Automation and Systems Issues in Air Traffic Control

NATO ASI Series

Advanced Science Institutes Series

A series presenting the results of activities sponsored by the NATO Science Committee, which aims at the dissemination of advanced scientific and technological knowledge, with a view to strengthening links between scientific communities.

The Series is published by an international board of publishers in conjunction with the NATO Scientific Affairs Division

A Life Sciences B Physics	Plenum Publishing Corporation London and New York
C Mathematical and Physical Sciences D Behavioural and Social Sciences E Applied Sciences	Kluwer Academic Publishers Dordrecht, Boston and London
F Computer and Systems Sciences G Ecological Sciences H Cell Biology I Global Environmental Change	Springer-Verlag Berlin Heidelberg New York London Paris Tokyo Hong Kong Barcelona Budapest

NATO-PCO DATABASE

The electronic index to the NATO ASI Series provides full bibliographical references (with keywords and/or abstracts) to more than 30000 contributions from international scientists published in all sections of the NATO ASI Series. Access to the NATO-PCO DATABASE is possible in two ways:

- via online FILE 128 (NATO-PCO DATABASE) hosted by ESRIN, Via Galileo Galilei, I-00044 Frascati, Italy.
- via CD-ROM "NATO-PCO DATABASE" with user-friendly retrieval software in English, French and German (© WTV GmbH and DATAWARE Technologies Inc. 1989).

The CD-ROM can be ordered through any member of the Board of Publishers or through NATO-PCO, Overijse, Belgium.

Series F: Computer and Systems Sciences Vol. 73

Automation and Systems Issues in Air Traffic Control

Edited by

John A. Wise
Center for Aviation/Aerospace Research
Embry-Riddle Aeronautical University
Daytona Beach, FL 32114-3900, USA

V. David Hopkin
RAF Institute of Aviation Medicine
Farnborough, Hampshire GU14 6SZ, United Kingdom

Marvin L. Smith
Airway Science Simulation Laboratory
Embry-Riddle Aeronautical University
Daytona Beach, FL 32114-3900, USA

Springer-Verlag
Berlin Heidelberg New York London Paris Tokyo
Hong Kong Barcelona Budapest
Published in cooperation with NATO Scientific Affairs Division

Proceedings of the NATO Advanced Study Institute on Automation and Systems Issues in Air Traffic Control, held in Acquafredda di Maratea, Italy, June 18–29, 1990.

ISBN-13: 978-3-642-76558-2 e-ISBN-13: 978-3-642-76556-8
DOI: 10.1007/978-3-642-76556-8

Softcover reprint of the hardcover 1st edition 1991

45/3140-543210 – Printed on acid-free-paper

Preface

In recent years, increases in the amount and changes in the distribution of air traffic have been very dramatic, and are continuing. The need for changes in the current air traffic systems is equally clear. For example, the impact of air traffic systems on the economic well-being of a country can be demonstrated by the fact that, in the United States alone, delays due to air traffic problems were responsible for an estimated loss of over five billion dollars in 1988 (for excess fuel, time losses, etc.). Losses of over ten billion dollars per year are expected by 1998 if dramatic changes are not made. As a result, members of the NATO Alliance are all in various stages of modernizing and automating their air traffic control systems.

While automation is generally accepted as a method of improving system safety and performance, one of the conclusions of the 1986 NATO Advanced Research Workshop on information system failure (Wise and Debons, 1987) was that high levels of automation in complex human-machine systems can have a negative effect on total system performance and have been identified as contributing factors in many accidents and failures. Examples of this effect can be found in the operation of most complex systems such as aircraft and nuclear power plants. As a result, there is a growing body of theoretical and systems knowledge that attempts both to explain these events and to provide guidance on how to avoid similar problems in new systems. For example, at the 1989 International Symposium on Aviation Psychology, several researchers identified critical issues and problems associated with the proposed increases in automation in air traffic control systems throughout the NATO countries. Morrison and Rosenthal, in separate papers, identified communications and human information processing loads as highly correlated with air traffic control system errors. Woods, in a counterpoint, presented a wide range of current examples where automation introduced to address those problems actually generated an entirely new set.

In addition, significant advances have been made in understanding the different types of cognitive errors that people make in complex, time-constrained, real-world decision making that is typical of air traffic control. This knowledge is critical not only to the design of the system hardware and software, but equally to the selection and training of operators. Likewise, the understanding of the impact of of automation-induced stress on system effectiveness has also advanced to the stage where it can assist in the analysis of failures and the design of more effective systems.

Those responsible for designing the advanced air traffic control systems to be implemented throughout the Alliance during the next decade need to be aware of recent progress concerning the most effective application of automation and artificial intelligence in human-computer systems. This need is critical. Design decisions made in the next few years will affect the safety and vitality of air transport in and between the nations of the Alliance. The application of the above information to those decisions will result in safer, much more efficient air transportation.

Acknowledgements

The editors would like to acknowledge the work of those individuals whose untiring effort and dedication to the Advanced Study Institute made possible the publication of these proceedings. In a certain sense, each of them should have his or her name on the cover, because without any one of them, this volume would not exist.

For the success of the Institute, we must thank a large number of people and organizations. First, we must thank our sponsors, without whom the Institute could not have taken place. These included:

- NATO Scientific Affairs Division
- U. S. Federal Aviation Administration
- Embry-Riddle Aeronautical University
- Research Institute for Information Science and Engineering

We also must thank the U. S. National Science Foundation which supplied financial support for several participants, and Transport Canada which assisted the Institute in every way it could.

We must thank the Institute's staff who worked hard before, during and after the meeting, particularly Barbara Gibson, whose administrative skills impressed everyone (and without whom, chaos could have easily broken out). We want to thank Jan Goodrich, whose skill with the written word helped everyone produce a paper to meet the goal of high information transfer. We also want to thank Kevin Norris for his outstanding work in recording the Institute, and Mr. Peter McAlindon for his technical and administrative support.

All of the participants owe a debt of gratitude to the staff at the Hotel Villa del Mare, who provided the ideal environment for exchange of current ideas and the generation of many new ones.

We also must thank a great number of people who assisted in a variety of ways to get the papers into the final electronic form that allowed their publication. At Embry-Riddle Aeronautical University, they include: Sandy Oliver, Mike Graves, Joan Sullivan, Tammy Clem, Kevin Minton, Carol Liptak, and Laura Hammond, Brian Pape. At the RAF Institute of Aviation Medicine, they include: Natalie Mitchell, Margaret Fawcett, and Jon Coldwell.

While our names are on the cover, we respectfully share whatever credit arises from this volume with all of the above.

John A. Wise
V. David Hopkin
Marvin L. Smith

Table of Contents

Editors' Summary

Introduction

The ASI, as most meetings of this type, raised more questions than it answered. For example, it was agreed that we need to identify the types of errors that are predetermined by the design process. While there is a tendency to see problems as system-driven, most discussions were about things that are human-driven. And while there was a feeling that design is biased toward technical factors, most discussions were about human factors.

We talked about the adaptability and flexibility of software. Adaptability and flexibility are emotive words that sound great. But at some point they change into instability and unpredictability. Where?

We discussed what should be taught in air traffic programs besides knowledge and procedures. What other kinds of attributes should be taught? We also mentioned the teachability of automation. One may have to determine not just that an automated aid can be made to function, but also whether it can be taught. Should we be selecting forms of automation based on their teachability?

We need further to address performance measures. Have we been measuring the right things? Are there things, like understanding or memory, which may be more important than costs, that we don't measure? We need tools to measure the effects of automation.

We addressed the concept of over-control, which implies an optimum level of sharing responsibilities between ground and air personnel within the air traffic control process. It is not necessarily all one or the other, but an optimum level is somewhere between. How do we recognize that we have it?

One criticism of the ASI would be that we concentrated too extensively on controllers and technology. We did not devote much time to the needs and contributions of the managers, instructors, engineers, maintainers, pilots, and passengers. They will all be affected by automation, some more than controllers. For example, there is a marginal, but definite, advantage to the air traffic control system to bring large aircraft in as a batch. But as a result, hundreds of people will be waiting at immigration and the baggage hall, trying to get through customs, and trying to get onto ground transport all at the same time. We need techniques and tools for comparing air traffic control advantages with the sort of chaos that might be resulting in other parts of the air transportation system.

Topics Addressed

Automation's Impact

A brief description and explanation of the structure of this text may help the reader to gauge the range of its contents and locate papers of particular interest. The Keynote Paper adopts a historical perspective to review previous forms of automation in air traffic control, some of the advantages and difficulties in applying them, and the future consequences of automation for air traffic control in the context that the envisaged future traffic-demands make the provision of more automation inescapable. Therefore, the debate is on the forms automation should take and the timescales for it, rather than on whether there should be further automation. Wise asks that the readers reassess their assumptions about both automation and air traffic. He attempts to assist the readers by providing a number of antitheses to commonly held assumptions.

Current Perspectives

The first section contains four papers which consider perspectives on the impact of automation on air traffic control, generally by relating envisaged future developments to current practices. The first paper, by Simolunas and Bashinski, describes the largest and perhaps most ambitious envisaged application of computerization and automation to air traffic control, in the example of the American Advanced Automation System, by indicating some of the underlying procedures adopted and the reasons for them; by describing the ways in which the user has been involved in the system evolution process; and by indicating some of the problems that have been envisaged and allowed for. The paper by Stager in this section, noting that Canada is marking fifty years of air traffic control, describes Canadian plans for replacing existing air traffic control systems, the intended uses of automation in the evolved replacements, and the rationale behind the project. The paper by Lane is a reminder that air traffic control must cover both military and civil requirements; it describes how military and civil air traffic control needs are coordinated in the United Kingdom, and suggests how future automation will affect military air traffic control with reference to data processing. Levesley's paper, giving the air traffic controller's point-of-view, indicates ways in which controller opinion is coordinated internationally and represented nationally in the planning and evaluation processes, to confirm that proposed forms of automation will be operationally satisfactory and acceptable.

Systems Issues

The section on systems issues begins with a paper by Debons, which provides a review of the kinds of thinking applied when systems issues are discussed, of the relevance of these to air traffic control, and of assumptions which underlie the impetus to extend automation in air traffic control. The remaining papers in this section exemplify the breadth of issues covered when automation in systems is considered. Bradbury's paper is a timely reminder that air traffic control is international, and that the sensed data, the equipment, and the procedures and policies of automation in air traffic control in the future have to be the subject of international coordination and agreement; agreement is perhaps particularly essential in relation to navigation and the ways in which more technically advanced and more automated navigation aids may be employed. The paper by Lindgren focuses on the harmonization problems of Western Europe, where differences in national practices, often confined to quite small geographical areas, have to be reconciled; common practices have to be developed and proved which are politically acceptable, can be translated into common regulatory practices, and will accommodate future changes, including increased air traffic and closer integration with Eastern Europe. The paper by McAlindon is a reminder that the effects of automation on air traffic control have to be predicted well in advance; both their nature and the timing of the need for them depend not only on technological advances but on forecasts of future traffic demands and resultant communications requirements many years ahead; these forecasts also must be linked to the mechanisms for international agreements and standardization.

The papers by Alpaslan and Harmancioglu draw parallels between water resources, engineering and air traffic control, demonstrating that automation being considered for application in both systems has a great deal in common, even though the purposes and products of the systems seem initially very different. These papers were a timely reminder that, in considering the impact of automation in relation to air traffic control, much can be learned by looking at its impact in other fields of application, in terms of both the problems likely to arise and the kinds of solutions that may be practical.

Automated Aids and Their Evaluation

This section considers some ways in which automated aids, particularly those related to artificial intelligence, are being considered or applied now in relation to air traffic control problems in various countries; it also looks at the facilities available for discussing and evaluating such issues by various forms of simulation. The very rapid and continuing developments in artificial intelligence raise the issue of how air traffic control might benefit from these advances. Edwards approaches this issue by considering the interactions between the controller and the system as a dialogue; he suggests that, while much existing functionality may be automated, and many of the traditional functions of the air traffic controller (particularly in en-route air traffic control) can be automated in the future, the issue arises of not just whether they *can* be but whether they *should* be. The paper by Benoît describes some of the work conducted by Eurocontrol to provide automated aids to the controller, in order to improve the efficiency of the on-line management of air traffic control; the emphasis is on evolving efficient means of controlling traffic within many of the existing constraints. Benoît also describes a new planning system for greater efficiency in controlling approaching aircraft, in order to increase the traffic handling capacity, improve the service offered, and not incur penalties in terms of excessive workload; this paper describes the expected benefits from a practical form of automated assistance for air traffic control, not in the distant future but being evaluated now. Sargis discusses the potential use of a multi-layer queuing model as an aid to check ATC system effectiveness; the logic -ased problem-solving approach discussed would allow an ATC system to learn and adapt to environmentally induced changes. The paper by David describes some of the practical steps being taken to integrate within Eurocontrol the approach to evaluations intended to answer both general and specific questions, dealing with particular airspaces chosen as exemplars; evaluations are being used both to furnish answers to practical issues and to try out and prove various measurement techniques and methods. The paper by Sandiford is based on the work of the Air Traffic Control Evaluation Unit in the United Kingdom, and pays particular emphasis to the relationships between real-time and fast-time studies and to the role of simulation in relation to valid findings which will extrapolate to real-life environments.

Human Attributes and Automation

The impact of automation on air traffic control from a variety of perspectives, ranging from the theoretical to the applied and from the general to the particular, is addressed in this section. Numerous perspectives are brought to bear on the central issue of the reconciliation of human characteristics and attributes with progressive automation and technological advances, in the furtherance of safe and efficient air traffic control. Hancock's paper takes a fresh look at the fundamental aims of human factors in the context of complex systems, bringing many theoretical constructs to bear to further an understanding of the relationships involved and how progress may be made on reconciling human and machine. Day, who also addresses the role of human factors as a discipline, advocates that human factors should return to its earliest roots through specialist involvement in the system-design cycle; he notes that those who commission air traffic control systems might learn from military experience about the procurement cycles and the integration of the user's point-of-view during various stages of system procurement. Garland makes more explicit a widespread implicit view, present throughout many of the deliberations of this ASI, questioning whether the availability of automation is a sufficient reason for introducing it; he advocates an approach which has a better philosophical base for what automation seeks to achieve and a more human-centered approach, to avoid the most adverse human factors consequences of automated systems and provide a better-planned progressive introduction of automated aids in step with user needs. Lind considers the various means of modelling the control task in complex systems, indicates that air traffic control is (in this sense, an example of broader kinds of systems, in that the modelling methods developed for them are applicable to it), and shows how models can not only describe the complex interacting dynamics of systems but also account for the end products of systems. Mogford also addresses the issues of modelling but from the viewpoint of mental models; he outlines their salient characteristics, how they can be applied to air traffic control, and how they may promote insight into various aspects of the controller's thinking and information processing. Tonner's paper lists, from the point of view of the air traffic controller, themes relevant to the application of automation in air traffic control which seem currently insufficiently recognized or unanswered, as well as the controller's view of essential requirements for successful air traffic control which did not seem to be recognized sufficiently elsewhere. Tonner emphasizes a theme which recurred often throughout the ASI: the need for everyone to pool their expertise and work with a common purpose, since ultimately all the work has the common purpose of ensuring safe and efficient air traffic control. Seidel notes, as do others, that controllers generally find air traffic control intrinsically satisfying; he also reviews European developments in the automated provision and manipulation of data in relation to air traffic control as it is — and as it is expected to become — with particular reference to controller acceptability of proposals.

Stress and Workload

The next section deals with some aspects of the impact of the air traffic control system on the human being, with particular reference to stress and workload and the changes that may be wrought by increased automation. A paper by Tattersall provides an overview of methods to assess the controller's ability to manage work demands, with emphasis on the effects of failure to manage the demands sufficiently and on the measures for identifying when this has occurred, he reviews the various kinds of responses to stress, mechanisms for workload regulation, and a recent extensive survey of factors relevant to stress and workload in air traffic control. Costa's paper focuses primarily on the effects of shiftwork and circadian rhythms on human performance, and on the consequences of such work for the human being in terms of medical indices, performance indices, and subjective assessments, drawing on both literature and survey data. Jorna, in considering workload problems of air traffic control, employs the concept of mental workload to help to explain some of the more unexpected human consequences of the introduction of more automation; he brings out the complexities of "workload" revealed both by the very high number of factors to which it appears to be related and by the fact that the relationships do not always accord with what would be expected. Goettl discusses the different kinds of automation that can be applied to air traffic control and some of the consequent human factors issues, particularly those related to task performance.

Error and Failure

This section of papers addresses issues related to human error or failure. Bradbury traces the connection between particular failures and the emergence and agreement of international regulations and practices, and discusses methods of disseminating literature to promote safety and minimize failure; sources of potential confusion, such as insufficient familiarity or discrepancies between military and civil practices or liaison, are identified in terms of their implications for the future safety of air traffic control. Fokas notes that an automated air traffic control system has to handle a mixture of old and new procedures; he reviews typical kinds of human error within air traffic control systems, particularly those related to communication, as a means of recognizing and minimizing such errors. Stager considers various kinds of models of human errors, and various error taxonomies, in relation to both the kinds of error which typically occur within air traffic control systems, and to the impact of automation on the nature of the errors that can be expected; from this point of view, he considers what the most appropriate levels of automated air traffic control operations might be. Empson, in looking at cognitive failures in air traffic control, provides a classification of failures and specific examples of the occurrence of various identified error-types within air traffic control communications or procedures; he considers the implications not only for automation but also for selection and workspace design. Horne uses Bailey's classification of seven broad categories of human error in systems to speculate on how they could be applied in air traffic control to resolve human-error problems.

Organizational and Cultural Issues

The papers in this section address aspects of automation in air traffic control related to its organizational structure and culture. Westrum addresses two issues in his paper: how automation may help resolve problems of information flow and consciousness, and how the organizational structures within air traffic control and the limitations imposed may lead to ignoring problems; he concludes that future systems, if they employ automation successfully, should involve a more effective consideration of consciousness and a better recognition of the kinds of management and decision making that are possible, given the prevailing organizational structures. Kaplan draws attention to those human factors areas related to air traffic control that are susceptible to cultural influence and to the need to recognize how and under what circumstances such cultural influences result in changes relevant to air traffic control; he deals not only with obvious cultural influences, such as language, but also with subtler issues, such as cross-cultural differences in stress aetiology. Harss also considers organizations and their structure, but traces their influence on workspace design and on conditions of employment; her paper argues persuasively that it is essential to view human-machine interactions within their broader organizational contexts.

Selection

The papers in this section deal with the selection of air traffic controllers. Della Rocco considers the desirable impact of more automated systems on the selection procedures for controllers — by looking at the evolution of the present selection procedures historically, reviewing the current selection criteria and programs, considering how automation may be applied to the selection procedures themselves, and then extrapolating what the future selection needs are likely to be. Wing, in her comprehensive review of the state-of-the-art of air traffic control selection, also discusses the issues of validation and brings out the complexity of air traffic control in relation to the selection procedures for it, since most jobs involve a considerable variety of tasks. Nyfield describes a program of selection and research to choose appropriate controllers for more automated future air traffic control systems, setting this in the context of preselection procedures, interviews, and test and simulation measures which may be used to assess various relevant factors, the aim being not only to adapt selection procedures more appropriate for future systems with more automated jobs, but also, in the process, to develop and validate better selection tools. Eißfeldt reviews the basic traits in the psychological selection procedures that have evolved in West Germany for air traffic controllers and, from this point of view, considers the changes needed to match controller selection with envisaged change requirements of more automated systems.

Training

The ninth section is concerned with air traffic control training. Baldwin's paper addresses the definition of training requirements in air traffic control and the methods of ensuring that requirements have actually been met. This paper views the problem from the Eurocontrol perspective where controllers from both the Eurocontrol and other nations are trained; he also considers other related issues, such as on-the-job training and the provision of stand-alone supplementary training devices to enable controllers to practice and hone their skills. Smith considers the effects of automation in air traffic control in the United States, first discussing predictable factors which must be taken into account (such as new training requirements, training costs, expected shortages of controllers, and demographic changes), then describing the Federal Aviation Administration's future policy on training, including ways in which automation is intended to help learning assessments during training and to adapt the training methods and techniques to the individual learning requirements of each student. He provides a list of individual differences that affect training and anticipates changes of emphasis in training requirements where automation could have a role. Herschler also identifies the need for new training approaches as a consequence of greater automation, placing particular emphasis on the need for more controller training in resources management, since this would be an implicit consequence of the changes automation would be expected to bring. Svátek emphasizes that resource management has become much more important and that trainers of cockpit crews are already paying more attention to the need to include cockpit management; she identifies objectives of this training, many of which clearly are applicable also to air traffic control. The paper by Chobot identifies future critical issues that must be addressed by those concerned with the planning and management of air traffic control training, discussing how training requirements should be integrated with the system acquisition process. Chobot also raises the issue of specifications and standards applicable to training and how they could be evolved and employed, as well as the roles of professional trainers during system procurement processes and the validation of training procedures.

Speculations on the Future

A final series of papers offers unbridled speculations about the probable evolution of air traffic control during the next fifty years or so from a variety of points of view. One theme which many of them have in common is that automation might have advanced *less* than many would guess. In this regard, these musings are perhaps more cautious and less adventuresome than they may have been, had a similar question been posed only a few years ago.

Final Thoughts

The closing remarks attempt to summarize the flavor of the meeting and its main broad issues, points of view, and patterns of agreement, rather that picking up any of the numerous issues raised in the specific papers and discussions. They emphasize the need for caution, lest we take the effects of automation for granted and fail to identify the whole range of their consequences before introducing them.

Introduction

The Impact of Automation on Air Traffic Control Systems

V. David Hopkin

RAF Institute of Aviation Medicine, Farnborough, Hampshire, UK

Introduction

When air traffic control began about fifty years ago, its main promptings were associated with the Second World War and the consequent need to fly at night or in poor visibility (Adair, 1985). Before then, aerial navigation relied on distance and time estimations, and on the judicious selection of distinctive features and landmarks along the intended route in order to verify position and progress. Ground support, mainly in the form of radio communications, did not constitute true control over air traffic, but was a service confined to regions around airports, as accounts by pioneer aviators make clear (e.g., Saint Exupery, 1939).

The main original reasons for air traffic control, all of which still apply, were to keep aircraft safely separated from each other; to ensure that pilots always knew where they were; to allow safe flying to continue in conditions of poor visibility when other aircraft could not be seen; to help flights to be efficient and expeditious; and to bring order, rules, and patterns out of potential chaos and arbitrariness, particularly near airports and other regions with dense traffic. All these needs were recognized when the air traffic was light and slow by modern standards. From them, the familiar traditional objectives of air traffic control can be derived: the safe, orderly, and expeditious flow of air traffic. These fundamental objectives must not be compromised, but other implicit objectives have been added in the interim: air traffic control must also be cost effective, impartial, noise abating and fuel conserving, and it must not harm those whom it employs.

The kind of air traffic control service that can be provided, and the applicable separation standards between aircraft, have always been critically dependent on the nature and quality of the available navigational data. Procedural methods of air traffic control have to allow for occasional major navigational errors, or initially small but cumulative deviations from track, even in well-equipped aircraft. Transoceanic flights beyond radar coverage typically require at least ten minutes' longitudinal separation between consecutive aircraft at the same height on the same route, and sixty miles' lateral separation between routes. By contrast, the navigational data about aircraft near an airport may be so good that the separation standards between consecutive aircraft on final approach path depend on other factors, such as the permissible distance between parallel runways and the wake turbulence or vortex regulations applicable to various aircraft weight categories. Because the advent of radar greatly enhanced the quality of the navigational information, the safe separation standards for much en-route flight could be reduced. In the future, satellite-derived information is expected to reduce the applicable separation standards for transoceanic traffic for the same reason. The quality of navigational data depends on several factors, including accuracy, precision, reliability, serviceability, frequency of updating, and the consequences of possible errors.

NATO ASI Series, Vol. F 73
Automation and Systems Issues in Air Traffic Control
Edited by John A. Wise et al.

Early Automation

Many of the human factors problems associated with air traffic control predate the introduction of automation as we now know it. The comprehensive proposals to apply human engineering to air traffic control, made by Fitts and his colleagues in 1951, had a major influence on human factors. Their paper referred to many concepts that are still current though with different connotations — both human factors concepts, such as the division of responsibility between humans and machines, decision making, information coding, and information flow, and air traffic control concepts, such as the air traffic control tower and center, the approach controller and en-route controller, and the flight plan. Much of the vocabulary of air traffic control dates from its beginnings when it relied on spoken interchanges between pilots and controllers.

Although air traffic control is now among the pioneering applications of automation, it was not always so. An authoritative review of automation (HMSO, 1956) scarcely alluded to aviation and never mentioned air traffic control, but discussed extensively economics and management, the engineering industries, process control, data processing, and large human-machine production systems. More surprisingly, it also discussed implications of automation for labor relations; the compatibility of human and machine; and the effects of automation on shift work, on women workers, on job satisfaction, on criteria for payment for work, on teams, and on opportunities for promotion. Why were so many of these human factors issues, recognized as relevant in 1956, ignored thereafter?

Increases in Demand

Probably the availability and potential benefits of automation would alone be sufficient to ensure consideration of its applicability to air traffic control — as to many other contexts, from offices to the entertainment industry. But in many parts of the world, automation in air traffic control is treated as a necessity, rather than a potential boon, for another reason. Although current air traffic control systems may be remarkably safe and efficient within their design limits, many are handling traffic levels at or near their planned maximum capacity and must evolve to accommodate more air traffic.

Although the actual size of the impending increase in the demand for air traffic control services is debatable, and projections from discernible trends in air traffic are subject to gross distortion by factors beyond the influence of air traffic control, there is agreement that the increase in demand will be substantial and will exceed the capacity of existing systems. There are two broad consequences. One is that the quality of navigational information, and of computations based on that information, must be enhanced so that more aircraft can be controlled efficiently in the same airspace with no loss of safety. The other is that each controller must spend less time dealing with each aircraft under control, again maintaining efficiency and safety. Both these consequences point towards more automation. The apparent option of increasing the number of controllers becomes self-defeating because of the extra communications and liaison entailed.

Automation is, in principle, beneficial. It can render practical and safe what could not otherwise be done. Its main objective must be to maintain and enhance the safe, orderly, and expeditious flow of air traffic, as traffic increases. The ways in which proposed forms of automation could meet this objective are usually clear from the proposals, although quantification of the benefits may be less certain. Expected benefits may not always materialize, or they may be canceled by associated unforeseen disadvantages. Only in recent years has the full range of the potential effects of automation on air traffic control begun to be recognized, and even now some its indirect effects can incur unanticipated problems, which have to be solved retrospectively after the automation is in service.

Basic Forms of Automation

Technological advances, more and better data, enhanced facilities for computations based on those data, greater system complexity, and additional demands for air traffic control services combine to make increased automation of air traffic control in regions with high density traffic inevitable. The issue is not whether there should be automation in air traffic control, but when, how, and how much.

The ways in which automation has been introduced into air traffic control parallel its progressive application elsewhere. The first functions to be automated concerned the gathering, storage, compilation, and presentation of data. Next came functions concerned with data refinement, synthesizing, and labeling: for example, the progression from primary to secondary radar was followed by the inclusion on secondary radar displays of labels containing additional information about each aircraft. Label content also expanded: aircraft identify was essential, but the label could also show current state, such as height, and intentions, such as destination. Further information based on calculations, such as predicted position at a future time or conflict alert, could also be displayed automatically. The benefits of label expansion encountered the constraints of excessive complexity of content, label size, and label overlap.

Most of the information in early forms of automation was quantitative. Often, various forms of automation changed or removed qualitative information — used for judgments of significance, accuracy, or trustworthiness — although this was seldom a specific intention but rather an incidental consequence. Primary radar displays could and did show considerable information about the signal-to-noise ratio, the incidence of clutter, and the propensity of information on the display to fade, but secondary radar displays normally lack such information, although predictions of the future positions of aircraft may attempt to depict probabilistic information.

Developments in Automation

From capabilities to sense, store, compile, condense, and present relatively large amounts of information, and to handle larger amounts of more frequently updated data, automation progressed to computations based on those data. Sometimes these were relatively simple, from the point of view of the controller using the information. Examples were conflict detection of a predicted future violation of the separation standards between two aircraft, and track monitoring, with an indication whenever any aircraft deviated from its planned track by more than a pre-defined distance. In such circumstances, the automatically detected impending conflict or the track deviation was displayed to the controller for action.

In human factors terms, some of these earlier forms of automated assistance were sensible and well matched with human strengths and limitations. Human beings are not efficient at continuous searching or monitoring, and the fulfillment of such functions automatically can, in principle, be far more efficient, reliable, and consistent. The human controller might also do the same tasks independently, but the system did not have to rely on adequate human performance of tasks for which people are unsuited, and this seemed a sound principle to follow.

Since all the information needed for such computations was in the database, the question arose of how else it could make air traffic control more efficient and enable the controller to handle more traffic and spend less time in dealing with each aircraft, without lowering safety standards. In particular, how might automation assist such functions as problem solving, decision making, prediction, and task scheduling, and such requirements as the structuring of menus and dialogues and the provision of memory aids?

If properly designed, such applications of automation can undoubtedly enhance efficiency and safety and increase system capacity. However, some of their human factors implications can be obscure and can engender problems that are difficult to resolve, partly because, potentially, there are so many of them. For example, it seems relatively straightforward to replace a human solution to a computer-detected conflict by an automated conflict resolution facility. The conflict having been detected automatically, the computer can also resolve it automatically, or at least offer a hierarchy of possible solutions, with or without a computer recommendation to the controller on which solution should be chosen. This seems sensible, and, in system description terms, there has been little significant change of function. In human terms, however, there has been a gross change. Deciding whether to accept or reject a computer solution of a problem presented on a display is very different cognitively from devising and implementing the solution of that problem oneself, particularly in terms of what is subsequently understood and remembered. Thus the introduction of computer assistance into such functions as decision making, problem solving, and prediction can affect human tasks significantly. It can make different task demands, change the knowledge and skills required to perform tasks, affect the levels of human involvement in them and human control over them, and alter the controller's understanding of them (Hopkin, 1982).

If these forms of assistance become very extensive, they introduce a different kind of problem, especially if the computer not only makes proposals but also goes some way towards implementing them. A computed solution to a problem may be adopted by a simple controller action such as pressing a button. The responsibility for the solution or decision thereby remains with the human, at least nominally. Automated aids to problem solving and decision making would never be introduced into air traffic control unless they were known to be very safe, efficient, and reliable. The controller is encouraged and instructed to accept computed solutions and decisions. Failure to do so defeats the purpose of providing such forms of automated assistance. However, at some point it is no longer apparent to anyone else or detectable by the system whether or not the controller performing such computer-aided functions retains sufficient understanding of the actions taken, the decisions reached, or the problems being solved. When all that is involved in accepting a computer solution is to press a button, it is not possible for the system or anyone else to tell what has been understood. In fact, the controller who always accepts the preferred computer solution promptly and never questions anything may be praised for using the system as it was designed to be used. The reasons for never questioning anything may not be that the controller is satisfied that all is well, but could be inadequate understanding. Ultimately, if the computer can do all but implement solutions, the question must arise: why is the human being present and why cannot the computer implement the solutions also? That raises the issue of where the legal responsibility lies.

Extended Applications of Automation

Although there can be debates about how far automation should be extended, its influence on modern air traffic control systems is such that it is now difficult to conceive how dense traffic might be controlled efficiently without computer assistance. Certainly manual reversion, if it remains possible at all (which is debatable in some circumstances), usually entails very substantial decrements in the efficiency of the system since the whole point of these forms of computer assistance is that they can fulfill functions more quickly, accurately, consistently, reliably, efficiently, and safely than a human being can. If for any reason they fail, it follows that the human controller cannot operate the system in manual mode with high traffic levels as efficiently without them: otherwise, there would be no justification for them.

According to future plans for air traffic control systems, automation will have a widespread though not ubiquitous role. Many functions will be performed wholly or partly by machines. The applications of automation will extend into traffic management systems, planning and prediction, artificial intelligence, expert systems, human-machine interface design, and the recording of system performance. They also will affect system architectures and the processing and use of sensed data, including data from satellites. Various advances in communications; in the transponding of data to, from, and within the system; in the prevention of collisions and avoidance of hazards; in the resolution of problems; in decision making; and in flow control and scheduling will rely on advances in automation for their efficacy. Automation is also expected to have major effects on air traffic control through its impact on associated disciplines, such as meteorology.

Automation is being applied more in the selection and training of controllers. Skills related to automation will be a desirable attribute of future controllers, and should therefore be a feature of selection and training procedures for controllers. But the administration of those selection and training procedures can also be substantially automated in the interests of efficiency, impartiality, and standardization. Automation is also influencing maintenance procedures, both in their concepts and in their practice. Many of these developments were mentioned in a recent Special Issue of the Proceedings of the IEEE on Air Traffic Control (Pozesky, 1989), which provides a comprehensive description of the plans for the future Advanced Air Traffic System in the United States, and incidentally shows how extensively such plans rely on automation.

Automation and the Controller

Many future forms of automation may possess characteristics hitherto thought of as uniquely human; these include intelligence, adaptability, flexibility, and innovation. An issue is how to apply these advances to air traffic control optimally to promote safety and efficiency. It is not yet known what the optimum human-machine interface designs, procedures, instructions, or divisions of responsibility should be, when such characteristics as intelligence, innovation, adaptability, and flexibility are possessed by both human and machine.

Further advances in automation can be conceived, for example towards a system that can be either highly manual or highly automated, without these being mutually exclusive alternatives. At one extreme: when traffic is light, the system may be almost entirely manual, and controllers may be able to maintain their skills through opportunities to practice them. At the other extreme: the system could be capable of functioning wholly automatically, for example in the event of total human incapacitation. Both of these extremes may seem outlandish and theoretical, but their technological feasibility is in the offing, and advances in technology will offer many alternatives, options, and ways of using automation which have not yet been considered. Such flexibility in the extent to which available automation is actually used could have major human factors advantages, in terms of job interest, professionalism, and effective human intervention in unusual circumstances.

To achieve optimum future systems, we must broaden our thinking beyond conventional assumptions. Before we can challenge them, we have to recognize what some of our present assumptions are: that functions should be allocated between human and machine; that every definable function can be fulfilled somehow; and that there must be an answer to every human factors problem, and generally a single optimum answer. The solutions of some problems which seem primarily technical or human factors issues may be dictated mainly by legal or financial considerations. As technology makes more and more options available, it may tend to obscure what would really be best. We need valid means to define what would be best before we can achieve it. We also need means to recognize when we have achieved the best, so that we do not waste resources in trying to improve the unimprovable. This is not as unlikely as it seems. Sometimes we have been inveigled into such a situation in the past, by failing to recognize the crucial limits or constraints on what can be achieved.

Relationships Between Human and Machine

As automation advances, the nature and range of human machine relationships within more automated systems are potentially enriched and extended. The notion that automation must be restricting and reduce human options, though often true of some of the earliest forms of automation, is not necessarily true of more recent ones, and will seldom be true of future applications. Many technological advances extend what is feasible and enhance the diversity of choice, without necessarily removing existing options.

The initial attempts to aid the human by automation often relied on machines that extended human capabilities by providing power. This induced mechanistic ways of thinking and implied that machines were inherently inflexible, so that a human role associated with automation was adaptation to the machine. Early measurement techniques, such as time and motion study, emphasized this mechanistic approach and also linked efficiency to output. At this time, close human-machine relationships were not sought, and human and machine roles and functions were not designed collaboratively. The efficiency of human roles was judged in machine terms, a point of view neatly encapsulated and criticized in Dieterly's (1981) concept of the "John Henry syndrome." Man was competing with machines and would lose the competition.

This kind of thinking was supported by many of the earliest ventures of human factors into applications of automation, including those related to air traffic control. The classic compilation by Fitts (1951) and his colleagues of functions suitable for human and functions suitable for a machine listed them under the heading of "Man Versus Machine," a clear testament to the underlying thinking. This approach excluded certain choices altogether. The independent fulfillment of the same function by both human and machine in parallel was not an option. The notion that some functions were suitable for neither human nor machine was not addressed. An implication was that everything could be done somehow, and the whole approach was technology driven. The question of whether a function should be performed by a human or a machine can be asked only about those functions for which a machine can be devised, since otherwise the question does not arise. The more technology advances, the longer the list therefore becomes of functions about which the question can be asked. Specifically, human attributes with no machine equivalents have no influence on the list and seem irrelevant to decisions about it. Such attributes therefore tended to be ignored.

As technology advanced, not only did more functions appear on the list but more were assigned to the machine, solely because of technological progress. Functions tended to be assigned to humans not because they could fulfill them well but because machines could not fulfill them at all. The notions implicit in the concept of human versus machine can be discussed only in machine terms — for it seems ridiculous to ask, for example, if a machine can tolerate enforced idleness or likes to demonstrate its efficiency to other machines. The fact that such concepts seem inapplicable helps to reveal the limitations of the approach, but human-machine function allocation as a kind of competition has a long history.

When the limitations of this competitive approach became obvious, the view that human and machine were complementary took its place. Attractive in principle, this view seemed suspect in practice. There was no fundamental reason why they should be complementary, although it was possible to design systems where they would be. Complementarity of human and machine was most in vogue in an era when any form of automation was hailed as progress, and innovations were introduced that often promised more than they delivered. Changes were made because they were technically possible, seemed to be cost-effective, were of proven reliability, and offered enhanced efficiency or safety — but not generally because they complemented human attributes, particularly when forms of automation were introduced retrospectively into systems originally planned without them. Successful complementarity proved to be somewhat elusive; it was not serendipitous, but had to be planned.

The notion that human and machine were complementary when using automated aids in air traffic control gradually evolved into the notion of symbiosis. In retrospect, this seemed rather strange, since one of the envisaged roles for the human was to exercise sole control in manual mode if the automation failed, a contradiction of true symbiosis. The idea that neither human nor machine could function without the other was in some respects attractive, but its influence was stronger in principle than in practice. Successful yoking of human and machine was not a main criterion of machine design, and there seemed no particular reason why machines with major technological constraints must be symbiotic in their relationship with optimum human roles. Recently, references to symbiosis have dwindled.

The concept now in fashion seems to be human centered automation. In principle, this represents the opposite thinking to the earlier notions about automation, with their implication that the human would do whatever the machine could not. Again, there may be a gulf between the principle and its practical forms. However, the notion of human centered automation does reflect a genuine change of emphasis which it has been possible to trace for some years. An invited paper in 1985 had the title "Fitting Machines to People in Air Traffic Control Automation" (Hopkin, 1985), and that paper began by noting the change from six years earlier, when the given theme of an invited paper had been "New Workforce Roles Resulting from New Technology" (Hopkin, 1979). Adaptation was a human function in the earlier paper, a machine function in the later one.

Although many of the above options no longer appear as attractive as they once did, they are still available. Technological advances in the future offer further options. In addition to human centered automation, the notion of the electronic crewmember, developed mainly for military cockpits, offers further possible applications of automation to air traffic control. Technology also can bring flexibility and fluidity, to the extent that the ideas of the machine adapting to the human or the human adapting to the machine need not be mutually exclusive.

Hitherto, individual differences between people have been an unwanted source of variability and unpredictability, to be curtailed by training which has sought uniformity among individuals. But if individual differences are viewed as a potential strength, and if automation is viewed as an adaptive form of assistance applied selectively to take full advantage of individual strengths, then different kinds of human-machine relationship become available. In particular, the human-machine relationship may not have to be fixed and general, but can be fluid and individualistic. Functions may be fulfilled by either human or machine or jointly by both, according to prevailing circumstances, and do not have to be allocated to one or other in advance. Exclusively human and machine roles do not have to be pre-defined. Thus a basic impact of automation on air traffic control can be to expand greatly the kinds of machine and human roles that are feasible, and the kinds of interaction between them.

Workload

In the past, controller workload has often seemed to be a main determinant of system capacity. Either the tasks required directly for air traffic control — or associated but essential communications, co-ordination, and liaison functions — have strongly influenced the number of aircraft each controller can handle safety. In air traffic control systems functioning at or near their planned maximum capacity, the individual controller can be continuously burdened by high task demands throughout most of the working day, with no prospect of respite. Sometimes co-ordination sets the limits, and the burden of communications and co-ordination has to be reduced if the capacity of the system is to be increased. Further partitioning of the airspace alone cannot offer a general solution, because of the extra communication and co-ordination tasks entailed by the more frequent handover of responsibility resulting from greater partitioning. The effects of automation have to include a reduction of the communication and co-ordination burden, as well as of the time required by the individual controller for the actual control of air traffic.

When data transponded automatically between air and ground are used for calculations and predictions, this can reduce the planning of traffic routings, the co-ordination tasks associated with the handover of responsibility, and the amount of speech between air and ground. Aids to achieving this include data links, satellite derived data, and the automated correction, updating, and prediction of data. A simple familiar example is that the updating of an expected time at one reporting point automatically updates the expected times at subsequent reporting points for that flight. Routine chores are avoided, and controller's time is saved. Such applications of automation are clearly beneficial on balance. If they can help the controller to spend less time dealing with each aircraft, without any impairment of efficiency or safety, then the limitations imposed by controller workload on the traffic handling capacity of the system should become less severe.

In practice, the effects of such a simple automated aid can be much more complex. When routine actions are performed automatically instead of by the controller personally, the controller's understanding of what is happening may be impaired. The controller reports a loss of picture, and spends time on the extra task of topping up human memory. The controller builds a picture of the traffic which contains a great deal of information about it, considerably beyond what would be predicted from theory, in the form of readily accessible images of the air traffic. These are sometimes, though not necessarily, visual: they may be mainly in terms of times. This loss-of-picture associated with some forms of automated assistance seems genuine and explicable because less detailed levels of processing are required to accept a computer decision than to work out that decision for oneself (Narborough-Hall, 1985). The associated quest for evidence to build the picture can impose additional task demands, which partly or wholly counter the savings made by automating routine functions.

Other consequences can arise, for example, from new or unforeseen circumstances. A controller who has worked out the solution to a problem knows at once if unusual circumstances or new evidence warrant its reappraisal. The controller may not know whether a preferred automated solution has or has not taken account of these new circumstances or evidence, and it can be difficult to design a human-machine interface to allow the human to ask the machine about the evidence being used in automated solutions to problems. Such dialogues may be a prerequisite for optimum human-machine relationships.

There are two ways of resolving such dilemmas. One is to design the system so that the controller's loss-of-picture is acceptable and the safety of the system is not predicated on its maintenance. This kind of solution is envisaged, in the long term, in some United States air traffic control systems. The other is to keep the controller sufficiently within the active control loop to prevent significant erosion of the mental picture; this latter seems to be preferred in some user centered systems, but it can reintroduce, in new guise, the problem of high controller workload.

Automation can fail to capture the full range of functionality of the apparently equivalent manual functions. This problem has become apparent in relation to flight strips. Most of the attempts to automate flight strips have led to a form of automated presentation which physically bears a close resemblance to the traditional paper flight strips. These electronic flight strips have then been found to be very different in their functionality, their manipulation, and their cognitive aspects from the traditional paper flight strips, to the extent that the approach has had to be revised and a greater cognitive understanding of all the roles of flight strips has had to be gained, in order to automate them successfully.

Some theories of workload in the past have given the impression that there is a fixed workload capacity, so that it is possible by saving work of one kind to devote the time saved to work of a different kind, as long as the fixed total capacity is not exceeded. A recent review of workload measures (Lysaght et al, 1989) still leans towards this approach. Recent theoretical research and practical experience alike suggest that it is inadequate. Specifically, only certain kinds of workload can be added without excessive interference with the work already in progress, and the ways in which new forms of work relate to the work already being done are crucial. Further research is needed to gain sufficient evidence on how controller workload can be optimized by an amalgam of suitably designed tasks, based on a detailed knowledge of the kinds of work which they entail. Traditional secondary task measurement, whereby it was thought that a secondary task could indicate spare capacity available, now seems at best an over simplification, and at worst misleading. This relates to another theoretical controversy, concerning the extent to which the human being can be a serial or parallel processor of information, a topic of much research, theorizing, and debate which also has implications for attention and memory.

Many air traffic control systems seem to have been designed in the past on the assumption that correctly specified visual information displays can attract and hold a controller's attention indefinitely. This is now at last being challenged: while every effort should be made to maintain attention, the safety of the system should not be predicated on its successful achievement — since it is of the nature of attention to wander, and no amount of exhortation can prevent this.

The traditional problem of workload as a limitation in air traffic control systems may, because of proposed forms of automation, gradually disappear. Many of the forms of automated assistance for tasks which currently impose most workload are intended to reduce both the work which the human has to do and the time required to do it. Solving a problem or taking a decision manually can involve the assembling of a great deal of information and various consultations. The acceptance of an automatically computed and recommended solution or decision can take much less time. Because automation in future air traffic control systems will help to guarantee that capacity limits for air traffic in regions are not exceeded, and will smooth the flow of traffic within those limits, current problems associated with local high traffic densities and bunching will diminish, along with the high workload in resolving the resultant air traffic control problems.

A further application of automation is to predict potential conflicts between aircraft, so that they are prevented long before they occur, rather than resolved as a matter of urgency. Much of the work in finding solutions to conflicts manually would be removed. Far fewer residual conflicts would remain to be resolved manually, if the system functioned as planned, and the time taken to resolve each would diminish, since automated solutions would be offered. Such factors mean that, if the forms of automated assistance planned for future systems are used by the controller as they are intended to be used, it should become almost impossible for the controller to be overloaded by excessive task demands. The controller can prevent the workload from becoming excessive by using the automated assistance provided. In that sense, excessive workload would be self-inflicted — the result of a controller decision to do some of the tasks manually, and not an inevitable consequence of high task demands. For this fundamental reason, the problem of workload is thought to be solvable. The cost may be reduced controller understanding of what is happening. It follows that conventional measures of controller workload may become much less useful as indicators of task demands or system capacities.

Manual Reversion

Often mentioned as a residual role for the human controller is the need to assume manual responsibility in the event of failure of some form of automation. It seems to be accepted that performance in manual reversion would often be less efficient than with automation, but checks are made to ensure that it would be feasible and safe. Manual reversion implies that the controller will have opportunities to maintain the necessary skills, and that the human-machine interface design has made provision for everything necessary for manual reversion. In practice, this may be very difficult to achieve although it sounds attractive in principle. It suggests a criterion for successful human-machine interface design: if the user has to ask a question that has never had to be asked before, the logic of that design should make it obvious at once how that question must be asked and whether there is an available answer to it. There is no time in such non-standard circumstances for trial and error attempts to discover whether the information needed urgently is available in the system, or for numerous futile attempts to access it. The extent to which the human can innovate and be adaptable and flexible in an emergency is set by the human-machine interface design, by the user's knowledge of how it should function, and by the help that it gives to a user in rare emergency conditions. These are some of the applications of automation in air traffic control which have not yet been adequately addressed.

Manual reversion raises a further issue. It seems often to be assumed that the main difficulty is to revert to manual functioning in the event of failure of automation; but this is only one aspect of the problem. A neglected aspect of many forms of automation is how to ensure that a human already functioning in the reverted manual mode can also prepare the system to bring it back into its normal automated state, while continuing to work in manual mode until it is safe to revert to the automated one. In many systems this seems difficult to do, but it must be done somehow, since a failure of automation should never be permanent. Successful manual reversion entails equally successful reversion from the manual to the automated mode.

Trust

Attention to some of the human consequences of automation has been very uneven. The introduction of automation can lead to certain difficulties. One concerns trust. The simpler aspect of this concerns the trust by the human of the automation provided. If there is a failure in the system, the user needs to know how far its effects extend. If a slight degradation in the system performance, for example because of poor weather or a failed radar, results in displayed information which, though usable, is less trustworthy than normal, the controller should be able to take account of it. Currently there are no satisfactory codings to denote how far visual air traffic control information should be trusted; the need to develop and prove such codings seems urgent.

Some of the roles of trust have not been thoroughly examined. Excessive task loading and workload have been considered as sources of stress, but other sources of stress can be working alongside a colleague who seems unpredictable enough to need watching, or having to use untrusted automation so that the user is forever trying to verify what the machine is doing. Training in the use of automation ought to include how far the user should trust it, and under what circumstances differences in trust are warranted.

Work on the electronic crewmember has suggested that the problem of human trust has its counterpart in machine terms. If a machine performs a collection of functions which seem, in aggregate, rather like having another person there (hence the electronic crewmember), the question arises how much the machine should trust its human counterpart, in this case the controller (if such anthropomorphic terminology is permissible for discussion purposes). As well as the human adapting to the machine on the basis of human trust, the machine is adapting to the human on the basis of machine trust. To some extent, this is already done. Certain human actions may be forbidden by the machine, on the grounds that they could be dangerous, but it is not clear how far this principle should be extended to actions which are not overtly dangerous but are non-optimum, or not really hazardous but not as safe as they might be. How far should the machine trust the human in the relationship between automation and the human controller? How can this be reconciled with the possible need to resolve an emergency by human actions which, under normal circumstances, would be forbidden by the machine?

Human Stress

Although many of the possible effects of the air traffic control system on the human being have not been studied, an exception is the extensive work on stress. For a long time, air traffic control has been deemed to be a stressful occupation, despite considerable evidence to the contrary (Melton, 1982). This does not mean that there is no stress induced by air traffic control, but rather that many jobs in life nowadays engender stress, and that air traffic control is not exceptional. Certainly, air traffic control as a profession does not seem to differ grossly from many other professions in the amount of stress it can generate. Stress need not, however, originate from the air traffic control itself but from management practices, from unjustified media disparagement of air traffic control, from the wide prevalence of shiftwork and the resultant impact on family life, from inadequate equipment, or from rostering or other aspects of working conditions.

On the whole, the evidence does not point to a strong relationship between stress and safety. Therefore, the alleviation of stress, though highly desirable on humanitarian and medical grounds, might not necessarily lead to significant enhancements of safety, although it is hoped that it would. A less welcome consequence of emphasizing stress in air traffic control so much is the comparative neglect of another problem which may be more serious, namely boredom. The relationships between boredom and safety are not known either. It is time they were. The problem is not new (Hopkin, 1980).

Teams

One striking aspect of automation applied to air traffic control systems is that most of the forms of automation for the controller to use, as distinct from those which sense or process or compile data automatically, are for one controller at a human-machine interface. They are aids to an individual controller's decisions, problem solving, or predictions, yet they are being introduced into contexts where many of these functions have previously been performed by teams. Many tasks in air traffic control are collaborative team tasks.

An air traffic control workspace has always been open, in that supervisors and colleagues can tell by observation a great deal about what the individual controller is doing and can form a view about the competence of the individual by observing control actions. There is usually a consensus of peer judgments about who the best controllers are. Such views can form, and such judgments can be made, only when the tasks are open enough to be observable by others. The more that air traffic control is performed by dialogues through a human-machine interface, the less transparent the processes of air traffic control become. Most forms of automation associated with the man machine interface render the individual controller's activities much less observable and interpretable by others.

As a result of the progressive replacement of team functions by individual dialogues through a human-machine interface, many team functions taken for granted in manual air traffic control systems will be absent from more automated ones. Positive planning is necessary to retain these team functions if they are still considered to be necessary. Teams fulfill many roles. They form the basis of supervisors' and colleagues' judgments of competence and confidence. They provide opportunities to gain the esteem of colleagues and to have self-esteem through the respect of colleagues. They provide the mechanisms by which professional norms and standards evolve and become entrenched and maintained. They allow various particular practices at local facilities to evolve, and be refined, agreed upon and implemented. They provide criteria to assist the fair choice of people for career advancement. They suggest requirements for individual training or retraining. They help to imbue a professional ethos among controllers. They allow incoming controllers to gauge what is expected of them if they are to be acknowledged as fully accepted members of their group and profession.

The transparency of manual air traffic control workspaces allows predictions to be made of how busy other controllers are going to be, and offers opportunities to help colleagues because of the flexibility of doing manual tasks. The openness of teams in manual air traffic control means that, if an individual controller is not performing as well as normal for whatever reason, colleagues notice this and take active steps to help, so that the efficiency and safety of the whole system are not impaired. This detectability of below-par performance can be used either to help a colleague through a temporary crisis or to reveal those who are no longer able to fulfill their duties to the required standard. Teams are the source of group identification, of camaraderie, of morale, and of job satisfaction. They help to provide opportunities to develop and exercise skills. Most members of each team believe that they belong to the best team, and have a personal stake in maintaining its high standards.

All of these team factors and many more are changed by the progressive introduction of automation into air traffic control systems, insofar as those changes in automation have a direct bearing on the tasks of the controllers and the manner in which they can be done. If this has not been acknowledged, then decisions about introducing automation into air traffic control have been taken in ignorance of the full range of their consequences.

Selection and Training

Automation influences the selection of air traffic controllers. It provides means to streamline the selection procedures themselves by automating the selection processes and tools, and it can change the procedures, for example by providing immediate feedback to an interview board on the performance by candidates of tests which they have just completed. Automation can also influence selection in another way, by raising the question whether compatibility with computers is a desirable attribute of future air traffic controllers — which should therefore be a feature of any procedures for selecting them. A current research issue is whether compatibility with computers is associated with poor ability to work closely within teams, given that the latter attribute may not remain as important in future air traffic control if team roles diminish.

Automation also affects the training of controllers. The impact of technological advances on future training of air traffic controllers has recently been discussed, and many of the implications of automation for training have been identified (Hopkin, 1988). Computers can be used to familiarize students with the forms of computer assistance and make them more comfortable in using them. They may also aid training directly by offering demonstrations, opportunities to practice and polish skills, practical examples of human-machine dialogue design and structures, and different forms of explanation.

Uneven Applications

Because the range of feasible options is being continually extended, as more forms of automation are devised and proven, many issues which have been decided in the past should be reconsidered from time to time. Automation is seldom applied uniformly throughout an air traffic control system. Traditional divisions between human jobs may need to be revised, if some have been largely automated out of existence while others remain substantially in their manual form. The choice, groupings, level of detail, and nature of various human tasks may need to be reexamined.

Any form of automation raises the issue of the extent to which the machine or the controller should initiate actions. This has been debated since the earliest stages of automation because all the arguments do not support one point of view or the other. It must be absolutely clear where the responsibility lies. Air traffic control is not a context where it is safe to wait indefinitely until uncertainty about the locus of responsibility for an action has been resolved. But reliance on the human may mean dependence on fallible human memory that needs to be supported by forms of automation initiated by the machine that can serve as reminders, whereas reliance on the machine may mean that the procedures which can be adopted in an emergency are excessively rigid and provision has to be made for human initiative to be able to override at least some of them.

Optimization for Human Use

One aspect of the impact of automation is of critical importance but is not developed in this paper, because most of the associated problems are relatively straightforward and can be solved on the basis of existing knowledge. Any new form of technology, such as a new display or input device, has to be evaluated and optimized for human use. As a part of this process, possible misuses by the human being which could lead to errors have to be identified, so that decisions can be taken as to how they can be prevented or their consequences circumvented. The procedures and measures for trying to optimize equipment for human use are relatively orthodox; they include design procedures; task analysis; ergonomic and anthropometric evidence; selection and training content and processes; and often data from experiments, simulation, and modeling. This essential work produces recommendations for display formats and contents, for the location and sensitivity of data input devices, for the provision of appropriate communications facilities and for the matching of displays, input devices and communications in appropriately designed air traffic control workspaces within optimized physical environments. Such work, generally orthodox in its methodology, can be very demanding in terms of resources. Proof that optimization, as distinct from efficiency or high safety levels, has been achieved may be impossible, since there are no tools to define what is optimum. The correct weighting of the many factors involved may not be agreed, and the implication that there is a single optimum may be wrong.

Some Possible Problems

Some proposed applications of new technologies appear premature. A recent example has been direct voice input, which offers great promise for the future but not yet. Other applications of advances in technology seem self-evidently helpful, but do not always produce the expected benefits. An example is color coding which will probably become universal on air traffic control displays in the future, but has not always led to demonstrable improvements in terms of efficiency or safety, often because it has replaced or been combined with monochrome codings which were themselves highly efficient for the job, even though they did not look nearly so attractive. Color coding or other innovations may be in response to strong user preferences, because they are well-liked or have become status symbols. A worrying feature is the common combination of strong subjective impressions of improved performance as a result of color coding, with equally strong evidence from actual performance measures that there has been no significant improvement. Perhaps what has actually been improved has not been measured — memory, for example.

Some forms of automation can apparently replace the real task with a surrogate. The controller may perform some representation or replica or mimic of the real task, rather than the task itself — may maintain minimum separation between blips or labels on a radar display, rather than between aircraft in the sky. This is not to be construed as a lack of professionalism or as reduced concern for safety. The more automation intervenes between pilots flying aircraft and controllers in charge of the air traffic control system, the more probable it becomes that some aspects of the controllers' tasks will become dissociated from their true nature. Precipitate judgments should not be formed about this. It may be safer to perform mimics of tasks than the real ones. It would be too hasty to conclude that the representative task is necessarily inferior to the real one, but the propensity for dissociation from the real task is always present as automated aids increase.

One consequence of automation, aggravated if team roles are reduced, is that it can become over-protective of human inadequacy. With some automated aids for decision making and problem solving, a controller would do the task very well and in accordance with instructions, by always accepting the preferred computer solutions of problems. It might never become apparent if a controller accepted a computed decision without understanding it. At some point, automation starts to be capable of being overprotective of human inadequacy and of concealing human incompetence. This is not necessarily an argument against automation. To be alert to the possibility may be enough to resolve the problem. It may become necessary in the future to test controllers regularly in a simulation of air traffic control, in which various emergency, non-standard, and standard procedures are presented to them, so that they are required to demonstrate that they have retained their skills, knowledge, and professional competence, since such evidence can no longer be deduced directly from the way they do their jobs. This is an example of consequences of automation in air traffic control that reach far beyond those which are immediately apparent. Air traffic control and other large human machine systems share a propensity to generate such problems.

Any new form of automated assistance that affects human activities may introduce new possible kinds of human error. On the whole, human and machine do not make the same kinds of mistake, and it is therefore worth trying to use each to compensate for or counter errors by the other. An error that is plausible to one is not usually plausible to the other. To convert this potential advantage to an actual one, both human and machine must be extensively involved in the same task. In principle, human and machine combined have greater potential to prevent human and machine error than either has alone. One issue concerns what the human user of the system needs to know about the ways in which automation can fail, before being able to recognize that a failure has occurred. A corresponding question is how far it may be feasible to detect human errors or prevent their consequences, by automated checks based on evidence stored in the computer about the kinds of human error that are typical.

Safety of Current Air Traffic Control Systems

In considering the past and future impact of automation on air traffic control, it is important to recognize how safe and efficient many air traffic control systems currently are. Because of the publicity which attends any lapses in air traffic control, the impression can be gathered that the system is inherently dangerous and seriously inadequate, yet this is misleading. Given human fallibility, it is a great tribute to the competence and professionalism of controllers that incidents are so few and that those with serious consequences are rarer still. An important perspective which has major effects on the practical applications of automation to air traffic control is that the purpose is not to improve a system which is thought to be highly inefficient and potentially dangerous; it is to increase the traffic handling capacity of a system which is already highly efficient and safe within its limits, but which must be able to handle more traffic without compromising that efficiency and safety. Because an incipient problem with any new form of automation is that it can inadvertently change functions which hitherto have been efficient and satisfactory, it is vital to identify beforehand the full range of the effects of automation, so that its introduction is successful and does not bring any unwelcome surprises.

When a system is already functioning manually at or near its optimum and maximum, any benefits from changes in the form of automated assistance may be limited, unless they affect the true system constraints directly. If the capacity limits are set by wake vortex regulations, minimum separation standards, runway capacities, route structures or the serviceability of navigational aids — and not by the factors which the automation actually improves — then automation may not increase capacity. In manual air traffic control systems functioning at maximum traffic handling capacity, controllers may become very busy and tired, experience stress, and believe that they are unappreciated. But such conditions also provide opportunities to maintain and exercise skills, to rise to challenges, to gain the esteem of colleagues, and to have job satisfaction. Automation changes all these factors, and the balance of advantage and disadvantage among them has to be assessed in deducing what its full implications are for future air traffic control systems. To get the best from automation, it must remove the disadvantages and limitations without losing any of the advantages.

References

Adair, D. (1985) *Air Traffic Control.* Wellingborough: Patrick Stephens.

Dieterly, D. L. (1981) Automation in Organizations: Eternal Conflict. Moffett Field, Ca.: NASA Ames Research Center, NASA Tech. Memo. 81290. AFHRL-A-81-808.

Fitts, P.M. (Ed.) (1951) Human Engineering for an Effective Air-Navigation and Traffic-Control System. National Research Council, Washington, DC.

HMSO (1956) Automation. A Report on the Technical Trends and their Impact on Management and Labour. Department of Scientific and Industrial Research. London: HMSO.

Hopkin, V. D. (1979) New Work Force Roles Resulting from New Technology. In *Proceedings of 24th Annual Meeting of the Air Traffic Control Association.* Atlantic City, NJ.

Hopkin V. D. (1980) Boredom. *The Controller. 19*, 1, 6-9.

Hopkin, V. D. (1982) Human Factors in Air Traffic Control. Paris: NATO, AGARDograph No. 275.

Hopkin, V. D. (1985). Fitting Machines to People in Air Traffic Control Automation. In *Proceedings of Seminar on Informatics in Air Traffic Control. EEC*: Capri, Italy. 147-163.

Hopkin, V. D. (1988) Training Implications of Technological Advances in Air Traffic Control. Keynote Address at FAA Symposium on Air Traffic Control Training for Tomorrow's Technology. FAA: Oklahoma City.

Lysaght, R.J., Hill, S.G., Dick, A.O., Plamondon V.D., Linton, P.M., Wierwille, W.W., Zaklad, A.L., Bittner, A.C. & Wherry, R.J. (1989) Operator Workload: Comprehensive Review and Evaluation of Operator Workload Methodologies. Fort Bliss, Texas. US Army Research Institute. Tech Report 851.

Melton, C. E. (1982) Physiological Stress in Air Traffic Controllers: A Review. FAA Office of Aviation Medicine, FAA-AM-82-17.

Narborough-Hall, C. S. (1985) Computer Assistance - Implications for Memory Performance as a Function of Control Responsibility Assigned to Man. In *Proceedings of 2nd IFAC/IFIP/IFORS/IEA Conference*, Italy. 67-73.

Pozesky, M. T. (Ed.) (1989) Special Issue on Air Traffic Control. Proceedings of the IEEE, 77, 11, 1603-1775.

Saint-Exupery, A. de, (1939) *Wind, Sand and Stars.* London: Heinemann.

Assumptions and Automation in Air Traffic Systems

John A. Wise

Center for Aviation/Aerospace Research, Embry-Riddle Aeronautical University, Daytona Beach, FL, USA

Assumptions

One of the goals of the Institute was to help the participants to question their assumptions about air traffic control (also see Hopkin, 1991). Assumptions are critical to all decision making and problem solving because they have a hidden, yet dramatic, impact on conclusions. Assumptions determine which data are accepted or rejected as relevant to the decision process. They determine the rules that will be used to process the accepted data. They determine the relative weight each datum gets. Thus, assumptions, hidden as they may be, directly impact whether a solution is accepted or rejected.

An interesting example of the significance of assumptions on decisions can be demonstrated by an analysis of medical diagnosis (Payer, 1990). Accepted medical data appear to show that Germany has a significantly lower rate of coronary artery disease than the U.S. Therefore, we conclude that there must be a difference in German diet, air, or environment to cause the difference. However, if we look at the total heart disease statistics, we find that they are essentially the same in the U.S. and Germany. The difference in coronary artery disease is in the criteria used to diagnose (i.e., sort) the various types of heart disease. The German criteria are slightly different from those used in the U.S. The assumption that we usually make, however, is that everyone uses the same diagnostic criteria. Thus, we assume there is a medical difference, when it is really only a semantic one (also see Westrum 1986 and 1991, for other examples)

The reality of the world is that assumptions are usually only made explicit to students in introductory classes on a subject, or possibly on the first day of a design. More often than not, assumptions are taken for granted even then, because they are usually *assumed* to be *unquestionable*. Even when assumptions were addressed early in design, as the system gets older, the assumptions slowly become matters of "faith and morals." And those who do question the assumptions are seen as radicals, or perhaps crazy.

Air Traffic Control: Key Assumptions

There is a crowded sky.

The popular press and even many people in aviation itself talk often of the crowded skies. They speak as if there are simply too many aircraft for the amount of airspace we have

NATO ASI Series, Vol. F 73
Automation and Systems Issues in Air Traffic Control
Edited by John A. Wise et al.

available (the proof can be seen in the take-off queues and holding patterns at most major airports). Therefore, some types of limits are needed.

However, if every single aircraft in the U.S. (private, corporate, airline, and military) were airborne at exactly the same time, they could all fly over the state of Texas and each have one square mile of protected airspace (from the surface up). The other 49 states would have completely empty skies. Could you imagine driving your car on a road where the closest car was a mile away? We would believe we were in heaven.

If we are not at the physical limits of airspace, then it must mean that the ATC systems are limited in their capacity to meet the demand being placed on them by the society (e.g., business, individuals) that they support. In fact, I might argue that, because ATC has traditionally been a government (i.e., bureaucratic) activity,[1] there is a natural tendency to ignore customer demand and try to regulate demand to meet capacity, rather than change capacity to meet demand.

Air traffic control provides universal separation.

ATC was originally intended only to assist pilots who were flying in instrument meteorological conditions (i.e., meteorological conditions that made it impossible for pilots to see the horizon or other aircraft and thus maintain their own separation). The U.S. requirement for airlines to use ATC services on all flights was not a decision based on analysis of ATC system performance or a safety analysis; it was purely political. It was an attempt by the Congress to show the public that they were aggressively doing something to improve aviation safety after the mid-air collision that occurred over the Grand Canyon in the 1950's. This requirement was politically acceptable to most Americans because pilots are often seen by the general public as adventurous, daring, and in need of control (especially when the public is in the back of the airplane).

While the general public assumes that ATC provides separation between all aircraft, in the U.S., ATC is not even required to provide guaranteed separation to all aircraft on instrument flight plans. In the U.S., separation is only *guaranteed* to those aircraft 1) flying under instrument flight rules, 2) in controlled airspace, and 3) in instrument meteorological conditions. This point can be clearly shown in the FAA's response to the 1978 San Diego midair that occurred between two aircraft under ATC control:

> Within the rules specifying controller and pilot responsibility and visual meteorological conditions, there was no requirement to provide any form of standard separation between the aircraft involved. (FAA,1979)

Airline greed is the cause of congestion.

Another popular assumption, especially in the U.S., is that airline deregulation and the airlines' innate desire to depart in the same time periods are the reasons for the current "air traffic congestion." As such, it is quite natural to assume that the best way to deal with the

---

1. It should be noted that, in fact, in some places ATC has been turned over to industry. New Zealand is a current example. These attempts deserve to be watched to see how they well they fare relative to government-based systems.

problem is control of the selfish airlines by government based flow control.[2]

But why would any airline want to depart at those times when large amounts of fuel will be wasted in taxiing and holding patterns, especially when fuel is such a large part of the operating budget?

One can also argue that the cause of congestion is really the airline consumer. As each airline tries to maximize its market share of passengers, they study the desires of their customers to determine the times at which customers want to fly. The airlines then match their flight schedules to demand (the appropriate thing for a free enterprise corporation to do). Thus, the airlines schedule their flights at approximately the same time. If this is the case, then perhaps the best method to minimize congestion associated with customer demand is not limiting actual flights, but an economic incentive for the passengers to use others time slots. For example, we might ask, how could we modulate air fares in order to get a flatter demand curve? This is what telephone companies do to spread demand over more of the day.

The government is the necessary guarantor of safety.

One of the basic assumptions of current systems is that there is always a need for someone outside of the activity of flying itself to guarantee the safety of the flying public. When we talk to true believers in the current approach to ATC, it is clear that they believe that without direct ground control, there would be carnage in the air. Pilots would be hurling themselves through the sky with disregard for their own safety. All we have to do is talk to controllers to believe that, if they were not on the job, everyone in the sky would quickly be involved in a midair.

In the U.S., this assumption is based in part on the American cultural tendency to regulate that which we do not understand or feel we personally do not have much control over. For example, while most Americans are completely for ground-based control of all aircraft, they go crazy anytime the government infringes on their "right" to drive their cars where and when they want.

But U.S. air regulations have a deadly Catch 22. While pilots have the ultimate responsibility for the safe conduct of their flights (after all, they are the ones who will die if something goes wrong), ATC (in order to be efficient) operates on the principle that pilots are subservient to the controller.

While expanded ATC authority has a certain face validity, one can in fact argue that reliance on the ATC system is more dangerous in visual meteorological conditions than is proceeding under visual flight rules. Remember, in visual meteorological conditions, there is no *requirement* for ATC to provide the pilot operating under instrument flight rules with separation or notification of *all traffic*. But pilots are lulled into a false sense of security, believing that ATC is providing them separation from everyone (even though they have formal knowledge that it is not true).

As a result, the pilot spends more time maintaining exact altitude and airspeed (i.e.,

2. It is particularity ironic, at the time in history when the government-controlled economies of eastern Europe have failed and are charging head long into free enterprise economies, that the world's leading advocate of free enterprise, the USA, is applying the communistic techniques of high government control of commerce in the form of "flow control." In flow control, the government says who can take off (read: build widgets), when they can take off (read: how many they can build), and the route they can take (read: who the widgets can be sold to).

head-down in the cockpit), instead of head-out of the cockpit looking for other traffic. When an aircraft slips by ATC (due to ATC overload or the ATC system's not interrogating its particular transponder code), there is a reasonable probability that it will be missed because the pilot[3] will not be looking for it.

As an U.S. Air Force instructor pilot, I watched student after student stare at the gauges trying to maintain exact altitude and airspeed and never look out through the beautiful bubble canopy the taxpayers provided him. No matter how much I preached "look outside," they would continually go back inside, checking to make sure they were following ATC directions to the letter, secure in the belief that ATC would clear the path for them.

Controlled airspace is safer.

One of the basic assumptions for the existence and application of air traffic control systems, as they are now designed, is that the most effective way to avoid midair collisions is to use positively controlled airspace (i.e., airspace where all aircraft are under direct ground based control). But when we look at the data (see Table 1) collected by NASA's Aviation Safety Reporting System (ASRS), we find an unusually high proportion of near-midairs in positively controlled airspace, as opposed to similar airspace that is not positively controlled.[4]

Table 1
ASRS reports of near midair collisions in terminal airspace (May - November 1978)
(Reported by Billings et al., 1980)

Category	TRSAs	TCAs	Other terminal airspace
All near-midairs	120	76	231
Estimated total 1978 near-midairs	222	140	426
FAA airport traffic volume (FY '77)	16.2M	7.8M	52.2M
Reported near-midairs per million ops	14	18	8

3. This is even true of fighter pilots in bubble canopies, who are supposed to be the best at visual clearing.
4. It should be emphasized that these are voluntarily reported incidents, subject to the biases that affect such data.

The above data indicate that there are more than twice the number of reported near-midairs in the positively controlled airspace of a TCA that in other types of terminal airspace. Perhaps a smaller percentage of near-midairs are reported in less controlled airspace because people expect other aircraft to be closer. If that is true, the question should then be "How close is unsafe?"

Controlled flight is more efficient.

If ATC systems were in fact providing the fastest, safest, and most efficient service possible, it would stand to reason that most pilots would take advantage of the service. After all, in the U.S. at least, the service is "free" and it is "easier" to use ATC than to do all of your own navigation. But most flights in the U.S. do not use the ATC system.[5] They simply go from point A to point B, using their own navigational and see-and-avoid skills.

The U.S. Air Force requires its solo student pilots to fly formation with one meter of wingtip-to-wingtip clearance. This clearance is expected to be maintained even in maneuvers and attitudes the FAA would define as acrobatic. If the Air Force considers two meters' separation safe for students with less than 200 hours total time, why is anything less than 8064 meters considered unsafe for high experience airline crews flying straight and level in TCAs? What requires the additional 8,062 meters in a supposedly safer, more efficient system? And what, if anything, can be done to reduce that "wasted" separation?

The best example of a highly efficient ATC operation is the Experimental Aircraft Association's Fly-In at Oshkosh, WI. This is only true, however, because during the fly-in, the usual control rules are thrown out the window by the FAA. As a result, that simple low-tech system can handle twice the volume of the complex and very high-tech Chicago O'Hare. And when we consider that, during the fly-in, Oshkosh is open for less than half of each day, and that a significant number of aircraft don't have radios, the performance is even more impressive.

So What?

This is not meant to be a condemnation of ATC or of controllers. Nothing could be further from the truth. My intention has been to demonstrate that at least some of the basic assumptions generally held about ATC may not match well with reality. If we design expensive new systems based on those assumptions, the results will be at best inefficient, and at worst failures. The design of future ATC systems needs to be based on the best and most accurate set of assumptions possible.

To determine what the best assumptions are will require a considerable amount of work on the part of both the funding agency and the designer. Perhaps they may only really be achieved by a disinterested third party. Those actively involved in ATC (both pilots and controllers) may be too close to the system to be ever really able to identify and eliminate their own biases.

---

5. Personally, even though I am an instrument rated pilot, I only use the ATC system when meteorological conditions require it. The remainder of the time, I proceed totally on my own.

Conclusions?

As Disraeli once noted, it is much easier to find fault with something than it is to develop and implement solutions to those problems. I freely admit that I have taken the easy path. I have only sought to identify some flaws in current system assumptions, and only in a few cases have I even provided a hint of a solution.

What I have tried to do in this paper is the same thing I attempted to do at the Institute; to remind us to recognize that all of our assumptions about ATC were not handed down directly from God, written in stone. ATC systems rest on the same shaky human determined assumptions as every other human designed system. As a result, no matter what technology we throw at ATC, it will only work as well as the underlying design assumptions allow it to work. The best we can hope for is to

- Use the best set of assumptions, given the state-of-the-art
- Design the optimal system that the assumptions allow.

In my mind, the papers in this book can be divided into two categories, based on these goals. The categories are:

- Those that address design, using current assumptions (i.e., all the papers except those in the section on "blue-sky" concepts)
- Those that address systems that might result if different assumptions were used (i.e., the papers addressing "blue sky" concepts[6])

As you read each paper, please try to identify the assumptions each author has used and try to generate an antithetical set, and then determine how the conclusions of the paper (and the system issue discussed) would change if your antithetical assumptions were closer to reality than the author's. Test your antithetical assumptions against all the authors' arguments (including my own) to see whether, if they were applied to future systems design, they might improve system performance. We might all be surprised at the outcome.

References

Billings, C. E., Grayson, R., Hetch, W., & Curry, R. E. (1980). A study of near midairs in U. S. airspace. In. *NASA Aviation Safety Reporting System: Eleventh quarterly report.* Moffett Field, CA: NASA Ames, NASA Tech Memo 81225.

Federal Aviation Administration. (1979). FAA practices, procedures, and actions at San Diego, California, September 25, 1978. Washington, D.C.: U. S. Department of Transportation, December 21.

6. It should be noted that all of the blue sky papers were generated after the Institute and were voluntarily submitted by the authors as a result of discussions held at the Institute.

Hopkin, V. David. (1991). The impact of automation on air traffic control systems. In: J. A. Wise, V. D. Hopkin, & M. L. Smith (Eds.) *Automation and Systems Issues in Air Traffic Control.* Berlin: Springer-Verlag.

Payer, L. (1990). Borderline cases. *The Sciences*. July-August. 38-42.

Westrum, R. (1977). Management strategies and information failure. In J. A. Wise & A. Debons. *Information Systems: Failure Analysis.* Berlin: Springer-Verlag. 109-128.

Westrum, R. (1991). Automation information and consciousness in air traffic control. In J. A. Wise, V. D. Hopkin, & M. L. Smith (Eds.) *Automation and Systems Issues in Air Traffic Control.* Berlin: Springer-Verlag.

Perspectives on Air Traffic Control Automation

Computerization and Automation: Upgrading the American Air Traffic Control System

Arthur A. Simolunas[1] and Howard S. Bashinski[2]

[1]Federal Aviation Administration, Washington, D.C., USA
[2]CTA Incorporated, Denver, Colorado, USA

Introduction

Implementation of the FAA Advanced Automation System will be the first major human interface change in the air traffic control system since the original automation program. Replacing flight progress strips with electronic displays will require significant human engineering effort. The implementation of more complex safety-related functions also will require considerable design efforts due to the system's inherent reliability and availability requirements. New commands and procedures, in addition, will be necessary. This report describes the FAA developmental approach of using user teams to ensure an effective computer-human interface.

Since the 1960's, the United States Federal Aviation Administration (FAA) has been implementing a series of automation upgrades for the Air Traffic Control (ATC) System. This effort is currently known as the Advanced Automation System (AAS) (Benel, Dancy, Dehn, Gutmann, & Smith, 1989). These developments were formally initiated in 1988 and will continue into the next century. They will result in a major re-alignment of the ATC system's functions and locations, including implementation of advanced systems for both the en-route and terminal functions.

Generally, any large-scale undertaking involves unexpected issues and challenges; the AAS is no exception. Particular challenges to AAS include the enhancement of complex command-and-control systems through automation, the large scale of the project (AAS will eventually require over one billion lines of software source code), and the system's accuracy and reliability requirements.

Although many specific design areas are affected, the change from paper flight-progress strips to electronic flight-data entries is worth particular attention. This process is critical to the controller's job and has required much effort. This paper addresses the following electronic flight-data issues associated with developing automated ATC systems:

- Application of automation to those controller tasks currently requiring complex cognitive processing (automated reasoning)
- Problems associated with the "hidden" complexities of ATC automation
- Role of the users (controllers) in the development of ATC automation
- Prospects for increasing automation in the future

NATO ASI Series, Vol. F 73
Automation and Systems Issues in Air Traffic Control
Edited by John A. Wise et al.

Overview

The FAA always has implemented changes in the ATC system progressively to minimize the impact on operations, and the AAS continues to use this approach. Training and transition requirements are important to these progressive changes.

The first segment of the AAS program, the Initial Sector Suite System (ISSS), will be the first significant change to the controller computer-human interface (CHI) since the implementation of the original National Airspace En-Route Plan View Display (PVD). In ISSS, the situation display (containing the processed radar returns and heading, speed, and altitude information) will remain basically the same as in PVD. The controllers, however, will face new data-entry devices, including new keyboards and trackballs. Most significantly, electronic displays, which will replace paper flight-process strips, will require new commands and operational procedures. Extensive adaptive-display capabilities and intra-sector operations also will affect the controller CHI significantly.

The initial NAS automation did not extensively change the controller functions, and the improvements focused on relieving the controller of routine duties. Processing flight-plans and combining separate data sources onto a single situation display, for example, relieved the controllers of the task of manipulating multiple data bases, and aided their thinking and perceiving processes. Although these improvements did automate some decision processes, they were routine and easily implemented; the number of variables was constrained, the data were unambiguous, and safety implications were minimal. The controllers, however, still performed the basic task of separating aircraft.

Since the initial implementation, a number of functional, primarily safety-related improvements have been added to help the controller detect such problem situations as the impending violations of separation standards. Identifying false alarms and problems undetected by the system, however, continues to be the controller's responsibility.

These new functions will help create a system that can apply automated reasoning across a large number of "fuzzy" variables with complex databases. Several factors challenge the adequacy of operational concepts and procedures; for example, response times are more critical for traffic-separation functions than for the earlier controller-aiding functions. As automation is applied to aircraft separation, the algorithms become more complex, data-reliability requirements more critical, and system fault-tolerance requirements more exacting.

The AAS program provides the platform for the next generation of automation improvements. In the past, detailed-specification developments in routine controller automation were flawed; thus, the FAA and its prime contractor, the International Business Machines Corporation (IBM®), are attempting to enlist the expertise of the controllers themselves in the development of the system. An operational user team (the Sector Suite Requirements Validation Team (SSRVT)) will work closely with the contractor and will be involved actively in all phases of the system-development process.

Although many automated flight-processing functions resulted in a control decision, such as posting flight plans or handing-off aircraft between sectors or facilities, most were limited to computerizing basic manual functions and composing data for video display. Very few of the complex aircraft-separation functions were truly automated for making control decisions, and automated reasoning using imperfect real-world information was minimal (Post & Sage, 1990). The addition of new computerized tools and services for the controllers increased productivity and presented a more integrated picture of the air traffic situation; however, it did not produce the commands and instructions that separate aircraft.

Basically, the FAA intended to reduce the controllers' data-entry load and aid their interpretation of the resulting display. First, a computer database was developed and flight strips were printed in an effort to computerize the flight plan processing system. Next, radar "blips" were electronically labeled with a data block displaying speed, heading, and altitude. This improvement, however, did not reduce the controller's responsibility for separating aircraft. Subsequent upgrades to the system also used this workload reduction approach.

The AAS will extend this computerization. In the place of paper flight strips, the flight data display will project electronically the recorded clearances and other control instructions. The system will automatically update the presentation in real time and allow manual changes as necessary. Other new features include techniques that minimize clutter and reduce data density, such as the time-sharing of data field and logical filters for routine information on both graphical and tabular displays. The controller, in addition, will have access to an electronic display of maps and charts. The chart information will be layered and filtered to minimize the cluttered appearance of paper charts. Features such as color, variable brightness, shading, blinking, and inverse video, as well as new technologies such as icons, menus, and pointers, will improve the display interpretation (Hopkin, 1989). The controllers will be able to adjust these features individually and save them as "preference sets."

These advanced features, however, cannot replace the controller decision process. In general, they are tools to aid the controller in the basic ATC task of separating aircraft. In fact, although the purpose of computerization is to improve safety and enhance controller productivity, it sometimes results in additional work for the controller. Electronic flight data displays, for example, will require new controller commands and interactive techniques that are very different from manual sorting and marking paper flight progress strips. It cannot be taken for granted, therefore, that computerization will cause a reduction in the controller's tasks.

Arguably, some complex automated decision processing has already been instituted. Conflict Alert (CA), Mode-C Intruder (MCI) detection, and En-Route Minimum SAfe Altitude WArning (MSAW), which calculate possible airspace utilization conflicts, are presently being developed and implemented. These "attention tools" strictly aid the controller in the decision-making process.

As with former new systems, the FAA will enhance AAS functionality in increments. The enhancement will take into consideration safety issues as well as minimize training and transition problems. Accordingly, the new electronic flight data display will approximate the current paper/manual protocol and syntax. An interesting exception to this computerization is the "implied commands" in which the controller will select a displayed object and the system automatically will proceed to the object's next logical state. For example, if the object is in a "handoff" state, the system will assume that the implied command is to "accept handoff."

Automated reasoning probably will be first applied to the basic controller tasks in 1998, with the Automated En-Route Air Traffic Control System (AERA). The initial implementation, AERA-1, will predict problems based on known intentions and assist with control and coordination tasks (Elsaesser, Gisch, Hines & Swedish, 1984). True automated reasoning probably will not be accomplished until the completion of AERA-2 and AERA-3 after the turn of the century (Coleman, Guntherse, Murphy, Rooney, Shannon, Sheppard, Stewart & Tuttle, 1986; Rockman, 1989). AERA-2 and AERA-3 will add the complex resolution processes required to process ambiguous, imprecise, incomplete, and inconsistent information.

Automation Complexity

Converting the paper flight progress strips to a computerized display has required major human engineering effort. Programming any basic function is more complex than it seems, and the AAS programs is not an exception. AAS's complexity, however, is not surprising, since its safety-critical environment requires extensive failure modes. Automating a manual function, in addition, can be very deceptive from an overall design perspective. Developing a basic program to replicate a function merely represents a subset of the total programming effort. The complexity of the relationships between functions, the establishment of failure modes, the design of software that recognizes failures and alerts the controller, and the fault avoidance and fault tolerant approaches must be considered as well. These ancillary problems are seldom obvious to the software designer and are not usually resolved until late in the testing phase of a procurement. While developing a specification and reviewing design documentation, the operational user often overlooks this sophistication. The software complexity appears not only in the "application" programs, which are the basic controller functions, but also in the software which controls the process.

One method of fault avoidance is to use commercially available software which, after extended use, has been thoroughly debugged. Unfortunately, commercial software may not have the fault tolerant and detection features necessary for the ATC environment; thus, additional monitor and mode management software will be required, resulting in further software complexity. These programs, in addition, affect air traffic control response times, which must operate at a very high functional availability; for example, critical functions must have availabilities approaching 0.9999999, which amounts to about 3 seconds per year of down time. To achieve the required fault tolerance, the system must be capable of fault detection and, in turn, fault isolation. The system must then reconfigure into a fail-safe mode if redundant elements are available, or, if necessary, a fail-soft mode retaining the critical ATC functions. During these fault situations, the system must notify the controller and retain the information necessary for timely maintenance actions. These recovery procedures require significantly more software than necessary to replicate air traffic control functions.

Obviously, computerization of routine functions, even in the formative stages, is a highly complicated operation. In order to perform controller reasoning functions, the system would require more reliability, better response times, and better alerts to allow the "traffic manager" to perform problem resolution verification. This confirmation procedure, of course, will result in additional system complexity and software development magnitude.

Another complicating factor is the enormous human database storage and retrieval capability necessary to perform the resolution process. The machine speed and capacity required to replicate this capability may be even greater than anticipated. In addition, the human logarithms used to perform this function are not well understood.

Recognizing their complexity, the FAA instituted an early evaluation of developing systems. An independent test system will be implemented close to IBM's development facility. In addition, IBM has agreed to furnish early versions of software builds, so that FAA operational personnel, both controller and maintenance personnel, can exercise the system in an unstructured manner. In previous automation procurements, unstructured environments were necessary to discover system flaws. Scripted factory testing did not explore the areas often traveled in the real world, and discovery processes were delayed until after the operational and test evaluation phase or even after field implementation. During these latter stages, problems were difficult to correct. The FAA, therefore, considers it imperative to involve operational personnel at the earliest date possible.

Controller Involvement in the Development Process

To ensure operational suitability and controller acceptance of AAS, the FAA has employed controllers as advisors to the development process. The controllers are divided into "user teams" which specialize in a broad technical or operational area, such as CHI and testing issues or en-route and terminal operations. The user teams have been involved from the beginning and will continue to provide technical support throughout system implementation.

There are several important issues associated with user input in the development of automated systems; for example, although the controllers will use the system, the program specialists will actually manage the development. The benefit of user input, therefore, must be weighed against the constraints imposed by budgets and schedules. Also, the ultimate audience is the design engineer, and the information must be presented in an engineering context. Finally, considering the complexities of the system, the user teams must maintain a high level of proficiency throughout its development.

The AAS User Teams are service organizations to the AAS Program Office; thus, the Program Office must manage the interaction between IBM and the Air Traffic organization. The user teams develop the operational and functional requirements and present them first to the Program Office, which assesses their programmatic impact. Requirements that are clearly safety issues are readily incorporated and passed on to IBM which also receives user input in the form of design guidance.

Because the controllers' operational and functional input eventually affects system design, this information must be presented in a useful way to the design engineer. The substance and the format of the information is important, for a controller discussion of operational requirements can sometimes resemble sheer anarchy. In order to focus these efforts, initial user teams developed detailed concepts of operations for the various segments of AAS. The resulting "ops concepts" are rigorous, systematic, and detailed descriptions of the controllers' job in each of the AAS segments (Ammerman, Fairhusrt, Hostetler, & Jones, 1988). In an hierarchical manner, air traffic activities are traced through sub-activities, tasks, and task elements. The ops concepts describe the criticality and frequency of all tasks, and include cognitive and sensory attributes for highly critical tasks; in addition, they describe all inputs and outputs necessary for the controller to perform each task. In effect, the ops concepts are detailed CHI task analyses that completely describe the controller's job (Phillips, Bashinski, Ammerman, & Fligg, 1988).

The operational concepts provide a framework for controller input and feedback on the various AAS segments. With respect to electronic flight data, the task requirements are specified comprehensively, and can be referenced to functional and software specifications. In this way, the system designers have a reference point for user team input. The ops concepts also offer a basis for developing operational test requirements for electronic flight data and determining operational suitability (Davies & Bashinski, 1987).

The controllers' commitment to long-term involvement is important to the FAA's user team program. Although the specific makeup of the various teams will inevitably change, some members have served since the inception of the teams, a period of over eight years. This consistency has allowed the teams to maintain a high degree of expertise in their respective AAS design areas. Expansion of the teams to meet the increasing demands for controller input, however, has freshened the overall team perspective. Furthermore, as the design matures and enters the test phase, the controllers must have a broader exposure to the systems.

The need for a high level of expertise in the area of electronic flight data is especially critical. As software and other design constraints arise, the controllers must continually re-evaluate the operational requirements of electronic flight data. In some cases, special intensive working groups are formed to resolve design issues. This approach is used often with respect to controller interaction with the electronic data. The controllers must determine whether the electronic system meets all requirements of today's manual system and takes advantage of automation benefits.

Ultimately, the goal of controller involvement is to help mitigate risks associated with operational suitability and controller acceptance. These risks are especially great in areas undergoing major changes in current operations, such as in the transition from paper to electronic flight data in the ATC environment. User involvement must be somewhat systematic, however, with the procuring agency in clear control of the interaction between the users and the development contractor. Finally, trade-offs must occur between the depth and the breadth of user involvement.

Future Shock

We have discussed some of the issues associated with automating the U.S. ATC system, particularly with respect to electronic flight data. Unfortunately, the future outlook shows that these issues will become even more difficult to resolve as the system moves from "computerization" to true automation. Additionally, new issues, particularly those concerning the changes in the air traffic controller's role, undoubtedly will arise.

The previously discussed AERA system was conceived as a follow-up to AAS. Gradually, AERA will replace the cognitive controller tasks with automated performance. In comparison to AERA, the early AAS segments represent "simple" computerization of manual tasks. AERA will provide capabilities, such as prediction and planning, that are now the sole responsibility of the controller.

In AERA, flight data ultimately will become superfluous; that is, controllers will not need any visual history/status on the aircraft because the system will access specific data after informing the controller that a decision must be made. As systems become more and more capable of accurately making safety-critical decisions, the controller's job may become an example of "management by exception."

Such advances will further increase the accuracy and reliability requirements for the future. Systems that are nearly 100% reliable are reasonably expected, especially when air safety is an issue. Managers and planners must understand the technical challenges of meeting this type of goal so that they can implement reasonable development plans. Providing significant resources for front-end research, development, and technical validation will be increasingly important.

Because the controller's role is likely to undergo significant change, the controller must maintain a high level of involvement in the development of future advances. Clearly, the controller is the single most important resource in determining the operational requirements of controlling air traffic.

All of the challenges discussed in the paper must be resolved before more advanced automated systems are implemented. Considering the past difficulties of computerization, we can expect that many risks to the automation process will not be fully identifiable during the planning phases. The planners and managers, therefore, must locate potential problem areas as early as possible and apply enough resources to advance planning and technical analyses.

Conclusions

Automated reasoning, implemented in the latter stages of AAS, will produce new and possibly unanticipated levels of system complexity. These functions will require an extremely high degree of fault tolerance not already designed into the basic system architecture. Previously, requirements were recognized and detailed only during development; thus, the contractor and the user need to establish new levels of understanding. It will be tempting, however, either to reduce the necessary requirements or to add requirements that are "nice to have." In order that the contractor avoids penalty for imperfect specification and the user receives an acceptable operational system, they must exert exceptional understanding and management.

To achieve this goal, the design and development phases must include the operational user. The customer, in addition, must periodically perform independent evaluations during the development phase. Regardless of specification requirements, the operational suitability and stability of the system must be established prior to the operational test and evaluation phase.

Field controllers in the Sector Suite Requirements Validation Team (SSRVT) have been highly productive. The formation of this team has led to the establishment of user groups in control areas besides the en-route. Also, employing user-experts for the entire course of development significantly helps to mitigate programmatic and technical risks in the CHI area. A clear technical framework for user involvement, such as detailed operational concepts, and long-term participation by individual team members is important to such an effort.

The issues discussed in this paper are only a subset of those encountered by developers of large, complex command-and-control-type systems. The problems are programmatic as well as technical, and will require innovation and ingenuity to achieve true automated decision capability.

References

Ammerman, H. L., Fairhurst, W. S., Hostetler,C. M., & Jones, G.W.(1988). FAA Air Traffic Control Operations Concepts: Vols. I-VIII (Report No. DOT/FAA/AP-87-01). Washington, DC: Federal Aviation Administration.

Benel, R. A., Dancey, R. D., Dehn, J. D., Gutmann, J. C., & Smith, D. M. (1989). Advanced Automation System Design. *IEEE Proceedings: Air Traffic Control, 77*(11), 1653-1660.

Coleman, B., Gunthersen, N., Murphy, E., Rooney, P., Shannon, R., Sheppard, S., Stewart, L., & Tuttle, B. (1986). AERA 2 Operations Concept (Contract No. DTF-A01-85-Y-01034). Englewood, CO: CTA INCORPORATED.

Davies, D. K., & Bashinski, H. S. (1987). Human Engineering Evaluations in Real Life: Operational Suitability Assessment During ATC Design Development. *Human Factors Society Bulletin, 30*(5), 1-3

Elsaesser, C., Gisch, A. H., Hines, A. L., & Swedish, W. J. (1984),). Description of AERA 1 Capabilities (MitreTech.Report No. MTR-84W88). McLean, VA: The Mitre Corporation.
Hopkin, V. D. (1989). Man-machine Interface Problems in Designing AirTraffic Control Systems. *IEEE Proceedings: Air Traffic Control, 77*(11), 1634-1642.
Phillips, M. D., Bashinski, H. S., Ammerman, H. L., & Fligg, C. M, Jr. (1988). A Task Analytic Approach to Dialogue Design. *I. M. Helander (Ed.), *Handbook of Computer-human Interaction* (pp. 835-857). North-Holland: Elsevier.
Post, S., & Sage, A. P. (1990). An Overview of Automated Reasoning. *IEEE Transactions: Systems, Man and Cybernetics, 20*(1), 202-224.
Rockman, M. J. (1989). Automated Decision-making in the Air Traffic Control Process: Vol.1 System Overview (Report No. DOT/FAA/ADS-89-01). Washington, DC: Federal Aviation Administration.

The Canadian Automated Air Traffic System (CAATS): An Overview

Paul Stager

York University, Toronto, Canada, and CAATS Project Office, Transport Canada

Introduction

Exactly fifty years after the initiation of air traffic control (ATC) in Canada, Transport Canada has launched a comprehensive and innovative program, the Canadian Automated Air Traffic System (CAATS) Project, to completely revamp the national ATC system. This paper provides a brief overview of the background, objectives, design, and implementation of that project. The intention of the paper is to present a general description of the CAATS concept, without detailing either the engineering requirements or the developmental program. Excellent reviews of some of the issues, particularly those concerning human factors engineering, that must be addressed in a comprehensive ATC automation project already have been provided by Benel, Dancey, Dehn, Gutmann, and Smith (1989) and by Simolunas and Bashinski (1991). An introduction to the concept of human error in the context of automation has been outlined by Stager (1991).

Until the present time, the elements of the Canadian flight data processing infrastructure have evolved in a piecemeal and non-integrated manner, with each element performing a particular function and frequently duplicating the functions of other components. The flight data processing functions for the national ATC service are scattered through many systems and subsystems, some involving obsolete technology. However, the increasing complexity of the modern air traffic environment requires advanced automation technology to aid the controller in performing ATC functions more efficiently and effectively.

In recognition of this requirement, Transport Canada proposed to modernize the processing and handling of flight data, and in 1985 received approval for what was called the Flight Data Systems Modernization Project (FDMP). In addition, the Canadian Airspace Systems Plan (CASP) (Transport Canada, (1988) identified the need for the development of a Common Controller Workstation (CCWS) which would increase the efficiency of ATC operations at the controller interface. The CCWS project would apply the principles of human factors engineering to provide a systematic integration of the various formats and sources of flight data. In 1988, FDMP was enlarged to include the CCWS project and became the Canadian Automated Air Traffic System (CAATS) Project. In November 1989, Hughes Aircraft of Canada Limited was contracted to build and install CAATS in conjunction with MacDonald Dettwiler and Associates and a number of other Canadian companies.

NATO ASI Series, Vol. F 73
Automation and Systems Issues in Air Traffic Control
Edited by John A. Wise et al.

The CAATS Project

The CAATS Concept

The CAATS project is the third in a series of Major Crown Projects under the CASP, a plan which is intended to guide the modernization of Canada's air navigation system. The Radar Modernization Project (RAMP) and the Microwave Landing System project are currently underway.

The overall objective of the CAATS project is to replace the existing infrastructure of Air Traffic Services (ATS) for the processing and display of flight data with a modern integrated system that will meet the future operational requirements of Transport Canada. Through CAATS, Transport Canada's Aviation Group intends to reduce significantly the projected long-term growth in personnel, operations, and maintenance costs and, at the same time, to meet future growth-related demands for air traffic services. These objectives are to be met while maintaining, or enhancing, system safety.

A fully automated real-time *Flight Data Processing and Distribution System* (FDPS) will be the key element in CAATS. The FDPS, using route analysis, will automatically acquire, process, and distribute flight plan data to appropriate ATS components (e.g., units, sectors, etc.). A dynamic air traffic model, incorporating a meteorological model, will be maintained by the FDPS. This model will be updated by position information derived from the *Radar Data Processing System* (RDPS), Automatic Dependent Surveillance (ADS) sources, VHF Air Ground Link, input by Flight Service Stations, and the Aeronautical Fixed Telecommunications Network/Automated Data Interchange System (AFTN/ADIS).

CAATS will also provide controllers with enhanced capabilities to handle air traffic both in Canada's northern airspace and in oceanic airspace, areas in which surveillance and communications are limited. For example, CAATS will display extrapolated flight plan tracks in areas of non-surveillance (e.g., radar or ADS) by means of a dynamic depiction of the air traffic situation. This capability will apply in geographical areas where radar coverage may be discontinuous , such as in mountainous terrain. Enhanced capabilities will include the determination of Minimum Time Tracks (MTT), optimum track structures (oceanic or continental), potential conflicts, and options for conflict resolution.

The CAATS system availability requirements are stated in terms of the probability of having at least a given level of functional capability available to the operators. The requirements call for high system availability, ranging from 0.9997, for a full service mode, to 0.999997 for an emergency mode of operation. However, these requirements address the availability of functions to ATS staff and do not necessarily represent the availability of specific equipment configurations.

Ada, a higher order programming language, will be used for all newly developed software in CAATS, including all newly developed data communications software, air traffic control software, and simulation, training, and support software. Ada was chosen because it will promote sound software engineering principles and support the concept of reusability for applicability within CAATS and in future systems. For example, Ada will increase the capability of incorporating new operational requirements. It is also anticipated that Ada will reduce life cycle costs by ensuring ease of maintenance and higher reliability.

Traffic Management in CAATS

CAATS will provide a *national* strategic air traffic flow management capability which will increase the controllers' ability to handle the expanding volume and complexity of air traffic. The CAATS Flow Management System will provide the capability to automatically predict demands on system components (e.g., airports, runways, airways, and sectors). In practice, flow management will be implemented from the Area Control Centers which will have direct access to national information. This nationally integrated system will provide both strategic and tactical solutions to achieve the required traffic flow and metering, to reduce air traffic delays, and to provide more energy-efficient routes. Carriers will be advised of anticipated delays and allocated metering times prior to loading aircraft.

The implementation of CAATS will result in a streamlining of the nation-wide transfer of flight data on all known aircraft movements among ATS installations. Automated interfaces will replace existing manual links for data transfer with other agencies. These links will provide flight data and aircraft movement information for a variety of purposes, including air defense, Search and Rescue operations, and the compilation of aviation statistics. Direct communication links will be provided for the ATC systems sharing joint airspace boundaries with Canada (e.g., the United States, the United Kingdom, and Iceland). Specified users and interfaced systems will be provided with a controlled query capability.

Conflict Prediction and Resolution and Flight Conformance Monitoring

Conflict-free flight profiles will be produced for all IFR and CVFR aircraft outside radar coverage areas. The FDPS will provide *conflict prediction* messages only between known radar and non-radar traffic, and between all known non-radar traffic. The FDPS will also present controllers with safe and efficient *resolution options* for non-radar traffic. However, in radar airspace, the RDPS will provide the critical functions of *conflict alert* and the *Minimum Safe Altitude Warning* (MSAW).

The FDPS will also perform conformance monitoring for all enroute aircraft to determine whether given aircraft positions are within calculated tolerances for longitudinal, horizontal, and vertical limits of the current flight-planned profiles.

Task Analysis for CAATS and the Controller Interface

It is intended that increased levels of productivity can be achieved within CAATS through the use of an advanced workstation, like the CCWS, together with the system's flight and radar processing capability. Several aspects of the CAATS concept which have already been alluded to illustrate the fact that CAATS will be able to automate many time-consuming and data-intensive tasks now performed manually by controllers. However, a significant human factors engineering program is currently underway in order to achieve the most effective hardware and software interface. The program will focus on the development of the graphics requirements for situation displays and the formatting requirements for flight data displays as well as other interface characteristics.

The approach to be taken to interface design will involve, in part, the exhaustive analysis of controller tasks and behaviors in a system context. A systematic and detailed decomposition of controller tasks (e.g., Lenorovitz and Phillips, 1987) for the CAATS environment will yield sets of tasks which can be validated by experienced operational personnel. The resulting data will be used to guide the iterative design of the controller interface and to address the semantic, syntactic, and lexical design requirements for software development.

Day (1991) has outlined some of the human factors concerns surrounding human-computer interaction and interface design, especially when structured design methodologies are used. These concerns include the role of experimentation, human factors standards, and usability testing. There is a need, for example, to promote human factors experimentation and user involvement during the design process. One of the objectives of the CAATS Human Engineering Program Plan will be to ensure active participation by operational controllers in the developmental work.

Systems Perspective

Replacement of the Current Systems

CAATS will integrate the functions of several systems which comprise the current national ATC service, including the National Flight Data Processing System, the Gander Automated Air Traffic System for oceanic control, the Runway Utilization and Automated Departure System, and the Northern Airspace Control System. Under the CAATS project, the existing flight data processing infrastructure will be replaced with an FDPS at each of the seven regional Area Control Centers (e.g., Vancouver, Edmonton, Winnipeg, Toronto, Montreal, Moncton, and Gander), the Transport Canada Training Institute (TCTI), the Research and Evaluation Center, and the Technical Systems Center. Peripheral equipment will be provided for the two Terminal Control Units (e.g., Calgary and Ottawa), the Altitude Reservation Unit, and sixty control towers.

The Visual Information Presentation System (VIPS) and the Operational Information Display System (OIDS) are the primary peripheral displays at the current workstations in the Area Control Centers. The VIPS contains airways charts, approach plates, and other publications. The OIDS presents NOTAMs and airport, weather, and other operations information. The planning for the CCWS has called for the visual displays of both of these systems, as well as the current situation (radar) display and the flight data display, to be presented on 20 in. x 20 in. Sony DDM-2800 high-resolution (e.g., 2048 x 2048 pixels) monitors. Up to three of the CCWS units can be combined to achieve an optimal configuration for a particular sector. The CCWS will be able to merge data from the RDPS, either to generate a mosaic display that combines data from different radar sites or to generate a composite display that depicts both radar and non-radar airspace by merging data from the RDPS with dynamic display of flight data. This use of flight plan data will effectively replace the paper flight progress strips. The CCWS will not be used in the control towers, but each tower workstation will consist of two high resolution displays.

System Interfaces

Several systems that will interface with the FDPS, such as the RDPS, have already been alluded to in earlier paragraphs. In addition, the FDPS will interface with the Transport

Canada Operational Communications Network (i.e., ADIS). Phased implementation will provide the interface with various units and systems outside of Transport Canada, including Department of National Defense (DND) Radar Terminal Control Units, Aerodrome Control Tower Units, Flight Planning Centers, and Maritime Command Mission Operation Centers. Interfaces will also be included for the NORAD Sector Operations Control Centers, the Automated Communications and Reporting System of the two national airlines, and the Atmospheric Environment Service weather processing system. The Altitude Reservation Unit will interface directly with the automated Central Altitude Reservation Facility and with the European Central Air Reservation Facility (EUCARF).

Implementation of CAATS

Implementation Plan

The objective of the implementation plan for CAATSs is to adopt a transition strategy which will enable existing systems to be replaced in a manner which is virtually transparent to the aviation community. CAATS will be introduced gradually, and in a manner that is not disruptive to either those using or those providing the traffic service.

In order to ensure that the new system is implemented successfully, system integrity of the operational environment must be maintained. All operational ATC staff will be trained prior to the commissioning of each phase, to facilitate a smooth transition from the current equipment to that of CAATS. For example, CCWSs will be installed in each Area Control Center for system testing and the operational training of the controllers prior to system commissioning. All of the various ATS training units, including TCTI, will have a simulation capability for operational training.

A phased implementation strategy has been planned for CAATS. The following description outlines the proposed strategy for a two-phase implementation, but as the project proceeds, it is anticipated that the boundaries of each of the phases may become less distinct. To achieve system integrity, there will be a full system review prior to Phase I.

Phase I. Phase I implementation is planned to start with installation and testing of CAATS at the Technical Systems Center, TCTI, and the Research and Experimentation Center. Installation and testing at six Area Control Centers will follow. The seventh Center, Gander, will not be updated during Phase I because of the advanced capabilities in the Gander Automated Air Traffic System; it will be replaced during Phase II.

During Phase I, the FDPS of CAATS will replace the current National Flight Data Processing System. On-line interfaces to such agencies as the Federal Aviation Administration (FAA), the Department of National Defense (DND), flight service stations, and control towers will be established. The FDPS will generate and distribute flight data and flight progress strips, based on a real-time meteorological model, to appropriate ATS units in Canada and abroad. All air traffic control tower systems will be installed, and aircraft operators with ground systems capable of linking directly with the flight data processing system will be connected "on-line."

Phase II. Gander Area Control Center will be brought on line during Phase II. Major new functions will also be implemented; these new functions will include the national flow management system; conflict prediction, detection, and resolution; and full altitude reservation capability with automated data communications between the Altitude Reservation Unit and all Area Control Centers, for the filing and plotting of altitude reservations and military missions. The implementation of Phase II will also enable additional statistics on flight movements to be automatically collected, analyzed, and distributed to appropriate agencies. This capability will provide for the distribution of raw data to the Transport Canada Network (TCN) host system in Ottawa.

Implementation Schedule

At this time, Phase I of CAATS is scheduled for commissioning at the non-operational support centers in late 1994. Phase I for operational units, including the Area Control Centers and the control towers, is scheduled for late 1994 and early 1995. Phase II is scheduled for late 1995 and early 1996. The commissioning dates will vary slightly from one operational unit to another.

Upon completion of CAATS Phase II implementation, controllers, supervisors, and support personnel will be provided with a modern, reliable, and versatile work environment, with the capability to fully exploit the features of state-of-the-art automation.

Acknowledgements

The preparation of this paper was supported by Transport Canada through the Department of Supply and Services Contract No. T8080-9-5072/02-ZA. The views and opinions expressed are those of the author and not necessarily those of Transport Canada. This paper is based, in part, on material prepared for Transport Canada publications by members of the CAATS Project Office. Their assistance in preparing this paper is gratefully acknowledged.

References

Benel, R. A., Dancey, R. D., Dehn, J. D., Gutmann, J. C., and Smith, D. M. (1989). Advanced automation system design. *Proceedings of the IEEE, 77*(11), 1653-1660.

Day, P. O. (1991). Human factors in system design. In J. A. Wise, V. D. Hopkin, & M. L. Smith (Eds.) *Automation and Systems Issues in Air Traffic Control.* Berlin: Springer-Verlag.

Lenorovitz, D. R., and Phillips, M. D. (1987). Human factors requirements engineering for air traffic control systems. In G. Salvendy (Ed.), *Handbook of human factors* (pp. 1771-1789). New York: Wiley.

Simolunas, A. A., and Bashinski, H. S. (1991). Computerization and automation: Upgrading the American air traffic control system. In J. A. Wise, V. D. Hopkin, & M. L. Smith (Eds.) *Automation and Systems Issues in Air Traffic Control.* Berlin: Springer-Verlag.

Stager, P. (1991). Error models for operating irregularities: Implications for automation. In J. A. Wise, V. D. Hopkin, & M. L. Smith (Eds.) *Automation and Systems Issues in Air Traffic Control*. Berlin: Springer-Verlag.
Transport Canada. (1988, September). *The Canadian Airspace Systems Plan - Facilities, Equipment, and Associated Development*. Third Issue. Ottawa, Canada: Aviation Air Navigation System Directorate. Transport Canada Publication TP-4334E.

The Implementation and Impact of Automatic Data Processing on UK Military ATC Operations

S. E. J. Lane

OC Data Processing Wing, Royal Air Force, West Drayton, UK

Introduction

In this paper I will give you a very brief description of the UK ATC system, with special emphasis on Military Air Traffic Operations. I will describe the important landmarks in development and stress the implementation aspects and the lessons learned.

National Air Traffic Services

Air Traffic Services in the UK are provided by a joint military-civil organization called National Air Traffic Services (NATS) and staffed by CAA Staff, personnel of the Services, and civilian MOD Staff. It is headed by a Controller (CNATS). Directly below the Controller is the Deputy Controller (DCNATS); the Director General (DG) (Projects & Engineering); the DG Strategic planning; and the DG Air Traffic Operations, who is responsible for operations in the field. CNATS, DCNATS, and the Director General Air Traffic Operations are appointed jointly by the CAA and the Secretary of State for Defense; the appointment of a Controller is additionally subject to the prior approval of the Secretary of State for Transport.

The appointments of CNATS and DCNATS reflect the principle of civil/military integration; when one is a military officer, the other is a civilian. Within the lower echelons, posts are designated civil or military as the task dictates. The joint field headquarters comprise the co-located military staffs of Civil Air Traffic Operations (CATO) and Military Air Traffic Operations (MATO), respectively.

The civil element of NATS is responsible for civil and military aircraft flying in regulated, controlled air-space, on recognized air routes and at civil aerodromes; while the Air Officer Commanding MATO (AOC MATO) is responsible, through DG Air Traffic Operations to CNATS, for the supervision and control of military ATC operations throughout the UK, other than those at military airfields (who answer directly to their appropriate Command Air Traffic Control Organization). Civil and Military Controllers hold separate licenses; however, both work to common practices and standards. That is to say: the *Civil Manual of Air Traffic Services* is mirrored by the equivalent military Joint Service Publication.

In simple terms, civil controllers control civil and military aircraft flying along the route

NATO ASI Series, Vol. F 73
Automation and Systems Issues in Air Traffic Control
Edited by John A. Wise et al.

structure whereas military controllers provide an area radar service to military and civil pilots who do not wish to use the air routes system. The vast majority of area ATC Service in the UK is provided from two Joint Air Traffic Control Centers. The London Air Traffic Control Center (LATCC) is situated at RAF West Drayton, two miles from Heathrow Airport. The Scottish Air Traffic Control Center (ScATCC) is situated at Prestwick.

LATCC Military Development

The main thrust of ATC systems development in the UK has occurred at LATCC, and thus, I shall now confine my presentation to military developments at that center.

Interestingly, the first attempt at automation, although only a partial success when introduced in the early 70s, is now considered to have been quite an achievement and has only just been replaced with a more modern system. Originally intended to provided both Radar Data Processing (RDP) and Flight Data Processing (FDP) for the whole of LATCC, the MYRIAD consisted of three Marconi 32K computers forming a single disc-based system. Each prong individually processed the total amount of data simultaneously, thus providing constant software accuracy checking and hardware redundancy. The concept was clever, as its longevity proves.

The history of the system raises many points which are still relevant today. First, because of memory constraints (only 32K), the system was found to be incapable of providing full RDP and FDP processing, and it was eventually relegated to providing an FDP System in the military control room only.

There is often a tendency by system designers to underestimate the size of new systems.

Specifications change, even during the formative design stage, and operational systems will always be subject to a constant stream of suggested improvements from users. This point is re-enforced by a relatively recent experience when a system's memory requirement grew initially from three 1.5 megabyte processors to four 4.5 megabyte processors in the space of five years.

The role of Flight Data Processing is to provide controllers with timely, accurate, easily amendable flight details — that is to say, aircraft type, callsign, flight level and route of aircraft under service. Thus, the initial aim was to remove the need for controllers to write down and amend what their aircraft were doing. Indeed, in the very early days, military controllers were banned from using china-graph pencils or pens to maintain a log. Despite this, it was soon discovered that an electronic data display and a touch-sensitive input device were not wholly the equivalent of a piece of paper and pencil. Eventually, use of the system gravitated to a level where the controller's assistant became completely responsible for inputting changes to the flight datalines, leaving the controller free to carry on with the primary task of separating aircraft.

The failure to impose a new system on normally intelligent human beings raises other worthwhile points, which are disregarded at the System Designer's and implementer's peril:

Controllers are conservative animals who generally do not like change.

ATC is a difficult, demanding job, and controllers are often working under stressful conditions at the very limit of their ability. Therefore, anything new, which initially may require additional conscious thought, adds precious micro-seconds to decisions and actions, leaving controllers feeling that they are getting behind the situation.

Don't assume that evaluation will solve everything.

This point leads on from the first. Even the very best system designers find it very difficult to design systems in offices because that which is acceptable on the "drawing board" may not be so under stressful operational conditions. Some form of evaluation is therefore essential; however, no matter how good the artificial environment, it may not be sufficient to prevent implementation problems when the system is exposed to real controllers in the real environment. There is no substitute for careful staff involvement and consultation at every phase of the project.

Automatic systems must be easy to use and give controllers what they want.

Specifically, in this case (and as we shall see later), the input device must be fast and easy-to-use. (Remember that it is attempting to replace the function of writing on a paper strip or perspex plaque.) Furthermore, the electronic data display which presents the information to the controller must be within his cone of vision and be easy to read and assimilate. Controllers do not like to continually have to look away from the radar screen and refocus on another electronic data display — all sorts of nasty things could be going on while they are looking away!

Radar Data Processing

The failure of MYRIAD to provide full Civ/Mil RDP and FDP resulted in NATS requiring, fairly quickly, a replacement system. The decision was taken to purchase the FAA NAS 9020 En-route system based upon IBM 360/365 computers. MYRIAD would continue to provide an FDP system in the military operations room, and the 9020 FDP program was successfully modified to provide an FDP service to the civil operations room. At first, the intention was to modify the RDP program to accommodate the differences between the American and UK ATC environment, and between the different operating methods of civil and military controllers. Progress was slow, involving several steps which initially modified the FAA RDP system to a UK variant and then to a more radical UK variant which, although acceptable to the civil element, still did not meet the requirements of the military off-route organization. Therefore, despite much joint work to overcome the compatibility problems, it was decided that RDP would be used in the civil operations room and a Code Callsign Converted single-radar-source processed picture would be used in the military operations room. That is the situation that we have today and it raises new points.

It is important to analyze the ATC requirement carefully before building the software system rather than modifying an existing system to fit the automation.

It may well be that, given careful analysis, system designers can demonstrate to operational staff that there is a better way of doing things; however, unless the advantages can be shown, one leaves oneself open to the criticism that "the tail is wagging the dog," and the job of implementation becomes that much harder.

Code Callsign Conversion is the automatic conversion of a four-digit octal secondary surveillance radar code to a pre-paired callsign for display in the aircraft label on the radar screen. The advantage is, of course, to provide the controller with constant identification of aircraft. The concept is not new and is used widely in ATC systems and within the new military control-room at LATCC. However, prior to its current successful usage, there were two abortive attempts at introducing it into the old military operations room. On the first occasion, Code Callsign Conversion implementation failed for a reason already mentioned: that is, because there was no operational input; no consultation; only the declaration that the new system would be introduced on the appropriate date. Frankly, it deserved to fail. On the second occasion, operational staff were consulted, a series of explanatory courses were laid on, implementation was gradual, and at least half of the control staff thought it to be a positive advantage — and yet it was again withdrawn from service because of considerable resistance from a noisy minority of controllers.

The control staff's innate and natural resistance to change needs, on occasion, to be matched by determined leadership and political will.

Interestingly, it was the new controllers who quickly accepted the system and saw its advantages, while the controllers who had to make the change from the old system had the most difficulty. Even so, one experienced controller admitted, later, that within six months he would have accepted the new system.

The Processed Radar Display System

Similar operational acceptance problems were encountered while implementing the new processed radar display system (PRDS). The new display system had many advantages over its predecessor: greater clarity, greater map selection, more filtering options, and greater flexibility; however, it also had some disadvantages in that the aircraft target and trail symbology was different, and, more significantly, the picture took longer to build after a radar or range change. (There were also a number of other observations, which I need not go into here). PRDS was successfully implemented, despite the initial problems, because the political will was strong and strenuous attempts were made to speedily improve the perceived shortfalls with the system. This may be an appropriate place to raise another point.

When existing systems are replaced, the new system should be at least as good as the one it is replacing.

The impression of going backwards (or "negative progress") can arise because the specification of new systems is often fixed some time in advance of the implementation date. In the meantime, the old system is further enhanced, so that the facilities provided by the old system are superior to the replacement system. The implementer in this situation will need to enhance the replacement system as soon as possible and use the often-heard statement "The old system had no more development potential whereas the replacement system will ultimately be better" (the "jam tomorrow" approach).

The introduction of digitized processed radars raised another valuable point:

> *Do not assume that the digitization process is naturally superior— it will only be so if careful attention is paid to the advantages and disadvantages.*

This again is the careful analysis/consultation/operational-evaluation argument restated in another form. Digitization will often raise problems while solving old ones. For example, with an analog radar response, controllers when separating aircraft, were not allowed to let the radar returns merge with each other. Thus, as the range of the aircraft from the origin increased and the azimuth definition became less precise, the radar returns became bigger and the safety margin also increased. However, despite the same radar definition constraints, the processed radar 'decides' where the center of the large imprecise radar return is, and then displays a uniformed sized synthetic symbol. The problem can be partially overcome on a single radar display by arbitrarily picking a range, say 80 miles, from the radar head to increase separation from 5 nms to 10 nms; however, the problem becomes greater in a multi-radar, mosaic scenario.

Military Airfield Remote Entry Terminal System

In 1985, the Military Airfield Remote Entry Terminal System (MARETS) was introduced to enable military airfields within LATCC's area to directly input Flight Plans and Prenotes into the proposed Electronic Data Display Update System — an intended FDP replacement for the aging MYRIAD Computer. The MARETS project was very successful, illustrating many of the points already discussed: careful analysis; user consultation; modification of the requirement to reflect user comment; and straight-forward, logical, easily-usable, menu-driven input devices. However, the great strength of MARETS is that it fulfills an important need — it is perceived by the users as a useful system which is far more efficient than using the telephone or the slow, cumbersome teleprinter system. At the risk of stating the obvious:

> *Automated systems must provide a recognizable benefit to the person doing the imput, rather than merely demanding information for a third party benefit.*

The point is illustrated by describing what happened when an additional weather-input facility was added to MARETS. The facility has two functions: first, all MARETS equipped airfields are required to use it to tell-in their weather to LATCC every hour, regardless of whether they were required to do this before; secondly, it provides an automatic method of keeping diversion airfields aware of each other's weather. It will come as no surprise that airfields which previously had to pass the weather by telephone or which regularly book diversions are much more likely to input the weather than those who see it as an extra task with no benefits to them.

Just one year after the successful introduction of MARETS, the need arose to produce a civil networked system to enable a number of airlines to request Flow-Control slot clearance times. Because the basic functionality was broadly similar to MARETS, the technical decision was taken to incorporate this new system within the same hardware. From a purely technical point of view, the decision was sound; however, the operational effect was quite significant, in that there were now civil and military users with differing operational requirements. Thus, if one user required a system outage to correct a software problem, the other user also suffered. Planned outages were therefore far more difficult to negotiate, because each side had the power of veto. Furthermore, because of the natural protectionist attitude of users, changes or enhancements to one system were resisted just in case they might adversely impact on the other system's performance. In this particular case, the systems were separated after a period of intense lobbying from both operators. Thus:

What makes sense in narrow system terms may not necessarily make sense from the operational ATC's and engineering manager's viewpoint.

The New Military Operations Room

And so we reach the current situation. The old (some would say ancient) Myriad computer has been relegated to a short-term subsidiary role, and both Eastern Radar and London Radar have been successfully co-located in a new military operations room using the Electronic Data Display Update System to provide FDP and high resolution 23 inch vertical radar viewing units with Code Callsign Conversion to replace the aged 16 inch unconverted units.

Someone once said that there has never been a significant software system delivered without bugs. — a sweeping statement perhaps, but certainly borne out from personal experience. The Electronic Data Display Update System was no exception. In fact, despite exhaustive contractual and operational testing the new system crashed within the first hour of operational use. The simplistic response is to say that the testing was therefore not extensive enough; however, the hard fact is that with complicated, multipath, real-time ATC applications, it is just not possible to test every single condition.

It is therefore important to realize that, failures will occur; no matter how much we try to design them out. Hence, we must apply either the software recognition and control of the failure (fail safe) approach, or the gentle degradation (fail soft) philosophy. It is true to say that, while solutions to hardware failure are progressing satisfactorily by using duplicated or triplicated systems, there is still much work to be done in overcoming the software failure problems, and from experience, I can confirm that there is no more demanding test of a new system than its operational use.

Conclusion

This brief history of the significant events in the progress of UK military ATC data-processing has been used to discuss some of the significant problems encountered by systems designers and implementors. The onlooker may feel that there has been little progress — no multi-radar data processing, no radar tracking, no correlation between flight data and radar data or conflict alert — and yet the processed radar system and flight data processing system which have evolved are right for the current military ATC environment. Other options were available (and are being used by the civil element), but they were considered to be inappropriate to current military ATC operational requirements.

I leave you with the question:

> *When modernizing the current system, should the planners and designers merely enhance what is already in existence — or should they be more radical and change current operating methods to utilize fully the benefits of state-of-the art automated systems?*

From Under the Headset: The Role of the Air Traffic Controllers' Professional Association in Present and Future Air Traffic Control Systems Development

John Levesley

Guild of Air Traffic Control Officers, Wimborne, Dorset, UK

Introduction

The Guild of Air Traffic Control Officers is a purely professional body with no political or trade union affiliation. It is pledged to defend and improve the technical and professional standards of the air traffic control (ATC) service within the United Kingdom. Its 1500 members include civil and military operational and non operational controllers working for various employers in the UK and other countries. Additionally, our life, associate, and corporate members include many electronics companies, airlines, private air traffic control agencies, and individuals.

Part I

Working within the Democratic Process

The Guild's permanent membership in the British Parliament's Advisory Group on Transport Safety supplies a valuable opportunity to make our opinions known to the representatives of both houses of the British Parliament.

Recent incidents have led to select committees of the British Parliament investigating various aspects of both United Kingdom and European air traffic control. The Guild has submitted written evidence to these enquiries and has also given evidence in person to one select committee.

In recent years the Guild has widened its horizons, and now maintains a dialogue with British members of the European parliament on ATC matters.

Representations to the State

In the United Kingdom, the provision of the civil air traffic control service is delegated by the state to a government agency, the Civil Aviation Authority (CAA). The CAA and the Ministry of Defence together staff an organization known as the National Air Traffic Services

NATO ASI Series, Vol. F 73
Automation and Systems Issues in Air Traffic Control
Edited by John A. Wise et al.

(NATS), which is responsible for providing most ATC services in the UK. The Guild meets regularly for discussions with both the chairman and managing director of the CAA, and the controller of the NATS and his senior managers.

The CAA organizes many special interest committees in the area of UK aviation. The Guild is represented on many of these. The most recent task has been as part of a committee to advise on the need for limiting controllers hours. We have provided input to a committee on microwave landing systems, and are members of both the flight safety committee and the general aviation safety committee. Airspace utilization in the United Kingdom is subject to a consultative process through the National Air Traffic Management Advisory Committee (NATMAC), which includes Guild members. Where losses of separation between aircraft have been known to occur, panels meet to investigate both pilot reported incidents (Airmisses) and controller reported incidents (Aircraft Proximity Hazard reports). Nominees from the Guild sit on the investigative committees.

Participation in the Aviation World

The British Airline Pilots Association runs an Air Traffic Service (ATS) study group. The Guild provides the controller input to the group. For many years we have arranged symposiums, study groups and conferences on many subjects from controller stress to pilot-controller forums, and have attended many venues to participate in debates or to give working papers.

Perhaps our best known effort in the aviation world is our biennial conference and exhibition, CONVEX. CONVEX has become a highly regarded ATC occasion, attracting many of the most influential people and organizations in both UK and (increasingly) European ATC and air transport.

Regional and Local Effort

The Guild is a regional based organization, with different regions having particular specialties. The Scottish region has, for example, a speciality in oceanic control whilst our Wessex region has a particular interest in training. Many of the pilot controller and other forums are organized on a regional basis. At an even more local level, each unit has a technical committee on which the Guild is represented.

Guild International

I cannot finish this introduction to the Guild without mentioning the international element of our work. We are members of an umbrella organization called the International Federation of Air Traffic Controllers Associations (IFATCA). For many years we have provided the chairmanship and the majority of members on IFATCA's standing committee 1, responsible for formulating IFATCA's technical policy. Our members of that committee have been the most frequent and regular representatives of IFATCA on the ATS study group of the International Federation of Airline Pilots Associations (IFALPA). Increasingly, we also find ourselves playing a role on other IFATCA standing committees dealing with legal and medical matters. We were proud to provide the official IFATCA delegate to the International

Civil Aviation Organization's Future European Air Traffic Systems panel (FEATS). Finally we are honoured to have been chosen to host the IFATCA conference in the UK in 1992, a pivotal year for those of us whose nations are members of the European Economic Community.

Part II

In the first part of this paper, I attempted to explain the role of an air traffic controllers' professional association in a modern democratic state, using the UK as an example. We would expect to maintain and even expand this role in a domestic and European context during the challenging years ahead, as we move towards greater European integration and interdependence in air traffic control systems. Each of you no doubt has a vision of the future, and we too have our hopes and fears. I do not intend to detail the perfect ATC solution for the future, but I will take the opportunity to place before you one or two things for you to think about, in the context of a future increasingly automated ATC system.

Some Non-Technical Considerations When Automating ATC

Much of the non technical work on air traffic control in recent years has been in the human factors field. The Guild is one of many associations that have been active in campaigning for the recognition that non technical human factors have a significant impact within the air traffic control service. My immediate intention, when I was planning this paper was to continue in that theme. On seeing the list of participants, I changed my mind, for gathered here are a considerable number of human factors experts who can present their own case better than I can. Instead, I intend to raise some non technical considerations which are not necessarily within the scope of human factors studies.

So What of the Future?

One frequently quoted vision of the ATC future is of a largely automated ATC service, with the remaining controllers involved as supervisors and managers of the airspace involved. We find it difficult to obtain a definition of what our task of "supervision and management" will actually entail. Does it mean that we have a role to intervene in the ATC operation when a situation exceeds the logic, rules, or priorities of the automated system? Are we still the final problem solver in the system?

If so, then there are some pretty important decisions to be made. Any controller who is not routinely carrying out his operational duties will suffer some loss of fine judgment and operational expertise. I say this with some confidence, having myself done a non-operational job in the past, and then subsequently returned to operational work. It takes time to develop your speed and judgment again. If we are to have an interventionist role within an automated system, we will need to retain our practical skills, presumably on a sophisticated simulation facility. This has obvious consequences for staffing and support facilities for an automated system. Additionally, controllers are expected to demonstrate their competence with an annual check and/or continual monitoring and supervision. Will a simulator serve for this function, or will controllers be required to demonstrate a "hand on" exercise to their checking examiner?

The second consequence of any need for the supervisor/manager to act as the problem solver, will have an effect on system capacity. Any system needing human intervention will be limited to a system capacity based on the human in the chain, not the potential capacity of the fully automated system. Believe me, few if any controllers are going to be stupid enough to accept an interventionist role in a system which exceeds their human capabilities. If anyone is mistakenly thinking that we can be nominated as the final authority in such a system, and hence the person who accepts the liability for error, forget it.

And so we come to a major problem for the future: *liability*. At the moment the controller and his employer carry much of the legal liability for actions of the ATC service. What of the future? If controllers cannot intervene in the new systems, due to a lack of practical continuity training or because the system speed exceeds human capability to intervene, he can no longer be liable, for they lack the ability to control the ATC situation.

Any employer who writes a specification for a new software driven automated ATC system will surely stipulate as follows:

Requirement 1: The system will maintain the legal minimum separation between aircraft at all times.

Requirement 2: The system will be flexible enough to cope with emergencies, unserviceabilities, diversions ,and disruptions to service.

Requirement 3: Both civil air traffic and military operational and training needs must be catered for.

Requirement 4: The system will expedite and maintain a safe and orderly flow of air traffic.

Requirement 5: The system will demonstrate compliance with reqirements 1 to 4 before acceptance.

Any software house or ATC system contractor will inevitably find that its customer will seek to impose these specifications in an automated system, for they are conditions which will absolve the supplier of an automated ATC service from most, if not all, of the liability for a system failure.

So who now will replace the controller and the service provider as being liable for this safety critical ATC system? Presumably the system designers and analysts who design the software, and the programmers who write the code. Are we looking possibly, at the introduction of an international licensing system for software personnel who write safety critical software? Problems in other areas with liability for safety critical software, may be experienced in the ATC industry, where software companies attempt to minimize or remove any liability for their work — or they withdraw from the project. Will we indeed find anyone willing to produce the software who will also accept liability for it? If not, how long will an ATC service provider have to test the system before introducing it into service, and thereby accepting liability themselves?

The final problem is a legal one, of international proportions. An automated ATC service will presumably initiate and issue ATC clearances. Although something of a simplification, it is not untrue to say that the ATC clearance is the basic technique of ATC: it clears an aircraft from, to, via, at. All ATC procedures are based upon instructions to aircraft in the form of a clearance. To issue an ATC clearance requires a number of qualifications to be held by the issuer. These vary state by state, but in the UK, the issuer needs an air traffic controller's licence, and a rating on the licence for the service being provided, (e.g., aerodrome control, area radar control etc.). Finally the issuer must have a validation for the unit at which they work, and the airspace for which they are responsible. Without any one of these qualifications a controller is not qualified to issue a clearance, except under the supervision of another suitably qualified controller. If the issue of a clearance requires these qualifications, what qualifications will be needed by the person who writes an algorithm to cover an ATC procedure, or a programmer who produces a piece of code to initiate the calculation of an ATC clearance? Will they have to be ATC qualified?

My brief from the directors of the institute was to introduce you to the work of a controllers' professional organization, using my own organization as an example, and to explain to you what role we are able to play in ATC systems development. However, no controller, given such an audience, would resist the opportunity to make a formal input on matters of concern to his association. In recent years, a fairly detailed agenda has been written for the technical aspects of automation and advanced ATC systems development. More recently still, that agenda has been expanded to incorporate all manner of human factors elements. I hope that the input from myself, and other civil and military ATC colleagues in this volume will place further items on that agenda for consideration by all our ATC services.

That there is much more to producing automated ATC systems than the technical challenge. There are also ethical, political, legal, and human considerations to be taken into account. One thing I can promise: that whenever any of you have to consider these problems, there will be professional air traffic controllers' associations throughout the world ready and able to participate in any discussions. If you do not approach them, don't be surprised if they take the initiative and approach you.

A Collation of Approaches to Systems

Air Traffic Control and Systems Issues

Anthony Debons

Robert Morris College, Coraopolis, Pennsylvania, USA

Introduction

In this essay, the focus is on systems issues as they relate to air-traffic control (ATC). The first issue concerns the extent to which automation should be applied in ATC. The second issue refers to the ability of the present ATC information system to provide efficient and effective control in a rapidly expanding, complex air space environment. How can our present capability in obtaining information be extended to satisfy the need for greater human cognitive competence demanded by the environment? The third issue is to determine how best we can reap advantage from our knowledge of ATC failures in designing future ATC systems.

Automation in ATC

The argument for the use of automation in ATC is based on the present state of ATC and what can be expected in the future. Given these expectations, a very basic issue is whether the current ground based ATC systems are really the appropriate model to be followed in the future. It may very well be based on out-of-date assumptions. For example:

1. More aircraft can be expected to occupy air-space in the future.

2. More sophisticated technology will be available to provide greater capabilities in directing aircraft through space.

3. Air travel will move to greater expanse. There will likely be space travel for the general population during the present decade.

4. There will be a changing population of pilots who will be flying aircraft—differently trained, younger, and expected to respond to different environments—changing faster, more complex circumstances, risks, etc....

5. There will be more international travel to be expected, due to the changing social environment.

These changes suggest a reanalysis of the actual needs of the air transportation system:

NATO ASI Series, Vol. F 73
Automation and Systems Issues in Air Traffic Control
Edited by John A. Wise et al.

(1) for the reduction of human error and equipment failure; (2) for the standardization of hardware, software, air procedures; (3) for reduction of staffing; and (4) for reduction in the cost of operation and maintenance.

Weiner (1988) sees the nature of the air-space environment as placing specific demands on human capabilities.

> The increasing complexity of the environment in which pilots fly — an environment congested with aircraft — demanded regulation and procedures; and (2) a vast number of computer-based devices at the pilot's fingertips replacing the demand for manual control and mental arithmetic. (p. 451)

In an extensive study of operational errors in the national air-space system for calendar year 1987, almost 98 percent of the errors were attributed to human factors (U.S. Dept. of Transportation, 1988). A broad range of human factors research points to significant limitations in both cognitive and motor performance with increasing environmental demands. Edwards (1975) cites the experimental work of Drew (1940) and Davis (1948), which indicated that when subjects were fatigued, their skilled performance disintegrated, leading to poor management of a complex task. Pilots overlooked fuel content and failed to distribute their attention to instruments. Undue demand on the pilot will induce stress, which predisposes the crew to a number of problems, including "misinterpretation of multipointer instruments, legibility problems, scale interpretation error, illusions, and accepting data from inoperative or faulty sources" (Edwards, p. 9).

From another human perspective, beliefs, values, and the language of a culture can be important in the analysis that precedes the design of the system. In a recent article, Payer (1990) has shown that the assessment or diagnosis of illness can be different for physicians of varying countries. There were significant differences both in the points of emphasis physicians applied to their interpretation as to the extent that physiological factors contributed to disease and in the drugs they prescribed.

Kaplan (1991) has pointed to differences in auditory, visual, and perceptual capacities resulting from cultural factors (language, background, instruction, etc.). For example, he notes that Gleiss and Renhall (1986) found that subjects judged quality of pictures (transmitted over different kinds of telephone lines differently) based on differences in their background. These differences extend to other sensory modalities as well. Kaplan also reported differences in speech quality judgments based on the properties of noise reference systems as a function of the culture of the listener. These influences cover a number of functions, including command and control, voice communication, and the differential response to stress.

The increasing complexity of the environment, the demands that are made on basic human capabilities, and the influences that the culture can play in ATC operations all seem to suggest the use of technology in present and future ATC systems. Yet many of the tasks of air traffic control can be expected to be subject to greater variability and less stability. *How* the specific technologies are to be used and distributed in the system are fundamental issues to be addressed. At the outset, a fully automated ATC system appears unlikely because of the human reluctance to surrender complete control of any system where human safety and life could suffer. At the moment, the air traffic control system is efficient in obtaining the data (from sensors and other sources) to maintain control. With greater variability in the environment, the need for operators and controllers to *understand* the data in the context of environmental circumstances will be of increasing importance. In the words of those concerned with air traffic control, the operators will need to understand the dynamics of the system, that is, how the system "thinks" and "acts." Many factors to be considered in these issues demand a systems approach.

General System Issues

Systems are all around us. We have transportation systems, health systems, command control systems, etc. There is general acceptance of what constitutes a system.

> A system is a set of objects together with relationships between objects and between their attributes. (DeGreene, 1970, p. 6)

According to Hall and Fagan (1956),

> ...objects are simply the parts, or components of a system—for example, stars, atoms, neurons, switches and mathematical laws. Attributes are the properties of objects... relationships to the system components together.... For a given system, the environment is the set of all objects a change in whose attributes affects the system and also those objects whose attributes are changed by the behavior of the system. (p. 6)

What a system "is" should not be belabored. The problem or issue is not in the acceptance of these concepts. The issue concerns the *practice* or application of the systems approach to the business of systems analysis and design. It narrows down to the fact that, when we consider automation in ATC as a system apart from the larger context in which it functions, the systems approach is violated. As a result, the question of role of automation continues to be begged, while the larger issues remain obscure. The understanding, acceptance, and application of the systems approach can be said to be vital to system analysis and design.

A major issue in systems thinking concerns the human predilection for sequential and reductionistic thinking — a carry over from the empiricistic philosophy of the eighteenth century. Because of this predilection, we are inclined to consider an element or component of the system (i.e., automation) as *the* system. This habit of thought does not conform to the system approach, wherein an element (object, entity) is considered to exist and function apart from the larger context the system serves. A system consists of a number of subsystems, which in unison contribute to specified objective(s) or goal(s). Each component or subsystem exists and functions in a hierarchy of levels, each level representing increasing complexity. J. G. Miller (1978) expresses this idea succinctly:

> The structure and processes of a given subsystem are more complex at a more advanced level than at the less advanced level. (Miller, p. 1)

This is shown in his hierarchy representative of all living systems (Figure 1).

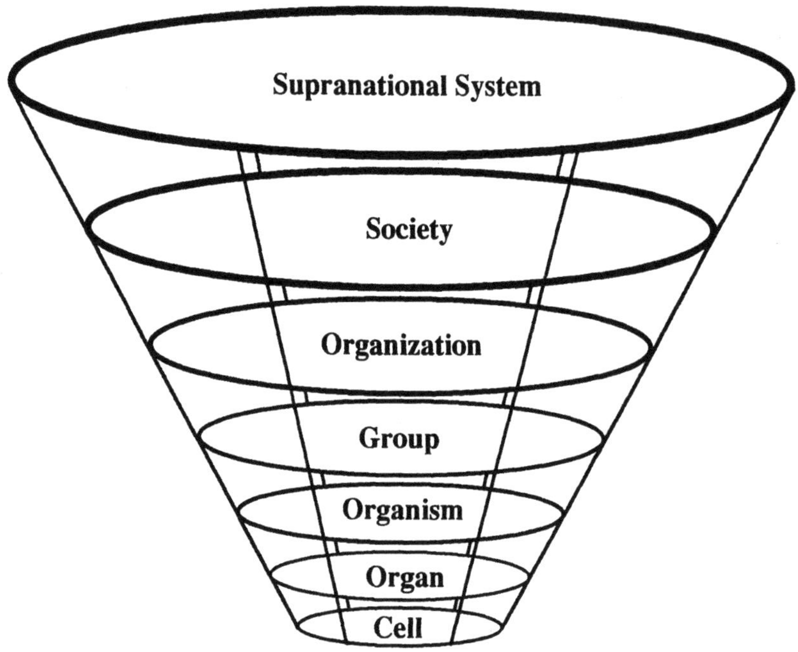

Figure 1. Miller's hierarchy

Sheridan (1988), in a graphic representation of the hierarchal nature of part of a National Air Space System, includes a description of a general pilot-aircraft system (Figure 2) and a system model of aircraft production (Figure 3).

As indicated previously, the air-transportation system (ATS) includes the presence and movement of aircraft and people on the ground and in the air. Human controllers, with available technology, guide the aircraft on the ground and in the air. Travel agents, ticket processors, and other agents move people and equipment through the system, both on the ground and in the air. Thus, the system can be said to consist of a number of subcomponents working in unison, interacting with other systems operating at other functional levels.

Basically, the question can be raised as to the nature of the cement or glue that binds these relationships enabling each of the subsystems to achieve their respective objective. Each component or subcomponent, as well as the overall system, represents a cybernetic environment wherein control is achieved through communication and feedback. It is suggested that the answer lies in the formal structure of all systems — namely input, throughput, output, and feedback functions—representing the constituents of the cybernetic model of a system. In this model, control is maintained through communication and feedback. The integrating force that is part of a system rests on a number of transformations occurring to energy and matter in the interrelationships between the formal structures that constitute the system. These structures are present in all living systems. In addition, these structures, the laws that govern their function, and the interrelationships between them, provide the basis for the analysis and design of artificial systems. The objective of artificial systems is to augment the capacities that are part of the basic structures of most if not all living systems. In Table 1, these structures are identified by Miller (1978) and Debons (1971); they are represented in Sheridan's (1988) depiction of an aircraft system in Figure 2. These six components are sketched graphically by Wise (1988) in Figure 4. The description of the basic structures and the functions they serve are detailed in Table 2. An attempt is made in this table to relate such functions to an ATC event to illustrate the respective function.

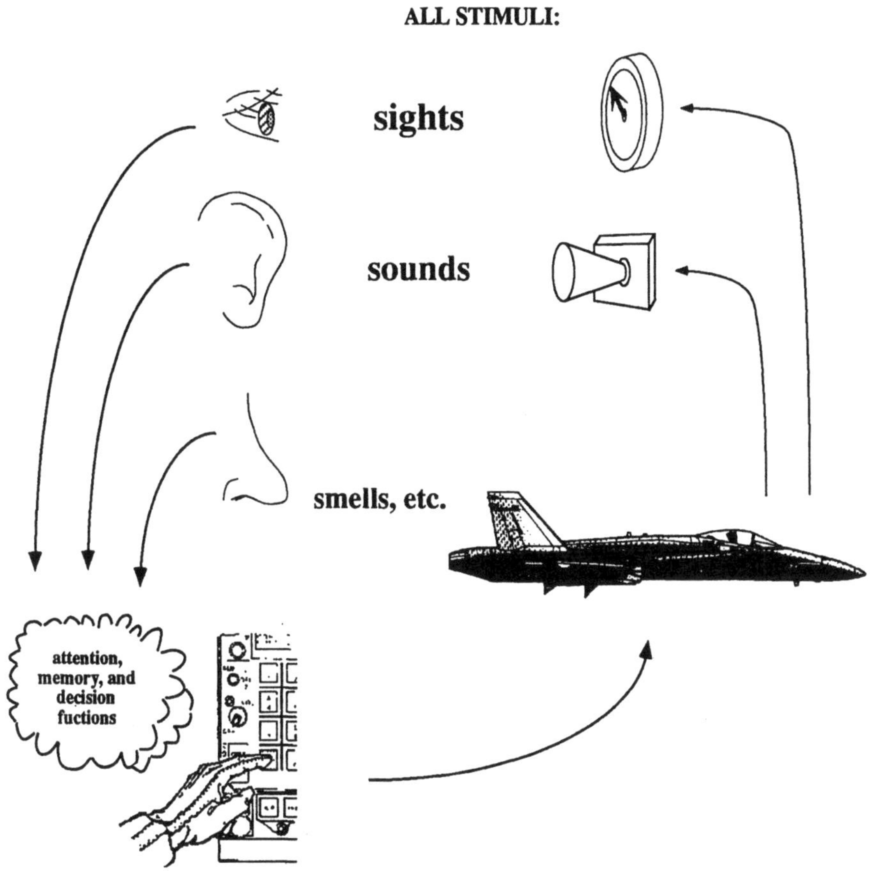

Figure 2. Sheridan's System Perspective

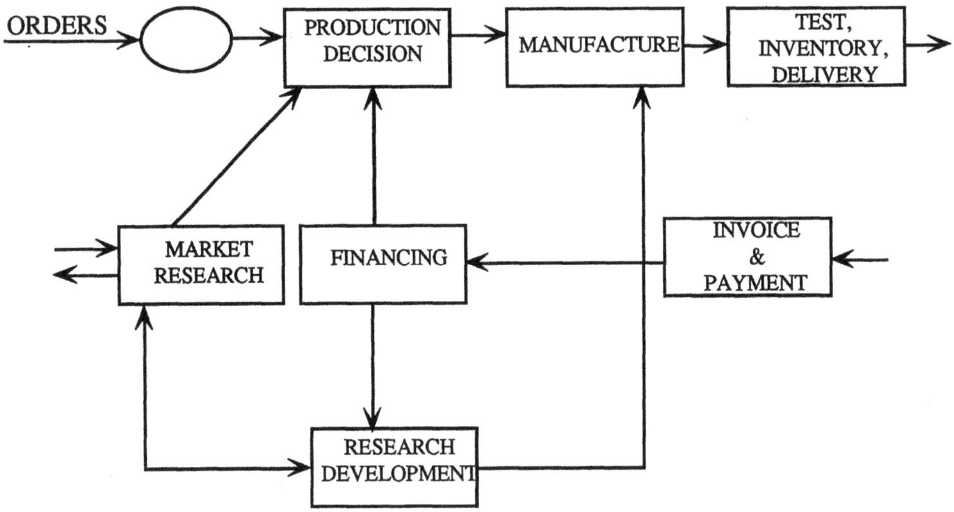

Figure 3. System of model aircraft production

Table 1. Aircraft systems analyzed using Miller's and Debons' models

Miller (1978)	Debons (1971)	Sheridan (1988)
	Event World	Aircraft in Space
Input Transducer	Acquisition	Sights, Sounds, Smell
Internal Transducer	Acquisition	
Channel and Net	Transmission	(Arrows in Figure)
Decoder	Transmission	(Arrows in Figure)
Associator	Processing	Attention
Memory	Processing	Memory
Decider	Utilization	Decision Function
Encoder	Transmission	(Arrows in Figure)
Output Transducer	Transfer	All Actions

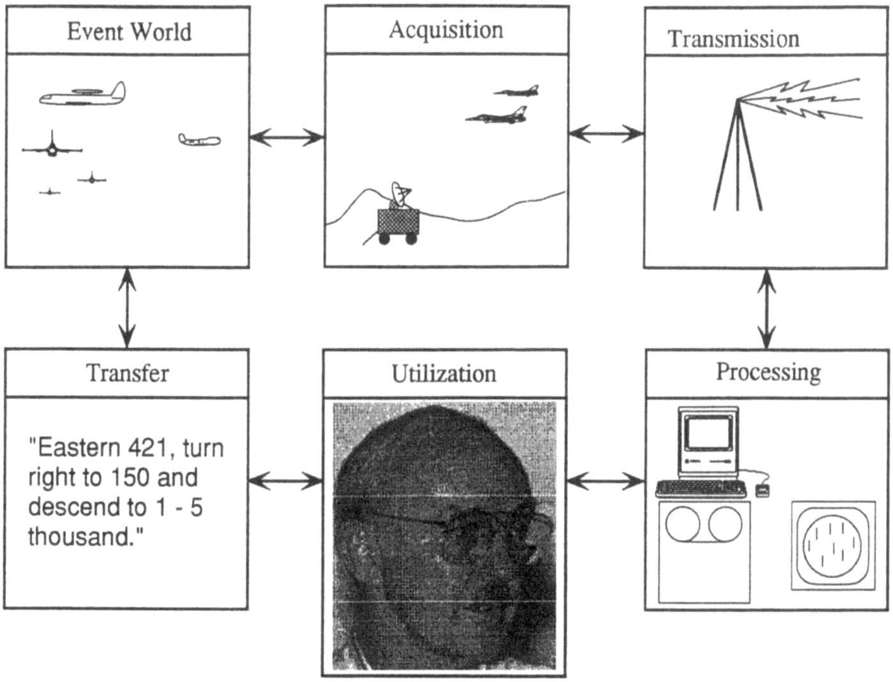

Figure 4. Graphical Analysis of the Airspace System

The Nature and Function of Data-Information-Knowledge in Systems

The question arises as to what cements or integrates the various components of a system. The life blood of any system is the movement and exchange of energy throughout the various constituent parts of the system. Energy constitutes the agent which innervates each of the components of the system. To deal effectively with energy and matter, humans devise symbols to represent states of energy and matter. Over time these symbols have been formalized, leading to the creation of language(s). For example, letters and numbers (alphanumerics) provide the basis for most of our data systems. These data systems provide us with an ability to deal with each of the interrogatives: *what*, *where*, *when*, and *who* — the basis of our awareness. Thus, data constitute the stimuli underlying our consciousness. Information systems are environments of people, technologies, and procedures which serve and augment out state of consciousness.

Table 2. Fundamental components of living systems and their representation

Debons	Miller	System Components (Human/technology)
Event World The occurrences that are relevant to the objective and function of the living system. It includes the process of classifying, and linguistic representation of the item in the form of symbols.	Not represented	Avianca airliner approaching JFK, relationship of all aircraft to ATC, priority rules, bad weather, large number of delays, pressure to meet the schedule, pressure to get passengers to destination
Acquisition The initial physical component of the system used to capture matter and energy (i.e., the event) from the event world	*Input Transducer* The sensory subsystem that brings into the system, changing them from new forms to transmit to t	Visual response to cockpit display of fuel level, controllers monitoring radar, flight strips
	Internal Transducer The sensory subsystem that receives data from other components and changes their form so they can be transmitted in the system	Response to auditory and other displays that indicate status of system; alarms
Transmission The actual movement of signals (data) within and between system components	*Channel and Net* Rock matter and energy in the system (interconnected components and parts)	Wire, cable, sound, light, lasers, smoke. Conversation between crew and ATC, language, human neural system
	Decoder Process the signal (data) it receives from the internal transducer from one form to another. So it can be used by the system.	

Table 2. Fundamental components of living systems and their representation (Continued)

Processing The ordering, storage, and retrieval of data for the purpose of applying it to problem-solving, decision-making	*Association* Ties one form of data to another, thereby creating significant relationships..	Crew's brains, expert system, ATC computer, weather computer
	Memory Stores data that are established at various times	
Utilization The component that represents the evaluative, interpretive requirements of the information system	*Decider* Regulates the flow of data among the components, thereby controlling entire system	The controller's decision process about landing and approach sequence, relative to Avianca's fuel state
	Output Transducer "Puts out markers bearing information..." and changes "markers within the system into other matter energy forms"	Muscles, voice (radio). language, motion-movement
Transfer The action component, the implementation of the utilization function through the system's medium. Transfer in this model can be seen as the communication or information transfer	*Encoder* Takes data obtained from the components of the system and attempts to make it understandable to other systems in its environment	Avianca pilot's decision to abort landing. The lack of use of the word "emergency" to ATC.

Although human awareness is a necessary concomitant of life, it is insufficient to deal with the exigencies that occur from time to time. Greater awareness due to greater availability of processed symbols can be debilitating. For example, in recent years, developments in radar: primary (reflection), secondary (computer enhanced) radar, and advanced automation systems, have substantially increased data acquisition component capabilities. From a general perspective, awareness has increased substantially since the turn of the century (de Solla Price, 1963). Some have referred to this growth as the information explosion (Toffler, 1970), others as the ignorance explosion (Lukasiewicz, 1972), suggesting the decay in the timeliness of data and the ability of education to catch up with the availability of information and knowledge.

Information systems that now provide us with data should also provide the potential for extending this awareness to knowledge, that is, meaning and understanding. Meaning and understanding are fundamental cognitive capacities. The knowledge system provides us with the ability to deal with the *how* and *why* questions of our existence. The how and why questions are essentially extensions of the what, where, when, who interrogatives. The gap between an information system and a knowledge system, of which it is part, is best understood through the acknowledgement of the six fundamental components that represent the structure of the knowledge system.

Event

The ability to account for the various aspects of an event is very much the function of the technology at our disposal. The "black box" placed on all aircraft is a technology that enables the reenactment of the event, enabling an identification of those factors that induced an error in flight or on the ground. But the capacity of this technology to reproduce the event is *virtual,* and thus, a limiting factor in the definition and acknowledgement of the event. The reproduction possible through the human or technology can never be absolute. It is always inherent in the characteristics of the receiving element, whether organic or technological. To the extent that language is an important tool used in representing events. its capacity to define and represent the event is also virtual and thus limited. Language, moreover, enables us to classify and categorize different properties and attributes of events. Because language is limited in event representation, categorization and classification of events could be significantly influenced. Windshear, a meteorological event, is a case in point. Windshear conditions are complex and often difficult to define and describe. Shifts in barometric pressure, wind velocities, and other attributes are at times difficult to acknowledge and describe by the language that we possess.

Acquisition

The ability of the aircraft to respond to different conditions on the ground and in flight is dependent on technologies such as radar, positioning satellites, and the crew who pilot the aircraft. Each sensor, whether human or machine, is limited in capacity to respond to energy (often referred to as the "Law of the Threshold"). The number of signals that can be obtained may be voluminous. The question remains as to the established criteria which efficiently filter the relevant from the irrelevant signals (often overlapping and in the presence of noise) from an event and that are required to command and control the aircraft.

Transmission

The movement of data among the various components of a knowledge system is a function of bandwidth and other perturbances that are part of the characteristics of ground lines and other technologies such as microwave, satellites, etc. When systems demand quick response to *how* and *why* questions, fast switching capabilities are demanded. When an aircraft completely loses its hydraulic power, the navigation of the aircraft requires quick access to information from various sources. The prioritization of signals is a critical aspect of this requirement, be-

cause of the loading of transmission lines that can, and often do, occur. This is an area where standardization of procedures could substantially benefit control.

Processing

The manipulation of data in machines is governed by rules and procedures which are part of the software. It should be clear, however, that some of the data representing the event may never reach the processor, either because of acquisition breakdown or for lack of standards established to process the priority of signals. Furthermore, the test of the adequacy of software is difficult(Bilinski, 1987). The acquisition and representation of ideas, concepts, and propositions (knowledge) are often difficult. This is the subject of considerable attention in the artificial intelligence literature. Our understanding in this area, although limited at present, is growing.

Human information processing, on the other hand, has received much attention since the turn of the century. Despite growing understanding of cognitive-affective processes, questions on sequential vs. parallel processing, satisficing, etc., remain to be matters of strong research interest. Of considerable importance is the ability of software to meet the demands of users for explanation as they process the data from the computers. Explanation is vital to support the interpretive, judgmental functions that are part of decision making and problem solving (Cosic, 1989).

In the final analysis, our ability to extend our cognitive competence could rest on our ability to establish technologies or procedures which go beyond the awareness made possible through data processing machinery. When something goes wrong, it is desirable to know how and why. This is in the nature of explanation. If control of events is to be achieved, the human operator must understand the dynamics of the system. Our understanding of explanation and the translation of this understanding to the design and application of automata is the heart of the matter.

Utilization

One area that has received considerable attention under the rubric of "information systems" concerns the ability of such systems to aid problem solving and decision making (functions under the continuous attention of behavioral scientists). Utilization means the exercise of evaluative judgment — perhaps the critical dimension of knowledge systems. Utilization is highly sensitive to the availability of data, but perhaps more in the manner in which data is organized and presented. Despite the availability of highly sophisticated transmission devices — such as electronic displays, diverse media, etc. — the question remains as to the ability of the human to use such data available through these technologies. Means for obtaining quantitative criteria to support display designs are being actively researched. New electronic displays now under testing provide pilots with spatial presentation of objects and targets in three dimensional perspective. The work of Tufte (1990) highlights numerous ways that visual perspectives can be achieved on flat surfaces. The work on quantifying the various perspectives on multivariate data by Huff (1981) and his colleagues is of interest. Huff notes:

> Three different major approaches have been proposed to display multivariate data (Fienberg, 1979). The first uses glyphs and metroglyphs, which are circles of fixed

> radius with rays of various lengths representing the values of different variables (Anderson, 1969). An example of this approach is provided by Siegel, Goldwyn, and Friedman (1971) who use K-sided polygons wrapped around a circle to represent multivariate data. The second approach displays multivariate data by using finite Fourier series (Andrews, 1972). A marketing application has been demonstrated by Darden and Flaschner (1974) in which they use this approach to group stimuli from a multidimensional scaling solution and to examine the uniqueness of groups from an overall cluster analysis. The third approach was developed by Chernoff (1973) and consists of representing a point in K-dimensional space as a cartoon of a face whose facial features (such as length of nose, curvature of mouth, size of eyes, etc.) correspond to components of the point. (p. 44)

Gestalt theory formulated during the early part of the present century, if expanded and tested, can be fruitful. But other theories can also be examined. Most certainly, composition and rhetoric reflect principles directly related to order which is part of logic and language in discourse (Young, et al, 1970). The editing process could also be useful in providing insight into the problem of data organization. Film theory represents another rich resource (Wise and Debons, 1987). Equally, principles of harmonics, as found in music, is suggestive. Creativity and art are relevant in providing principles in the organization of data (information) and knowledge on electronic mediums.

Transfer

In addition to our ability to classify/categorize and in general deal with the variance of events, the transfer component plays a critical role in the system's ability to achieve its objectives. Ample evidence in the literature on information system performance leads to the conclusion that the ability to translate judgment to action (which is the fundamental property of transfer) is critical. Problems generated because of the vagaries of language and human habits are classical. Tendencies to use terms incorrectly and inappropriately to reflect states of the environment and particularly the properties of events are abundant. Of course, it is also acknowledged that absence of action is also a means for transferring meaning, understanding, and related intentions. Furthermore, the organizational attributes play an important role in mitigating communications. Many organizational factors (Westrum, 1987) and cultural factors (Kaplan, 1990), international in scope, can influence communication and can be quite critical to ATC operations viewed from a global perspective.

System Failure

One way the effectiveness of a system can be determined is to account for its failure. A number of failures on record, associated with ATC operations, can provide insights as to the causes of such failures. For example, reports are available from investigatory teams assigned to isolate the causal factors.

A reasonable proposition is that many, if not most, organizations fail because of the breakdown of D-I-K systems designed to support them. Organizations both public and private do fail. By and large, the role of information systems in such failure is inferred, rather

than directly linked. At times organizational failure is considered a matter of management practices and the idiosyncratic style of the manager or decision-maker and problem-solver, rather than the information system design.

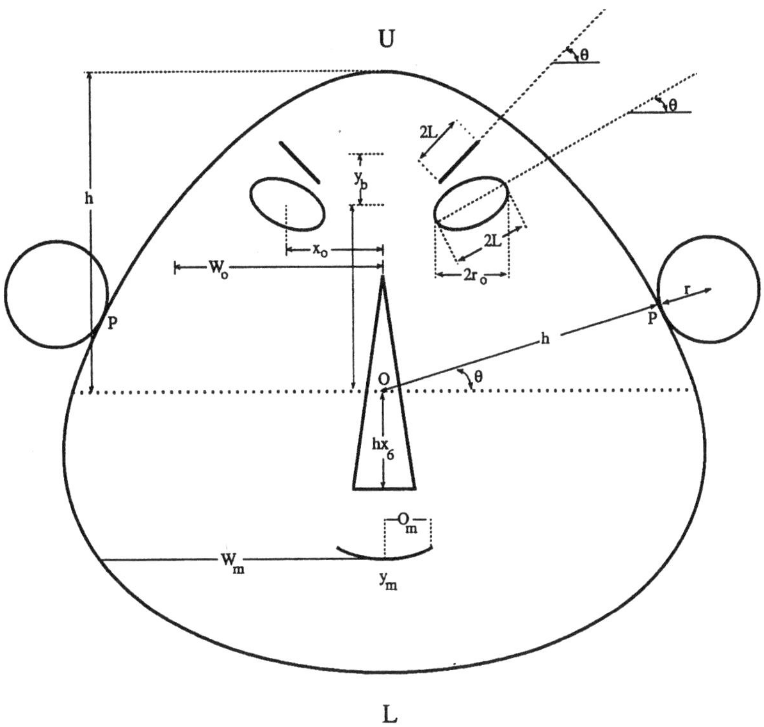

Figure 5. Huff: Basic features in facial construction.

How and why information systems fail cannot be determined without a carefully monitored account or discipline for documentation established within the organization. (Spiers, 1987)

- Information systems need to be audited to ensure that they are operating effectively and are likely to continue to do so.

- Information systems need an audit trail but do not in themselves automatically provide such an trail through their functional operation. The audit trail requires the introduction of controlled redundancy.

- The audit trail needs to be protected against damage during the failing of the information system.

• The components of a system should as far as possible be insulated from other components. This serves both to limit the consequence of a failure and to preserve as far as possible the evidence of the system state in the period leading up to failure.

From the system perspective, there are many reasons why systems fail. Certainly organizational culture can play an important role Westrum, 1987). The tragic case of the Challenger illustrated how information management in an organization can be a powerful determinant in decision making. Other factors are also important, as indicated by Hopkin (1988).

At present, it is safe to say that ATC has an effective information system, for no other reason perhaps, than its high safety record in contrast to the other forms of transportation humans now use. But with increasing environmental complexity, the need is to extend present information systems to knowledge systems. To achieve this, we need to extend our thinking from technology and more to the infinite capacity that the human brain possesses to process data while achieving meaning and understanding. The development of knowledge systems will permit us to achieve greater wisdom in the way we communicate with and control our environment.

Summary

At present air traffic operations are reputed to enjoy a high measure of safety in contrast to other forms of transportation available to people. Much of this acknowledgement is focused on air traffic control and technology that has been incorporated in ground and air operations. But the air traffic environment is expected to change dramatically in the years ahead. With increasing environmental complexity, the demands on personnel throughout the system will increase. Although automation may remain a viable design option, the question remains as to how this option is integrated into the air traffic system — a system international in scope, bridging a number of national boundaries and socio-cultural structures.

A systems approach is deemed essential in our attempt to understand the coming environment and our ability to design systems that are responsive to expected changes. What this means is that we should avoid the tendency to look at (or emphasize) one aspect of the total system while overlooking other factors of the system that may be more critical — or at least related. The language of air traffic control may be limiting our attention to more pertinent dimensions of the problem. For example, automation remains a viable design option, given the expected increase in control. But the use of the term "control" implies an outside guarantor of safety. It implies that it cannot be achieved through a more comprehensive view of the system at large. Moreover, use of automata in the system may be a moot issue. Total automation of the human function doesn't present a realistic option, given certain human attitudes and inclinations. The risk for human error in dealing with event contingencies is too high to allow the human to surrender completely control of the environment to automata.

If we are to understand how we can make the air traffic system human fault tolerant, acknowledge the role of stress factors, and develop a comprehensive training program for operators in the system, then it will be necessary to acknowledge the role and function of data, information, and knowledge (D-I-K) in the system. D-I-K are states that are created by processes part of all living systems. All living systems consist of subsystems that induce these states. Artificial (technological) systems serve to augment the capacities of each of these subsystems. For example, sensor technologies augment the viewing capacity of pilots to acknowledge impending weather disturbance. The autopilot augments the capacity of the brain and motor components to deal with the control of forces that impinge on the aircraft as it proceeds to penetrate a weather front, etc. On occasions data-information (those states that are part of our consciousness and awareness), may not be sufficient to deal with event contingencies. Meaning and understanding may be mandated by the situation. Such functions are aided by our ability to incorporate explanation into the design of present knowledge (expert) systems. D-I-K concepts can substantially aid us in the analysis and design of the air traffic system of the future.

Given this basic structure, we can benefit substantially by our understanding of D-I-K system failure. This requires a discipline for documenting the activities of the system over time, and for the generation of timely audits which identify the dynamic properties of D-I-K in specific, critical air traffic operations.

References

Ackoff, R. L. and F. E. Emery. (1972). *On Purposeful Systems..* Chicago: Aldine-Atherton.

Beniger, James R. (1986). *The Control Revolution.* Cambridge: Harvard University Press.

Bilinski, Andrew E. (1987). System Failure Models as a Result of Design Inadequacy. In J. A. Wise & A. Debons (Eds.). *Information Systems:: Failure Analysis.* Berlin: Springer-Verlag.

Churchman, C. W. (1971). *The Design of Inquiring Systems.* New York: Basic Books.

Cosic, Cynthia. (1989). *Issues in Explanation.* Unpublished paper.

Davis, D. R. (1948) *Pilot Error: Some Laboratory Experiments.*London: His Majesty's Stationary Office.

Debons, A. (1971). "Command and Control: Technology and Social Impacts." In M. Yovits (ed.) *Advances in Computers, Vol. 11.* New York: Academic Press.

Debons, A., E. Horne, & S. Cronenworth. (1988). *Information Science: An Integrated View.* Massachusetts: G. K. Hall. p. 69-68.

De Greene, K. B. (1970). *Systems Psychology.* New York: McGraw Hill.

De Solla Price, D. J. (1963). "Prologue to a Science." *Little Science, Big Science.* Columbia University Press.

Drew, G. C. (1940) *An Experimental Study of Mental Fatigue.* (FPRC Report 227). London: Medical Research Council

Edwards, E. (1975). "Stress and the Airline Pilot." Paper presented at British Airline Pilots Association Medical Symposium, London.

Fox, C. J. (1983). *Information and Misinformation: An Investigation of the Nations of Information, Misinformation, Informing and Misinforming.* Westport: Academic Press.

Hall, A. D., and R. E. Fagan. (1956). Definition of System. In L. von Bertalanfy and A. Rapoport (eds.) *General Systems Yearbook of the Society for the Advancement of General Systems Theory, I*, 18-28.

Harmon, Glynn. (1984). The Measurement of Information. *Information Process Management. 20*(1-2).

Hopkin, V. David. (1988). Air Traffic Control. In E. L. Wiener and D. C. Nagel (Eds.). *Human Factors in Aviation.* New York: Academic Press. p. 639-662.

Huff, D. C., V. Mahajan, and W. C. Black. (1981). Facial Representation of Multivariate Data. *Journal of Marketing, 45,* p. 54.

Kaplan, M. (1991). Cultural Ergonomics: An Evolving Focus. In J. A. Wise, V. D. Hopkin, & M. L. Smith (Eds.) *Automation and Systems Issues in Air Traffic Control.* Berlin: Springer-Verlag.

Lukasiewicz, J. (1972) *The Ignorance Explosion: A Contribution to the Study of Confrontation of Man with the Complexity of Science Based Society and Environment.* New York Academy of Sciences Transaction (Series II), 34(5), May, p. 373-391.

Miller, J. G. (1978). *Living Systems.* New York: McGraw-Hill.

Payor, L. (1990). Borderline Cases. *The Sciences.* July-August, pgs. 38-42.

Scheridan, T. B. (1988). The System Perspective. In E. L. Wiener and D. C. Nagel (Eds.). *Human Factors in Aviation.* New York: Academic Press. pp. 29-34.

Spiers, J. D. (1987). Failure of Business Information Systems. In J. A. Wise & A. Debons (Eds.). *Information Systems:: Failure Analysis.* Berlin: Springer-Verlag.

Toffler, Alvin W. (1970). *Future Shock.* New York: Random House.

Tufte, E. R. (1990). *Envisioning Information.* Connecticut: Graphic Press.

Westrum, R. (1987). Management Strategies and Information Failure. In J. A. Wise & A. Debons (Eds.). *Information Systems: Failure Analysis.* Berlin: Springer-Verlag.

Weiner, E. L. (1988). Cockpit Automation. In E. L. Wiener and D. C. Nagel (Eds.). *Human Factors in Aviation.* New York: Academic Press. p. 451.

Wise, J. A. (1988). Unpublished Sketches ,1988.

Wise, J. A. and A. Debons (Eds.). (1987). *Information Systems: Failure Analysis.* Berlin: Springer-Verlag.

Young, R. E., A. L. Becker and K. L. Pike. (1970). *Rhetoric: Discovery and Change.* New York: Harcourt Brace Jovanovich.

ICAO and Future Air Navigation Systems

John N. Bradbury

International Civil Aviation Organization (ICAO), Montreal, Canada

Introduction

During flight, the crew of an aircraft frequently communicates with stations on the ground. They may seek authorization for flight maneuvers, obtain information necessary to avoid collision, receive an update of weather conditions ahead, or question the operational status of navigation aids en route or at destination. There is a continuous invisible link between the aircraft and the ground stations, and among the ground stations themselves. Many ground facilities and supporting services are needed for the safe and efficient operation of aircraft. To achieve harmonious functioning of all these ground facilities and services, international standardization is necessary.

To ensure safety, regularity, and efficiency of international civil aviation operations, international standardization is essential in all matters affecting them, that is, all matters in the operation of aircraft; aircraft airworthiness; and the numerous facilities and services required in their support, such as aerodromes, telecommunications, navigation aids, meteorology, air traffic services, search and rescue, aeronautical information services, and aeronautical charts. A common understanding between the countries of the world on these matters is absolutely necessary.

To achieve the highest practicable degree of uniformity whenever this will facilitate and improve air safety, efficiency,and regularity, the ICAO Council adopts (or amends when necessary) International Standards and Recommended Practices and approves procedures for the safety, regularity, and efficiency of air navigation.

Although the Council has the responsibility for final adoption of these Standards and Recommended Practices and approval of these procedures, the principal body concerned with their development is the ICAO Air Navigation Commission. As provided for in the Convention, the Commission is composed of fifteen persons who have "suitable qualifications and experience in the science and practice of aeronautics." Its members are nominated by Contracting States and are appointed by the Council. The Commission reports to the Council and is responsible for the examination, coordination, and planning of all of ICAO's work in the air navigation field. It is assisted in this work by an internationally recruited technical secretariat of the Organization's Air Navigation Bureau. In the advancement of solutions to specialized problems requiring up-to-date and specialized expertise, the Commission is assisted by panels. These are small groups of experts, nominated by Contracting States and international organizations and approved by the Commission. The panel members act in their personal, expert capacity and not as representatives of their nominators.

NATO ASI Series, Vol. F 73
Automation and Systems Issues in Air Traffic Control
Edited by John A. Wise et al.

Technical Work and Standardization

Since ICAO came into being in 1947, a main feature of the technical work of the Organization has been the achievement of agreement of the Contracting States on the necessary level of standardization for the operation of safe, regular, and efficient air services. In turn, this has resulted in high levels of reliability in all the many areas which collectively make up international civil aviation. This has particularly been so with respect to aircraft, the crews that operate them, and the ground-based facilities and services. The necessary standardization has been achieved by the Organization primarily through the creation, adoption, and amendment by the Council, as Annexes to the Convention on International Civil Aviation, of specifications known as International Standards and Recommended Practices (SARPs). A Standard is a specification whose uniform application is necessary for the safety or regularity of international civil air navigation, while the Recommended Practice is one agreed to be desirable but not essential. At present, there are eighteen sets of International Standards and Recommended Practices. Seventeen of these are within the technical field.

The basic criterion which decides whether a particular specification should be a Standard is an affirmative answer to the question "Is uniform application by all Contracting States necessary?" The applicability of the Standard may be made subject to the existence of certain conditions relating to the nature of terrain, density of traffic, states of flight, climate, etc., but the Standard must be applied equally by any Contracting State where the specified conditions are encountered.

The specifications usually originate with the ICAO divisional-type air navigation meetings. These divisional air navigation meetings and air navigation conferences are periodically convened to discuss important and urgent problems in one or more fields of air navigation. All Contracting States are invited to participate in them with equal vote.

Specifications proposed to be given the status of Standards or Recommended Practices are, after consultation with all Contracting States, finalized by the Air Navigation Commission and submitted to the Council for adoption by a two-thirds majority. Following their adoption, and provided a majority of Contracting States do not disapprove them, the SARPs become effective and subsequently applicable at dates set by the Council. These Standards and Recommended Practices are considered binding unless Contracting States have given notice to ICAO that they find it impractical to comply with them. The differences notified are then published by ICAO in Supplements to Annexes.

ICAO also formulates Procedures for Air Navigation Services (PANS). which do not have the same status as the SARPs. The PANS are developed by the Air Navigation Commission, usually on recommendation of divisional-type air navigation meetings. Following consultation with all Contracting States, they are approved by the Council and recommended to Contracting States for worldwide application. The PANS comprise, for the most part, operating practices as well as material considered too detailed for SARPs. PANS often amplify the basic principles in the corresponding SARPs to assist in the application of those SARPs. The PANS become applicable on a date set by the Council.

In addition to the various PANS (Aircraft Operations, Rules of the Air and Air Traffic Services, and the ICAO Abbreviations and Codes), there is a need for procedural specifications to meet needs peculiar to specific areas. To meet this demand, specifications known as Regional Supplementary Procedures have been developed, approved by the Council, and recommended to Contracting States for application in the flight information regions to which they are relevant. These procedures may indicate modes of implementing procedural provisions in SARPs and PANS; specify detailed procedural options for regional application, or promulgate a procedure of justifiable operational significance, additional to, but not in conflict with, existing provisions in the Annexes or PANS.

In keeping pace with the rapid development of international civil aviation, ICAO is conscious of the need to adopt in its specifications modern systems and techniques. In recent years, extensive work has been undertaken by ICAO in the areas of data interchange system; automation of air traffic and meteorological service; a review of the general concept of separation; and the application of space techniques, including satellites, for air navigation. Furthermore, a computer system called Accident/Incident Data Reporting (ADREP) has been installed at Headquarters to provide States with accident and incident reports for use in their accident prevention programs. For instance, requests from States for information relating to specific safety problems or studies can be processed and answered in 24 hours.

As part of its continuing effort to improve air safety, the organization has undertaken a comprehensive program involving a complete revision of Annex 5 (Units of Measurement to be used in Air and Ground Operations) designed to introduce the use of a single standardized system of units of measurement in all air and ground aviation operations. The standardized system based on the International System of Units (SI), and will eventually eliminate the use of different units of measurement for the same quantity and provide for the standardized application of all units of measurement for those quantities used in air and ground operations.

To facilitate the implementation and to promote the uniform application of SARPs and PANS, technical manuals or other guidance materials are issued by the Organization under the authority of the Secretary General. At present, there are more than 30 different publications providing this type of guidance.

The Organization also undertook a detailed review of the subject of interception of civil aircraft and civil/military coordination, and developed a comprehensive package of amendments and guidance material involving several Annexes and the PANS-RAC and PANS-OPS, as well as two new manuals. These amendments were adopted or approved by the Council on 10 March 1986 and 12 March 1990, respectively, and all of them have become effective. Technical standards are kept under constant review and updated as necessary, in response to the needs and requirements of international civil aviation.

Regional Planning

In dealing with international civil aviation, ICAO considers many subjects on a regional basis as well as on a worldwide scale. To facilitate detailed planning of the facilities and services and formulation of supplementary procedures necessary to support increases in traffic density, new air routes,and the introduction of new types of aircraft, regional air navigation meetings are held periodically to consider the requirements of air operations in specified areas.

These meetings result in the enumeration of the many facilities and services to be provided by States in the nine regions of ICAO. After review of the meeting's recommendations by the Air Navigation Commission and approval by the Council, they are presented in Air Navigation Plan Publications covering the nine regions. The requirements of these plans are so designed that, when implemented by the States, they lead to an integrated international civil air navigation system.

Although meetings of Contracting States are necessary to determine and express the needs for facilities and services, every State is individually responsible for providing the recommended facilities and services within its own territories. As financial and technical resources vary widely between States, there may be uneven implementation of parts of the air navigation plans. For this reason, ICAO makes a major effort to secure implementation of the plans and to help governments provide the facilities and services called for within their territories.

Implementation of Regional Plans

ICAO gives continuing attention to implementation. The regional meetings establish dates for provision of the required recommended facilities and services on a co-ordinated basis as appropriate, taking into account the regional plans and the ability of States to implement them. Following the meetings, the ICAO regional offices concerned consult with States for the purpose of co-ordinating implementation plans and provide all possible advice and assistance in this regard.

When States require it, help is available through ICAO. Seven ICAO regional offices are maintained — in Bangkok, Cairo, Dakar, Lima, Mexico City, Nairobi and Paris — each one accredited to a group of Contracting States. These offices have, as their main function, the duty of maintaining, encouraging, assisting, expediting and following up the implementation of the air navigation plans. Under certain conditions, help is also available from the Secretariat Headquarters in Montreal; technical consultants are recruited from time to time on a short-term basis, to undertake special implementation projects with the task of studying in detail serious shortcomings in specific technical fields and recommending to governments appropriate remedial action. Such projects have been undertaken so far in South East Asia, the Pacific, Africa, the Middle East, and Central and South America on problems in communications, air traffic control, search and rescue, meteorological services, aerodromes, aeronautical information services, aeronautical charts, and personnel licensing and training. Technical assistance and joint financing are among the other potential solutions to difficulties in the implementation of Air Navigation Plans.

The Council gives direction to these implementation efforts. It has set up, *inter alia*, a Standing Group of Implementation to recommend solutions to implementation problems in cases where special measures are needed to cope with important deficiencies in the implementation of regional plans.

In general, difficulties in implementation can be divided into three main groups: those which arise through problems of organization and administration of departments of civil aviation; those which come from the lack of a sufficient number of trained people to install, maintain, operate, and supervise facilities and services; and those resulting from the scarcity of funds for construction and improvement of airports, for the purchase of equipment and spares. and for provision of adequate salaries and working conditions. Very often these difficulties are combined, for lack of personnel may come from shortage of funds (and shortage of funds from a lack of appreciation of the value and benefit of civil aviation to the country).

As the demand for air navigation services and facilities continues to grow with the expansion of international civil air transport and international general aircraft operations, it can be expected that implementation will continue to be one of the most important questions facing ICAO in the next decade.

Future Air Navigation Systems (FANS)

Background

Since the early 1960s, civil aviation authorities and airlines have taken an active interest in an aeronautical satellite service to provide and operate air navigation in international civil aviation. Such an interest was encouraged by specialists in space technology from all over the world, who believed that aviation was a field for which supportive air navigation facilities had to be developed. Research on this topic has continued within the framework of co-operation among user communities, and, as a result, serious market surveys and studies have been performed by responsible international organizations and private institutions.

In 1968, ICAO established a panel of experts to deal with the application of space technology to aviation (ASTRA). The results of the panel's work were reviewed by the ICAO Seventh Air Navigation Conference in 1972, which recommended that interested States and international organizations initiate and carry out international programs for research, development, and evaluation of aeronautical satellite systems.

In June 1983, the ICAO Council established a Special Committee on Future Air Navigation Systems (FANS), in support of its continuing policy to develop a projection of air navigation requirements for international civil aviation.

FANS is a Technical Committee of the ICAO Council. It forms part of the special machinery of the Council to deal with important, complex, and urgent problems involving technical, economic, social, and legal aspects, whose resolutions require worldwide expertise.

The terms of reference, approved by the Council in November 1983 for the establishment of the FANS Special Committee, read as follows:

> To study technical, operational, institutional, and economic questions, including cost/benefit effects, relating to future potential air navigation systems; to identify and assess new concepts and new technology, including satellite technology, which may have future benefits for the development of international civil aviation. including the likely implications they would have for users and providers of such systems; and to make recommendations thereon for an overall long-term projection for the co-ordinated evolutionary development of air navigation for international civil aviation over a period of the order of twenty-five years.

Composition of FANS and Work Program

FANS membership was drawn from twenty-two Contracting States and international organizations, together with a further ten participants having the status of observer.

The first meeting of FANS, held in Montreal in July 1984, was devoted to development of the work programme. The work was structured to formulate, through a series of steps, a co-ordinated development plan up to the year 2010. The objectives and scope of the study were based on the terms of reference and on the techniques upon which future work should be focused. Although the terms of reference of FANS indicated that the Committee would have to identify and assess new concepts and new technology, including satellite technology, it was clear from the beginning that the study would be concentrated mainly on the use of satellites for the three basic functions of future air navigation systems:

- Communication (voice and data)
  - Air-ground
  - Ground-ground
  - Air-air

- Navigation
  - Self-contained
  - Radio navigation
  - Other

- Surveillance
  - Independent
  - Co-operative independent
  - Dependent

At its second meeting in April 1985, the Committee 1) concluded that the present ground-based, line-of-sight air navigation systems could not technically, operationally, or economically provide an efficient global communication, navigation, and surveillance (CNS) system; and 2) addressed the potential that the use of satellites offered to overcome the present shortcomings, while at the same time providing for future requirements. The Committee also defined scenarios to represent various air traffic conditions in different parts of the world. These scenarios would be used to assess the comparative ability of candidate systems to evaluate efficiently the aviation requirements in terminal, continental, and oceanic airspace.

At its third meeting, in November 1986, the Committee defined the minimum satellite-air-ground communication systems characteristics necessary to ensure system interoperability and the concept of automatic dependent surveillance to a point where, with the proper ground automation capability, a complete pseudo-radar environment could be created.

The FANS Committee concluded its task during the fourth meeting, held in May 1988. The report of FANS/4 contains a consolidated set of material developed by all four meetings, and therefore represents substantially the final report of the Committee.

Although the technical work of FANS was completed at the fourth meeting, the detailed work is only beginning; as a result, existing or new ICAO technical bodies will have to finalize the key issues identified by the Committee, including those related to the institutional aspects associated with the implementation of a global CNS system. These technical bodies are the

- Special Committee on Future Air Navigation Systems
- Aeronautical Mobile Satellite Services (AMSS) Panel
- Automatic Dependent Surveillance (ADS) StudyGroup/Panel

Present Air Navigation Systems Capabilities

At its second and third meetings, the FANS Committee made a comprehensive assessment of the characteristics and capabilities of the present air navigation systems and their implementation in various parts of the world. The review of this accurate profile of the air navigation infrastructure, on a regional and sometimes sub-regional basis, has shown that the shortcomings and the needs to be satisfied vary considerably, due to the types and densities of traffic, topography, and social and economic conditions. However, close analysis ascertained that the shortcomings of the communications, navigation, and surveillance (CNS) systems around the world essentially amounted to three factors.

- The propagation limitations of current line-of-sight systems, and/or accuracy and reliability limitations imposed by the variability of propagation characteristics of other systems

- The difficulty, caused by a variety of reasons, to implement present CNS systems and operate them in a consistent manner in many parts of the world

- The limitations of voice communication and the lack of digital air-ground data interchange systems to support automated systems in the air and on the ground

Although the effects of the above limitations are not the same for all regions of the world, it is evident that one or more of these factors inhibit the desired development of air traffic management almost everywhere. New CNS systems should enable the flexibility, efficiency and safety of global air traffic management to be considerably improved on a global basis. Therefore, the more important needs to be met are:

- Global communication, navigation, and surveillance coverage up to at least 70 000 feet, including for off-shore and oceanic regions, as well as at low altitudes over much of the populated areas of the world

- Digital data interchange between air and ground systems

- Navigation/landing services for runways and other landing areas which need not be equipped with precision landing aids

Air Traffic Management

Components

Air traffic management (ATM) consists of a ground part and an air part, both of which are needed to ensure the safe and efficient movement of aircraft during all phases of operation. ATM is comprised mainly of the air traffic control (ATC) system.

Air traffic management is comprised of ATC, air traffic flow management (AFTM) and airspace management. These three elements contribute to achieve the ATM objectives during different phases of flight.

Objectives

The primary objective of ATM is to expedite and maintain a safe and orderly flow of traffic. The efficiency of the system is an important consideration, which relates, in particular, to aspects such as costs of implementation and operation of both the air and ground parts.

The principles necessary to satisfy the above objectives include the following:

- The ATM system should offer users the maximum flexibility in airspace utilization, taking into account their operational and economical needs as well as the ground system capabilities

- Functional compatibility of the data exchanged between the airborne and the ground components is essential to ensure the global efficiency of the system

- The sharing of airspace between different categories of users must be organized as flexibly as possible, taking into consideration different levels of aircraft equipage

- The various ATC components of the overall ATM systems must be designed to work together effectively to ensure a homogeneous, continuous, and efficient service to the user from take-off to landing. International harmonization is needed to provide for consistency in operations across national boundaries.

The realization of a system capable of satisfying the above principles will require the adaptation of existing procedures and facilities, as well as the development of new ones.

Present ATM

Until now, automation in air traffic control, which is necessary to handle the continuously increasing volume of traffic, has been based mainly on a better utilization of the limited information available on:

a) the intention of the pilot as provided in the flight plan and its updates; and

b) the position of the aircraft as provided by current surveillance means.

Despite the introduction of secondary surveillance radar (SSR), the burden of voice communication between pilot and controller has not been appreciably reduced. There still remains a wide gap in the information flow between air and ground. In addition, large areas of the world lack reliable coverage by CNS systems. Without proper measures, efficiency will further decrease in the future because of the predicted growth in air traffic and the widening gap between airborne and ground capabilities.

The examples presented below are applicable to even the most developed ATC system environments.

- The information flow within and between ATC units (ground-ground communication) and between ATC units and aircraft under their control (air-ground communication) is insufficient to support further significant improvements. However, it is recognized that the utilization of flight plan data processing and radar processing is fairly widespread, with various degrees of refinement, and automated assistance to traffic flow management already exists

- ATC needs improved data and procedures for surveillance, prediction, and optimization of the air traffic flow

- The most advanced ATC systems are still working on the basis of data representing the aircraft performance and environment conditions, which approximate reality only poorly

- Route structures are generally inflexible, although increasingly direct routings are allowed as controller workload permits

Evolution of ATM

The improvement in airborne and ground systems capabilities will be used in a complementary fashion to maximize the efficient use of aerodrome and airspace resources. In this respect, the foreseen implementation of an air-ground data link will play an essential role.

The evolution of ATM will have to take place under a number of conditions, in particular:

- A future system can only be implemented by means of an evolutionary process

- Pilots and air traffic controllers must be kept involved in the ATC process

- The new system should not be overly sensitive to random disturbances, such as outages, emergencies, errors in forecasting, etc.

- The system must be easily expanded to accommodate anticipated future growth in air traffic demand

To support the evolution of ATM, CNS services need to be improved. Notably, they should provide better coverage and include an efficient air-ground data interchange capability. In order to obtain the maximum benefit from new CNS systems, it is essential that these systems are matched with the required ATM services. Although the objectives and principles of ATM are basically the same for any area, the required comprehensiveness of an ATM environment is highly dependent on the type of airspace, and, therefore, its detailed implementation requirements will vary around the world. In addition, the degree to which those ATM requirements can be met is heavily influenced by the practicability of providing CNS services.

CNS System Concept

General

The envisaged changes in air traffic management, and the shortcomings and future needs identified, constitute the basis which underlies the FANS approach to a future CNS system concept. Essentially, the Committee concluded that the application of satellite technologies is the only now viable solution that would enable international civil aviation to overcome the shortcomings of the present CNS systems and to fulfill the needs and requirements of the foreseeable future on a global basis. Indeed, the Committee sees very little prospect of substantial improvement in air traffic management if it were to continue having to reply on the present CNS systems. Therefore, in response to its terms of reference, the Committee developed "an overall long-term projection for the co-ordinated evolutionary development of air navigation for international civil aviation over a period of the order of twenty-five years," in which, complementary to certain terrestrial systems, satellite CNS systems would be the key to worldwide improvement of air traffic management. The overall CNS system concept proposed is presented function-wise as follows:

- Communications
- Horizontal navigation
- Vertical navigation
- Surveillance
- Emergency location
- Airborne collision avoidance

Mobile Communications

- Safety communications:
  - Safety-related communications carried out by air traffic services (ATS) for air traffic control, flight information, and alerting services
  - Aeronautical operational control communications carried out by aircraft operators, which also affect air transport safety, regularity, and efficiency (AOC, aeronautical operational control)

- Non-safety communications:
  - Private correspondence (AAC, aeronautical administrative communications)
  - Public correspondence (APC, aeronautical passenger communications)

Horizontal Navigation

Required Navigation Performance Capability (RNPC)

Modern aircraft are increasingly equipped to utilize newer techniques, generally referred to as area navigation, the use of which is appropriate and inevitable because it facilitates a flexible route system. A new concept, primarily for use in airspace where adequate surveillance is available to ATC, was developed and called "required navigation performance capability" (RNPC). RNPC is broadly defined as the maximum deviation of aircraft from assigned track within which the aircraft can be expected to remain with a given degree of probability. The RNPC concept has been approved by the ICAO Council and assigned to the Review of the General Concept of Separation Panel (RGCSP) for further elaboration.

Present Radio Navigation Systems

The present radio navigation systems serving en-route navigation and non-precision approaches will be able to meet the RNPC conditions and co-exist with satellite navigation systems. However, it is foreseen that satellite systems will eventually become the sole means of navigation, and that the present systems will be progressively withdrawn. The timing of such withdrawal will depend on many factors, such as the implementation and quality of the new systems (which will probably differ in various parts of the world). Because of its impact on international civil aviation on a world-wide scale, the implementation of transition/withdrawal needs to be planned by ICAO.

Approach and Landing Guidance Systems

In accordance with the ICAO ILS/MLS transition plan, MLS will be used for precision approach operations, and no other means of providing final all-weather landing guidance will exist after the year 2000. It is foreseen that global navigation satellite systems (GNSS) will eventually provide adequate information to support non-precision type approaches and, possibly, certain parts of the initial approach fix for all-weather landing procedures.

Vertical Navigation

The ICAO RGCS Panel is studying the relationship between barometric altimetry and vertical separation of aircraft at altitudes above FL 290, with the objective of reducing the presently applied separation criteria of 2 000 feet. It is recognized, however, that barometric altimetry is no longer functional at the very high altitudes at which multimach aircraft will operate. In those cases, geocentric altitude measurements from satellite navigation systems may have to be used.

Surveillance

Automatic Dependent Surveillance (ADS)

The introduction of air-ground data links, together with sufficiently accurate and reliable aircraft navigation systems, gives the opportunity to provide surveillance services in areas which lack such services in the present infrastructure, particularly in oceanic and other areas where the current systems prove difficult, or even impossible, to implement. Automatic dependent surveillance (ADS) is a function for use by air traffic services in which aircraft automatically transmit, via a data link, data derived from on-board navigation systems. As a minimum, the data include aircraft identification and three-dimensional position. Additional data may be provided as appropriate. The ADS data are used by the automated ATC system to present a "pseudo-radar" traffic display to the controller. In addition to areas which are at present devoid of traffic position information other than pilot-provided position reports, ADS will permit beneficial application in other areas, including high-density areas, where ADS may serve as an adjunct and/or back-up for secondary surveillance radar, and thereby reducing the need for primary radar.

Secondary Surveillance Radar (SSR)

Secondary surveillance radar is in wide use in many parts of the world where terrestrial surveillance systems are appropriate. In several countries, the accuracy of position information is improved by the application of monopulse techniques. By enlarging SSR with Mode S, the selective address and data-link capabilities will further enhance the beneficial role of SSR for surveillance purposes. Mode S is expected to come into use in some areas during the 1990s. However, because of the limited capability of the Mode S data link, and due to its line-of-sight constraints and its lack of potential worldwide coverage, the Mode S data link cannot be considered as a universal data link system.

Primary Radar

Although the use of primary radar is already declining, its need will continue to exist in those airspaces where there is a mix of SSR-equipped aircraft and non-SSR-equipped aircraft, with compatible services provided to both categories. However, as such circumstances decrease, the need for primary radar will be reduced. This reduction will be further advanced by the introduction of ADS in its role as an adjunct/back-up for SSR. It is foreseen that secondary radar, through the Mode S system and its data link, and ADS through satellite communications, will be of such high integrity that the justification for primary radar in air traffic services for international civil aviation will greatly diminish. It is recognized, of course, that primary radar for other purposes, including weather detection, will continue to be needed.

Satellite-aided Search and Rescue (COSPAS-SARSAT)

An aircraft under emergency conditions can be detected, and an approximate position established, using its last messages and last positions as recorded on the surveillance system. In this respect, AMSS and ADS will extend that possibility to a worldwide coverage. The transmission of signals by emergency location transmission (ELT), which starts operating after an accident, allows a more complete detection of the emergency by giving an actual location of the grounded aircraft. The satellite-aided search and rescue (COSPAS-SARSAT) system operates on the standard ICAO frequencies for ELT of 121.5 MHz and 243 MHz. It also uses a new frequency, 406 MHz, which allows a better detection and location service.

Airborne Collision Avoidance Systems (ACAS)

Whereas, for the prevention of collisions, air traffic management should continue to rely primarily on air traffic control services, airborne collision avoidance systems (ACAS) will play a useful role as a last resort back-up system to prevent collisions in the event of a large error or undetected blunder.

While recognizing that ACAS would offer safety benefits, it is not to be included as an integral part of the future air navigation concept. Where applied, ACAS will be a system by itself supplementary to, but separate from, ATC systems.

Conclusions

In conclusion, the global concept of CNS systems for the future global needs of international civil aviation, proposed to be implemented within a period of about twenty-five years to eventually succeed the present concept, may be summarized as follows:

Communications

- Satellite data and voice communication capability over most of the world. Initially, HF may have to be maintained over polar regions

- VHF will remain in use for voice and certain data communication in many continental and terminal areas

- The SSR Mode S data link will be used in high density airspace

Navigation

- Navigation accuracy has to concur with required navigation performance capability (RNPC) criteria

- Global navigation satellite system(s) will provide worldwide coverage and will be used for aircraft navigation and for non-precision type approaches

- Microwave landing system (MLS) will have replaced ILS for precision approach and landing

- ADF/NDB and VOR/DME will be gradually withdrawn

- The use of Omega/Loran-C will progressively decrease

Surveillance

- SSR Mode A/C or SSR Mode S will be used in terminal areas and in high density continental airspace

- Automatic dependent surveillance (ADS) will be used in other airspace and may eventually replace some SSR

- The use of primary radar will diminish

An evolution towards the above concept over a period of the order of twenty-five years would result in an air navigation environment free of the present CNS limitations.

Automatic Dependent Surveillance (ADS)

In the absence of long-range radar surveillance of air traffic operations over extensive oceanic and certain land areas, the procedural air traffic control system was necessarily based upon position reports provided by pilots via voice communications channels. The concept of ADS, where aircraft-derived positional information would be transmitted automatically or on request to ATC units at frequent intervals through a suitably developed air-ground data link, is defined as follows:

Automatic dependent surveillance is a function for use by air traffic services in which aircraft automatically transmit, via data link, data derived from on-board navigation systems. As a minimum, the data include aircraft identification and three-dimensional position. Additional data may be provided as appropriate.

The system elements needed to implement an ADS function include: air traffic control, communication elements, and c) avionics.

ADS Functional Objectives and Areas of Application: The functional objectives of ADS, which could vary in different areas of application, may be characterized as follows:

- the enhancement of flight safety through the ability to provide a means of surveillance on aircraft operating outside surveillance radar coverage

- timely detection and correction of deviations from the cleared track

- reduction in separation made possible by improved communications, and improved ATC data processing and display capability

- an increased level of tactical control enabling a more flexible use of airspace

Consideration of specific areas for the potential early application of ADS indicates that the most beneficial is likely to be in areas with transoceanic operation or:

- oceanic areas with off-shore fixed-route operations

- areas with low-altitude operations (e.g. helicopter oil rig operations)

- low traffic density land areas

- high-density areas where automatic dependent surveillance can eventually serve as a substitute for secondary surveillance radar

ATC Automated System Based on ADS

The functional capabilities of an ATC automated system based on ADS will include the aircraft position, tracking, air-ground data message processing, flight plan data processing, and graphical display for traffic situation and tabular information. The system will also include:

- communication system through data link via satellite with ground systems and ATC units

- aircraft position report message containing the aircraft position, flight projection intentions, heading, and on-board navigation systems status

- smoothing process based on historical position report, ADS messages, and aircraft position display

- aircraft position display with flight plan information

- automatic generation of alerting messages for non-receipt of position reports

- air traffic controller capability to request and obtain aircraft position independent of ADS communication system

- processing capability for the projection of flight trajectory and conflict detection

Evolution and Transition

In considering the overall CNS system concept, the questions of evolution and transition are most important. For instance, careful planning will be necessary to ensure that aircraft of the future are not required to carry a multiplicity of existing and new communications, navigation, and surveillance equipment. In addition, there is a close relationship between the required CNS services and the desired level of air traffic management. And finally, for reasons of both economy and efficiency, there is a need to ensure that differences in the pace of development around the world do not lead to imbalance and incoherence in the global air navigation system. In particular, due to the wide coverage of satellite CNS systems, the above considerations call for conscientious worldwide coordination of the planning and implementation if such systems are to be optimized.

Table 1 presents a suggested evolution of the CNS systems for four different types of airspace. Such evolution would lead to meeting the future needs and to reducing the complement of avionics required for international worldwide operations.

Following development of the FANS global concept, the Council established a special committee to monitor and coordinate development and transition planning for the Future Air Navigation Systems (FANS-Phase II), with the following terms of reference:

- To identify and make recommendations for acceptable institutional arrangements, including funding, ownership, and management issues for the global future air navigation system

- To develop a global co-ordinated plan, with appropriate guidelines for transition, including the necessary recommendations to ensure the progressive and orderly implementation of the ICAO global, future air navigation system in a timely and cost beneficial manner

- To monitor the nature and direction of research and development programs, trials,nd demonstrations in CNS and ATM, so as to ensure their co-ordinated integration and harmonization

Tenth Air Navigation Conference

A divisional-type air navigation meeting makes recommendations, within the scope of its agenda, for new Standards and Recommended Practices (SARPs), Procedures for Air Navigation Services (PANS), and guidance material, or for amendment to the existing SARPs, PANS,and guidance material. Apart from this primary objective, the meeting prepares, as necessary, other recommendations on individual agenda items for action by Contracting States or the Organizations.

Table 1. CNS systems evolution.

Type of Airspace	Current			Proposed		
	Communications	Navigation	Surveillance	Communications	Navigation	Surveillance
Oceanic/continental en-route airspace with low traffic density (Note 4)	VHF voice HF Voice	OMEGA/LORAN-C NDB VOR/DMF Barometric altitude INS/IRS	Primary radar/SSR Voice position reports	VHF voice/data AMSS data/voice HF over poles only (Note 5)	RNAV/RNPC GNSS Barometric altitude High altitude GNSS altimetry (Note 2) INS/IRS	ADS
Continental airspace with high density traffic	VHF voice	OMEGA/LORAN-C NDB VOR/DMF Barometric altitude INS/IRS	Primary radar SSR Mode A/C	VHF voice/data AMSS data/voice SSR Mode S data link	RNAV/RNPC GNSS Barometric altitude High altitude GNSS altimetry (Note 2) INS/IRS	SSR Mode A/C or SSR Mode S ADS
Oceanic airspace with high density traffic	HF voice	MNPS OMEGA/LORAN-C Barometric altitude INS/IRS	Voice position reports	AMSS data/voice	RNAV/RNPC GNSS Barometric altitude High altitude GNSS altimetry (Note 2) INS/IRS	ADS
Thermal areas with high density traffic	VHF voice	NDB VOR/DME ILS Barometric altitude INS/IRTS	Primary radar SSR Mode A/C	VHF voice/data SSR mode S data link	RNAV/RNPC GNSS MLS NDB (Note 3) VOR/DME (Note 6) Barometric altitude INS/IRS	SSR Mode A/C or SSR Mode S ADS (Note 1)

<u>KEY</u>

AMSS - aeronautical mobile-satellite service
MNPS - minimum navigation performance specifications
RNAV/RNPC - area navigation/required navigation performance capability
GNSS - global navigation satellite system
ADS - automatic dependent surveillance
INS/IRS - inertial navigation system / inertial reference system

Note 1 - The need for primary radar is reduced
Note 2 - To be used where barometric altimetry is not functional
Note 3 - NDB will be progressively withdrawn
Note 4 - Includes low-altitude, off-shore and remote areas
Note 5 - Until such time as satellite communication is available
Note 6 - VOR/DME will be progressively withdrawn

Purpose and Scope of the Tenth Air Navigation Conference

After receiving the FANS/4 Report (Doc 9524), the Air Navigation Commission believed that the global concept of the future CNS system was a significant departure from the present air navigation system, and that States should be consulted on the matter in order to obtain a clear mandate for such an important project. Such an approach would avoid the risk of States' opposition or regional fragmentation at the planning and/or implementation stage. In line with the above, the Council agreed to the convening of an air navigation conference and requested the newly formed Special FANS (Phase II) Committee to assist in the preparation of such a conference.

The Air Navigation Commission, on 15 and 17 May 1990 (at the sixth and seventh meetings of its 124th Session), discussed the subject and agreed that States and interested international organizations be consulted on the convening of a Tenth Air Navigation Conference to discuss subjects in the field of future air navigation systems which are expected to be mature for worldwide consideration by September 1991. In agreeing to such a conference, the Commission recognized that the new air navigation systems concept, which was developed using satellite technology, would overcome the shortcomings of the present air navigation system in addition to yielding additional benefits. It also recognized that the full system benefits could be realized only through global implementation of the future CNS system. It was essential in the view of the Commission that the new concept, which is a significant departure from present air navigation systems, receive worldwide acceptance through the air navigation conference.

The purpose of the Tenth Air Navigation Conference is to introduce the CNS concept and obtain endorsement from States of a plan for the implementation of the future system. The theme for the conference is "Future Global CNS Systems." Tentative dates for the Conference are 3 to 20 September 1991.

List of tentative subjects

The following list of tentative subjects expected to be mature for examination by the Tenth Air Navigation Conference has been identified by the Commission and sent to States for comment.

- Subject 1: Introduction to the CNS concept
- Subject 2: Consideration of the FANS concept for the future air navigation system, and its capability of correcting the shortcomings of the present CNS system
- Subject 3: Technical development of the future air navigation system
- Subject 4: Consideration of institutional aspects of the future air navigation system
- Subject 5: Safety, technical, operationa.l and economic advantages in adopting the future air navigation system, including cost-effectiveness and penalties associated with non-implementation

- Subject 6: Consideration of air traffic management (ATM) aspects
- Subject 7: Global transition planning
- Subject 8: Coordination of the implementation of the future air navigation system through regional activity
- Subject 9: Recapitulation of allocations for aeronautical services in the L-band portion of the electromagnetic spectrum
- Subject 10: Endorsement of the CNS/ATM concepts by States and international organizations

Summary

To ensure safety, regularity, and efficiency of international civil aviation operations, international standardization is essential in all matters affecting them — that is, all matters in the operation of aircraft, aircraft airworthiness, and the numerous facilities and services required in their support, such as aerodromes, telecommunications, navigation aids, meteorology, air traffic services, search and rescue, aeronautical information services, and aeronautical charts. A common understanding between the countries of the world on these matters is absolutely necessary. To achieve the highest practicable degree of uniformity whenever this will facilitate and improve air safety, efficiency, and regularity, the ICAO Council adopts (or amends when necessary) International Standards and Recommended Practices and approves procedures for the safety, regularity, and efficiency of air navigation.

In keeping pace with the rapid development of international civil aviation, ICAO is conscious of the need to adopt in its specifications modern systems and techniques. In recent years, extensive work has been undertaken by ICAO in the areas of data interchange systems; automation of air traffic and meteorological service;, a review of the general concept of separation; and the application of space techniques, including satellites, for air navigation.

In dealing with international civil aviation, there are many subjects which ICAO considers on a regional basis as well as on a worldwide scale. To facilitate detailed planning of the facilities and services and formulation of supplementary procedures necessary to support increases in traffic density, new air routes and the introduction of new types of aircraft, regional air navigation meetings are held periodically to consider the requirements of air operations in specified areas.

When States require it, help is available through ICAO. Seven ICAO regional offices are maintained — in Bangkok, Cairo, Dakar, Lima, Mexico City, Nairobi and Paris — each one accredited to a group of Contracting States. These offices have, as their main function, the duty of maintaining, encouraging, assisting, expediting, and following up the implementation of the air navigation plans. Under certain conditions, help is also available from the Secretariat Headquarters in Montreal, and technical consultants are recruited from time to time on a short-term basis to undertake special implementation projects with the task of studying in detail serious shortcomings in specific technical fields and recommending to governments appropriate remedial action.

In June 1983, the ICAO Council established a Special Committee on Future Air Navigation Systems (FANS) in support of its continuing policy to develop a projection of air navigation requirements for international civil aviation. FANS is a Technical Committee of the ICAO Council. It forms part of the special machinery of the Council to deal with important, complex, and urgent problems involving technical, economic, social, and legal aspects, the resolution of which requires the representation and participation of worldwide expertise.

The terms of reference, approved by the Council in November 1983 for the establishment of the FANS Special Committee, read as follows:

> To study technical, operational, institutional and economic questions, including cost/benefit effects, relating to future potential air navigation systems; to identify and assess new concepts and new technology, including satellite technology, which may have future benefits for the development of international civil aviation including the likely implications they would have for users and providers of such systems; and to make recommendations thereon for an overall long-term projection for the co-ordinated evolutionary development of air navigation for international civil aviation over a period of the order of twenty-five years.

FANS developed a global concept of the future communications, navigation, and surveillance system and addressed the potential that the use of satellites offered to overcome the shortcomings of the present air navigation system. As a follow-up to FANS, Council established a special committee for the monitoring and coordination of development and transition planning for the Future Air Navigation Systems (FANS II) with the following terms of reference:

a) To identify and make recommendations for acceptable institutional arrangements, including funding, ownership, and management issues for the global future air navigation system.

b) To develop a global coordinated plan, with appropriate guidelines for transition, including the necessary recommendations to ensure the progressive and orderly implementation of the ICAO global, future air navigation system in a timely and cost beneficial manner.

c) To monitor the nature and direction of research and development programs, trials and demonstrations in CNS and ATM, so as to ensure their coordinated integration and harmonization.

The ICAO Council and Air Navigation Commission have agreed to consult States on the convening of a worldwide air navigation conference, in order to obtain a clear mandate for the global CNS system concept and obtain endorsement of a plan for the implementation of the future system. Tentative dates for the Tenth Air Navigation Conference are 3 to 20 September 1991.

References

ICAO (1989a). Procedures for Air Navigation Services. ABC - ICAO Abbreviations and Codes. (Doc 8400). 4th edition.
ICAO (1989b). Report of the Fourth Meeting of the SSR Improvements and Collision Avoidance Systems Panel (SICASP/4). Montreal, 10-28 April 1989. Yellow cover report.
ICAO (1988a). Annex 13 - Aircraft Accident Investigation. 7th edition.
ICAO (1988b). Annex 16 - Environmental Protection. Volume I - Aircraft Noise. 2nd edition. Volume II - Aircraft Emissions. 1st edition.
ICAO (1988c). Report of the Fourth Meeting of the Special Committee on Future Air Navigation Systems (FANS/4). Montreal, 2-10 May 1988. (Doc 9524).
ICAO (1988d). Report of the Sixth Meeting of the Review of the General Concept of Separation Panel (RGCSP/6). Montreal, 28 November - 15 December 1988. Doc (9536).
ICAO (1987a). Annex 11 - Air Traffic Services. 8th edition.
ICAO (1987b). Annex 15 - Aeronautical Information Services. 7th edition.
ICAO (1986a). Annex 2 - Rules of the Air. 8th edition.
ICAO (1986b). Annex 6 - Operation of Aircraft. Part III - International Operations - Helicopters. 1st edition.
ICAO (1986c). Procedures for Air Navigation Services. OPS - Aircraft Operations. (Doc 8168). Volume I - Flight Procedures. 3rd edition. Volume II - Construction of Visual and Instrument Flight Procedures. 3rd edition.
ICAO (1986d). Report of the Third Meeting of the Special Committee on Future Air Navigation Systems (FANS/3). Montreal, 3-21 November 1986. (Doc 9503).
ICAO (1985a). Annex 4 - Aeronautical Charts. 8th edition.
ICAO (1985b). Annex 10 - Aeronautical Telecommunications. Volume I (Part I - Equipment and Systems; Part II - Radio Frequencies). 4th edition. Volume II (Communication Procedures including those with PANS status). 4th edition.
ICAO (1985c). Procedures for Air Navigation Services. RAC - Rules of the Air and Air Traffic Services. (Doc 4444). 12th edition.
ICAO (1985d). Report of the Second Meeting of the Special Committee on Future Air Navigation Systems (FANS/2). Montreal, 10-26 April 1985. (Doc 9458).
ICAO (1983a). Annex 6 - Operation of Aircraft. Part I - International Commercial Air Transport - Aeroplanes. 4th edition.
ICAO (1983b). Annex 6 - Operation of Aircraft. Part II - International General Aviation - Aeroplanes. 3rd edition.
ICAO (1983c). Annex 14 - Aerodromes. 8th edition.
ICAO (1980). Convention on International Civil Aviation. Signed at Chicago on 7 December 1944. (Doc 7300). 6th edition.
ICAO (1979). Annex 5 - Units of Measurements. 4th edition.
ICAO (1972). Report of the Seventh Air Navigation Conference. (Doc 9004). Montreal, 5-28 April 1972.

Social, Political, and Regulatory Issues Concerning Harmonization of Interacting Air Traffic Control Systems in Western Europe

Randahl N. Lindgren

Systems Control Technology, Inc., Arlington, Virginia, USA

Introduction

The advent of unprecedented worldwide growth in air traffic, together with forecasts for continued increases at a rate which will double today's levels by the year 2000, have jolted governments and the aviation industry — airspace users, air traffic authorities, and fare paying passengers — into the realization that today's air traffic management mechanisms can cope neither with today's air traffic demand nor with that of tomorrow.

Long the neglected link in the transportation chain, the air traffic management system has been identified as the tool which will — incorporating advances in communications, navigation and surveillance — begin to build air traffic capacity to a level where air traffic demand is satisfied with a safe, orderly, and expeditious air traffic service.

As a result, governments and air traffic control authorities worldwide are engaging in a recovery effort intended to increase total capacity and relieve the costs and inconvenience of delay.

Nowhere is the problem more acute than in western Europe, where the realization of a lagging infrastructure has created increased activity and a charged atmosphere of challenge.

Atop the challenge list for Europe is the issue of fragmentation and lack of integration of its regional air traffic management machine. Western Europe's air traffic management mechanism consists of 22 independent systems run by autonomous governments and agencies which historically addressed their air traffic problems on a national basis.

However, in recent years these authorities have come to realize that Europe's air traffic future must be attacked on a regional basis, and that their separate mechanisms must be made to operate as a single integrated system. A harmony must be designed into the hardware and software which is being, and which will be, procured (and, it must be added, to that which currently exists). Some notable examples of forward thinking exist. Especially pertinent is The 4 States/EUROCONTROL Integration Project to harmonize air traffic services involving Germany, Belgium, the Netherlands, and Luxembourg.

Much has been written of the technical aspects of the harmonization process, of the pros and cons of certain technical proposals and the difficulties in implementing them.

It is the purpose of this paper, however, to illuminate some of the tangential issues associated with this harmonization process and to show that these associated issues can greatly influence the level of success enjoyed by any technical solutions which are implemented.

NATO ASI Series, Vol. F 73
Automation and Systems Issues in Air Traffic Control
Edited by John A. Wise et al.

Harmonization in western Europe could be realized either through an effort to make the currently installed mechanisms play together as a single integrated system, or it could come as a complete replacement of the current infrastructure with an entirely new pan-European design.

The first is the more likely scenario, for it could be done more quickly and less expensively than the second. It could also be a logical forerunner to the latter, allowing costs of current equipment to be amortized out over a natural equipment life span while a new pan-European system is designed and procured. But for the purposes of this paper, either scenario presents the same subset of social, political, and regulatory issues — issues which cannot be considered as secondary, but which must be attacked with the same political will and determination, and within the same implementation time-scale, as the well known technical side. If this is not accomplished, the technical solutions become white elephants, unable to fulfill their intended purpose to the appropriate level of efficiency.

What are these issues, and how do they affect the end product? Broadly, they can be divided into social, political, and regulatory issues, but first let's lay some groundwork.

Harmonization

Technical harmonization occurs when the many hardware and software devices which comprise the air traffic management systems of the various national facilities, are made compatible. That is, they are so designed as to completely interact, communicate, and negotiate automatically, or at controller direction, with each other. Harmonization, from a user's point of view, results in a high level of air traffic service which is so uniform throughout the system as to be transparent. The pilot cannot identify any difference in service between the various ATC facilities. Coordination and work practices are also uniform within and between the control facilities, because controllers perform duties to precisely the same standards throughout the system. Coordination with a controller seated two meters away is no different than with one 200 miles away, and the expectations, one with the other, are the same. When equipment harmony is achieved, an air traffic control center in France, for example, will be directly interfaced with one in the UK, and together they would share respective responsibilities for traffic passing over the English Channel. With the controllers on either side of the Channel sharing exactly the same radar data, flight plan data, hardware and software capability, the potential for greatly improved capacity is achieved because separation requirements can be reduced and coordination requirements can be both lessened and simplified.

For example, when two unharmonious radar data systems are in use on either side of a control area boundary, the radar displays of the two involved controllers may show slight differences in the plan position. Because the controllers may not have high confidence in this presentation and in their ability to provide assured minimum separation across such borders, they may sometimes increase the separation slightly just to compensate for any error that may be present between the two systems. That increase of two or three miles in separation (in some cases more), when multiplied by the number of aircraft operating across the border area, can lead to a substantial loss of capacity (i.e., fewer aircraft per hour). All the links along the chain should have access to exactly the same information about each aircraft. However, the controllers on both sides must also be doing the same thing, understanding and interpreting the like data in the same way. This is total harmonization and it encompasses the social, political, and regulatory issues which are the focus of this paper.

Harmonization could well mean facility consolidation. Currently, some 42 air traffic control centers, vastly more than is required for the geographic area of Europe, provide western Europe's enroute ATC services. By comparison, the United States uses 20 centers to control an airspace area about double the size. The correct number of centers needed for Europe could be debated at length, but some consolidation could and should be achieved.

Europe's centers could be consolidated if its airspace configuration were changed to a design based upon traffic flow, efficient routing, and traffic density — rather than on arbitrary national borders. With fewer centers responsible for larger areas, the sovereign airspace once controlled by a national entity might be delegated to a center located out of the country, a situation not unlike that of Luxembourg today. Such a situation would raise several interesting issues.

Social Issues

For example, part of the personnel required to run the consolidated center would come from those facilities which become redundant. The result of this could be a rather disrupting move of many families from familiar national surroundings to a neighboring country. Some controllers, maintenance staff, and administration specialists would doubtless not make such a move, choosing instead to retire or change employment fields. The losses from the field could either result in additions to the European controller shortage of 1500 controllers as estimated by the United Kingdom Guild of Air Traffic Control Officers (Domogala, 1989), or it might be that the consolidation will actually reduce the need for controllers, and in that case they would become offsets and sources of needed attrition.

Consider also the language issue. Even though English is considered the international language of air traffic control, many non-native English speaking controllers utilize their native language for certain control and coordination tasks. Such language usages are not really a problem in a "national" ATC facility, but they take on added significance in a multi-national environment.

Dealing with unions would become a new challenge. It may occur that the unions representing controllers would themselves wish to harmonize in some way. But certainly dealing with them uniformly in a multi-national setting, would become a challenge. Salary levels would somehow have to be indexed to location cost-of-living, so that salaries become consistent throughout the entire system yet not out-of-step with the local area. Some form of international equal-opportunity program might need to be implemented to assure that controllers of all nationalities have equal opportunity for employment and advancement. Many other issues, including national holidays, vacation time, medical benefits, and retirement programs would require adjustment.

Political Issues

One of the most serious detriments to harmonization of national air traffic services has been the reluctance of national authorities to relinquish control over sovereign airspace. This is one of the prime issues which caused the original charter of EUROCONTROL to falter. As a result, only a very small portion of control authority was ever handed over by the states. In order for any new attempts at harmonization to be successful, this prominent hurdle must be tackled straightaway.

Fisher (1989) states,

> The principles on which European airspace is organized were established to cope with the problems which arose 30 years ago with the advent of jet aircraft. The EUROCONTROL ideal of upper airspace sector boundary determination by operational and technical considerations alone was never fully achieved because of the politics involved. The situation has changed now, and worries about national sovereignty can no longer be used as an excuse to block moves towards a centralized system.

In most western European countries, the military is both a significant user and a significant provider of air traffic control service.

As users of these services the military has special requirements that remain classified. Special provisions for handling these issues on a multi-national basis would have to be worked out. This would seem to be quite feasible for most cases, as large scale military exercises encompassing multiple states in Europe are already routinely undertaken. As providers of air traffic services the military plays a large role. Typically, military controllers control not only military aircraft, but in fact make significant contributions to the control of civil aircraft as well. For example, in the United Kingdom, military controllers are largely responsible for off-airways enroute traffic, much of which is civilian. Thus, in any technical and procedural harmonization which is undertaken, the military cannot be ignored. Interestingly though, with the relaxation of tensions in east-west relations, and with the continuing draw down in military activity, there may soon be a desire to reduce defense spending, especially in pseudo-defense roles like air traffic control. There could well be a serious movement to civilianize portions of the military sector in coming years.

Regulatory Issues

If controllers are going to have basically the same tools to do their jobs, it will be necessary to standardize the procedures they use as well. For example, little gain will be realized if controllers, given the capability to make automatic transfer of control handoffs, cling to old manual methods, or if integrated surveillance capability gives them the opportunity to provide five-mile separations, yet they persist with ten or twenty. Of course, there is an intricate web of training, tradition, management emphasis, and professionalism dictating how these behaviors will emerge, but the issue remains that all of the control facilities must be using the same basic procedures, or full benefit will not be gained.

Training systems will require a complete renewal. Not only must controllers be trained on the new equipment they are receiving, but they also must be indoctrinated on new control concepts that can be implemented because of the new equipment they have at their disposal. Training may need to be centralized, so that controller vacancies are filled with a stream of consistently prepared air traffic personnel. An expansion of the current EUROCONTROL training facility at Luxembourg may be an answer.

Differences in licensing requirements between the various states would require some standardization. Some states require very stringent examinations for licensing, while some require no license at all. Also, differences regarding the types of organizations providing the air traffic service, government or private, could someday become an issue. Currently, Switzerland employs the independent enterprise swisscontrol™ to handle its ATC service, unlike most of Europe, where government agencies provide most ATC services.

Many regulatory issues in the financial and budgetary arena will require attention. Much of the revenue earned through service to enroute aircraft is acquired simply because an aircraft is flying through sovereign airspace. With airspace consolidations, new criteria for division of revenues may be required to properly align income with costs of services provided. Likewise, expenditures for capital improvement needs to be carefully coordinated, so that scarce financial assets can be used in ways that will best benefit the entire system. In that regard, calls for a supra-national agency, like the current EUROCONTROL, but with added authority and responsibility as granted by the states, should be considered.

Eastern Europe's Impact

As if the integration and harmonization problems already identified for western Europe were not enough, now the stage is set for much more interaction with eastern Europe, as the result of the political changes which have recently occurred. For example, a new agreement was signed by Presidents Gorbachev of the USSR and Bush of the USA, at the May 1990 Superpower Summit, which quadruples air service between the two countries. It is not hard to imagine over whose airspace those new flights will fly, and equally what impact the opening the east, in general could have on traffic levels of the West. Suffice it to say, it could be very, very significant — and it leads to an equally strong argument for eventually including eastern European states as fully integrated partners in a European air traffic management system. The implications of that are considerable and must be the subject of separate work, but briefly stated, all of the social, political, and regulatory issues already identified for western Europe alone, would apply and be compounded exponentially because of the economic and ideological differences that are also manifest. However, we must realize that if the current political climate in Europe is maintained and nourished, the day will come — probably much faster than we could ever imagine — when we must expand the integration and harmonization effort through all of Europe...and beyond.

Conclusion

Technical harmonization in Europe will be achieved as means are procured and installed. But these harmonizations cannot achieve their full potential unless the social, political, and regulatory issues are concurrently resolved.

There are some who would say that this cannot be done, and they would cite the stalled efforts at achieving military equipment inter-operability between the establishments of the various NATO countries. It is agreed that it will be a very difficult task. In the words of I.R. Jan P.F. de Konging (1989), regarding regulatory and procedural harmonization of The 4-States/EUROCONTROL Integration Project,

> Use of the same equipment and software of all ATC centers concerned should imply optimum collaboration. However, national regulations, civil/military relations, working methods and procedures are in each case different for historical reasons. This means that many details are dealt with differently in each system. [Attempting to]...harmonize regulations and procedures...proved to be a virtually impossible task. But it is quite clear that if the intention is to achieve joint procurement of systems, then harmonization and standardization are essential in many fields...genuine results will be achieved only with great difficulty and in the long term.

This statement of experience, from an integration project well underway, shows us that the challenge to achieve this harmony is great, the task is essential, and the payoff crucial to European air traffic control integration.

References

de Koning, I.R., Jan P.F. (1989). "The 4 States/ EUROCONTROL Integration Project," EUROCONTROL CONTEXT.

Domogala, P.R.V. (1989). "The Future European Air Traffic Controller," *Guild of Air Traffic Control Officers CONVEX 89.*

Fischer, F. (1989). "Wanted: One ATC System for Europe.," *Janes Airport Review.*

Janes Information Group, Ltd. (1990). "World Survey of Air Traffic Control Projects," Janes Information Group, Ltd.

Aircraft Traffic Forecast and Communications Requirements in the Year 2000

Peter J. McAlindon

Embry-Riddle Aeronautical University, Daytona Beach, Florida, USA

Introduction

Aviation activity during the coming decade is expected to grow at about the same rate as the general economy. Aviation will continue to dominate all other transportation modes in the commercial intercity passenger market. Regional/Commuter aircraft activity and the business use of general aviation is expected to experience greater growth than the larger, established airlines and personal use of general aviation (Federal Aviation Administration, 1988).

In the U.S., the continuing growth in the number of aircraft operations, diversity of operations, and number and sophistication of aircraft will place unprecedented demands on the national airspace system, from now through the turn of the century. In addition, the full effects of the Deregulation Act of 1978 and the resulting increase in competitive scheduling, lower prices, new airline entries, and airport hub use will continue to place extra demands on the system. The safe and efficient operation of the national airspace system will require improved services, new facilities for system expansion, orderly replacement of aging facilities, and provisions for adequate airport capacity (Rodgers, 1988).

Similarly, limits in the current air traffic control system seriously disrupts air travel in Europe. Air travel remains safe, but it is inefficient and subject to intolerable delays. In Germany, for example, nearly one-third of all flights are delayed (German Airspace Users Association, 1989). The European air traffic control system will require improved facilities for its system expansion; implementation of new, more efficient technologies; and integration of an air traffic control system for all of western Europe.

Aircraft Traffic Forecast

Aviation activity at FAA facilities is expected to continue the upward growth pattern which began in 1983. The demand for FAA operational services is anticipated to increase over the forecast period as a result of continued strong growth in aviation activity. Total aircraft operations at FAA-towered airports are forecast to increase to 81.4 million in 2000, a 2.4% annual growth rate over the 61 million operations achieved in 1987 (Rodgers, 1988).

Aircraft operations, including takeoffs and landings at all towered and non-towered airports in the National Airport System Plan (NASP), were anticipated to grow by 54 percent

NATO ASI Series, Vol. F 73
Automation and Systems Issues in Air Traffic Control
Edited by John A. Wise et al.

between 1982 and 2000, with itinerant operations increasing 70% (Table 1). Itinerant operations refer to those aircraft departing to or arriving from an area outside an airport's local operating area (Federal Aviation Administration, 1988).

Table 1: Total National Airspace System Activity
(from the *National Airspace System Plan*, 1988)

	1982	1985	1990	2000	% Growth 1982-2000
Airport Operations (millions)					
Aircraft Operations	127.6	143	155	196	53.6
Itinerant Operations	68	78.8	89.4	115.6	70.0
Instrument Operations	31.7	38.7	49.1	60.6	91.2
Towered Airport Operation	50.6	57.9	66.1	82.2	62.5
ARTCC Operations (millions)					
IFR Aircraft Handled	27.9	32.7	39.2	48.0	72.0
ACF Approach Control Ops	----	----	----	60.6	----
Hours Flown (millions)					
Air Carrier	6.1	7.4	11.3	14.7	137.1
General Aviation	37.7	36.2	34.5	37.1	-1.6

For the period from 1986 to 1990, Eurocontrol projected increases in air traffic of 2.4% annually. However, the European airlines warned that the increases would be more than twice that figure. The actual increase has been more than three times the Eurocontrol projections. Eurocontrol is now projecting annual air traffic growth of up to 6.5% through 1994 (German Airspace Users Association, 1989).

Civil Aviation Growth

Civil aviation is expected to remain a high growth industry in the foreseeable future, partly because of continuing advances in technology, and partly because of people's discoveries of new aviation needs. If the jet revolution of the 1960s is any indication of the traffic growth potential of advanced technology, then the introduction of advanced supersonic transport should once again lead the industry through double digit growth (Taneja, 1987).

The increased use of avionics by regionals/commuters and general aviation, plus further implementation of additional airport radar service areas, will contribute to the high growth expected in instrument operations at FAA-tower-operated airports. Instrument operations are forecast to increase from 43.4 million in 1987 to 59.5 million in 2000, a 2.7% annual growth rate (Table 2) (Rodgers, 1988).

Table 2: Aviation Activity Forecasts, Fiscal Years 1988-2000
(Rogers, 1988)

	Historical		Forecast			% Average Annual Growth			
Aviation Activity	1980	1986	1987	1988	1989	1999	80-87	87-88	87-99
Air Carrier									
Enplanements (Millions)									
Domestic	278.2	385.2	415	431.7	450.8	713.7	5.9	2.1	4.6
RPM's[1] (Billions)									
Domestic	203.2	294.4	322.0	333.3	348	566	6.8	3.5	4.8
Commuters/Regionals									
Enplanements	12.9	24	27.2	29.1	31	57.9	11.2	7.0	6.5
RPM's (Billions)	1.6	3.6	4.2	4.7	5.1	11.0	12.3	16.7	8.3
Fleet (thousands)									
Air Carrier	2,394	3,168	3,401	3,528	3,658	4,651	5.1	3.7	2.6
Commuter	1,413	1,538	1,604	1,648	1,711	2,252	1.8	2.7	2.9
Gen. Av.	210.3	210.7	220.0	219.0	218.5	220.9	0.6	-0.5	0.0
Hours Flown (Millions)									
Air Carrier	6.5	9.3	10.0	10.4	10.9	14.3	6.4	4.0	3.0
Gen. Aviation	41.9	34.5	34.5	34.4	34.3	36.6	-2.8	-0.3	0.5

[1] Revenue passenger miles

Large Air-Carrier and Commuter Operations Growth

The large U.S. air carriers enplaned 415 million passengers domestically in fiscal year 1987, an increase of 68% over the 247 million enplaned in fiscal 1978 (the last year of a regulated industry). The FAA forecast is projecting a 72% increase over the next twelve years, reaching 714 million by 2000.

In 1978, the large air carriers performed 10.1 million operations at FAA towers. With the development of a hub-and-spoke route network, operations increased by almost one-third, to 13.1 million in 1987. The forecast is for continued strong growth, reaching 17.3 million operations by 2000. During this same period, the commuter air carriers experienced spectacular growth. In 1987, 10 million passengers were carried by the commuter industry. In 1987, they were carrying almost three times that number (27 million) and the forecast predicts strong growth to 58 million passengers by 2000. Equally impressive is the increase in commuter operations from 3.8 million in 1978 to 7.3 million in 1987 (forecast to reach 11.2 million by 2000) (Rodgers, 1988).

The number of passengers on European flights has doubled in the last ten years. Current projections are that passenger volume will double again by the year 2000 (German Airspace Users Association, 1989).

General Aviation Growth

General aviation has not experienced the expansion enjoyed by commercial aviation. Hours flown in 1987 (34.5 million) were unchanged from 1986. Aircraft shipments in 1987 declined for the ninth consecutive year. Turbine powered aircraft shipments showed some improvement over 1986 totals. Since the single-engine piston aircraft market is the base upon which general aviation activity builds, the trend in aircraft shipped is cause for concern for the general aviation industry. Three fourths of the aircraft fleet are single-engine piston (Rodgers, 1988).

Preliminary figures for the most recent FAA "Aviation Forecasts" indicate that the number of aircraft using the system will double by the year 2000. Accommodating this anticipated demand growth has been a primary justification for proposed investments in system improvements (Office of Technology Assessment, 1982).

Assumptions and Methods Used for Forecasting

FAA predictions are based on a combination of econometric modeling, trend extrapolations, and expert judgment. Forecasts of key economic indicators are prepared by Wharton Econometric Forecasting Associates, Inc., using their long-term industry and economic forecasting model. Specific assumptions include the following:

General aviation– continued rapid growth of business and commercial GA (i.e., larger turboprops and jets used as corporate aircraft or air taxis) and continued availability of aviation fuel, although prices rise more rapidly than the consumer price index.

Air carriers– additional mergers, resulting in route optimization and more efficient fleet utilization, and continued replacement of older equipment with larger, quieter, more fuel-efficient aircraft.

Commuter carriers– a decrease in the number of carriers as competition leads to mergers, no loss of competitiveness with the personal automobile, and increases in average aircraft size and stage length.

FAA workloads– increases in the number of FAA-towered airports and terminal control areas, which will tend to increase the number of IFR operations and flight-plan filings, and greater utilization of flight services due to increased convenience and improved services. (Federal Aviation Administration, 1988)

The third phase in the en-route automation program provides an advanced capability for the centers which will be focused on improved communication between pilot and controller, between controllers, and between the controller and the computer. Also included is provision of an accurate surveillance base for a computer generated conflict prediction function and a computer generated conflict resolution function.

See Figure 1 for comparison between the United States' basic third generation and the proposed upgraded third generation, in terms of principal air traffic management functions and communications (Gilbert, 1973).

Aircraft Separation Sequencing, Spacing

Third generation	Upgraded third generation
Primary radar	**Improved beacon/secondary radar system**
Radar beacon (ATCRBS)	**Upgraded ATCRBS**
See and avoid	**Discrete address beacon system (DABS)**
Procedural	**Beacon/secondary radar backup**
Manual radar vectoring	**Primary radar**
	Procedural
	See and avoid
	Manual radar vectoring
	ARTS II and III data acquisition (terminal)
	NAS Stage A data acquisition (en route)
	Automated metering and spacing
	Automated conflict prediction and resolution
	Expanded positive control areas (PCA)
	Airspace organization
	VFR only
	IFR only
	Performance segregation

Communications

Third generation	Upgraded third generation
Voice communication	**Data link**
Teletype	**Position**
Manual FSS advisory	**Heading**
Manual preflight briefing	**Altitude**
	Voice communication backup
	Discrete frequency
	Satellite
	Oceanic
	Contiguous
	Automated FSS Advisory
	Automated mass preflight briefing
	Intermittent positive control (IPC)

Flow Control

Third generation	Upgraded third generation
Manual computations	**Automated central traffic flow management**
Manual central-flow-control	**Gross prediction of high-density terminal**
Management	**demand and capacity.**
	Flow-rate control
	Automated local (center area) flow control
	Short-term demand and capacity prediction
	Flow-rate control

Figure 1: Principal air traffic management functions and communications (Gilbert, 1973).

Communications

Communication is a key element in the present ATC system, and advances in communication technology may open new options for configuring the system in the future. Historically, voice radio has been the primary and almost exclusive means of communicating between aircraft and the ground. In the future, it is expected that an air ground digital data link will play an increasingly important role, as the automation of ATC functions requires more direct communication between airborne and ground based computers (Office of Technology Assessment, 1982).

Future Communications

In the future, with the introduction of a digital data link capable of selective address, two distinct modes of communication can be expected. Broadcast, the mode now used, will continue for voice or digital transmissions of general interest, such as weather, airport status, and traffic advisories. Other transmissions, pertinent only to specific aircraft, will be sent by a discrete address digital data link that allows isolation of specific receiving stations. However, to the extent that communication relative to position and intent uses a discrete address data-link rather than broadcast, the side benefits of the "party line" would be diminished (Office of Technology Assessment, 1982).

Communications will shift more to data-link applications. Voice communications will continue as a backup, with discrete frequencies being used for specific communications purposes. Satellites will be introduced for oceanic communications and, perhaps to some extent, over the continental United States. Weather information and flight plan distribution will be expedited via high speed teletype circuits (Gilbert, 1973).

Highly automated NASP equipment scheduled to come on-line over the next ten years includes the controller voice switching and control system, central weather processor, and Mode S data-link communications. An FAA request for bids on the controller voice switching system is due by year's end (*Aviation Week & Space Technology*, 1988).

Future Air Navigation System Committee

The task of the Future Air Navigation System (FANS) Committee is to define an updated, improved, and modernized global air navigation infrastructure covering a time period of the order of 25 years into the future. It has been asked to complete its work in three to five years, and it is now completing its third year.

The FANS Committee has made the following preliminary postulations on the nature of a future civil aviation communication infrastructure:

1) A satellite-based system is envisaged, in which digital data transmissions predominate, but in which voice service also is available for emergencies and other non-standardization types of communication.

2) In the vicinities of airports, especially the world's highest density facilities, there will be a continuing requirement for non-satellite, ground-based communication techniques.

3) The emergence of a global satellite navigation service also is envisaged as the basis for all but precision approach and landing. Such a service is seen ultimately to replace most ground based navigation facilities, of which there are thousands.

4) The full service coverage of communication, navigation, and surveillance systems will need to extend from anywhere on the Earth's surface to an altitude of 70,000 feet.

5) Integration of satellite based communication, navigation, and surveillance systems may be possible, but will be acceptable only if sufficient surplus systems can be maintained to prevent a complete air navigation services failure from the failure of a single element.

6) The FANS Committee sums up its preliminary postulation of a future system by indicating that the provision of communication services for air traffic control purposes will be the most important application of satellite technology in aeronautical mobile communications. (Freer, 1986)

What Are the Aviation User Industry Requirements?

There would appear to be at least four fundamental classes of communication traffic, each embracing several sub-classes. The four fundamental categories are:

Air traffic services: Safety of life communications which involve direct pilot or aircraft to air traffic controller communications of both voice and data form, including surveillance. In addition, air traffic information services (meteorology/system status/traffic, etc.) and perhaps navigation information are seen as sub-sets of the ATC requirement.

Aircraft operational control: This involves the means by which the aircraft operating company exercises control on its aircraft, wherever they are. This class of communications also embraces voice and data offering safety, and expeditious and economic use of its highly capital intensive equipment.

Aircraft operator business communications: All aircraft operators have general business communications requirements which do not fall under the umbrella of safety as defined within ITU and ICAO regulations, indicated in the annexes to the Convention of Civil Aviation. These communications represent the provision of passenger booking services, sales, and information for passengers, etc.

Public correspondence: Facilities provided by the aircraft operator to enable its customers to access the public telephone and data networks from any point, worldwide. These services embrace voice telephony, telex and even computer access of ground computer/data base systems. (Clark, 1987)

The Radio Technical Commission

The Radio Technical Commission for Aeronautics is a long standing association of U.S. government and private aeronautical organizations. The Commission does research to develop technical solutions to aeronautical problems involving the application of electronics and telecommunications. At the FAA's request, the Commission established a special committee, in July 1983, to develop representative civil user views of the requirements and forecasts of airspace conditions for technological applications, including satellite technology, for at least the next 25 years. The Commission concluded that future airspace management systems should, with more flexible real-time monitoring, permit each flight to operate on its most favorable and fuel-efficient path without intervention, except as necessary to avoid severe weather or collision with another aircraft. This represents a substantive change in control philosophy, from either the current close control to fixed altitudes and preset routes or the currently planned automated control to preset flight plans (Palmer, 1986).

Global Positioning System

The Global Positioning System is a space based navigation and time transfer system. It provides worldwide navigation service for an unlimited number of passive users. It has the potential of enhancing the safety of air travel and increasing the effectiveness of air operations (Kane, 1982).

Because of its absolute accuracy, its modest cost, and its simplicity of operation, GPS will be the optimum area navigation system. Because GPS navigation requires a competent computer, many attractive and useful features can be added at very low cost, by a simple expansion of the computer program. Route planning, multiple waypoints, and great circle tracks are but a few of the many additional capabilities (Kane, 1982).

GPS evolved from Navy and Air Force satellite research of the 1960s. With its eighteen satellites (and three spares) up, it will provide 60 foot horizontal and 90 foot vertical accuracy, regardless of vehicle speed, plus groundspeed within one-half knot. The GPS system can support a practically unlimited number of passive on-board navigation systems (Scott, 1986).

GPS System Improvements for the 2000s

Three improvements affecting the civil use of GPS are being investigated for future satellites. They are clock technology, system integrity, and coverage. One or more of these improvements could be a part of the GPS program in the year 2000 and beyond (Baker, 1987).

Satellite Technology

Defense Department officials expect by year's end to define requirements for next generation communication satellites to provide reliable command and control in the late 1990s and beyond. Milstar will be the Pentagon's first operational space communications system to operate in the extremely high frequency (EHF) part of the spectrum, at approximately 45 GH. for the uplink, and approximately 21 GHz for downlink in the super high frequency (SHF) band. At these higher frequencies, radio signals are less vulnerable to the aftereffects of nuclear explosions. Additionally, these higher frequencies make it easier to provide smaller satellite spot beams for increased communications privacy (Klass, 1989).

Mode S

Another key element in the FAA's overall plan for improving the ATC system is the Mode S data link, an improvement to the secondary surveillance radar that allows properly equipped aircraft to be interrogated selectively by ground stations. Mode S provides greater surveillance accuracy than the present Air Traffic Control Radar Beacon System (ATCRBS) equipment and avoids the problem of "synchronous garble" that occurs when more than one aircraft respond simultaneously to interrogation. The discrete address capability also provides a two-way ground to air data link to transmit clearances, weather information, traffic advisories, control instructions, and flight data automatically in digital format, without using VHF voice channels. The Mode S data link feature provides the basis for automation of ATC functions and other system improvements in the years beyond 1990 (Office of Technology Assessment, 1982).

Conclusions

Aviation activity is forecast to increase substantially over the next few decades. Continuing growth in the number of aircraft operations, number of aircraft, enplanements, diversity of operations, and sophistication of aircraft will place unprecedented demand on the air traffic control system. Meeting this challenge requires improved and expanded services, additional equipment and facilities, improved work force productivity, and the orderly and rapid replacement of aging equipment.

The present air traffic control system has evolved over several decades from the one that was first put in place in the 1930's. The operational characteristics and organization of the original system were determined largely by the technologies then available — radio for navigation and air/ground communication, and telephone and teletype networks for distribution of information among ATC ground facilities. New technologies, such as surveillance radar, Air

Traffic Control Radar Beacon System (ATCRBS) transponders, microwave relays, and electronic data processing, were added as demand increased and the state of the art progressed after World War II; but they did not change the essential characteristics of the earlier generation of air traffic control, which was a ground based, labor intensive, centralized system.

The evolution of the ATC system will be influenced by changes in user demand, market forces, and regulatory policy, as well as the availability of new technologies. The new technologies may help achieve greater effectiveness of the ATC system through higher levels of automation. There are several ways of meeting specific air traffic control needs, and the choice of which path to take will reflect a combination of technological, economic, and policy considerations. Prospective changes in the system will be dictated by the ability to increase system capacity to accommodate growth when and where it occurs, adding new capabilities to support improvements in efficiency and productivity by automating more functions and by introducing features to take advantage of technological improvements.

Solving the numerous automation and system issues in air traffic control will require interdisciplinary cooperation. Professionals from many disciplines have very valuable input in solving these issues. The following represents general fields presently addressing the issues.

Researchers have studied various aspects of:

- Attention, workload, and working conditions (Harss et al., 1991; Goettl, 1991)
- The role of the air traffic controller (Levesley, 1991)
- Automation's effect on the selection of controllers (Chobot, 1991; Della Rocco et al., 1991; Eißfeldt, 1991; Nyfield, 1991; Smith, 1991)
- Stress and workload on performance (Tattersall, 1991)
- Acceptance of new systems (Seidel, 1991)
- Training (Baldwin, 1991)

Large government organizations are needed to financially and politically support such large projects, and to perform research only large, resource laden organizations can perform. Numerous government agencies, including the FAA and ICAO, have been addressing automation and system issues since the inception of air traffic control. The following represent some of the issues currently being addressed:

- Cultural ergonomics (Kaplan, 1991)
- Computerization and automation (Simolunas & Bashinski, 1991; Lane, 1991; Sandiford, 1991)
- Dynamic management (Benoit, 1991)
- Selection (Della Rocco et al., 1991; Wing, 1991)
- ICAO and future air navigation systems (Bradbury, 1991)
- Air Traffic Control design principles (Volckers, 1991)

Independent research firms and universities also play an important role in solving the automation and system issues in air traffic control. Universities, like government organizations, have the necessary resources for research, but aren't as active politically. Universities and independent research firms help provide non-biased, non-political solutions to highly political issues. They also provide most of the foundation on which the higher levels of research are performed. Examples of these views include:

- Perception of automation (Horne et al., 1991; Westrum, 1991; Harmancioglu, 1991; Alpaslan, 1991)

- Resource management training (Herschler, 1991; Svátek, 1991)
- Mental models (Mogford, 1991)
- Impact of automation (Debons, 1991;Hopkin, 1991: Lindgren, 1991)
- Queuing Models (Sargis et al., 1991)
- Operator Workload and Performance (Costa, 1991; Jorna, 1991; Empson, 1991)
- Assumptions and Air Traffic Control Systems (Wise, 1991a)
- A Modest Proposal for Future Systems (Wise, 1991b)

Working with the numerous technical aspects of such an endeavor, the engineers and human factors specialists concern themselves with the actual functioning of the air traffic control system. Human factors specialists and engineers analyze the capabilities, limitations, and needs of people and systems in the air traffic control system. The following exemplify the issues being addressed:

- Communications (Fokas, 1991; Garland, 1991)
- Artificial Intelligence (Edwards, 1991; David, 1991)
- Human factors engineering (Day, 1991; Hancock, 1991; Tonner, 1991)
- Models in Automation (Stager, 1991; Lind, 1991)

The future of aviation lies in integrating modern technologies into the present air traffic control system without jeopardizing the safety of the system. Modern equipment and technology are imperative to meeting the challenges of the system. Through the cooperation of many individuals from various backgrounds, a complete plan that focuses on meeting user needs, addressing human concerns, and providing the impetus for a 21st century air traffic control system can be devised and successfully implemented.

References

Alpaslan, N. (1991). Automation in environmental engineering. In J. A. Wise, V. D. Hopkin, & M. L. Smith (Eds.) *Automation and Systems Issues in Air Traffic Control.* Berlin: Springer-Verlag.

Aviation Week. (1988). FAA Predicts airspace system sales overseas could reach $30 billion. *Aviation Week & Space Technology*, Nov. 21 p. 102.

Baker, P. J. (1987). GPS in the year 2000 and beyond. *Journal of Navigation*, pp. 175-182.

Baldwin, R. (1991). Training requirements in ATC. In J. A. Wise, V. D. Hopkin, & M. L. Smith (Eds.) *Automation and Systems Issues in Air Traffic Control.* Berlin: Springer-Verlag.

Benoit, A. (1991). Close ground/air cooperation in dynamic air traffic management. In J. A. Wise, V. D. Hopkin, & M. L. Smith (Eds.) *Automation and Systems Issues in Air Traffic Control.* Berlin: Springer-Verlag.

Bradbury, J. (1991). ICAO and future air navigation systems. In J. A. Wise, V. D. Hopkin, & M. L. Smith (Eds.) *Automation and Systems Issues in Air Traffic Control.* Berlin: Springer-Verlag.

Bradbury, J. (1991). ICAO and civil/military coordination. In J. A. Wise, V. D. Hopkin, & M. L. Smith (Eds.) *Automation and Systems Issues in Air Traffic Control.* Berlin: Springer-Verlag.

Chobot, R. B. & Chobot, M. C. (1991). Critical issues for decision makers in providing operator and maintainer training for advanced air traffic control systems. In J. A. Wise, V. D. Hopkin, & M. L. Smith (Eds.) *Automation and Systems Issues in Air Traffic Control.* Berlin: Springer-Verlag.

Clark, J. (1987). An overview of aeronautical satellite communications. *Telecommunication Journal,* pp. 384-389.

Costa, G. (1991). Shiftwork and circadian variations of vigilance and performance. In J. A. Wise, V. D. Hopkin, & M. L. Smith (Eds.) *Automation and Systems Issues in Air Traffic Control.* Berlin: Springer-Verlag.

David, H. (1991). Artificial Intelligence and human factors in ATC current activity at eurocontrol experimental center. In J. A. Wise, V. D. Hopkin, & M. L. Smith (Eds.) *Automation and Systems Issues in Air Traffic Control.* Berlin: Springer-Verlag.

Day, P. (1991). Human factors in system design. In J. A. Wise, V. D. Hopkin, & M. L. Smith (Eds.) *Automation and Systems Issues in Air Traffic Control.* Berlin: Springer-Verlag.

Debons, A. (1991). Systems issues in air traffic control. In J. A. Wise, V. D. Hopkin, & M. L. Smith (Eds.) *Automation and Systems Issues in Air Traffic Control.* Berlin: Springer-Verlag.

Della Rocco, P., Manning, C., and Wing, H. (1991). Selection of air traffic controllers for automated systems: applications from today's research. In J. A. Wise, V. D. Hopkin, & M. L. Smith (Eds.) *Automation and Systems Issues in Air Traffic Control.* Berlin: Springer-Verlag.

Edwards, J. (1991). Intelligent dialogue in air traffic control systems. In J. A. Wise, V. D. Hopkin, & M. L. Smith (Eds.) *Automation and Systems Issues in Air Traffic Control.* Berlin: Springer-Verlag.

Eißfeldt, H. (1991). Automation in ATC- how does it affect the selection of controllers. In J. A. Wise, V. D. Hopkin, & M. L. Smith (Eds.) *Automation and Systems Issues in Air Traffic Control.* Berlin: Springer-Verlag.

Empson, J. (1991). Cognitive failures in ATC. In J. A. Wise, V. D. Hopkin, & M. L. Smith (Eds.) *Automation and Systems Issues in Air Traffic Control.* Berlin: Springer-Verlag.

Federal Aviation Administration (1988). *National Airspace System Plan.* Federal Aviation Administration/Washington, DC: U.S. Government Printing Office.

Fokas, I. (1991). Human factors in ATC communication. In J. A. Wise, V. D. Hopkin, & M. L. Smith (Eds.) *Automation and Systems Issues in Air Traffic Control.* Berlin: Springer-Verlag.

Freer, D. W. (1986). Applications of satellite communication technologies of international civil aviation. *Telecommunication Journal,* pp. 710-714.

Garland, D. J. (1991). Automated systems: the human factor. In J. A. Wise, V. D. Hopkin, & M. L. Smith (Eds.) *Automation and Systems Issues in Air Traffic Control.* Berlin: Springer-Verlag.

German Airspace Users Association. (1989). *The Crisis of European Air Traffic Control: Costs and Solutions.* The German Airspace Users Association.

Gilbert, G. A. (1973). *Air traffic control: The uncrowded sky.* Washington, DC.: Smithsonian Institution Press.

Goettl, B. (1991). Attention, workload, and automation. In J. A. Wise, V. D. Hopkin, & M. L. Smith (Eds.) *Automation and Systems Issues in Air Traffic Control.* Berlin: Springer-Verlag.

Hancock, P. (1991). The aims of human factors and their application to issues in automation and air traffic control. In J. A. Wise, V. D. Hopkin, & M. L. Smith (Eds.) *Automation and Systems Issues in Air Traffic Control.* Berlin: Springer-Verlag.

Harmancioglu, N. (1991). The role of automation and control systems in water resources engineering. In J. A. Wise, V. D. Hopkin, & M. L. Smith (Eds.) *Automation and Systems Issues in Air Traffic Control.* Berlin: Springer-Verlag.

Harss, C., Jitzchack L., Kastner M., and Jan Goodrich (1991). Air traffic controller working conditions and organizations: suggestions for analysis and improvements from a psychological point of view. In J. A. Wise, V. D. Hopkin, & M. L. Smith (Eds.) *Automation and Systems Issues in Air Traffic Control.* Berlin: Springer-Verlag.

Herschler, D. (1991). Resource management training for air traffic controllers. In J. A. Wise, V. D. Hopkin, & M. L. Smith (Eds.) *Automation and Systems Issues in Air Traffic Control.* Berlin: Springer-Verlag.

Hopkin, D. (1991). The impact of automation on air traffic control systems. In J. A. Wise, V. D. Hopkin, & M. L. Smith (Eds.) *Automation and Systems Issues in Air Traffic Control.* Berlin: Springer-Verlag.

Horne, E., Pierce, J., and White, D. (1991). Results of a perceptual study on human error in computer systems based on bailey's research. In J. A. Wise, V. D. Hopkin, & M. L. Smith (Eds.) *Automation and Systems Issues in Air Traffic Control.* Berlin: Springer-Verlag.

Jorna, P. (1991). Operator workload as a limiting factor in complex systems. In J. A. Wise, V. D. Hopkin, & M. L. Smith (Eds.) *Automation and Systems Issues in Air Traffic Control.* Berlin: Springer-Verlag.

Kane, F. X. (1982). Navstar. *Artificial Satellites in Navigation*, pp. 331-341.

Kaplan, M. (1991). Issues in cultural ergonomics. In J. A. Wise, V. D. Hopkin, & M. L. Smith (Eds.) *Automation and Systems Issues in Air Traffic Control.* Berlin: Springer-Verlag.

Klass, P.J. (1989). Gains in satellite technology shape trends in C3 development. *Aviation Week & Space Technology*, Mar. 22 pp. 251-253.

Lane, J. (1991). The implementation and impact of automatic data processing on UK military ATC operations. In J. A. Wise, V. D. Hopkin, & M. L. Smith (Eds.) *Automation and Systems Issues in Air Traffic Control.* Berlin: Springer-Verlag.

Levesley, J. (1991). From under the headset: the role of the air traffic controllers professional association in present and future air traffic control systems development. In J. A. Wise, V. D. Hopkin, & M. L. Smith (Eds.) *Automation and Systems Issues in Air Traffic Control.* Berlin: Springer-Verlag.

Lind, M. (1991). Modelling control tasks in complex systems. In J. A. Wise, V. D. Hopkin, & M. L. Smith (Eds.) *Automation and Systems Issues in Air Traffic Control.* Berlin: Springer-Verlag.

Lindgren, R. (1991). Social, political, and regulatory issues concerning harmonization of interacting air traffic control systems in western europe. In J. A. Wise, V. D. Hopkin, & M. L. Smith (Eds.) *Automation and Systems Issues in Air Traffic Control.* Berlin: Springer-Verlag.

Mogford, R. (1991). Mental models in air traffic control. In J. A. Wise, V. D. Hopkin, & M. L. Smith (Eds.) *Automation and Systems Issues in Air Traffic Control.* Berlin: Springer-Verlag.

Nyfield, G. (1991). Automation issues for the selection of controllers. In J. A. Wise, V. D. Hopkin, & M. L. Smith (Eds.) *Automation and Systems Issues in Air Traffic Control.* Berlin: Springer-Verlag.

Office of Technology Assessment. (1982). *Airport and Air Traffic Control System.* Washington, DC: U.S. Office of Technology Assessment. (NTIS No. PB82-207606).

Palmer, C. R. (1986). *Use of satellite technology for air traffic control and navigation.* Washington, DC: U.S. Government Printing Office.

Rodgers, J. M. (1988). FAA forecast. *Thirteenth Annual FAA Aviation Forecast Conference Proceedings* (pp. 14-30). Washington, D.C: Office of Aviation Policy and Plans.

Sandiford, W. (1991). Meeting the ATC challenge through simulation. In J. A. Wise, V. D. Hopkin, & M. L. Smith (Eds.) *Automation and Systems Issues in Air Traffic Control.* Berlin: Springer-Verlag.

Sargis, J. (1991). Systems issues in air traffic control. In J. A. Wise, V. D. Hopkin, & M. L. Smith (Eds.) *Automation and Systems Issues in Air Traffic Control.* Berlin: Springer-Verlag.

Scott, V. (1986). Navstar GPS for general aviation. *Avionics News,* 23, pp. 40-51.

Seidel, E. (1991). Acceptance of new systems by air traffic controllers. In J. A. Wise, V. D. Hopkin, & M. L. Smith (Eds.) *Automation and Systems Issues in Air Traffic Control.* Berlin: Springer-Verlag.

Simolunas, A., and Bashinski, H. (1991). Computerization and automation: upgrading the american air traffic control system. In J. A. Wise, V. D. Hopkin, & M. L. Smith (Eds.) *Automation and Systems Issues in Air Traffic Control.* Berlin: Springer-Verlag.

Smith, M. L. (1991). Adaptive training to accomodate automation in the air traffic control system. In J. A. Wise, V. D. Hopkin, & M. L. Smith (Eds.) *Automation and Systems Issues in Air Traffic Control.* Berlin: Springer-Verlag.

Stager, P. (1991). The canadian automated air traffic system (CAATS): an overview. In J. A. Wise, V. D. Hopkin, & M. L. Smith (Eds.) *Automation and Systems Issues in Air Traffic Control.* Berlin: Springer-Verlag.

Stager, P. (1991). Error models for operating irregularities: implications for automation. In J. A. Wise, V. D. Hopkin, & M. L. Smith (Eds.) *Automation and Systems Issues in Air Traffic Control.* Berlin: Springer-Verlag.

Svatek, N. (1991). Resource management at canadian airlines. In J. A. Wise, V. D. Hopkin, & M. L. Smith (Eds.) *Automation and Systems Issues in Air Traffic Control.* Berlin: Springer-Verlag.

Taneja, N. K. (1987). *Introduction to Civil Aviation.* Massachusetts: Lexington Books.

Tattersall, A, Farmer, E., and Belyavin, A. (1991). Stress and workload management in air traffic control. In J. A. Wise, V. D. Hopkin, & M. L. Smith (Eds.) *Automation and Systems Issues in Air Traffic Control.* Berlin: Springer-Verlag.

Tonner, M. (1991). The controller in human engineering. In J. A. Wise, V. D. Hopkin, & M. L. Smith (Eds.) *Automation and Systems Issues in Air Traffic Control.* Berlin: Springer-Verlag.

Volckers, U. (1991). Application of planning aids for air traffic control design principles, solutions, results. In J. A. Wise, V. D. Hopkin, & M. L. Smith (Eds.) *Automation and Systems Issues in Air Traffic Control.* Berlin: Springer-Verlag.

Westrum, R. (1991). Automation information and consciousness in air traffic control. In J. A. Wise, V. D. Hopkin, & M. L. Smith (Eds.) *Automation and Systems Issues in Air Traffic Control.* Berlin: Springer-Verlag.

Wing, H. (1991). Selecting for air traffic control: The state -of-the-art. In J. A. Wise, V. D. Hopkin, & M. L. Smith (Eds.) *Automation and Systems Issues in Air Traffic Control.* Berlin: Springer-Verlag.

Wise, J. (1991a). Assumptons and air traffic control systems. In J. A. Wise, V. D. Hopkin, & M. L. Smith (Eds.) *Automation and Systems Issues in Air Traffic Control.* Berlin: Springer-Verlag.

Wise, J. (1991b). A modest proposal for future systems. In J. A. Wise, V. D. Hopkin, & M. L. Smith (Eds.) *Automation and Systems Issues in Air Traffic Control.* Berlin: Springer-Verlag.

Automation in Environmental Engineering

Neçdet Alpaslan

Dokuz Eylül University, Environmental Engineering Department, Bornova, Izmir, Turkey

Introduction

The field of environmental engineering encompasses a wide spectrum of subjects and problems. Among these issues, water and wastewater treatment and water distribution systems are the areas where automatic control is extensively used. Actually, water or wastewater treatment plants can be operated manually with no need for any level of automation. This may also be true for landing or departing of a single aircraft from a simple airport. When the number and sizes of the aircraft are increased, the need for automation becomes eminent. Similarly, the growing complexity of water and wastewater treatment and distribution systems, and the need to economize steadily-declining water resources, have led to the continuous and automatic monitoring of these processes. This means that the need for automation in the operation of such systems increases as the dimensions or the number of components involved in these systems increase.

Treatment plants consist of various pools into which water (or wastewater) enters, goes through certain processes, and then exits with a desired quality. These processes require the automatic measurement of a number of parameters which depend on many factors, such as the type of the process, operation policy, and effluent standards. Although such parameters vary from plant to plant, a few are common to all treatment systems. Among these, one may name flows, levels of liquids or solids, pressures, and temperatures.

Treatment processes may be dominantly physical, chemical, biological, or a combination of all three. The typical schematic flow diagram for water and wastewater treatment is given in Figure 1.

The Importance of Automation in Environmental Engineering

The purposes of automation in treatment processes are to (ICA, 1986):

- Increase the reliability of the system
- Upgrade the effluent quality
- Provide cost-effective operation of the systems

Reliability of a water or wastewater treatment plant depends on the performance of each operational unit, as depicted in Figure 1. Improvement of the control and operation of these units

NATO ASI Series, Vol. F 73
Automation and Systems Issues in Air Traffic Control
Edited by John A. Wise et al.

by automation may increase their efficiency and therefore their reliability. The other aspect of reliability is related to human error. Automation decreases the number of failures due to operator errors. This leads to improved reliability of the system. On the other hand, the situation may be the contrary for developing countries, where maintenance may be a significant problem since it requires special expertise. This is true especially when the required equipment has to be imported from abroad. Therefore, in such countries, the major concern seems to be the failure of the equipment rather than the failure of the operator.

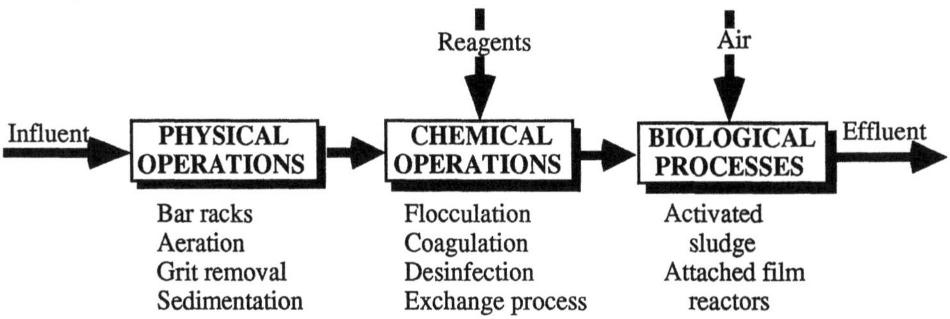

Figure 1. Flow Diagram for Water and Wastewater Systems

If the treatment plant is properly designed and constructed, the quality of effluent is a function of both the influent characteristics and the capability and skill of the operators. In practice, influent characteristics are not under the control of operators or engineer., so that the operational skills of the personnel generally determine the success or failure of the system. Under these conditions, the operators need to be selected from highly qualified and trained personnel, and such personnel may not be available at all times. On the other hand, an automated system can be successfully operated by a high school graduate with appropriate training (Alpaslan, 1987).

Furthermore, automation contributes to the reduction of the physical work load for operation and to the establishment of better working conditions. There are some tasks, especially in wastewater treatment plants, which are pretty unpleasant for the employees to accomplish (e.g., cleaning bar racks or the bottom of sedimentation tanks). The automation of such procedures leads to better working conditions.

Finally, automation leads to savings in energy as well as in chemicals used, which eventually results in a cost-effective operation. Furthermore, automated plants require only a few operators, so that significantly reduced human labor also decreases the fixed costs.

Parameters Measured and Automatically Monitored in Treatment Plants

The most common parameter is one that describes liquid flows, which are measured with conventional equipment such as turbine or piston, type meters, flow-type flowmeters, or negative pressure devices connected to transmitters, which may be either electrical or pneumatic. Liquid levels are measured by pressure or floating apparatus. Pressure measurements are generally done by piezometric tubes, differential manometers, and similar pressure gauges. The temperature of water is usually measured with a resistance-bulb thermometer, so that it can be transmitted for remote readings.

Besides the above mentioned common parameters, monitoring of specific parameters may be important. Various standard methods of analysis are applied automatically in the apparatus for measuring and monitoring specific parameters of water. These methods include: nephelometric analysis (measurement of turbidity), resistivity analysis (measurement of salinity), potentiometric analysis (measurement of pH), amperometric analysis (measurement of oxidizing agent content), photocolorimetric analysis, and titrimetric analysis (measurement of solute concentrations) (Degremont, 1979).

Forms of Automatic Operation in Treatment Plants

With respect to the automatic operation of water and wastewater treatment plants, the terms "automation" and "computerization" are interchangeably used and understood, although they essentially have different meanings. "Automation" relates to the automatic operation of the equipment, whereas "computerization" refers to the operation of the whole plant with respect to algorithms (or models) loaded on the computer. Computerization may provide real-time operation when it is associated with dynamic mathematical operation models.

As far as automation is concerned, three categories of operation may be listed (Degremont, 1979):

- Remote manual sequential control
- Manually-started automatic sequential control
- Automatically-started sequential control

In case of treatment, these operations consist mainly of closing or opening electrically- or pneumatically-actuated valves and starting or stopping pumps.

Recently, however, some modern plants have incorporated automated equipment which monitors and controls both machines and processes, besides improving their performance. This facility is provided by programmable automatic units; the choice of such equipment depends on a wide range of possibilities, extending from electromechanical units via static logic control devices to systems employing mini-computers (Kuleli & Boveri, 1988).

Finally, the use of process computers which, in some cases, are relatively cheap, is convenient for plant management and supervision.

They provide the means for (Degremont, 1979):

- Centralized data gathering and monitoring of plant performance
- Calculation and recording of parameters affecting plant operation
- Printing out selected historical data for purposes of checking or inspecting after the event
- Maintenance programs

Where automation is concerned, computers offer the advantages of:

- Greater flexibility in compiling automated programs
- Optimization of processes
- Integration of several automatic systems controlling a number of processes

Automatic Control in Water Distribution Networks

A water distribution system comprises pipelines, reservoirs, storage tanks, and pumping stations. In fact, water abstraction and treatment are also the components of a full network. However, since automation in water treatment is presented in previous sections, it is not included here again.

The automation in such waterworks covers both the relay of information for storage and processing in the control center, and the implementation of operator command. The main functions of automation may be summarized as follows (Martini, 1981):

- Supervisory control of pipeline pressure, flow, reservoir level, pump operation, treatment plants, storage tanks
- Monitoring of quality and flow of water, of designated functions at all plants, and of burst pipes in consumer areas (leakage detection)
- Remote operator control of all elements (valves, pumps, etc.)
- Optimization of water procurement and distribution according to predetermined economic criteria
- On-line calculation of valve settings based on a) reservoir filling rates and expected demand, and b) inlet sewage flow rates
- Calculation of reservoir filling rates, hourly reserves, demand and supply predictions
- Daily / weekly / monthly printout of reports of all events and operating values

Conclusion

Automation is highly significant in water and wastewater treatment systems, due to the complexity of these systems. The large number of parameters to be continuously controlled and the precision essential at each control require the use of automatic control systems. Automation in treatment systems has the advantages of reduced human labor, increased efficiency and reliability of treatment processes, and decreased operational costs.

In environmental engineering systems, automation is basically needed for better and more efficient operation. Therefore, its role in such systems may be described as "optional" rather than "essential" (Harmancioglu, 1991). In air traffic control systems, for example, automation is essential, especially with respect to system failures. Air traffic control has no tolerance for human errors, since the failures in these systems have immediate and drastic effects on human life. Therefore, automation is required particularly to reduce human errors and minimize failures in environmental quality and the resulting health risks. The impact of such failures is seldom immediate and seldom disastrous. Nevertheless, automation in such systems serves to minimize human errors or possible failures for a more effective operation.

It must also be emphasized that the use of automation in environmental engineering system is a matter of cost minimization. The decisions as to whether automation should or should not be applied — or if it is to be use, what its extent (full or partial) should be — are based on economic considerations apart from the technical aspects discussed earlier. If significant decreases in operational costs can be achieved by automation, then engineering judgment calls for its application. This is also one of the basic reasons why automation is considered to be optional rather than essential, in environmental engineering systems.

References

Alpaslan, N. (1987). Assessment of the operation of a sewage treatment plant: A case study for the City of Utrecht-Netherlands, *Proceedings of the International Symposium on Environmental Management,* Bogazici University, Istanbul, Turkey.

Degremont, (1979). *Water Treatment Handbook.* New York: John Wiley.

Harmancioglu, N. (1991). The role of automation and control systems in water resources engineering, in J. A. Wise, V. D. Hopkin, & M. L. Smith (Eds.) *Automation and Systems Issues in Air Traffic Control.* Berlin: Springer-Verlag.

Kuleli, O., Boveri, A.B. (1988). Automation in the water industry for optimum plant management. *Environment'88: Environmental Science & Technology Conference, 5-9 June,* Izmir, Turkey.

Martini, P. (1981). The remote control plant of the Rome water supply system, June 1981.

Seminar notes on Instrumentation Control & Automation-ICA (1986). IHE, ICSE, Delft, The Netherlands.

The Role of Automation and Control Systems in Water Resources Engineering

Nilgün Harmancioglu

Civil Engineering Department, Dokuz Eylül University, Bornova, Izmir, Turkey

Introduction

It is the responsibility of water resources engineering to control and regulate water to serve a wide variety of purposes. Certain fields of water resources engineering — such as flood control, land drainage, sewerage, and highway culvert design deal primarily with the control of adverse effects of water. Water supply, irrigation, and hydroelectric power development are the three major fields which serve the utilization of water for beneficial purposes. It is basically in these fields that the largest water resources systems are developed. In recent years, pollution control or water-quality management has become an important phase of water resources engineering.

The development of water resources systems is accomplished in two phases. The first one comprises the planning, design, and construction of the system, which require interdisciplinary expertise and judgment. The second phase, the operation of the system, is equally important as the first. It is only by means of efficient operation that the system can fulfill its expected purpose with respect to technical, economical, social, political, and legal criteria.

Operation is extremely significant, especially in large water resources systems that are developed for the utilization of water (i.e., water supply, irrigation, and hydroelectric power production). In the operation of such systems, automation is called for by two factors:

- The significantly large dimensions of the system, together with numerous components involved, each of which needs to be controlled
- Temporal variation of either the inputs or the outputs, which require continuous control

In addition, each control within the system has to be realized with utmost precision, to prevent either total or partial breakdown of the system. Under these conditions, it is not wise to leave the whole job to the operators, although this is often the case in many systems, especially in developing countries. Automation helps to solve most of the operational problems to achieve higher efficiency. Consequently, either total or partial automation has entered the field of water resources engineering.

It may be of interest to note here that quite a number of parallel aspects exist between water resources engineering (WRE) and air traffic control (ATC), although the two are related to completely different fields of application. Both deal with systems (i.e., water resources systems and ATC systems) which require controlled operation. Hydrologic flow can occur

NATO ASI Series, Vol. F 73
Automation and Systems Issues in Air Traffic Control
Edited by John A. Wise et al.

naturally within an existing solid medium; however, controlled operation is needed so that water flows in desired amounts and in selected directions. On the other hand, air traffic flow occurs in a controlled form where automation plays an essential role. Within this framework, automation in WRE may be considered "optimal" rather than "essential," since the control of water flows does not necessarily call for automation. The dimensions of both the ATC and WRE systems are significantly large; therefore, they can be described as "complex systems." The latter have become highly complicated, especially within the last two decades, so that the operation of such systems puts a growing demand on automation. In addition, precision is extremely significant in the operation of both ATC and WRE systems, to avoid system failures. Automation helps to preserve a certain level of precision by reducing human errors and the associated uncertainties. However, despite all its advantages, automation has not yet become as essential in WRE systems operation as it is in ATC systems.

This paper discusses the role of automation in irrigation and hydroelectric power production systems which show the greatest need for automatic control. Water supply and pollution control systems also use automation in various applications, and these are considered in another paper as issues of environmental engineering (Alpaslan, 1991).

A part of the paper is devoted to discussion of automation in data processing systems, which provide hydrological data as inputs to WRE systems at both the design and the operation phases. Data processing is an essential component of WRE systems, since all decisions are based on available data. Management of vast amounts of such data also requires automated processing. Otherwise, data handling is a highly time- and energy-consuming activity, subject to human errors.

Automation in Irrigation Systems

Often, irrigation systems have large area dimensions with numerous components (e.g. canals, gates, checks, valves, pumping stations, etc.). To give an example, the irrigation system designed within the Southeastern Anatolian Project in Turkey is going to supply water for more than one million hectacres of irrigatable land by means of four large dams and power plants, two dozen small dams, four small power plants, and numerous pumping plants. The design of this system is a perfectly elegant one, but the operation will be a huge job, especially if it is realized by manpower. Full, or at least partial, automation is definitely required for this project.

A good example for an automated irrigation system is the Salt River Project in Tucson, Arizona, where a 138 mile canal network can be monitored and controlled by a single operator.

This sophisticated water-handling system, known as Supervisory Control, is made possible through the use of electronic equipment coupled with a small process computer. The equipment receives telemetered data from 80 canal stations and displays water levels and gate positions at the control console. The computer automatically calculates the flow rate for each gate and will add the flow for all gates at any given station. The operator is able to control 331 radial gates, 46 deepwell pumps, and eight large booster pumps.

The first phase of the Supervisory Control construction was begun in 1970. This phase consisted of the installation of the master control console, the system controller, automatic data logger, data printer, and the Supervisory Control electronic logic equipment. In addition to the master control equipment, 11 canal stations were placed under control in the first phase. The remainder of the system was installed in four additional phases, at a rate of one phase per year.

In this system, the operator can monitor a gate by selecting the site from a push button panel which connects the console to the canal site through a commercial telephone network. When the contact is made, all the site information is displayed on a television screen in the center of the console. Included in the information shown are gate openings, water levels upstream and downstream from the gate, and computed flow rates. The electronic system is designed to display water levels and gate positions with an accuracy to the nearest one-hundredth of a foot.

An automatic printer near the console prints out the data for each site when it is first selected and the changes that the operator makes. This provides a permanent record of conditions at the site. The operator can raise or lower gates as desired to meet changing water delivery requirements. As this is happening, the gate movement and the changing water flow rates are displayed on the television screen.

The 46 electrically-driven pumps and eight booster pumps connected to the Supervisory Control system can also be started and stopped by the operator from the console. The operator selects the site with the desired pump from the push button panel and sends an electronic start or stop command. If a start command is sent, the pump will start after a predetermined time delay. Attached to the pump discharge is a water sensing device that will send a signal to the console, causing a blue light to come on. This indicates to the operator that the pump is running and pumping water. To stop the pump, the operator can depress the pump-stop switch, de-energizing the pump control circuit.

The Supervisory Control system continuously scans all 80 sites on a 30.8 second cycle, to monitor either normal or irregular conditions. Emergency conditions include canal high water, power failure, or communications failure at a site. When any one or more of 23 abnormal conditions occur at one of the canal stations, a warning horn and flashing light will attract the operator's attention. The operator can silence the horn and acknowledge the irregularity by depressing a push button on the console. The lights will stay on until the abnormal condition is corrected at the site, but will cease flashing when the condition is acknowledged. If the operator cannot solve the problem by manipulating various console controls, then help is summoned from technical or field personnel by radio or telephone.

The Supervisory Control system has made it possible for the Project to save water, manhours, and electricity. Because the system permits the operator to monitor the entire canal system at one time, water control is more precise. The operator can follow a change in flow rate down the canal much faster then when canals were controlled manually and someone would have to drive from gate to gate. As a result, excesses and shortages of water have been minimized.

Prior to the installation of the Supervisory Control system, the changes at canal gates were made by Project field personnel who operate the distribution system. Many times the work load of these people conflicted with the canal operations and made timely reports or gate changes impossible. Now, field personnel can devote almost full time to the distribution system. The manhour savings for the transmission system was equal to approximately 4.5 full-time employees.

Now that it is possible for one operator to quickly start and stop pumps from the control center, pump operation has been reduced during the summer peak load periods. This reduces power costs and improves the efficiency of the Project's power operations.

Automation in Power Production Systems

Today, nearly all power systems are served by several plants, and some systems are interconnected with each other. In general, a combined power system with both hydroelectric and thermal plants is much more economical and reliable than a power system with a single plant. During periods of low streamflow, thermal plants in a combined system are operated almost continuously near capacity to carry the base load, while hydroelectric plants with storage (dams) are used intermittently to generate power for peak loads. During periods of high streamflows, the functions of the thermal and hydroelectric plants are reversed. In each case, the operation of both types of plants is realized, often in an interconnected manner, to cover the total power load. Hydroelectric power plants with storage are coupled by dams as hydraulic structures. Therefore, the operation of the power plant is also accompanied by the operation of the reservoir to produce energy.

In general, each hydroelectric power plant is an automated system by itself. In the control unit of the system, the operation of both the reservoir and the power plant is controlled automatically. The incoming and outgoing (to-power-production and to-the-stream) flows to and from the reservoir are time-variant. By means of automatic control systems, real-time operation of the reservoir is realized both for the safety of the structures and for power production. In addition to the continuous observation of inflows and outflows, the operation of water control structures (e.g., inlet valves, pressure valves, spillway gates, inlet gates, and so on) is also accomplished by remote control. Next, the automation system provides data on power loads and power production. Furthermore, the operation of power production units (e.g., turbines, generators, etc.) can be realized again by remote control.

Some hydroelectric plants require a few operators, so that they are partially automated; however, some are operated totally by remote control. One example is the Mammoth Pool powerhouse of the Southern California Edison Company, where a microwave link to Big Creek No. 8 powerhouse provides channels for supervisory control, telemetering, and visual and audible control of operation by means of closed-circuit television and sound monitoring (Linsley and Franzini, 1964).

In large power systems with both thermal and hydroelectric plants, the operation is often interconnected, with the former running continuously and the latter operating intermittently with low load factors. Such an interconnected operation is especially important at times of repair or breakdown of certain units. Hydroelectric plants can be kept in standby, to start production whenever a unit in the system is out of operation.

In such interconnected systems, real-time operation is also desired, since power demands vary temporally (within both a day and a season). This is where automation is called for, to follow variations in power demands and to adjust the system of plants to meet the demand. Often, the supervision of the system to determine loads and production at any time is automated. However, the decision as to which plant goes into operation, when, and for how long, is often made by the responsible personnel at a central control unit where all data are put together. Then, this decision is communicated to specific plants where a change in operation has to be made. Such an operation may be called semi-automatic, since it involves human beings.

In Turkey, power production is realized by a semi-automatic interconnected system (thermal and hydroelectric plants) to meet the power requirements of the entire country. In Ankara, the National Power Distribution Center automatically controls the power production system to evaluate loads, production, and distribution of energy. Data collection and evaluation are automatic; however, decisions as to which units should start or quit operation are made by the Center personnel and communicated to the relevant units by either phone or energy-link lines. However, in cases of emergency, like the breakdown of a certain unit, the operation is totally automated, so that the load to be met is immediately reduced by a predefined program. Often, authorities seem to favor full automation in emergency cases but are still reluctant to apply the same under normal operating conditions. This situation holds true especially in developing countries like Turkey. The major reason for this attitude appears to be the fear of equipment failures. This may be associated with human psychology; however, it definitely has economical aspects. Concerns about the maintenance of an automatic control system discourages full automation, since it requires highly-trained personnel and seems to be more costly, especially when such expertise and equipment have to be imported. The authorities at the Turkish Electricity Agency indicate that full automation is to be realized several years from now, basically because it doesn't look feasible at present.

Again in developing countries, the reluctance for full automation is also associated with social factors such as the concern about unemployment. In most cases, two or three personnel are employed for a duty that can be taken care of by a single employee; this is contradictory with one of the purposes of automation, i.e., the reduction in human labor.

Automated Data Transmission and Processing

Hydrologic data are collected either manually or by continuous automatic recorders. Both practices produce vast amounts of data to be transmitted to the users and processed. Transmission is realized by a number of ways (Starosolszky, 1987):

- Manual, with the observer sending messages or data to a central bureau, or calling the bureau by phone or radio

- Automatically timed, with self-timed automatic equipment at the field station initiating radio transmission of instantaneous observations

- Automatic event indicators at station, which communicate automatically by radio any specific change

- Automatic, with station transmitting and central office recording data continuously through electrical wires (analog transmission) or through special radio or telephone lines (frequency modulation)

- Fully automatic, with station sensors, encoder, or radio transceiver transmitting coded information to a central office where data are decoded for entry into a computer; the central office's computer initiates sequential calls via telephone, telex, or by radio (digital transmission)

The replacement of observers (due to increasing costs of manpower) by automatic sensors is usually combined with the installation of automatic data transmission lines. Basically, there are two types of transmission systems:

- Off-line transmission, with a break in data flow between transmission and processing
- On-line data transmission and processing, with no break in data flow between transmission and processing (this system is now possible as a result of advances in telemetering technology, computer techniques, and mathematical modeling for processing and forecasting hydrologic variables)

In the second case, telemetering is a highly beneficial means of transmitting real-time data (with warnings or forecasts), especially when the presence of a continuous observer cannot be assured. The telemetering equipment consists of sensors, transmitters (amplifiers, converters, encoders), the transmission links, and the receiver.

Next, with the advent of high-speed computers, data processing is also realized automatically. A wide variety of computer programs are available for continuous processing of field data. Figure 1 shows schematically the computer hardware for automated processing of hydrologic data. Figure 2 summarizes the computer-controlled flow regulations used in the operation of water resources systems.

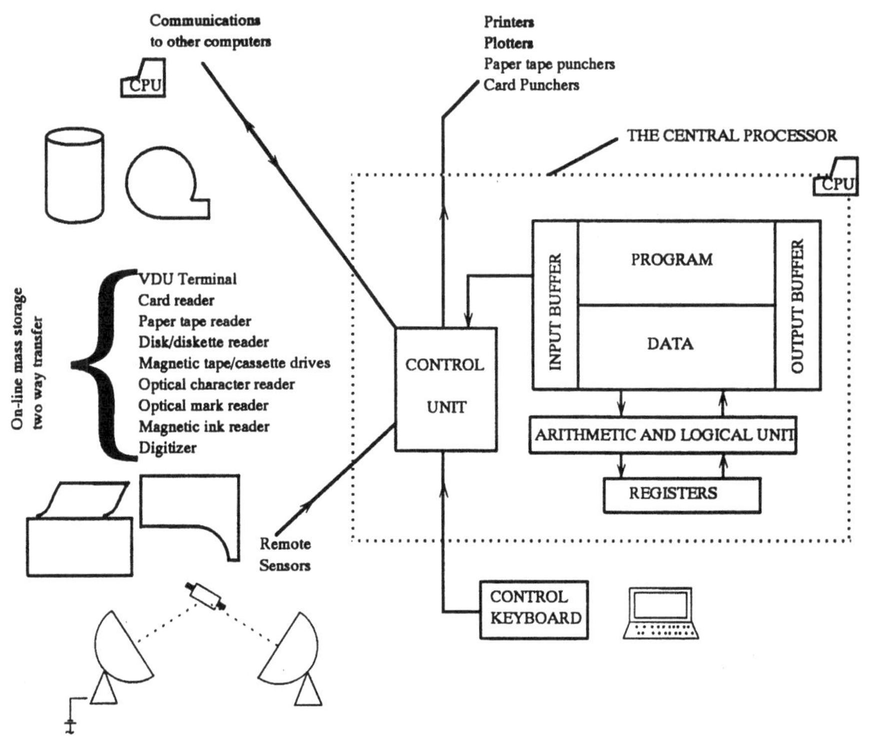

Figure 1: Schematic picture of computer hardware for processing of hydrologic data (Starosolszky, 1987).

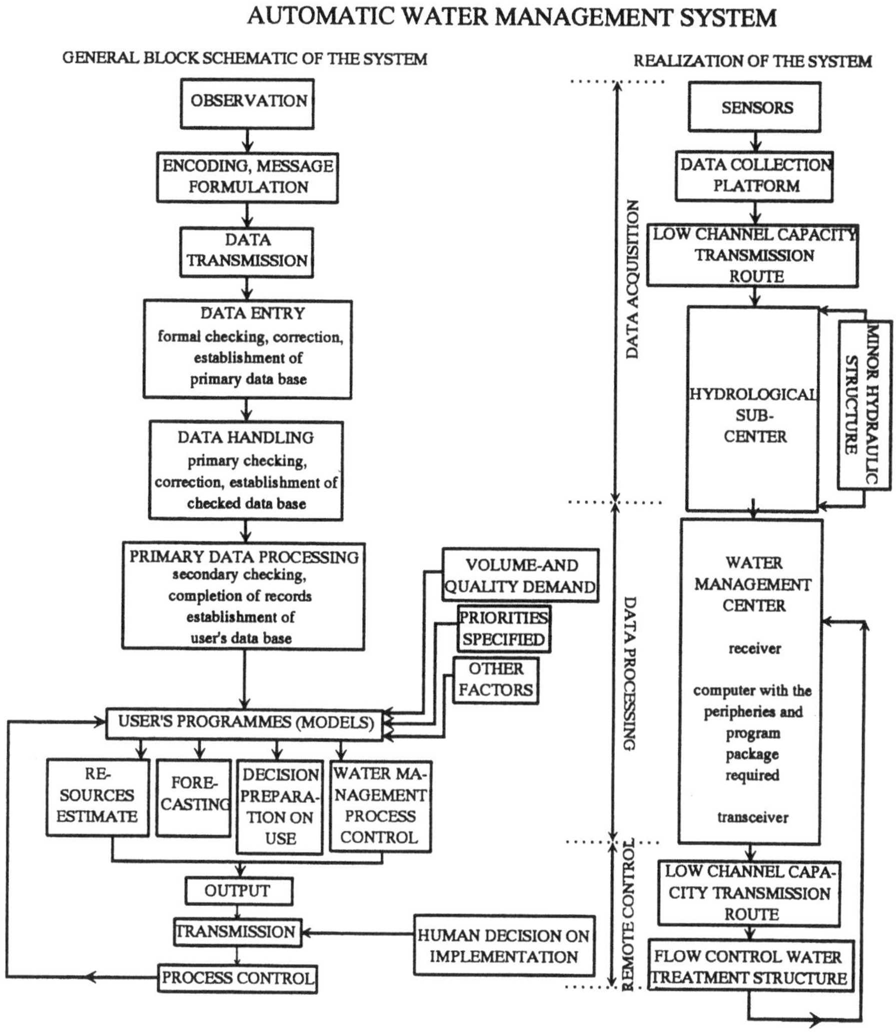

Figure 2: Outline of automatic computer-based water management system (Starosolszky, 1987).

Conclusions

Automation plays a significant role in the operation of water resources systems as required by

- the significantly large dimensions of these systems, with numerous components involved, each of which needs to be controlled
- temporal variations of either the inputs or the outputs, which require continuous control
- the need for precision, both in collecting or evaluating data and in operating each component of the system

While developed countries have initiated full automation in most systems, developing countries favor semi-automation at present, due to economical and social considerations.

On the other hand, hydrological data acquisition systems seem to be the area where full automation is most widely accomplished. Other fields of water resources engineering reflect varying degrees of automation.

In WRE, the basic approach, at both the design and the operation phases, is to reach optimal solutions. Therefore, the decisions to be made as to when, where, and how much automation is required are based on comparisons between its advantages and disadvantages; that is, the application of automation in WRE is a matter of optimization. This is actually why automation is described as "optional" in WRE system operation.

Automation is essential in safety-critical systems, such as those of ATC, to reduce human-induced errors. In ATC systems, there is actually no tolerance for errors, and the goal is to keep safety factors as high as possible. This is because any failure of the system has a direct and immediate effect upon human life. In this case, failure means disaster, and automation is required to minimize failure probabilities.

In WRE, system failures range from correctable errors to complete disasters which may or may not have a direct impact upon human life. Often, the results of system failures show indirect and slow effects on human beings. On the other hand, system safety is also a critical issue in WRE, and an optimal operating policy requires possible errors to be kept at a minimum. However because of uncertainties inherent in nature, there is no 100% reliability in WRE systems. Here, automation is called for by other considerations, like saving energy and labor or minimizing operational costs. As these considerations have gradually become significant in the operation of complex WRE systems, automation is entering the field of WRE at a growing pace.

References

Alpaslan, N., (1991). "Automation in environmental engineering", In: J. A. Wise, V. D. Hopkin, & M. L. Smith (Eds.) *Automation and Systems Issues in Air Traffic Control.* Berlin: Springer-Verlag.

Linsley, R. K. and Franzini, J. B. (1964). *Water-Resources Engineering.* New York: McGraw-Hill. pp. 378-468.

Starosolszky, O. (1987). *Applied Surface Hydrology.* Water Resources Publications. pp. 314-374.

Automated Aids and Their Evaluation

Intelligent Dialogue in Air Traffic Control Systems

Jack L. Edwards

Artificial Intelligence Management and Development Corporation, Toronto, Ontario, Canada

Introduction

Air traffic controllers are currently engaged in a dialogue with various automated subsystems: they access databases of textual information, examine and interact with radar displays, receive and transmit flight information on hardcopy, and communicate with pilots and other controllers over audio links. Many of these activities do not have the look and feel of dialogue, but convey the impression of working with a set of tools to perform a variety of tasks. Advances in computer technology over the last five to ten years reflect the increasingly successful use of this "tool metaphor," and it is likely that future, near-term Air Traffic Control (ATC) systems will continue to reflect this approach.

In spite of the prevalence of "point-and-click," direct-manipulation environments, conceiving of a controller's interaction with sophisticated system software as "dialogue" has its advantages. Currently, the tools available in those environments are tied very closely to a user's behavior and, when activated, produce an immediate response of relatively short duration. The evolution in the behavior of those tools, however, will gradually lead users to view many of them as sophisticated, partially autonomous agents, and to describe interaction with them as having more of the "look and feel" of dialogue.

The purpose of this paper is to explore efforts to bring together some of the technological developments in the areas of Artificial Intelligence (AI) and ATC systems. The paper covers a sample of the work in North America, in which AI is already being used to address problems of ATC, and explores the potential of a relatively new area of AI to ATC systems: the incorporation of explicit models to support intelligent dialogue capabilities (Edwards & Mason, 1989; Kass & Finin, 1988; Mason & Edwards, 1988). Models that might be incorporated include those of the task, individual controllers and the system itself.

In the sections to follow, some of the challenges of advanced automation systems for ATC will be identified and some of the forces that are already at work to bring ATC and AI together will be mentioned. Also, work in North America on advanced automation that incorporates AI technology will be reviewed; this includes work on AERA 2 and AERA 3 in the U.S. and similar work in Canada. The review is not meant to be exhaustive but should give a flavor of current contributions being made to ATC through AI and expert systems technology.

After looking at some of the current efforts to combine AI with ATC, the research on explicit-model building will be described. A brief history of this field, its current status, and its potential contribution to ATC in the post-2000 era will be presented. The discussion on modelling will focus both on recognizing the goals and plans of a controller and on capabilities for characterizing the goals and plans of the system, so that they can be presented

NATO ASI Series, Vol. F 73
Automation and Systems Issues in Air Traffic Control
Edited by John A. Wise et al.

to a controller in a way that allows a quick grasp of the key elements of the situation. Finally, the way in which explicit-models design provides a natural way of incorporating human factors concerns into advanced Air Traffic Control systems will be described.

Combining Air Traffic Control and Artificial Intelligence

Several assumptions motivate the move to increasingly automated systems in Air Traffic Control. One key assumption is that there are tasks that controllers perform that can be automated in a way that will improve the performance of those controllers and the effectiveness of the total ATC controller-computer system. Another assumption arises from tradeoffs that have to be made; for example, one may be willing to accept a slight decrement in overall performance introduced by an automated system for a considerable savings in costs, especially if it is likely to be a temporary trade-off and as long as it does not impair safety. A third assumption is that the amount of world-wide air traffic will show a substantial increase over the next few years, and that in order to cope effectively with that increase, more sophisticated support systems will have to be provided to controllers. A final assumption here is that support by automated systems will allow controllers to handle increased work loads as effectively as, or more effectively than, they handle their present ones.

There are a number of challenges in developing effective automated systems for ATC, including the development of a more complete understanding of what controllers actually do and an appreciation for the things they might do that they do not do today. In addition, there is a challenge to understand more deeply the differences between the capabilities of controllers and machines. Meeting these challenges will mean, first, acquiring the relevant knowledge, and then, using it to evolve and refine the tasks handled by controllers and their machines. That includes the intelligent distribution of those tasks. Along with the goal of sensible task distribution is the challenge of how best to engineer the interaction between the controller and the machine: how is the nature of this new interaction to evolve, and what are to be the elements of the "dialogue"?

Aspects of these challenges involve formulating what air traffic controllers do in such a way that makes automation of ATC tasks possibe;, developing criteria to determine which ATC skills should be automated and which should be left to controllers; determining which tasks, or aspects of tasks, can be successfully automated in the near term and which are possible only in the next 10-20 years; and finally, ensuring that the interaction ("dialogue") between controllers and machines is a harmonious one.

A number of forces at work in both ATC and AI will help decision makers meet these challenges. In Air Traffic Control, task analyses, such as those produced for the FAA (Computer Technology Associates, 1987-1989), help to make explicit the goals, plans, and supporting actions of controllers for the entire range of activities in which they are engaged. Any specific task analysis may be more or less helpful in this regard. In part, such efforts should be seen as evolving a language for representing ATC tasks.

The analyses of the activities of controllers help to foster the application of technologies like AI to the ATC area. For example, one clear advantage of task analyses, from the point of view of commercial AI, is that they may save considerable time for knowledge engineers in their attempts to elicit expert knowledge from controllers. Such analyses also provide a basis for formulating goal-plan-action hierarchies that could be used to help recognize what a controller is doing at any given moment, based on his or her low-level actions at the interface. Generating such hypotheses in real time can be seen as tracking the goals of a controller who is pursuing some task. Designing such an activity for an automated system would allow designers and developers to begin asking such questions as, "If the system is able to track the controller's goals for some well-defined task, with some acceptable degree of accuracy, how might that knowledge be used to support more intelligent dialogue with the controller?" (One example might be the presentation of information that a controller would find useful at different stages in the elaboration of a task, without the controller having to ask for the information explicitly.)

In a recent exercise, this author back-inferred the goals for a user from the methods ("services") specified in an object-oriented analysis of a workstation layout design problem. The result was several hundred goals, organized hierarchically, that now will be used to build a model of a user's plans in that environment. That exercise, the majority of which was completed in an afternoon, shows how user goals implicitly residing in a software design may be made explicit. Once decisions are made about the interface elements to be incorporated into the layout design software, that goal-hierarchy can be extended to the level of action at the interface.

The preceding illustrates a few of the activities taking place in both AI and ATC that are helping to bring together researchers from both areas. In the next section, a sample of that combined research, as it is happening in North America, will be reviewed.

A Sample of Work in AI and ATC in North America

Research and Development in the U.S.

The application of AI technology to ATC in the U.S. began in earnest in the 1980s with projects supported by the Federal Aviation Administration (FAA). To meet the goals of the 1990's and beyond, the FAA has developed the National Air Space Plan, part of which calls for the implementation of an Advanced Automation System (AAS) to provide Air Traffic Controllers with new hardware, software, and an important group of system support functions known as the Automated En Route ATC (AERA) capabilities (Scardina & Ryberg, 1989).

AERA is to be implemented in three stages, characterized respectively by the goals of detecting and displaying problems that might arise from a planned flight route, resolving problems detected by AERA 1 capabilities and displaying its resolutions, and introducing automated decision making and control into the ATC process.

A key problem definition in AERA 1 for conflict prediction is the identification of potential violations of separation minima between two or more aircraft, or between aircraft and some protected airspace. Detecting such problems makes use of standard algorithmic procedures which do not involve the use of AI technologies, but AI has been applied in AERA 2 for rapid prototyping.

Work at MITRE Corporation in McLean, Virginia, has been directed at building an Automated Problem Resolution (APR) prototype that generates possible solutions for aircraft-aircraft and aircraft-airspace conflicts in both radar and non-radar environments. This work has provided for the evaluation of the prototype by FAA field personnel — the Air Traffic AERA Concepts Team (ATACT) — and has helped establish the APR requirements specifications for the Advanced Automation System. ATACT, a sixteen-person team consisting of senior controllers, has met at MITRE for periods of approximately two weeks every other month for the past six years (Fearnsides, 1990).

The APR prototype has three major functions: conflict and resolution information display, automated problem detection, and resolution generation. It has been developed on seven networked Symbolics 3650 Workstations, using Intellicorp's Knowledge Engineering Environment (KEE) Expert System Shell and the Common LISP programming language. For each conflict the APR prototype detects, it attempts to generate a set of resolutions for each aircraft involved in the conflict and then to use its problem-detection capability (a subset of AERA 1 problem detection) to determine possible conflicts that might arise from resolutions it has generated. All resolutions are generated in ways that attempt to minimize the number of primary and secondary conflicts.

The Knowledge Base of the APR prototype contains the rules to be used in generating possible resolutions to conflict situations, and it is those rules that MITRE has sought to validate. The prototype system attempts to generate resolutions in three Maneuver dimensions: (1) the Lateral Dimension (Left/Right); (2) the Vertical Dimension (Climb/Descend); and, (3) the Longitudinal Dimension (Increase/ Decrease Speed). More specifically, APR attempts to generate one resolution for each direction within each dimension for each aircraft.

In the lateral dimension, different types of resolution are generated based on an aircraft's navigational capabilities, e.g., six types of resolution for Very High Frequency Omnidirectional Radio (VOR) aircraft and three types for Area Navigation (RNAV). Single dimension resolution is tried first and, if that fails, APR tries "composite" maneuvers (manipulating lateral and vertical dimensions simultaneously) and then "multiple" maneuvers (two independent maneuvers for two aircraft in one resolution). If no lateral dimension resolution is found, the system proposes a single-dimension, lateral resolution with the most remote secondary conflict attached. If it can find no resolution for a given side, none is proposed for the selected aircraft (Fong, 1989).

In the vertical dimension, generating possible resolutions depends on whether an aircraft will be climbing, descending, or maintaining a stable altitude "...within a parameter distance beyond its position at conflict time..." (Fong, 1989, p. 3). For aircraft maintaining a level altitude, "initiate-climb" and "initiate-descent" resolution maneuvers are examined, and for climbing and descending aircraft, "alter-climb" and "alter-descent" resolution maneuvers are attempted.

The third dimension is the longitudinal dimension. Resolutions along this dimension are not attempted if the conflict detected is one of a head-on encounter geometry. If such is not the case, however, resolutions are attempted through increasing and decreasing the speeds of the aircraft concerned. If both resolution attempts fail, a multiple maneuver will be attempted which involves increasing and decreasing the speed of two aircraft, respectively (Fong, 1989).

Evaluation of the prototype has been in terms of both its technical feasibility and its operational acceptability, and a scenario approach has been used based on flight data associated with four ATC Centers in the US. The prototype has been developed so that different conflict resolutions can be interactively created, detected by an Automated Problem Detection (APD) capability, and resolved using the rules of the APR Knowledge Base. Each evaluation performed is targeted at some area of that Knowledge Base.

The APR prototype has been developed in four phases, or "Builds." In the first Build, completed in 1988, strategy and maneuver-type selection rules were evaluated in a simple scenario involving a pair of aircraft in a "windless" and radar environment. A total of 180 evaluation runs were designed, consisting of all encounter geometries and exercising all of the

Build 1 rules (Scardina & Ryberg, 1989). Build 2 incorporated multi-aircraft and multi-airspaces and evaluated resolution maneuvers that were generated for all those aircraft in all dimensions and all directions. Build 3 incorporated an APR Knowledge Base for non-radar environments, also incorporating wind layers, and Build 4 included resolutions that incorporated Flow Instruction (FI) and interfaces with the Traffic Management System (TMS).

The evaluations of these implementations have provided guidance for how to extend, in an incremental way, the application of AI technology to Air Traffic Control and have shown, in the later phases, that the Knowledge Base is stable and requires no major restructuring of its rules (Scardina & Ryberg, 1989). MITRE is currently writing the system-level and algorithmic specifications for an APR that will be delivered to IBM, the AAS contractor. The next activities in AERA 2 for MITRE will be to examine the operational procedures required for controllers operating within this new environment.

Perhaps the most important limitation to applying AI technology in ATC arising from MITRE's work has to do with the criticality of real-time systems. Although there are a number of commercial AI inference engines available, it is questionable whether any of those can currently generate solutions and provide feedback as rapidly as is required in ATC environments, at least in tasks that are as critical as conflict resolution. Certainly the KEE environment does not appear to have provided the required response time. Nevertheless, AI Expert Systems technology has provided a useful environment for developing and evaluating the APR prototype, even if it is not used in implementing the operational system.

AERA 3 Work at MITRE: AERA 3 is targeted at post-2000 ATC systems and work in that area is still largely in the concept stage. The goal is a relatively full automation of the en-route controller's task (that is, the development of automated systems to handle most en-route air traffic and communicate with pilots through some form of voice synthesis). The "Airspace Manager" (the controller in this new environment), although still in the loop, would not handle problems that arise with normal en-route traffic. Instead, he or she would deal with high-complexity and emergency situations such as airport closings, high-density airspace, on-board emergencies, and unexpected and severe weather changes (Ryberg, 1990).

Currently, no AI technology is being used with AERA 3 prototyping though it is expected to be part of those efforts. One area of AI research that may prove useful is the work in planning (see Barr, Cohen, & Feigenbaum, 1989; Georgeff & Lansky, 1986; and, for early work in ATC, see Cammarata, McArthur, & Steeb, 1983), but work at MITRE is focused, in this early stage, on more traditional algorithmic approaches and other techniques such as linear programming.

The goals of AERA 3 are ambitious, and designing and building prototype systems for this phase of the AAS will be a difficult task. In fact, it is not clear at this stage that the goals of AERA 3 are achievable, and final decisions about the feasibility of those efforts are still several years away.

Research and Development in Canada

Attempts to modernize the Air Traffic Control system in Canada are being guided by the efforts of Transport Canada, in part through the Canadian Automated Air Traffic System (CAATS). The contract for CAATS was awarded to Hughes Aircraft at the end of last year, after several years of preliminary work.

Some work that may eventually become part of the CAATS project is being done by a group called CompEngServ, Ltd., who have been exploring how to automate conflict resolution for conflicts detected in en-route air traffic. This project is called CORES for CONflict REsolution System. The objective of the CORES project is to assist Canada's Transportation Development Center, a branch of Transport Canada, in examining tools and training media that might be used in future ATC workstations. The specific goal is to develop a prototype conflict resolution advisory system for a non-radar environment using Expert System Technology.

Work on CORES began in the 1980s (Bowen, 1987), with the expressed purpose of demonstrating that AI, and specifically expert systems, could be applied to help resolve aircraft conflicts. The success of that effort led to another project, whose goal was to explore in more detail the problem of "convergence" (that is, the creation of new conflicts as a result of resolving those currently predicted). Similar to efforts at MITRE, this new project made use of Intellicorp's Knowledge Engineering Environment (KEE) Expert System Shell but was less ambitious in its hardware support, targeting a personal computer 386 as its platform. The goal of the CORES project was also less ambitious, in that the system was to be a demonstration system as opposed to a full prototype.

CORES is intended as an advisory system to a controller who retains the final judgment as to the appropriateness of the resolutions generated. Some of the principles of conflict resolution that guided the development of CORES include minimal perturbation of the airspace (the maneuvering of as few aircraft as possible in generating a resolution); consideration of ATC constraints such as airspace structure, the performance characteristics of aircraft, and whether an aircraft has been previously maneuvered in a resolution situation; and finally, consideration of factors *immediate* to the predicted conflict first, and then expanding that consideration in a widening field of view (Smith, Crabb, Ram, & Attisha, 1989).

Four stages of conflict resolution are supported by CORES. The initial phase seeks to classify conflicts into one of six basic types and then to select appropriate strategies to resolve them. Strategies that fail the initial stage are considered again in the second stage, in which some of the initial constraints are relaxed. The third stage looks at strategies that failed in the relax-constraints stage and, making the assumption that the problem space cannot be a "conflict-free" space, generates only partial solutions. Finally, if all attempts at resolution fail, the system seeks help from the controller.

Resolutions generated by CORES are ranked and presented to the user. Some of the criteria used in the ranking of the system's resolutions include the type of resolution maneuvers to be performed, the expected deviation in flying time as a result of adopting the resolution, which of the four stages produced the resolution,[7] and the ease with which the aircraft can be maneuvered. There are three modes of operation for CORES, including the creation and modification of scenarios used to demonstrate the resolution algorithm, the actual running of those scenarios with problem detection, and the generation of resolutions to conflicts as they are detected.

During creation and modification of the scenarios (Update Mode), the user sets variables that govern the behavior of the simulation. Various variables related to flight plans can be specified, and "events" such as the current estimated position of an aircraft and its ETA at the next flight plan position can be entered. In the second mode of operation (Run Mode), the simulation is activated. During this stage, scenarios can be altered, but only by halting this mode and reentering the Update Mode; after making desired modifications to the variables, the simulation can be resumed.

7. The earlier in the staged approach that a resolution is generated, the higher is its ranking.

Once a conflict is detected, the system moves into Solving Mode, where it generates resolutions, ranks them, and then displays them to a user. Once the resolutions are displayed, the system provides several options. Users may select a solution and examine its effects by entering a "what-if" mode; they may request details that show the original and modified versions of the aircraft's flight plans; or, they may request an explanation of how a resolution was determined and receive a trace of the logic used in generating the resolution.

To evaluate CORES, scenarios were developed that corresponded to each of the four stages of resolution: (1) the initial stage, (2) the relax-constraints stage, (3) the convergence stage, and (4) call controller. The resolution advice offered by the system as a result of running the scenarios was viewed as appropriate under the circumstances and conformed largely to those resolutions that had been generated by the controllers prior to the evaluations. The system was characterized, however, by time constraints that appeared to detract significantly from its performance. Those appeared to be due largely to the hardware and software platforms chosen. (CompEngServ Ltd., 1989).

A follow-on project, CORES 2, has been initiated to extend the CORES system and to address the time-constraint problems encountered in the previous project. Timing is a key concern in the new project, and a number of real-time issues will be addressed. Other objectives of CORES 2 include demonstrating conflict resolution in an en-route radar environment and building a "tactical" system that can address conflicts bounded by emergency situations, on the one side, and strategic problems that are more than twenty minutes away, on the other. A specific goal of the CORES 2 project is to control up to 50 aircraft in an airspace of 500nm x 500nm.

In order to overcome the time constraint problems encountered in building the CORES demonstration system, more powerful hardware and software will be used. The new platforms include a DEC 2100/3100 and Gensym's G2 Expert System builder. This latter choice indicates a continuing commitment to AI technology in the context of developing conflict resolution algorithms for ATC. It is hoped that the very fast inference engine provided in the G2 product will help CORES 2 successfully address a number of real-time operational issues. After six months of work, the prototype interface has been developed, and the design of the conflict prediction and conflict resolution algorithms are under way.

Voice Recognition for ATC Support: In addition to work on Conflict Resolution, Transport Canada is supporting efforts to examine the usefulness of speech recognition devices for future ATC workstations. This work began in 1988, and a presentation and demonstration of progress were given in Hull last summer (CAE Electronics Ltd., June 8, 1990) at the Transport Development Center's Research and Evaluation Center.

The configuration of the system is that of a controller communicating by voice commands to a Marconi MacroSpeak 640 speech recognition system. Responses to voice commands include displays of information from the Operational Information Display System (OIDS) and from the Visual Information Processor (VIP) onto an IRIS 30/30 Graphics Workstation monitor. OIDS contains information on weather and wind, digital altimeter display, navigation and landing aids. The current form of the OIDS display is a page format brought up on a screen as a consequence of a controller making a keypad selection. VIP is a data store of 35mm slides of area maps and operations procedures and is also activated by keypad selection. An important advantage to using speech recognition technology to retrieve information from these data bases is the fact that a controller can issue the command to retrieve the information without having to look at and manipulate a keypad, thus leaving hands and eyes free for other, concurrent activities.

In the demonstration, the displayed VIP data covered the entire screen, with the OIDS data occupying a window inset in the upper left corner of the IRIS display. A human "pseudo-pilot" simulated responses from pilots of the various aircraft in the simulation and controlled the movement of aircraft displayed on a simulation of a radar scope.

In addition to research into speech recognition technology for ATC, the contractor also looked at the feasibility of a voice-activated pseudo-pilot. In the second part of the demonstration, the human pseudo-pilot was replaced by a DecTalk speech synthesizer. The simulation was then fully automated, and responses from the simulated pilots were delivered to the controller by the DecTalk.

Results of the research appear to support the conclusion that a fully voice-activated pseudo-pilot is feasible, at least for simple exercises and for an entry-level trainer. The voice recognition system behaved reliably, only misunderstanding part of one command during the demonstration. As expected, the individual acting as the controller was easily able to maintain visual attention on the display screen, thus not having to locate and manipulate the keypad in order to bring up the required information.

A next step in this research will likely be exploring the use of voice in a training context. At some point, the voice-recognition capability will likely be considered for use by controllers in operational settings, but there was no evidence that this would be done in the near future.

Intelligent Dialogue and Modelling

Intelligent dialogue between humans and machines typically takes place in the context of some task environment. In order to support intelligent dialogue, as one thinks of intelligent dialogue with humans, a number of conditions must be met. The system must "understand" — that is be able to represent and reason about where it is in the execution of a task, where its dialogue partner is, and how the goals and actions of each are mutually complementary. Thus, it must have some way of monitoring and interpreting the flow of the interaction by building models of itself and its users. Software designers and implementors have, until recently, incorporated only implicit models of tasks and very limited, implicit models of the system and its users. There have been few attempts to provide representations of goals, plans, and actions about which a system might reason directly. In order to carry on intelligent dialogue with their users, systems of the future will need to have at least some of the components of those models explicitly represented, and the requirement that those models be explicit is the starting point for considerations of future ATC systems in what follows. Before considering the nature of those models, however, a few points are in order about entities with which controllers will be "dialoguing" in the ATC workstations coming on-stream in the next several years.

The Concept of Agents in Future ATC Environments

In the introduction to this paper, a contrast was drawn between a "tool" metaphor, represented by the direct manipulation environments that will characterize the advanced controller workstations of the next five to ten years, and a "dialogue" metaphor, represented by communication with "intelligent" agents, similar to human-human communication (for more information on interface design metaphors, see Hutchins, 1989, and Hutchins, Hollan, & Norman, 1986). It was suggested that one might view interaction in direct manipulation environments in a way similar to how one views environments that involve "intelligent" agents (that is, as environments in which interaction is characterized as "dialogue"). To support this claim, one only needs to consider the objects in direct manipulation environments and a few of the actions performed by a user on those objects.

Clicking on an icon typically highlights that icon; double-clicking "opens" a document or activates an application that it represents; clicking on an icon while holding down a mouse key and dragging, typically moves the icon from one place to another on the display screen. In a very real subjective sense, the responses of such objects are similar to passive objects encountered in the natural environment: when some action is performed on them they react. That interaction can be characterized as dialogue with reactive, or "tropistic," agents (Genesereth & Nilsson, 1987): that is, dialogue that involves simple messages of very narrow scope sent from the human to the machine, and reactions (returned messages from the machine to the human) that are simple, well-defined, and immediate.

On the other end of the spectrum are agents that are adaptable and can generate internal representations of themselves and other agents in their world; agents that can even model other agents' models of them. Those agents have typically been humans, but increasingly they are the "intelligent" machines *qua* agents that are coming to occupy our world.

The advantage to the view of direct manipulation environments that has been developed here is that the concept of agency and its complimentary notion of dialogue can be seen to unify the view of interfaces that are being developed and implemented in advanced automation systems, such as those of air traffic control. By viewing agency as a continuum from passive, reactive agents, through those with fixed or implicit models, to agents capable of generating explicit, adaptable models that can model the models of other agents, one can treat *interaction* in a consistent and coherent way — that is, as *dialogue with increasingly sophisticated agents* (see Figure 1). Such a view can help to unify characterizations of advanced automated systems and make it easier to understand and represent the interactions between controllers and machines.

Explicit Models: An Introduction

Aspects of agents (both human and machine), including "dialogue" exchanges, can be represented in computing systems by explicit models. Serious work in the area of explicit models is relatively new and is represented most coherently through research on user modelling. In spite of its relative recency, the study of user models is related to a number of other fields in AI, namely "intelligent computer-assisted instruction (ICAI), multi-agent planning systems, text comprehension and generation, intelligent help systems, game playing and expert systems" (Kobsa & Wahlster, 1989, p. 7). A user model has been defined as:

> ...the knowledge about the user, either explicitly or implicitly encoded, that is used by the system to improve the interaction. (Kass & Finin, 1988, p. 6)

and, in the context of natural language dialogue, as:

> ...a knowledge source in a natural language dialogue system which contains explicit assumptions on all aspects of the user that may be relevant to the dialogue behavior of the system. These assumptions must be separable by the system from the rest of the system's knowledge. (Kobsa & Wahlster, 1989, p. 6)

		Actor Sophistication			
		Reactive ("tropistic") actors	Actors with implicit or fixed models	Actors with explicit, adaptable models	Actors with adaptable models of other actors
User Support	Actors that help the user	simple direct-manipulation tools	versatile, well designed machines or programmable tools		supportive intelligent agents
	Actors that hinder the user	simple obstacles	complex obstacles or faulty machines or programs		adversaries or intelligent adversarial agents

Figure 1: The Representation of Intelligent Agency and Direct Manipulation in Terms of Dimensions of Actor Sophistication and User Support

The first definition admits of implicit encoding which, though not part of the notion of explicit models, needs to be considered in any comprehensive view of models and modelling.

Workshops on user modelling are held every two years, with the most recent in April,1990. At that Workshop, the first journal dedicated to user modelling was announced, the *Journal of User Modelling and User-Adapted Interaction (UMUAI)*, the first issue of which was scheduled for publication in January, 1991. Other than individual papers (see UM90, 1990, for example), the major sources of work on user modelling are the previously referenced book by Kobsa and Wahlster on user models in dialogue systems and a 1988 special issue of the *Journal of Computational Linguistics*. Bibliographies of user modelling are available in those references as well as in Kobsa (1988).

One of the key advantages of explicit-models design is that it is *a natural way of introducing human factors considerations into the design process*. If a designer designs a system by representing the characteristics, goals, and behaviors of a user explicitly and separately from other knowledge in the system, then issues related to the user must be dealt with directly. Further, by designing explicit models of the system, along with explicit user models, the designer is forced to address the interactions among each — that is, who is doing what, when, how, and why, relative to the other.

The explicit and separate nature of modelled knowledge allows a system to make inferences about itself and its users, and to combine the results of these inferences with its knowledge of its tasks. Even the "dialogue" itself needs to be represented explicitly so that the user and the system may refer to and discuss things that have taken place in the interaction. Results of inferences on those separate knowledge sources become part of the explicit knowledge stored in the system's models and support such things as intelligent help facilities and explanations for why the system performed some action.

Explicit Models in Air Traffic Control

A variety of components could be included in models a system might maintain of itself and a controller in order to support intelligent dialogue. Those of interest in the present context are related to the complementary goals and plans a controller and a system might be pursuing at any given moment.

Such components have been treated in a more mature area of AI, that of AI planning research. Planning representations, models of how some task might be accomplished, are often organized as goal-subgoal hierarchies, with the terminal nodes of the hierarchies specified as actions to be taken at the interface. By representing the actions of a controller in such a framework, it may be possible to build systems that can dynamically model where a controller is in a goal-subgoal hierarchy from an analysis of the low-level actions at the interface, much as humans model what goals other humans are pursuing by observing their behavior. Further, if the system is programmed to maintain a goal-subgoal hierarchy for itself, then, as indicated earlier, that should lead naturally into a consideration of how the controller's and the system's goals, plans, and actions might be coordinated — and thus into designing dialogue opportunities to support that coordination.

Some early work in AI planning and ATC was done by Cammarata, McArthur and Steeb (1983), in which aircraft were viewed as agents in a multi-agent environment that negotiated to determine which of them was best suited to develop a plan to resolve a predicted conflict. Unfortunately, neither the concept of planning nor negotiating ("dialogue") is well-developed in that article. A review of contributions that AI planning research can be expected to make to applied settings can be found in Bonasso (1988). Although Bonasso deals with a different problem domain, that of battle management, his conclusions are equally applicable to Air Traffic Control.

Intelligent dialogue can be important whenever there is a reciprocal dependency in the actions of entities. In Air Traffic Control, dialogue participants must be able to understand the "meanings" of the actions of one another in real time. This means that a participant must be able to have a good sense of what is "in focus" in the interaction, or have efficient ways of finding out what is in focus for the partner, and then adjusting his or her focus accordingly. As well, a dialogue participant needs to have ways of introducing items into focus and of being able to determine whether the partner understands what has been introduced and is following the focus as "dialogue" proceeds.

In suggesting that explicit-models design might benefit advanced ATC systems, it is assumed that the interaction between the controller and the system will continue to evolve in the direction of dependencies that necessitate intelligent dialogue. That is, it is assumed that both participants will be engaged in a cooperative effort, and that, at any given point in time, information vital to one partner may reside in the other, and vice versa. Conditions under which intelligent dialogue may be of little value are those in which one participant is able to perform all aspects of a task independent of another, or in which one participant's responses are highly stylized (that is, serve only to constitute a "reactive" agent in the sense defined earlier).

The next generation of automated ATC systems, incorporating components similar to APR and CORES for example, will be agents whose characteristics go beyond the passive and reactive nature of simple tools. They will appear to be more intelligent agents that possess elaborate models for the tasks of conflict detection and resolution. Although requirements for dialoguing with a controller will still be fairly constrained, coordination of activities ("dialogue") will be necessary to make the total system work effectively. For example, decisions will have to be made about the conditions and actions required to activate the conflict resolution components, the nature of the feedback they will provide to the controller, and any subsequent interaction. That feedback will include not only *what* is to be fed back but *when* and *how*. Also, procedures that allow a controller to inspect the resolutions proposed by a system will need to be identified, and similar procedures will be necessary to query the system as to its explanations for why one resolution is to be preferred over another.

In spite of their apparent sophistication, "intelligent dialogue" between the computer and the controller likely will remain limited in automated ATC systems that will be implemented in the next ten years. Work on the next phase of the AERA 2 project at MITRE, for example, will deal with the operational procedures a controller will use to interact with an APR, and those procedures will continue to limit the kinds of "dialogue" that can take place.

It is unreasonable to assume that explicit-models design can be part of any implementation of AERA 2, though that should not preclude consideration in the work that remains. Regardless of the limitations imposed to ensure that the system performs appropriately as a "tool," there is some scope for examining how explicit models might be applied in such an environment. As the protocols for interfacing with a controller are designed, researchers should be developing representations for where the system and the controller are in a goal-plan hierarchy at each potential interchange between the two. Those representations would create the two separate views of a task necessary to the exploration of how mutually complementary progress in its execution is dependent on "dialogue." They would also help focus attention on the nature of the information that must be exchanged and on the most effective manner of that exchange.

Such research could be of vital importance if the system and the controller are to evolve in ways such that tasks like conflict detection and resolution become increasingly shared activities. Explicit-models research and design can contribute to the intelligence in the dialogue when a controller is to communicate about such things as when an automated system is to become active, what things it is to consider, and when and how ambiguities and errors in the "dialogue" are to be resolved.

One of the possibilities recognized by research like AERA 3 is that all of the normal activities and decisions of an En-Route Air Traffic Controller may be able to be automated. Thus, if "dialogue" is to take place between a controller and such a system, it may be only in specialized areas of conflict detection and resolution, such as emergencies and situations of high complexity. Since "dialogue" would be at a minimum except for those situations, the kind of automated system required would be one that could maintain an explicit model of the critical elements of the situation and use information in that model when interacting with the controller. If the situation were critical enough to alert the controller, then it would be mandatory that the controller grasp the problem quickly in order to act effectively. The critical nature of such situations would be compounded if the controller could not immediately understand the full implications of the problem and had to look to the system for clarification. A system with the capability of understanding *why* a controller did not grasp its problem representation would have a much better chance of clarifying that problem than if it relied on simple restatements of the situation, or "canned" paraphrases, without taking into account the controller's expression of what he or she did not understand.

A system that could represent how its characterization of some task might be confused, identify the elements that were likely producing confusion for a controller, and integrate its contributions to the dialogue with those of the controller in order to refine those representations, would be an effective support in the highly automated environments envisioned in AERA 3.

One of the suggestions that arose repeatedly in this Workshop was that controllers should become more pro-active in determining their role in the new ATC systems — that is, that they should take initiative to define what *they* want their role to be, rather than waiting for advancing technology to define it for them. They should determine the activities they want to retain for themselves, the way in which automated systems should support those activities, and the "dialogue" that will be required. Design by explicit models could provide a natural support at the system design level to the on-going consideration of those issues.

Finally, decision makers in ATC who support moves to increasing automation but who also want to preserve an active role for controllers in the systems of the future, would do well to consider support for explicit-models research and design, as a way of helping to realize both of those goals.

Conclusions

Conceptualizing the interaction between a controller and a system as "dialogue" has its advantages. In this view, both are seen as agents whose acts, including dialogue acts, are mutually dependent on one another. This view can be applied even to those ATC systems proposed for the near term, which will continue to have the look and feel of tools. Although the metaphor is one of a "tool," the systems in fact can be seen to contain reactive agents that respond according to the behavior with which they have been programmed, and that are evolving. This evolution, which is currently far more rapid than human evolution, will be from simple, reactive agents to partially autonomous ones with increasing scope for executing tasks on their own.

Although there is evidence to suggest that advanced automation systems of the post-2000 era may be assuming more of the functionality associated with normal air traffic control, at least in en-route sectors, there will certainly be situations in which the intelligence of the dialogue between controller and system, even though infrequent, will be critical. In support of that intelligence, it has been suggested that work in the area of Artificial Intelligence (AI), dealing with explicit models of the task, user, system, and dialogue, be seriously considered in building the automated ATC systems of the future. Components of such models would include the goals and plans that a controller and system would follow in executing ATC tasks, as well as procedures for how "dialogue" about those goals and plans should be coordinated between the two participants.

Advantages to integrating explicit models into ATC systems include a deeper understanding of the current and future tasks of Air Traffic Control, how those tasks might be distributed more effectively among controllers and their automated support systems, and how the activities of the two participants might be more intelligently coordinated to maximize their value to the total ATC system. Finally, perhaps the key overall advantage to explicit-models design is that it provides *a natural way of building human factors concerns into ATC systems at the earliest possible stage in design.*

References

Ammerman, H. L., Fairhurst, W. S., Hostetler, C. M., & Jones, G. W. (1988). *FAA Air Traffic Control operations concepts, Vol. 1: Background and analysis methodology.* Federal Aviation Administration, Department of Transportation Report DOT/FAA/AP-87-01.

Barr, A., Cohen, P. R., & Feigenbaum, E. A. (Eds.) (1989). *TheHandbook of Artificial Intelligence.* Vol. 4. Don Mills, Ontario: Addison-Wesley.

Bonasso, R. P. (1988). What AI can do for battle management: A report of the first AAAI workshop on AI applications to battle management. *AI Magazine, 9(3),* 77-83.

Bowen, B. A. (1987). An expert system for conflict resolution in dense airspace. *Knowledge based concepts and Artificial Intelligence: Applications to guidance and control.* NATO Advisory Group for Aerospace Research and Development (AGARD) Lecture Series, AGRAD-LS-155.

Cammarata, S., McArthur, D., & Steeb, R. (1983). Strategies of cooperation in distributed problem solving. *Proceedings of the Eighth International Joint Conference on Artificial Intelligence,* Karlsruhe, West Germany, 767-770.

CAE Electronics Ltd. (June 8, 1990). Presentation of work on voice recognition in ATC display and pseudo pilot, Research and Evaluation Centre, Hull, Quebec.

CompEngServ Ltd. (1989). *CORES: Air traffic conflict resolution system.* Final report TP9990E, Transport Canada: Transportation Development Centre Policy and Coordination Group, Ottawa, Ontario, Canada.

Computer Technology Associates (1987-1989). *FAA Air Traffic Control operations concepts, Vol. 1-8.* Federal Aviation Administration, Department of Transportation Report DOT/FAA/AP-87-01.

Civil Aeronautics (1987). Toronto North EnRoute Qualification Training Program. Amendment #8, Civil Aeronautics: Air Traffic Services, Ottawa, Ontario, Canada.

Edwards, J. L., & Mason, J. A. (1989). The structure of intelligence in dialogue. In M. M. Taylor, F. Néel and D. G. Bouwhuis (Eds.), *The structure of multimodal dialogue.* Amsterdam: Elsevier Science Publishers.

Fearnsides, J. (1990). Personal Communication.

Fong, C. V. (1989) *Automated problem resolution prototype in automated En Route Air Traffic Control.* Report MP-89-W00027, MITRE Corporation, McLean, VA.

Fong, C. V. (June, 1990). Personal communication.

Genesereth, M. R., & Nilsson, N. J. (1987). *Logical foundations of Artificial Intelligence.* Los Altos, CA: Morgan Kaufmann Publishers, Inc.

Hutchins, E. (1989). Metaphors for interface design. In M. M. Taylor, F. Néel & D. G. Bouwhuis (Eds.), *The Structure of Multimodal Dialogue.* Amsterdam: North-Holland.

Hutchins, E., Hollan, J. D., & Norman, D. A. (1986). Direct manipulation interfaces. In D. A. Norman & S. W. Draper (Eds.), *User Centered System Design: New Perspectives on Human-Computer Interaction*. Hillsdale, NJ: Lawrence Erlbaum Associates, Publishers.
Georgeff, M. P., & Lansky, A. L. (Eds.) (1986). Reasoning about actions & plans. *Proceedings of the 1986 Workshop*. Los Altos, CA: Morgan Kaufmann.
Kass, R., & Finin, T. (1988). Modeling the user in natural language systems. *Computational Linguistics, 14(3)*, 5-22.
Kobsa, A. (1988). *A bibliography of the field of User Modelling in Artificial Intelligence dialog systems*. Memo Nr. 23, University of Saarbrücken, Federal Republic of Germany.
Kobsa, A., & Wahlster, W. (Eds.) (1988). Special issue on user modeling. *Computational Linguistics, 14 (3)*.
Kobsa, A., & Wahlster, W. (Eds.) (1989). *User models in dialog systems*. New York: Springer-Verlag.
Mason, J. A., & Edwards, J. L. (March–April, 1988). *Explicit models in intelligent interface design*. Paper presented at the Workshop on Architectures for Intelligent Interfaces: Elements and Prototypes, Monterey, CA.
Scardina, J. A., & Ryberg, P-Y. (1989). *Future ATC automation aid based upon AI technology*. Report MP-89-W00036, MITRE Corporation, McLean, VA.
Smith, I. G., Crabb, C. A., Ram, G. R., & Attisha, M. R. (1989). A unique approach to aircraft conflict resolution using artificial intelligence techniques. *Proceedings of the Canadian Conference on Electrical and Computer Engineering (CCECE)*, September, 1989.
Transport Canada (1988). Air Traffic Control manual of operations: Procedures. Report TP703, Transport Canada: Air Traffic Services, Ottawa, Ontario, Canada.
UM90 (March-April, 1990). *Proceedings of the Second International Workshop on User Modeling*. Honolulu, Hawaii.

Close Ground/Air Cooperation in Dynamic Air Traffic Management

André Benoît

Engineering Directorate, European Organisation for the Safety of Air Navigation, Brussels, Belgium

Introduction

The paper broadly follows the general structure of a lecture delivered at the Academie Nationale de l'Air et de l'Espace, Toulouse, France, in November 1989. The paper describes some of the work conducted by the Eurocontrol Engineering Directorate to provide automated aids to the controller and appreciably improve the efficiency of the on-line component of air traffic management and control.

The presentation is supported by work in six areas, in particular, (1) the air traffic management and aircraft guidance in and around a main terminal, (2) the basic principles of the control of inbound flights, (3) the methods, techniques, and tools developed for the generation of aircraft trajectories for on-line operation, (4) the integration of aircraft capability into air traffic handling simulations, (5) the integration of a realistic airline/aircrew/aircraft component in ATC simulations, and (6) an illustration of the quality of the ground-based 4-D guidance system when used in critical atmospheric conditions.

Framework of Contribution and Position on the Space-Time Horizon

Air traffic handling involves a vast system whose position in the space-time continuum differs greatly, depending on the component considered. It ranges from 20 to 40 seconds for collision avoidance action, up to some 20 years or more for the planning of large-scale facilities such as the construction of new airports or the introduction of a new surveillance/communication environment.

In this Section, I shall discuss specific practical requirements constrained by present capacity limitations (the human factor, airspace, runways).

Geographical Extent

In Western Europe, there is an extremely heavy concentration of major Terminal Manoeuvering Areas, or TMAs — such as London, Frankfurt, Paris, Amsterdam, Brussels, Luxembourg, and Strasbourg — in a small and highly-populated area. Brussels, for example,

NATO ASI Series, Vol. F 73
Automation and Systems Issues in Air Traffic Control
Edited by John A. Wise et al.

is within one hour's flight of a number of major airports — 30 to 60 minutes from Amsterdam, Paris, London, and Frankfurt, and 60 to 90 minutes from Zurich, Milan, Dublin, and Vienna. For such flights, and more especially for the shorter ones, any extension of the time or distance flown, because of either holding or vectoring, has an appreciable negative impact on flight economy.

Integration of Control Phases

At present, a flight departing from Copenhagen destined for Paris passes through four national or international jurisdictions. In two, or possibly three, of these jurisdictions, it will be handled by two different centres (e.g., en-route and approach, and upper and lower airspace). If we add to this situation the sectorisation effect (a requirement linked with controller capacity), we end up with a number of flight phases, each phase being conducted almost totally independently of the previous and subsequent ones. The contrast with the on-board potential for the planning and control of a flight is striking.

On-line Component of Air Traffic Handling

This description will be limited to one essential component of air traffic handling, namely the on-line component (i.e., air traffic control itself and the management of traffic conducted on-line). The time-horizon extent of the management will obviously vary with both the size of the geographical area covered and the aircraft type. The area concerned includes and surrounds a series of main terminals (e.g., Amsterdam, Brussels, Frankfurt, London, and Paris) and extends for, say, 200 nm around each main airport.

Accordingly, in view of the geographical scale of Europe, the on-line handling of air traffic should (a) cover areas within 1-2 hours' flight time from a fairly-central selected point, and (b) consider each flight as a complete entity, rather than a succession of phases. This means that management timescales would be of the same order. This clearly implies a need for tools to implement the subsequent control directives, while ensuring overall stability and still operating with predictions of the same order of magnitude.

Traffic Growth and Resulting Constraint

Air Traffic Evolution and Air Traffic Control Status

In June 1972, when welcoming the participants to the AGARD Air Traffic Control Systems Conference in Edinburgh, Scotland, Mr N. Coles, Deputy Director of Establishments and Research, concluded with the following observation "... we in the United Kingdom have yet perhaps to experience some of the most acute congestion in the Air Traffic Control field which is arising on the American continent. Nevertheless, on occasions, we experience stacking times equal to or greater than the associated flight time!"

This quotation is included simply to remind us that the existence of the problem we now face in Europe was announced some 20 years ago. So, have we done nothing to avoid it? In spite of tremendous developments in all branches of technology and the efforts undertaken by national and international institutions, a simple 45-minute flight leaving Brussels bound for London has no certainty of being allowed to land after this nominal time. Why is it so?

Air traffic control as a large scale system is unique. Hundreds of aircraft may be simultaneously present in an area. All the decisions on controlling these many elements are made by humans on a very short time scale, beyond which the behaviour of each of the elements exhibits a highly stochastic component. Furthermore, the set of elements constitutes a chaotic system, especially in the absence of any global view available to and controlled by a "central" unit.

Knowing the capability and limitations of the human being in the overall control loop, and given our present understanding of the complex process of air traffic control, it should and will be possible to provide automated aids to the controller to increase the integration of control phases and consider a flight as a single entity, instead of the juxtaposition of a series of disconnected phases.

The Saturation Barrier[8]

The saturation effect at both the en-route stage and at the runway will be illustrated, using two examples: London (Heathrow), UK, typifying a heavy-density traffic terminal, and Brussels, Belgium, representing a European medium-density airport. The potential and limitations of both the long-term regulation and the on-line handling of the air traffic will be discussed.

Stability of Air Traffic Management

For conducting the aircraft to a terminal, whatever its density, the present system is naturally stable, the stack constituting the regulating reservoir. However, any efficient management component avoids the systematic use of stacking facilities.

Instead, it determines the time of arrival and adapts the speed at entry accordingly for the aircraft concerned, or for another aircraft, if applicable. These two operations, which are associated with the traffic situation and the characteristics of the aircraft concerned respectively, are closely coupled. As a result, any management system appreciably more efficient than present operational ATC will clearly operate without the reliable, although very costly, buffers constituted by the stacks (see Benoît and Swierstra, in press).

The numerous, varied factors that affect the course of a flight are such that the uncertainty as to arrival time at the runway threshold could amount to several slots.

This fact was recognised and duly taken into account at an early stage in a project aimed at the management of traffic and guidance of flights in and around a main terminal. Indeed, following the first full-scale ATC simulation conducted at the Experimental Centre of the Eurocontrol Agency in 1982, an appreciable effort was devoted to the development of a ground-based 4-D guidance system designed to maintain the assigned time of arrival to within seconds (error less than 10 seconds). This system, which is referred to as CINTIA, has been tested under a variety of conditions, using full-scale flight simulators operated by airline crews. CINTIA's flexibility, adaptability, and accuracy ensure the stability of the overall

8. See "Le mur de la capacit" Jacques Villiers, *ITA Magazine*, December 1989

management and control system in an extended area including and surrounding a main terminal (Zone of Convergence, ZOC), from entry to touchdown.

Guidance of Flights

For those on-line management components which generate a set of time-of-arrival constrained trajectories, there is a need to provide a stabilizer to prevent the numerous perturbations that affect the conduct of a flight from upsetting the expected "organisation."

To this end, a specific prediction-guidance-and-control module was developed which makes it possible, for instance, to maintain a time-of-arrival assigned, say, one hour earlier to within seconds, if the need arises.

The basic principles and operation of the system are described in detail in Benoît and Swierstra (1990a). The results obtained to date in current operation and in critical atmospheric conditions (turbulence and strong winds) are illustrated in the first application outlined in the section "illustration of automated advisories."

Air Traffic Management Approaches

Air traffic management approaches comprise two essentially different aspects, depending on whether or not the initial time — in particular, the departure time — is included in the control variables.

Clearly, efficient air traffic management over an extended area embracing several main terminals should include the departure times in the set of control variables. By contrast, traffic management over one main terminal — and possibly secondary airports — will handle the entering flights as they come, the arrival times being the essential control variables.

The extent of such management will be discussed later. The case of efficient regulation over an extended area including and surrounding a main terminal is presented in Benoît and Swierstra (in press) and could be the subject of a thorough discussion.

Interfaces Involving Human Intervention

"The Air Traffic Controller Facing Automation: Conflict or Co-operation?"

This important topic was the subject of a separate paper presented at the Royal Institute of Navigation, London, UK, in September 1987.

Indeed, as was stated previously, the complexity of any management system should be kept within the computer.[9] This fact was recognised early in our development work. As a

9. *The Zone of Convergence (ZOC): Overview.* Video film produced by the EUROCONTROL Experimental Centre, Bretigny-sur-Orge, France, December 1986).

result, in the initial system (referred to as ZOC, using ROSAS and CINTIA as management and guidance components respectively — see references), the actual advisories are integrated into the aircraft label displayed on the controller's radar screen, without complementary tabular displays or any other additions.

To ease subsequent discussion, the main interfaces are summarised below. Two essential categories of interface can be envisaged: those involving automatic exchanges only, such as the generation, transfer, and reception of radar data; and those involving human participation. Only the latter will be considered here.

Computer / Controller

This interface involves two paths :

- Computer ➔ display ➔ controller: This path is used, firstly, for guidance directives generated and displayed automatically, and secondly, for providing information requested by the controller either directly on his own initiative or following requests from an aircraft.

- Controller ➔ keyboard/mouse ➔ computer: This path is required only to effect changes in information presentation, to express requests for additional data, or to alter a proposed arrival time; it is never used for actual aircraft control or guidance, even in cases of trajectory alteration.

Controller / Aircraft Crew

This radio/telephone (R/T) supported interface transmits two essentially different types of message, namely:

- Display ➔ controller/aircraft crew: 4-D guidance directives are displayed on the radar screen, translated into current ATC terminology, and sent to the aircraft, with or without acknowledgement.

- Controller/aircraft crew: For all other non-guidance messages, the crew and the controller use standard phraseology.

Aircraft Crew / Aircraft Control

In this context, this interface corresponds to the implementation of the 4-D guidance directives at the discretion of the aircraft commander, within, of course, the possibilities offered by the onboard equipment; depending on the flight phase and local/instantaneous conditions, this can be done in several ways:

- Manually
- Using the auto-pilot in the *ad hoc* mode
- Using the flight management computer system
- Using combinations of the above modes of operation

Line Generation of Aircraft Trajectories

The available knowledge concerning aircraft characteristics and operating procedures (preferred profiles and performance envelope) should be included in the ground-based system (for prediction, 4D-guidance, and control purposes) in a way consistent with on-line operation. The methods, techniques, and tools produced to achieve this aim were outlined in the presentation. Additional information is provided in Benoît and Swierstra (1990b) for further discussion.

Impact on ATC Simulation Capabilities

Air traffic controllers are well represented in traditional ATC simulations. However, specific requirements for ATC simulations result from the need to represent correctly the air component (comprising the pilot, the aircraft, and the avionics).

This subject is dealt with in some detail in Benoît and Swierstra (1990c) and Benoît, Swierstra, and Delnatte (1989) which emphasize the integration of aircraft capability into air traffic handling simulations and show how a reliable, low-cost experimental facility can be obtained.

Illustration of Automated Advisories

By way of illustration, three sets of experiments are presented.

The first pertains to the control of traffic and guidance of flights in strong winds and turbulence. It is shown how the controller can maintain landing time to within an accuracy of better than 10 seconds in such critical conditions (i.e., where there is a 100 kt variation in groundspeed over the last turn preceding localizer interception). The description of the experiments conducted with the cooperation of professional airline pilots (including the scenarios and detailed results) is presented in Benoît and Swierstra (1990c).

The second illustration relates to large-scale simulations and trials, including a full-scale ATC simulation of integrated control over the whole of Belgium (simulation conducted at the Eurocontrol Experimental Centre). The third covers plans for assessments in actual practice and preliminary tests of the use of the ground/air data link capability (as associated with Mode S) for ATC purposes.

Conclusions

The progressive introduction of automation in the ATC field calls for careful design of the aids envisaged. The references listed below present several aspects which derive from a simple initial requirement, namely that airspace users should be able to fly their preferred trajectories — possibly constrained by air traffic situations.

References

AGARD. (1990). "Aircraft Trajectories : Computation - Prediction - Control " AGARD AG-301, Vol. 3, May.

Benoît, A. and S. Swierstra. (In press). Air traffic management and aircraft guidance in and around a mian terminal. In M. Pélegrin and W. Hollister (Eds). *Concise Encylopedia of Aeronautical and Space Systems.*

Benoît, A. and S. Swierstra. (1990a). The control of inbound flights. *Journal of Navigation, 43.*

Benoît, A. and S. Swierstra. (1990b). Generation of aircraft trajectories for on-line operation. *European Journal of Mechanical Engineering.* September.

Benoît, A. and S. Swierstra. (1990c). Integration of a realistic airline/aircraft component in ATC simulation. 50th Synposium of the Guidance and Control Panel of AGARD, *Computer Aided System Design and Simulation.* Cesme (Izmir), Turkey, 22-25 May.

Benoît, A., S. Swierstra and Y. Delnatte. (1989). Integration of aircraft capacity into air traffic handeling simulators. Eurocontrol Doc. 892010-E, December.

Multi-layer Queuing Model to Check Automated Air Traffic Control Systems

J. C. Sargis and A. V. Sebald

University of California, San Diego, California, USA

Optimum performance of an automated air traffic control (ATC) system requires it to learn and modify itself to improve performance by practicing tasks over and over, solving problems, creating new terms and concepts, catching and exploiting explanations as a human air traffic controller would. Additionally, the system must check itself for errors, inconsistencies, unstable searching, identification of responsibility for decisions, and default reasoning. The multi-layer queuing model is a logic based problem solver to integrate an expert system, truth maintenance system, and queue model, to ensure efficient learning and checking are performed on-line in automated air traffic control systems. Many classes of parallel distributed processing models (also called neural network models) have been developed to address the learning issue. Expert systems have been used to organize knowledge, use rule-based and object-oriented reasoning models, and diagnose (or even execute) recommended repair plans to address problems where knowledge can be partitioned or organized with rules. Truth maintenance systems have been developed to perform indirect proofs, diagnosis, dependency-directed backtracking, dependency-directed searches, etc. to detect contradictions in conjunction with results from an expert system. The queue model and truth maintenance systems perform independently after calculations have been performed.

The Multi-layer Queuing Model is proposed to reduce error propagation and to optimize scheduling in an automated ATC system by using artificial intelligence (AI) tools. Implementation involves assignment of different tool components to conceptually varying tasks. Components are developed using a hierarchical structure ranging from information automation at the foundation to cognitive modeling at the top. The major components in the Multi-layer Queing Model (MQM) are an expert system, truth maintenance system, and a queuing model, as shown in Figure 1.

The purpose of the expert system is to manage data and rules where decision procedures exist. Examples include rules corresponding to weather conditions, coordination of aircraft, and air traffic controller language interpretation. The flow of the expert system is depicted in Figure 2. It is important to resolve conflict and maintain control in the expert system. Additionally, the expert system will contain a smart data manager to allow associative linkage and immediate cross-referencing retrieval of key concepts described in air traffic control documentation. The smart data manager greatly improves efficiency for tasks involving rapid information retrieval.

The truth maintenance system (TMS) is based on the work of Johan deKleer (Xerox PARC). It is designed to check the expert system in terms of rule inconsistency, default reasoning, and correct cache maintenance and updating. Additionally, the TMS delegates responsibility for decisions, guides backtracking (for search control in the expert system), and

NATO ASI Series, Vol. F 73
Automation and Systems Issues in Air Traffic Control
Edited by John A. Wise et al.

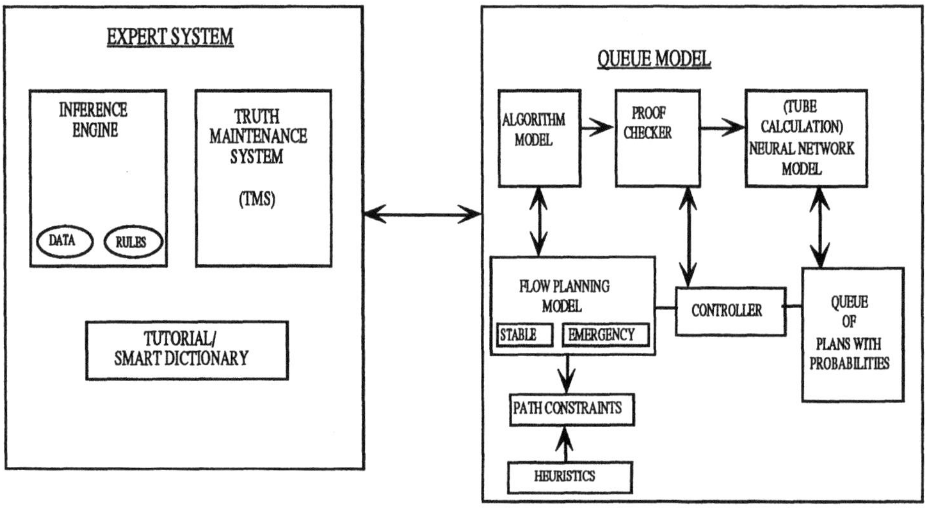

Figure 1. Major Components of Multi-Layer Queuing Model

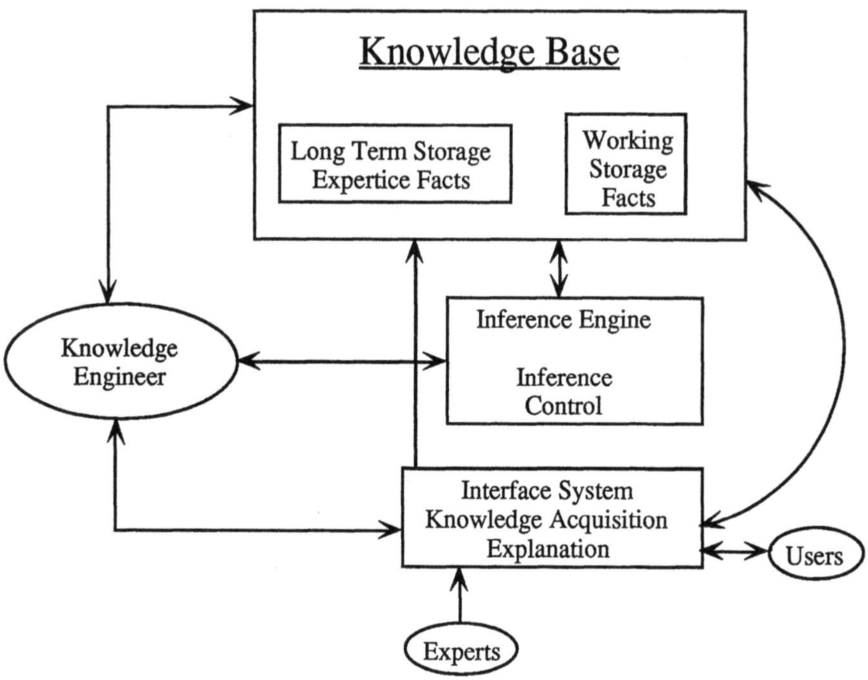

Figure 2. Expert System Architecture

interacts with the User Interface to provide path-audit capabilities and explanation facilities as shown in Figure 3.

Truth Maintenance System (TMS)

- Check for Inconsistent Rules
- Establish Decision Responsibility
- Guide Backtracking for Searching
- Provide Warnings, Audit Paths and Explanation Facilities

Figure 3. Truth Maintenance System Capabilities

The queue model is designed to manage information where effective algorithms are used to solve problems with decision procedures which are difficult to convert into if-then rules. In these uses, the decision procedures are condition, event, and strategic dependent. The queue model consists of an algorithm model, a proof checker, a tube calculation neural network model, a queue of plans, a controller, and a flow planning model.

The algorithm model provides inputs to the proof checker in the form of trajectory and scheduling calculations, flow planning for stable and emergency conditions, path constraints, and heuristics. The proof checker compares the initial conditions, the inputs from the algorithm model, and the goal states (minimum separation and no collision constraints). The proof checker uses the mathematical proofs — namely forward-backward, contra-positive, contradiction, construction, and choose to discard inputs which will never match the desired outputs. The most commonly used proof techniques are summarized in Table 1.

Table 1. Proof checker techniques

Proof Technique	When to Use it	What to assume	What to conclude	How to do it
Forward-Backward	As a first attempt, or when B does not have a recognizable form	A	B	Work forward from A apply the abstraction process to B
Contra-Positive	When B has the word "not "in it	Not-B	Not-A	Work forward from A & not B to reach a contradiction

Table 1. Proof checker techniques (Continued).

Proof Technique	When to Use it	What to assume	What to conclude	How to do it
Contradiction	When B has the word "not" in it, or when the first 2 methods fail	A & Not B	some contra-	Work forward from A & not B to reach a contradiction
Construction	When B has the term "there is," "there exists," etc.	A	There is a desired object	Guess, construct, etc. the object having the certain property & show that the something happens
Indication	When B is true for each integer, beginning with an initial one, say n_o	Sentence is true for n	Sentence is true n+1. Also show it is true for n_o	First substitute no every-where & show it true. Then invoke the induction hypothesis for n to prove it is true for n+1
Specialization	When A has the term "for all," "for each," etc.	A	B	Work forward by special-izing A to one particular object, the one obtained in the backward process
Uniqueness (A)	When B has the term "unique" in it	There are 2 such objects, & A	The two objects are equal	Work forward using A & the properties of the objects. Also work back-ward to show the objects are equal
Uniqueness (B)	When B has the term "unique" in it	There are 2 different	Some contra-dictions	Work forward from A using the properties of the 2 objects & the fact they are different
Either/or (A)	When B has the form "C or D"	A & not C	D	Work forward from A & not C ,& back from B
Either/or	When B has the form "C or D"	A &	C	Work forward from A & not D, & back from C
Max/min (A)	When B has the form "max S≤x" or "min s≥x"	Choose an s in S & A	S≤x or s≥x	Work forward from A & the fact that s is in the S. Also work backward
Max/min (B)	When B has the form "max S≥x" or "min S≤x"	A	Construct s in S so that s≥x	Use A & the construction method to produce the desired s in S

	AIRCRAFT			
Attribute	Plane A	Plane B	Plane C	Plane D
priority	3	1	3	4
speed-weight	4	4	3	3
turbulence-avoidance	4	1	4	3

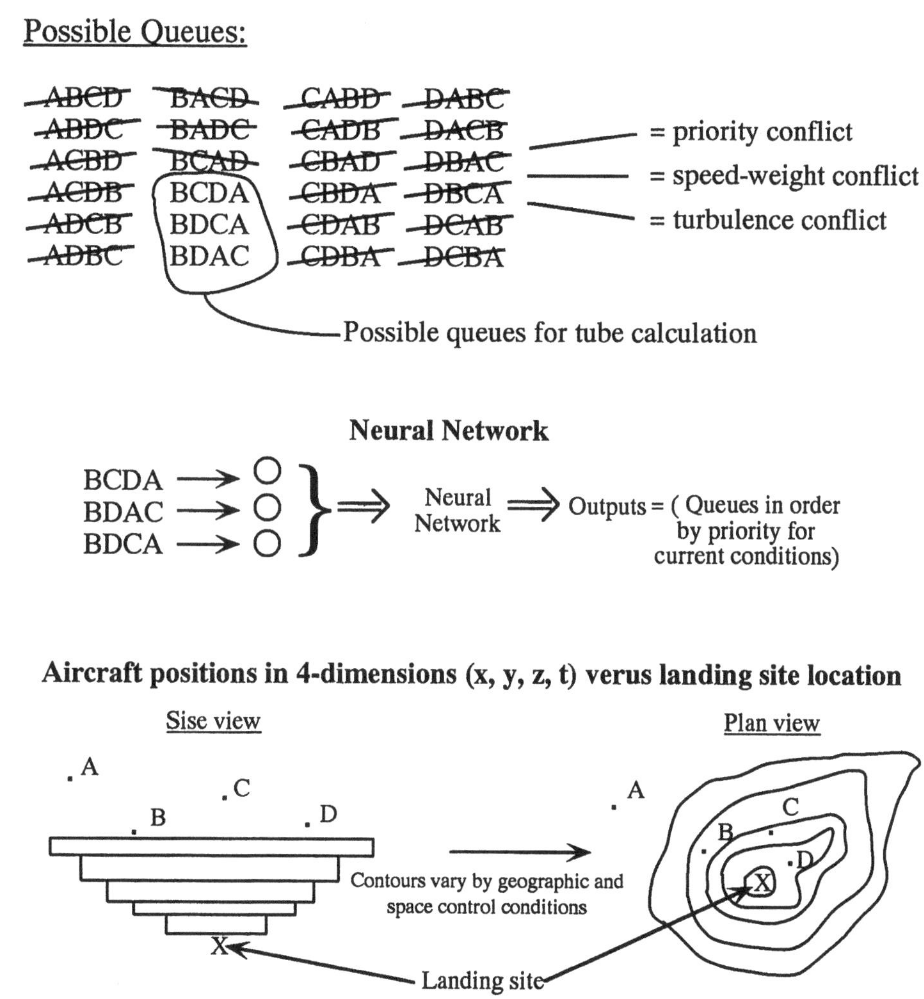

Figure 4. Queues for calculations

After inputs have been screened by the proof checker, the tube calculation (neural network) model is activated. Inputs are placed in queues by landing priority and current conditions, using object oriented programming. For example, let three of the queries be as shown in Figure 4.

The neural network model uses a set of processing units (each representing a possible queue) and activation values for each unit which are condition dependent. In this way, a pattern of excitatory and inhibitory connects are modified to produce rules to simulate learning. The patterns of activity can be analyzed and modified for new cases, using various neural network algorithms as depicted in Figure 5. This methodology is effective when inputs and desired outputs are known but algorithms correlating them are not.

The controller assesses the queues and their associated probabilities and compares them with a flow planning model. The flow planning model works with path constraints and heuristic variables to determine stable and emergency plans, then feeds back into the algorithm model to improve solution selection.

Neural net models are specified by net topology node characteristics and training or learning rules. These rules specify an initial set of weights and indicate how weights should be adapted during use to improve performance. For an ATC system, the inputs are the possible queries, and the outputs are trajectory, satisfiers, and minimum separation and collision-avoidance criteria.

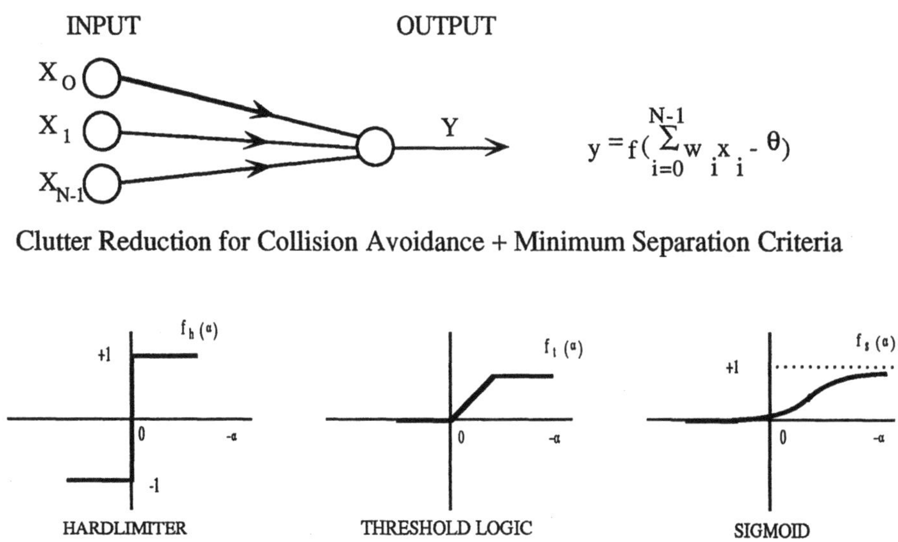

Figure 5. Computational element or node which forms a weighted sum of N inputs and passes the result through a nonlinearity. Three representative nonlinearities are shown.

References

DeKleer, J. (1988). "Truth Maintenance Systems." UCLA AI Short Course Notes. July .
Sargis, J. C. (1989). "AI/Expert Systems Tools." Jaycor Report, Nov
Partridge, D. (1988). "Expert Systems." UCLA AI Short Course Notes, July.
Solow, D. (1982). *How to Read APO Proofs.* New York: John Wiley
Rummelhart, D. E., & J. C. McClelland. (1986). *Parallel Distributed Processing: Exploration in the Microstructures of Cognition.* Cambridge; MIT Press.
Lippman, R.P. (1989). *Neural Networks: Artificial Neural Networks: Theoretical Concepts.* Computer Society Press of the IEEE.
Korf, R. (1988). "Problem Solving, and Search" UCLA AZ Short Course Notes. July.

Application of Planning Aids for Air Traffic Control: Design Principles, Solutions, Results

Uwe Völckers

DLR - Institute for Flight Guidance, Braunschweig-Flughafen, Germany

Introduction

The Institute for Flight Guidance of the DLR (Deutsche Forschungsanstalt für Luft- und Raumfahrt), together with the BFS and air traffic controllers from Frankfurt, have developed a new air traffic planning system called COMPAS. This new system is now installed at eight controller working positions in the Regional Air Traffic Control Center of the BFS in Frankfurt.

The aim of COMPAS is to assist the air traffic controllers in the planning and control of approaching aircraft in order to:

- Use the available runways to their optimum
- Recognize and reduce any congestion of incoming air traffic in time
- Reduce delay times
- Reduce controller workload

This is done by automatic input of radar data and flight plan information, taking into consideration other basic data such as aircraft performance, airspace, weather, separation standards, etc. This information is formed in an integrated process; the results are displayed to the controllers on a monitor and, if necessary, can be modified by using a keyboard. The air traffic controllers consider the proposed landing order given by COMPAS then direct the incoming planes in the usual way by radio transmitted commands.

Since September 1989, the system has undergone a six months' evaluation under real air traffic conditions in the Regional Air Traffic Control Center of the BFS in Frankfurt. The system was well accepted by the controllers and has been in all-day operations since then.

Problem

The highly frequented airport constitutes the most critical bottleneck for the aircraft approaching from various directions. Within the terminal area they are channeled into a few arrival routes and finally merged on the extended runway centerline(s).

NATO ASI Series, Vol. F 73
Automation and Systems Issues in Air Traffic Control
Edited by John A. Wise et al.

For safety reasons, minimum separation intervals between succeeding aircraft must be maintained by approaching aircraft. These intervals depend mainly on the type of aircraft, the runway configuration, and the prevailing weather conditions.

Each airport has a maximum possible take-off and landing capacity, which is determined by, amongst others, the runway and taxiway configuration and the installed equipment for monitoring and controlling the traffic on the airport surfaces.

Air traffic control has to guide approaching traffic to avoid jamming of the traffic flow, while at the same time achieving the maximum possible usage of the available landing capacity. This requires a thorough adjustment of the planned flight conduct of all arriving aircraft, which has to be established by appropriate coordination between all participating control sectors. The coordination effort required for this task increases rapidly with increasing air traffic volume.

These complex factors require central planning and control of all approaching traffic.

The numerous aircraft included in the planning process, the large number of static and dynamic parameters and marginal conditions to be considered, and the relatively short processing time available mandate a computer-supported planning system.

Solution

A concept was therefore developed by the DLR for the planning system COMPAS (Computer Oriented Metering Planning and Advisory System) with the aim of assisting the air traffic controllers in their planning and control of the approach traffic.

The computerized planning procedure is performed as follows: the entry of one or more approaching aircraft at the so-called ENTRY FIXES (about 100 NM before the airport), the anticipated, unaffected arrival time at the GATE (i.e., shortly before glidepath intercept) is initially computed on the basis of the current radar data and the flight intentions (these data are already stored in the ATC data processing systems). This is the arrival time which the aircraft would achieve under normal conditions, regardless of other traffic. With consideration of other traffic and the required wake vortex separation, each aircraft is given a place in the landing sequence and a nominal gate arrival time Starting with this time, the nominal times at other waypoints are retrospectively computed, taking into account further boundary conditions (weather, operating data, etc.).

Comparing the nominal time at the gate with the previously computed, unaffected gate arrival time, a time difference results (delay or advance). This difference should in future be compensated on the flight path between the ENTRY FIX and the METERING FIX, at the boundary of the TMA (i.e., during descent on appropriately shaped, economic flight profiles).

The planning described above looks ahead and is directed to a unified goal criterion. All essential boundary conditions, such as traffic loading of the sectors, wind data, aircraft wake turbulence categories, aircraft parameters, economic flight profiles, and current conditions, are taken into account.

The summarized planning results (e.g., arrival sequence, arrival schedule, time compensation) are presented to the controller. Under normal conditions, no interaction is necessary with the computer. If special traffic conditions make it necessary, the controller can modify the plan.

The executive function remains with the controller: an essential feature. COMPAS combines the individual planning tasks distributed over the different ATC sectors concerned. This reduces the coordination effort between the ATC sectors involved in the planning process.

Concepts aimed at similar goals have also been demonstrated recently in other European countries and the U.S.. In contrast to these concepts, which incorporate in part a higher degree of automation, the COMPAS system, after thorough studies of the DLR in the approach area of Frankfurt Airport, was deliberately designed only as a system to aid the controllers, following the principle of planning and proposal through the computer, decision and actuation by the controller.

Design principles

It is significant that in the acronym COMPAS the "A" stands for advisory: The computer simply provides advisories as to the sequence, schedule, and handling of each aircraft in the system. The controller is still in charge of the traffic situation, and is free to disregard or change the suggested plan if desired. The computer serves as an aid to judgment, but does not enter the primary control loop.

This system solution, therefore, can be classified as level-2-type, according to the categories of five levels of authority distribution between the operator and the machine given by Winter (1986):

Level 1: Human does the planning, makes the decision between suitable plans, and implements the decision (manual mode).

Level 2: Machine does the planning, offers a restricted set of suitable plans to the human, and human decides on one of these and implements it.

Level 3: Machine does the planning, suggests a plan to the human and implements it, if the human approves.

Level 4: Machine does the planning, decides on a plan, implements it, and informs the human after the fact.

Level 5: Machine does the planning, makes the decision between alternative plans, and implements the decision (autonomous mode).

The level of automation chosen for COMPAS may appear rather simple. However, the complex functional relationship between controllers and computers requires not only the computation of the correct planning results but also their representation to the controller in a understandable, reasonable, and acceptable manner. Furthermore, COMPAS must be able to modify the plan as specified by the controller. The computer proposes the medium-term frame of operation, while the controller makes decisions and is able to enter his ability to put the plan into effect and provide flexible short-term compensation for inevitable deviations and disturbances. The controller is therefore not the actuating assistant of the computer but remains the active and solely responsible link in the control loop.

In order to interact easily with the computer, the controllers use easy-to-handle displays and keyboards. The principle followed in the layout of these displays and keyboards was to "provide just as much as necessary" and to keep it "as simple as possible."

In the controller display, the various aircraft in the arrival flow are listed sequentially, with the earliest arrival at the bottom. Heavy or light aircraft are marked with an "H" or "L" respectively. The vertical position of each arrival is correlated with a vertical time-line which moves slowly downward with the aircraft listings. The bottom of the time line represents the time over the approach gate. The controller can move a cursor up or down on the time scale to identify a specific aircraft for a sequence change. The flow rate as well as the landing direction may changed by the controller and is displayed in the top right corner of the display.

The computer tracks each arrival individually, and continually updates its predicted gate time in relation to its assigned one. It then provides advisories to the controller (e.g., "expedite," "reduce," "hold").

Controllers are free to accept or reject the advisory. They can modify the computer-generated sequencing plan if desired, or if unforeseen events have occurred, using a functions keyboard. Ten functions keys to actuate the operational functions (clear and execute keys, as well as "-/+" keys to move the cursor up or down, to modify the plan for a specific aircraft, or to change parameter values. Example: press flow key; current value displayed at upper right can be changed with the "-/+" keys. The new value is entered by pressing the execute key).

Results

The results of the COMPAS evaluation will be published by DLR in the second half of 1990. They can, however, be summarized already in general. COMPAS:

- Has been developed *with* air traffic controllers *for* air traffic controllers
- Supports the controller in planning and control of incoming air traffic
- Reduces the coordination effort between controllers
- Reduces the amount of radio communication between controller and pilot
- Is widely accepted by controllers
- Helps to reduce jams and holdings of approaching traffic
- Causes the approach traffic to become more fluent and allows a more equal treatment
- Contributes to optimum utilization of available runway and air space capacity
- Can be incorporated into air traffic control systems without repercussions on already-existing ATC equipment

Artificial Intelligence and Human Factors in ATC: Current Activity at Eurocontrol Experimental Centre

Hugh David

Eurocontrol Experimental Centre, Brétigny-sur-Orge, France

Introduction

The European Organization for the Safety of Air Navigation is an international organization set up by a number of West European states in 1962, with the original intention of carrying out en-route air traffic control over Europe. This mission was later modified, so that Eurocontrol now provides, *inter alia*, advice and planning facilities for other ATC authorities.

The Eurocontrol Experimental Centre, situated at Brétigny-sur-Orge, south of Paris, was initially set up to provide real-time simulation facilities for the member states of Eurocontrol, and for other ATC organizations. Over the years, it has also developed the capacity for fast-time (arithmetic) simulation, and now provides a wide range of studies, tests, and trials facilities.

The activities of EEC are essentially pragmatic in orientation, intended to provide experience of, and familiarize ATC staff with, modern developments that may assist the controllers with their increasingly difficult task.

This paper describes the current activities related to the near and medium-range future, most of which are contained in the STAR (Studies, Tests and Applied Research) program, and will contribute to the ARC 2000 project, which will be a demonstration system for future highly automated air traffic control (Krella, 1989).

ASTA

ASTA (Air Traffic Management Strategic and Tactical Adviser) is an application of a production system for rule-based conflict resolution to on-line ATC. (A production system consists of a series of rules in the form IF [condition] THEN [action]. The conditions may be complex, and the action may involve setting values for further conditions.)

Two earlier studies furnished the groundwork for this study. The first was the development of a Computer Aided Planning Experiment (CAPE) which examined procedures for trajectory projection and conflict detection — leaving conflict resolution to the controller. This experiment, incorporated into a real-time simulation, provided for the examination of flights fifteen minutes before their entry into the simulated area, presenting any potential conflicts on a specially designed display. Controllers were able to make effective use of this aid during real-time simulation exercises (Braun, 1986).

NATO ASI Series, Vol. F 73
Automation and Systems Issues in Air Traffic Control
Edited by John A. Wise et al.

On the basis of the data and static environment of the CAPE exercise an off-line "production system" model was developed. This used a series of rules to produce "solutions" to the conflicts detected in a reasonable time and to present them to the controller. Written in PASCAL, this trial system involved a simple but rather rigid form of rule definition. Since it operated on recorded traffic, it could not simulate the changes induced by the solution of conflicts (Krella, 1986).

These exploratory studies were sufficiently promising to suggest the development of a more elaborate rule structure developed in PROLOG, and producing results on the traffic. This is the ASTA system proper, as described in Tumelin (1990), which provided a real-time system for providing "expert" solutions to potential conflicts.

Functional Organization

The ASTA demonstration system consisted of a "navigator" (which generated aircraft positions using a total energy model), an exit level manager,(which generated the necessary data and applied a knowledge-based reasoning system to assign an exit level five minutes before the entry of the aircraft to the sector), a conflict detector, a data collector which collected relevant data for aircraft involved in conflicts, and a conflict resolution monitor. The latter employed a rule-base with three main categories. These contained conflict resolution rules (which should be greatly expanded in future versions); a set of operational rules, relating to timing, aircraft attitudes, etc;, and a group of utility rules. Six work-months of expert controller effort were used in deriving the control-specific rules alone.

Operation

The system generated a group of potential solutions, calculated the resultant trajectories and estimated their cost (including a factor to discourage multiple interventions on one aircraft), and verified that the new proposals did not create further conflicts.

The information was presented on a SONY 2000 x 2000 pixel color display, on which was presented a radar-like image with three windows.

These contained a list of "strips" for the aircraft entering, and for any aircraft potentially in conflict; a list of proposed solutions, with the most economic presented in green; and a record of the solutions so far adopted. A further display could be called up to explain the reasoning behind the strategies employed.

Evaluation

The ASTA system was presented at the Le Bourget Air Show, where it attracted considerable interest from controllers. It was clear that more work needed to be done in smoothing the user-system interface, and in elaboration and proving safe the control rules. Work is continuing on the system, with a view to integration into the ARC 2000 advanced technology demonstrator. A real-time simulation is planned for October 1991.

ODID

Over the years, EEC has carried out a number of trials and evaluations of color displays and stripless systems (Clarke, 197;, Englmeier et al., 1980; David, 1981, 1984a, 1984b, 1986). In April 1986, it was decided that a series of operationally oriented studies should be carried out to combine the experience to date, and apply it to more realistic operational situations. Two simulations have so far been completed, and a third was released in September 1990.

The first ODID simulation, based on studies in the UK, took place in June 1987 (Prosser et al., 1988) and studied the use of colored EDD (electronic data displays) as substitutes for the traditional strips, with monochrome radar displays. Two ways of using color — to indicate flight direction, and to indicate the status of the flight — were contrasted (it was found that they could be combined). Space does not permit a full repeat of the 15 conclusions and 36 recommendations, but it was clear that the user-system interface was "input-heavy", and could be improved by the adoption of "default" procedures and handover conventions, and a multitude of minor adjustments to display characteristics.

The second ODID simulation, based on studies conducted in Germany, took place in May 1988 (Prosser et al., 1989) and studied the use of colored EED and raster-scan colored SDD (Synthetic Dynamic Displays). Two sector suites were simulated, representing sectors in Frankfurt and in Brussels. This simulation generally confirmed the conclusions of the first study, particularly the need to reduce routine workload by (for example) the introduction of pre-allocated levels for traffic passing from sector to sector, the need for flexibility in the allocation of tasks between controllers within a sector team, and the need for more sharply-defined and more closely integrated displays.

The third ODID simulation, based on preliminary experiments carried out by the French ATC services, will take place in September/October 1990 and will simulate sectors in France and Switzerland. This simulation will employ large 20-inch square displays, with 2000 x 2000 pixels containing all the displays for one controller. It will compare an approximation to the current "strip" in electronic form, with a set of analog displays for the Planning controller, using a WIMP (Window-interface-mouse-pointer) system for display. It will also compare the performance of two controllers working side-by-side with their performance when separated, communicating only through their displays and telecom links.

Common Operational Performance Specification

The Common Operational Performance Specification (COPS) is a Eurocontrol initiative to draw together requirements and ideas on the design of future ATC systems within the European context. COPS will identify which aspects are most in need of definite specification, and which may be adapted to special national requirements. The COPS group consists of a nucleus, based at Eurocontrol Headquarters in Brussels, and two members from each of the national administrations concerned. The nucleus works continuously on this task, filling out a skeleton Operational Performance Specification, which is supplemented and checked by all the participants in regular workshop sessions.

The skeleton COPS for controller workstations is divided into 13 chapters :

- General - background and scope
- Functions - time scale, ATC functions
- Requirements - workstation layout, sector layout, combination, aircraft performance assumed, system reaction times
- Architectural and functional components - working position processor, user interfaces, local communications, general communications
- Software architecture - language, operating system, functions
- Detailed functional specification - display presentation, communication and input processing
- Data formats and flows - radar server, plot, weather, static data
- Capacity and performance - displays, color, response time hardware characteristics.
- System monitoring - Availability, reconfiguration
- Environmental aspects - operating conditions etc.
- Software engineering aspects - maintenance, updating.
- Verification and validation - criteria, acceptance tests.

The COPS initiative is making steady progress, and has already defined many of the basic aspects of the specification and designated members of the group to elaborate the basic aspects of the system.

ATC Work Strain Measurement

As part of our continuing efforts to improve the value of real-time simulations, we have carried out a series of studies to evaluate different methods of measuring the "strain" on the controllers. We require methods that are minimally intrusive, cheap, easy to analyze, sensitive to "strain" and not so affected by irrelevant factors that the effects we are analyzing are swallowed up by noise. A preliminary survey (David and Noonan, 1984) identified 26 potential measures; others have since come to our attention.

The traditional method of assessing the effects of a simulated system on the controllers is the questionnaire. This has many obvious advantages, but it does not usually produce data that can be generalized. We have recently employed two techniques originating in the United States, the Subjective Workload Assessment Technique (SWAT) and the NASA Task Load Index (NASA/TLX). Both these methods employ subjective ratings on multiple scales, after a preliminary phase where the individual participant performs a ranking to determine the relative importance of the scales.

The Subjective Workload Assessment Technique (SWAT) was developed by Dr. Gary Reid at the Armstrong Aeromedical Research Laboratory, at Wright-Patterson Air Force Base (Reid. 1987). It employs three scales (time pressure, mental load, emotional strain) at three levels (low, medium, high). Before the simulation, the controllers are asked individually to rank all 27 combinations from best to worst. During the simulation, the controllers are asked to rate the task they are currently engaged on as high, medium or low on all three scales. These ratings are made "on-line" and may therefore be intrusive. Subsequently, the ratings are combined to form a score which is more reliable than the components.

The NASA Task Load Index was developed by Dr. Sandra Hart at NASA Ames

Research Center, at Moffet Field, California (Hart, 1987). It uses six scales (mental demand, physical demand, temporal demand, performance, effort, and frustration level). Before the simulation, the participants rank the scales by comparing each possible pair of scales; this ranking is used to develop a relative importance for each. After the simulation, the controllers (before leaving their positions) rate the run on a 1-20 scale for each of the six scales. These ratings are converted, using individual weightings, to a Task Load Index.

We have now employed each of these methods in at least two real-time simulations (Prosser et al., 1989, other reports in press), and have found that they produce statistically valuable results. We analyze the individual scales as well as the combined scores, finding that this provides useful, if more variable, diagnostic information. We are planning to install an automated data collection system for the SWAT test.

At the opposite end of the spectrum of methods, we have been investigating electrophysiological methods for assessing the mental activity of the controller.

We have carried out a series of studies which established, first, that EEG (electroencephalography — the recording of the electrical activity of the brain) was possible in the ATC environment, that brain activity was observably affected by ATC-like tasks (Vanwonterghem et al., 1988); and that certain characteristics of brain activity could be associated with particular types of ATC work, on the basis of our understanding of the general functioning of the brain (Vanwonterghem et al., 1989). These methods have been found to be relatively non-intrusive, although they require constant detailed supervision, and are too costly and slow in their present forms. These studies involved relatively few controllers, and considerable variation was observable between individuals.

To enable us to make a closer connection between control activity and psychophysiological measures, we have currently commissioned a feasibility study, by Professor Heinz Wolff of the Brunel Institute for Biomechanics, of a portable self-contained Work Strain Monitor. This will permit analysis of physical and psychological data in near-real time, in close association with data from the simulator (or, eventually, from real ATC systems).

AT30 Person-Machine Test-bed

EEC is developing, under this task, a Test-bed Controller Working Position for the development and evaluation of displays and control methods (Schmidt, 1990). This system will allow us to carry out realistic evaluations of the detailed interactions of display and control features, and to develop properly integrated and tuned control interfaces. Based on a 2000 x 2000 pixel display and associated windowing and pointer software systems, the system is designed to operate in a "stand-alone" mode, using its own resources to present a dynamic ATC image. It is also compatible with the main real-time simulator, so that it can be incorporated into future studies. It is at present being used to develop further the Conflict Risk display already used in ASTA (Tumelin 1990) in a more interactive form and will be used in the development of COPS. This display shows time on a horizontal axis, from the present to (say) fifteen minutes ahead, and separation distance on a vertical axis, from nil to (say) fifteen miles. For each pair of aircraft expected to infringe separation limits, a horizontal line is shown at the minimum distance, for the full period of the potential conflict. This line is numbered, and the corresponding aircraft are shown in an adjacent table. The Test-bench will be used to select potential conflicts using a mouse, displaying the corresponding aircraft positions in full color, and de-emphasizing (but not suppressing) those not involved.

AT42 Initial Sector Suite System Evaluation

Eurocontrol has obtained on loan two consoles designed for the Federal Aviation Authority's Initial Sector Suite System (ISSS), and arranged for software to allow the display of data from the Real-Time Simulator, although only display controls will be active. They will be demonstrated to, and evaluated by, European air traffic controllers during this year and next year.

Acknowledgements

I am indebted to the following colleagues for information on specific projects: M. R. Braun (COPS); M. G. Gabas (ISSS); M. P. Schmit (AT30); M. J-C. Tumelin (ASTA); Mr. D. Young (ODID III). I am also indebted to Mr. C. Massie (Head of Simulation Division) for his support and encouragement, to M. G. Maignan (Director of the Experimental Centre) for permission to present this paper, and to Mr. K. Mack (Director General of Eurocontrol) for permission to attend this Advanced Study Institute.

References

Braun, R. (1986) Simulation en temps réel d'un dispositif expérimental de C.A.P.E. EEC Report 185 February 1986
Clarke, L.G. (1976) Operational Evaluation of Colour Displays in Air Traffic Control EEC Report 102 November 1976
David, H. (1981) Study of a Colour SDD Display in an ATC Context EEC Report 140 July 1981.
David, H. (1984a) Background to the use of Colour in ATC Displays EEC Report 170 February 1984.
David, H. (1984b) Use of Colour in Electronic Data Displays (Preliminary Study) EEC Report 171 February 1984.
David, H. (1986) Use of Colour in Electronic Data Displays (Second Study) EEC Report 199, September 1986.
David, H. and Noonan, C.M. (1983) Potential Measures of Strain on Controllers EEC Report 186 August 1983.
Englmeier, H.P., Weis ,H., and Warns, O. (1980) Colour Television and Plasma Displays: Development and Integration into Experimental ATC systems. EEC Report 133 April 1980.
Hart, S. G. (1987) NASA Task Load Index V1.0: Paper and Pencil Package. Human Performance Research Group, NASA Ames Research Centre, Moffet Field, California.
Krella, F. (1986) Experimental Use of Artificial Intelligence for (future) ATC systems. EEC Technical Note 15/86 September 1986
Krella, F. (1989) ARC 2000 Scenario, EEC Technical Note 6/89 April 1989
Prosser, M. and David, H. (1988) Real Time Simulation in Support of the ODID (Operational Displays and Input Development) Group. EEC Report 215 April 1988
Prosser, M. and David, H. (1989) Second Real Time Simulation for the Operational Displays and Input Development Group. EEC Report 226 May 1989
Reid, G.B., (1987) Subjective Workload Assessment Technique (SWAT) : A User's Guide. Harry G. Armstrong Aerospace Medical Laboratory. WPAFB Dayton Ohio April 1967.

Schmidt, P. (1990) Test Bench for Controller Workstation Experiments, EEC Technical Note (In press).
Tumelin, J.C. (1990) ASTA EEC Technical Note 2/90 January 1990
Vanwonterghem, K. and Verboven, J. (1988) Brain Activity in a simulated ATC Task. EEC Report 219 August 1988.
Vanwonterghem, K. and Rabit, M. (1989) Activité Electrique Cérébrale et Charge Mentale Chez les Controlleurs du Trafic Aerien. EEC Report 228 August 1989.
Walraven, J. (1987) Colour Coding for ATC Displays in ATC. EEC Report 212 March 1988.
de Weert, C.H.H. (1987) Advisory Report on the Introduction of Colour in ATC. EEC Report 200 May 1987.

Meeting the ATC Challenge through Simulation

W. K. Sandiford

Air Traffic Control Evaluation Unit, Civil Aviation Authority, UK

Objective of this paper

This paper briefly describes the facilities used at the Air Traffic Control Evaluation Unit (ATCEU) for assisting airspace planners and decision makers to make sure that airspace capacity can keep pace with forecast demand.

Background

Air traffic in United Kingdom airspace is managed by the National Air Traffic Services (NATS), an organization jointly responsible to the Civil Aviation Authority and the UK Ministry of Defence. The ATCEU is a branch of NATS. The Unit is based at Bournemouth Airport in the south of England.

The Unit is established to evaluate and develop techniques and equipment for future air traffic systems; to investigate methods of control necessary to manage the growth in air traffic density; to undertake initial testing of new radar installations; and to collect and analyze data concerning aircraft flight profiles, including operations into and out of major airports in relation to minimum noise routes and flight paths over particular localities.

The Role of Simulation

As the number of aircraft under control increases, and they extend their performance capabilities, new methods of control using more sophisticated equipment are required to ensure a safe, reliable, and timely service. More automation is being steadily introduced to relieve the workload on human controllers and to improve their traffic handling capacities.

But, of course, all developments must first be thoroughly evaluated, so that their usefulness and cost effectiveness can be assessed. It is the task of the ATCEU to carry out such evaluations, and simulation plays a major role in these studies. Conclusions arising from simulation studies must obviously be treated with caution because simulators are not the "real thing," whatever level of sophistication and realism may be employed. Nevertheless, it is the

NATO ASI Series, Vol. F 73
Automation and Systems Issues in Air Traffic Control
Edited by John A. Wise et al.

UK experience that major changes to ATC equipment, airspace sectorization, and ATC procedures need to be evaluated thoroughly to reduce the risk of failure on implementation.

Fast-Time Studies

A relatively quick appreciation of the problems and benefits to be expected from new ATC structures can be gained by fast-time modelling. A fast-time model is a computer program which represents mathematically a given part of the ATC system and the associated ATC procedures. Samples of air traffic can be processed through the model, which calculates resultant aircraft movements. These can then be analyzed to indicate points of congestion and levels of loading within the system. The effects of alternative procedures can be compared by re-running the model.

Several examples of these models are to be found in the Directorate of Operational Analysis (Chief Scientist's Division, CAA). They include taxiway simulations, runway simulations, sector capacity models, and airspace congestion models. One such model (Richmond, 1988) attempts to estimate the workload on the controller, and so predict the traffic handling capacity of individual airspace sectors. Another (Richmond, 1989) represents the main features of the whole UK ATC system, and attempts to estimate the overall achievable throughput against different levels of traffic demand and sector flow constraints.

The ATCEU has developed a fast-time model to examine the impact of airspace and ATC restrictions on sector loading, holding stack utilization, and potential conflictions within defined regions of the UK airspace (Johnson, 1986). Further development is underway to incorporate a controller workload estimator. An important feature of the model is that it is fully harmonized with the ATCEU real-time simulator. Thus, a sequence of relatively quick applications of the fast-time model can prepare the way for a real-time simulation of the more promising options.

Real-Time Studies

The ATCEU mounts real-time simulations to examine plans for significant changes to ATC systems involving new equipment, modified control room configurations, and revised control procedures. The essential feature of a real-time study is that it will involve the active participation of professional controllers, usually on loan from the field units. These simulations fall into the following four broad categories.

Test-bench simulations are small-scale studies of one specific aspect of the ATC system, such as the merit of using color when displaying flight plan data electronically, or to assess the suitability of speech digitizers to support ATC voice communication via satellite.

Development simulations enable planners to refine complex proposals before submitting them to a formal evaluation. These development simulations allow changes to be made to the controller seating positions, airspace allocations, and ATC procedures during the course of a series of exercises.

Evaluation simulations are formally designed, controlled, and analysed, to compare in detail a proposed new system or technique with current operational practice. An evaluation aims to extract valid comparisons. The exercises to be played are carefully planned at the outset, and no changes to the program are allowed once the simulation is underway.

When major changes to operational practices have been evaluated by the Unit, it may be necessary to make the real-time simulator available to train operational controllers in the new techniques.

Fast-Time Simulation

The ATCEU fast-time model has given ATC planners an early insight into benefits associated with a radically new airspace in the southeast of England. The new airspace will provide each major airport in the southeast with its own separate "tunnels in the sky" for arriving and departing traffic. Projected traffic loadings and potential conflict points were noted, and the results helped to refine plans that were subsequently tested by real-time simulation.

The model has recently contributed to a study of the effects of providing additional runway capacity in the south of England. Implications for airspace utilization were assessed, and results will influence major policy decisions on how traffic levels in 2000 are to be accommodated.

Real-Time Simulation

The most important function of the ATCEU is the development and evaluation of existing and future ATC systems by real-time simulation. The complexity of these simulations is such that the successful completion of each simulation project requires effort from a mixed-discipline team involving project managers, air traffic controllers, telecommunication engineers, software specialists, human factors analysts, and scientists. Seven large-scale real-time simulations were mounted in 1989/90.

Operations within Manchester Sub-Centre were simulated to investigate the practicality of placing a support controller alongside each radar controller. Individual responsibilities were examined to achieve an equitable balance of workload, and to identify requirements for fundamental changes to ATC procedures, sectorization, holding areas, and aircraft routings.

Another simulation helped to develop new procedures for a new sector controller position within the present London Terminal Maneuvering Area. A further simulation was provided to enable controllers to be trained in these new procedures prior to implementation in spring 1990.

A series of real-time simulations were used to study various aspects of the plan to replace the London Terminal Maneuvering Area by a radically new airspace.

An important feature of many of these simulations was the need to evaluate the ergonomics of proposed new vertical radar furniture and the use of electronic devices to assist the communication and coordination between controllers.

Summary

The ATCEU provides a support service to NATS Directorates, primarily in the field of development and evaluation of ATC systems and facilities. Simulation techniques are fundamental to these studies.

Fast-time simulation is used in preliminary feasibility studies to gain a relatively quick insight into potential problems of overloading and loss of separation.

The Unit is becoming increasingly involved in testing and developing new ATC sub-systems and is devoting more time to test-bench activities.

System development and evaluation requires a much higher level of realism, in which large scale real-time simulations are used. These involve a mixed discipline team, including professional controllers from field units; human factors experts from the RAF Institute of Aviation Medicine (Farnborough); and substantial staff, technical, and computer resources at the ATCEU.

References

Johnson, A. E. (1986). A fast-time model comparison of Mediator and CCR Operations in the LTMA.ATCEU Memorandum 126.

Richmond, G. C. (1988). An interim description of the DORATASK methodology for Assessment of Sector Capacity DORA Report 8818.

Richmond, G. C. (1989). Outline description of the CAA Air Traffic Simulation (CATSIM) model. DORA Communication 8904.

Human Attributes, Representatives, and Requirements in Relation to Automation

The Aims of Human Factors and Their Application to Issues in Automation and Air Traffic Control

P.A. Hancock

Human Factors Research Laboratory, University of Minnesota, Minneapolis, Minnesota, USA

Preamble

The primary objectives of air traffic control have been described as "the safe, orderly, and expeditious flow of air traffic" (see Hopkin, 1991). At a fundamental level, this represents the aided retention of object dispersal in four dimensional space-time. As the objects under air traffic control are not within the range of "humanscale", (Hancock & Chignell, 1990) the recognition and manipulation of such objects can only be achieved via the use of technological prosthetics. Three major constraints are imposed on this general case. First, objects are limited with respect to resources; second, they are required, at some point, each to occupy essentially a common spatial location; and third, the objects themselves are under volitional control and can act independently. Separation distance covaries with object density, as does demand on communication capability and control action. The process by which such objects are controlled itself represents a record of iterative evolution to which proposed forms of automation will add an additional layering. The impact of such automation on the human operators within such a system is the major subject of this paper.

There is considerable literature on the questions of human factors as related to air traffic control (Hopkin, 1988) and it would be of little benefit here simply to re-iterate what has been said previously. Instead, this paper examines the fundamental aims of human factors and within a general framework, identifies critical commonalities subsuming the function of many complex systems. Principal among these are questions which are central also to an understanding of human actions. These include the problem of representation, intention, and the substitution of metaphor for direct experience. In understanding such questions, the role of automation as a prosthetic is examined. The emergent and nominally "preferred" role is in stark contrast to real-world systems, whose motivation for development is primarily technical and financial in nature. In illustrating contemporary incompatibility between human and machine, the paper examines potential, preferable, and probable solutions to the question of future interaction. Particular application to the questions of air traffic control is highlighted.

Introduction

Much has been made recently about the concepts of chaos and non-linear systems, and how an understanding of the patterns intrinsic to such systems provide a rich and varied

NATO ASI Series, Vol. F 73
Automation and Systems Issues in Air Traffic Control
Edited by John A. Wise et al.

understanding of multiple phenomena, ranging from the weather to human cognitive activity (Gleick, 1987). It has been asserted that the behavior of such systems, while not completely random, is also not completely predictable due to the acute sensitivity to minor variations in initial conditions, where the establishment of initial conditions is constrained by Heisenberg's principle of uncertainty (Thompson & Stewart, 1986). While the latter may be the subject of philosophical contention, the present status of practical knowledge confirms that predictability cannot as yet be achieved. Order opposes chaos, and stability is the enemy of uncertainty. Hence, when a system is designed, it is purposely created for stable states of operation.

Like nature, the design engineer is faced with a dilemma. How can a system be constructed so that it is able to oppose the vagaries of an uncertain environment while continuing to perform its desired function?

The history of technology is a handbook of the ways in which engineers have attempted to solve this question. Examples abound of strategies such as surrounding the system with layers of protection, or making the system itself partially adaptive, both of which are taken from nature's own solutions. Many systems require that the unique capabilities of a human be employed, and this is one unpredictable element that the engineer may view as "the enemy within."

To eliminate the undesired component of human spontaneity and unpredictability, the engineer of the past disenfranchised the worker, emphasizing repetitive motions and tasks of essentially negligible cognitive content. Problem solving endeavors such as breakdown and maintenance were separated from the line worker and were the province of a different group of individuals.

Both time and engineers themselves have defeated this sterile conception of the human operator. Any task as mindless as that described above has already become a prime target for automation. However, the role of the human in current systems retains much ambiguity. The changing nature of this role has been observed ad nauseam, but what actually emerges as the primary role for the human is still unclear. Are we further along the path toward an automated utopia, where the engineer can finally, and thankfully, usher out the last human operator from the system, close the door, press the on switch and retire?

For those with that form of vision the answer is still no. Humans retain a vital role in the vast majority of contemporary systems, at least on the basis of their abilities to respond to the uncertainties noted above. But the role the human operator is constrained to adopt today is often anathema to a variety of human performance characteristics. Operators are regularly required to sit for several hours watching the condition of a system which varies little in status, a task at which humans are notoriously poor (Parasuraman & Davies, 1976; Mackworth, 1950/1961; Warm, 1984). Further, in many complex operations, operators are required to change from a passive monitoring mode to assume active control of a system of whose contemporary status they may have little fundamental conception. Essentially reverting from audience to action.

In short, the human is new and will be required to act specifically in non-stable conditions. The processes of problem-solving, and error identification and correction are emphasized in these transient phases rather than steady-state modes of operation. Such conditions are often time-restricted and are attended by high levels of stress. How human operators respond under such conditions is a critical question (see Hancock & Warm, 1989; Tattersall, 1990).

Human Factors: Definition and Aims

Typically, definitions through historical references to human factors start with the first human use of tools and quickly skate several millennia to the late twentieth century and the nascent efforts of systems integration stimulated by the necessities of global conflict. However, it is important to examine this putative foundation more closely.

First, to constrain the use of tools to humanity alone is incorrect. Many animals use naturally occurring prosthetics to accomplish tasks, and some even modify such primitive tools to suit their required function. The human is certainly not the only animal or organism which orders its environment to promote self-welfare. Although Louis B. Leakey named a branch of the human family tree (Homo Habilis) after this tool-wielding capacity, it cannot be legitimately used as the differentiate for human factors, unless we choose to include all such tool users.

The essence of the difference lies not in the local and constrained manipulation of nature but in two fundamental ideas which also underlie the foundations of science itself, namely, the conception that nature can be controlled, and further, that there are principles that allow control by transcending individually-specific conditions. It is this affirmation of control, the expectation of progressively greater control, and the use of induction through reality-matching that lies at the very heart of Bacon's conception of the scientific method (Eiseley, 1973). In essence, the better differentiation is not as wielders of tools, but as wielders of what has become formalized as science. It is more than worth noting here, that the scientific method as conceived by Bacon was vitally, and "above all things, for the uses of life." While much of technology extends knowledge, it is human factors that critically focuses on science for the uses of life. In the long search for the leitmotif of human factors, the observation by Bronowski (1978) typifies the frustration of the user with an unknown and arcane technology, supposedly designed to assist, but which in reality resists. He noted that:

> Science and society are out of joint. Science has given to no one in particular a power which no one in particular knows how to use. Why do not scientists invent something sensible? Wives say it every time they stub their toe on the waste bin, and husbands say it whenever a fuse blows. Why is it the business of no one in particular to stop fitting science for death and to begin fitting it into our lives?

Although Bronowski's comment was voiced over technology of the 1950's, we hear continued and equally strident cries for help over four decades later. The following is taken from a newspaper, earlier this year, it reads:

> Infuriating gadgets:
> Sir, May I add my own "design" complaints to those of Geoffrey Wheatcroft. The off-set control panel on moderm cars is a maddening example of overdesign leading to front seat passengers having to crane necks and thrust elbows to operate the radio: a flurishing gear change by the driver produces a set of grazed knuckles.
>
> Infuriating in the home is the grill pan with the "easily detachable handle" - especially when tilting the pan to remove excess fat. But my most recent cause for despair was the attempt to buy a comfortable garden chair with hinged foot rest. There were dozens of them in all makes. But the foot rest can be raised only when the chair tips backward so that to read a newspaper with your feet up you have to be

lying on your back. When challenged, shop assistants say feebly: "They're all made like that now." Why?

Such comments reinforce the necessity for human factors and ergonomics, while simultaneously reflecting their importance in percolating into everyday society.

A Descriptive Structure

The question of the definition and aims of human factors is inextricably linked with the basic motivations of human action. To understand this primary linkage, we have to understand the changing scales of human capability. This can be accomplished by plotting a representation of the scales of time and space. In Cartesian coordinates, we can view our environment as ranging spatially from the very small to the very large, and temporally from the very brief to the very prolonged.

For anyone who has seen the spatial representation of size in *Powers of Ten*, this range effect is clearly illustrated (Morrison, Morrison, Eames, & Eames, 1982). In the center of the axes at 10° and 10°, we establish an arbitrary intersection. This is an arbitrary choice of location, as scales must always be specified relative to alternative entities. Here, and in common with human observers throughout the ages, we have put ourselves at the center and referenced everything else with respect to our own size and perception of immediate duration. From this perspective, at the center of these axes we have humanscale or the ecological scale, as it has been termed by Shaw (see Shaw & Kinsella-Shaw, 1988). We may visualize humanscale as an envelope centered at the intersection of these axes whose boundary connotes the region over which humans can exercise unaided influence.

For example, on the spatial scale of the very small, unaided, we may perceive and manipulate objects some millimeters in size. On the time scale, the lower boundary is represented by events that are separated by fifty to one hundred milliseconds (Stroud, 1955), a period that is projected to represent the "now," the perceptual moment, or in the terms given by Minkowski (1908), the specious present.

The boundaries at the upper end are somewhat more ambiguous. For simplicity, if we consider actions which occur with a brief time duration, a human might throw a spear or javelin and exercise unaided influence over an area of approximately one hundred meters in radius. However, with the aid of prosthetics, it is clear that the range over which a human may exert unaided action is vastly increased.

Therefore, surrounding this tiny region of humanscale is the vast expanse of space and time over which we exert influence when we act in conjunction with the technology we have created. This latter region is labeled prostheticscale. Spatially contemporary boundary markers to this prostheticscale are represented by elementary particle manipulation, at the lower spatial bound to Voyager and its physical presence beyond the edge of the solar system, at the upper end of the spatial scale. It might be argued that humankind has exercised influence over a much larger range when we consider the information intrinsic to radiowaves that have left this planet within the last century. Our choice of physical manipulation as a criterion is on this basis one that may be challenged; however, as this simply extends the envelope by some multiple, it is not a question of particular moment.

On the temporal scale, we have become familiar with nanosecond-based measures

(Rifkin, 1988) at the lower boundary, while storage and dynamic knowledge representation of expert systems promises the use of technology to preserve at least a small portion of our personal selves beyond our actual lifetime. Further, according to allometric scaling, humans should live on average to 23 years of age (Schroots & Birren, 1990; Yates, 1990). Already, our use of technology prolongs our expected life-span more than three-fold. Also, there is a trend with improvements in medical facilities for individuals to live longer.

Outside prosthetic scale, is perceptualscale. This represents the boundaries of what we may perceive. We could, as with action, differentiate perceptual scale into aided and unaided perception. Unaided, the human observer can see objects down to quite small sizes and from their actions infer the presence of even smaller particles, as for example in observing Brownian motion. However, it is the unaided eye directed toward the night sky that perceives the vastness of large scale spatial representation.

We should not neglect at this juncture, to point out that an individual looking out into space is also looking back in time. The interdependence of space and time has been recognized by philosophers and physicists for over two centuries, and more recently combination (space-time) has been explored with respect to human behavior (Hancock & Newell, 1984). With the aid of contemporary technology, our range of perception is vastly increased. The resolution of the Hubble telescope promises to expand our perceptions close to the boundaries of the known Universe, and our "perceptions" of smaller and smaller particles of matter is further expanded by tools such as the superconducting supercollider.

One of the major rationales for exploration is the desire to exercise influence over the things we are able to perceive. Words are small substitute for the illustration of William Blake. In his woodcut "I want, I want," he expresses the quintessence of the human desire to reach beyond such continual restrictions. We should note that Blake's example is taken from the large scale of space, where our manifest inability to exercise influence over far distant objects has been clear for some millennia.

Nor should we ignore the cultural bias of such desire, in that many cultures beyond the Occidental World express little interest in physically manipulating elements beyond their immediate reach.

With respect to this endeavor, technological innovation serves to expand the envelope of aided actions and so enlarge the area we have labeled prostheticscale. In so doing, however, technology often and purposefully serves to expand the region of perceptual scale. It is suggested that there is a continual tension between these two regions, as humankind seeks to physically control what they can perceive. It is indeed the purpose of technology to expand these envelopes.

However, in addition to the tension created by the dissonance between regions of perception and action, the further the envelopes expand away from the relatively fixed humanscale, the further divorced are actions from everyday experience. As we progress in our efforts to influence the very large and the very small, we begin to rely more on metaphorical representations of entities which we cannot sense directly and with which we have no direct empathy on a spatial and temporal basis. Time provides us with a most pertinent example here, as indicated by Rifkin (1988). Supposing two supercomputers of the near future are exchanging information and a user interrupts them with a line code of ASCII such as the enquiry "what are you doing?" In the time taken to enter such an enquiry, such computers can exchange more information than is contained in the sum total of spoken language for all time. Similar examples abound on the spatial scale. The point is simply that we are progressively operating more and more with metaphor, and the metaphors are becoming clearly more inadequate as technology progresses.

Technology is driven by forces which require that the boundaries we have identified are progressively enlarged. For market purposes, systems must be faster, they proliferate in physical size, and when connected with other units they grow in complexity. Often such progress is relatively "mindless," that is, driven by forces with few concerns whether such "progress" is beneficial.

One fundamental rationale for human factors is the dissipation of tension between ranges of perception and action, but a further one is to assume a leadership role in directing technological innovation. Often human factors is characterized as facilitating human interaction with machine systems; however, a more fundamental role is in the active direction of systems technology. The price of this dissonance is the spectacular system failures which adorn the news media. Recent insights suggest that micro-ergonomic manipulations can have only a limited effect in ameliorating the tension created. It is at the level of organizational structure that a more profound impact may be effected.

Yet the paradox remains that organizations are assemblages of individuals who each have disparate and sometimes opposing goals. The systems approach may be a vehicle through which to integrate the concept of adaptability in reconciling the internal pressures with the external tensions mandated by my above stated premise. Briefly, both individual and organization seek to increase their range of functional influence to reach desired goals, set future goals, and expand the horizon of future feasible goals. Each employs adaptive strategies to reconcile possible goal paths with imposed constraints. The critical role of macro-ergonomics is in harmonizing individual and organizational goals, by generating a functional synchronization between each, so that composite task and goal solution paths can be readily recognized and pursued as dynamic constraints act to continually change both personal and organizational solution spaces.

Vast efforts in all societies are given over to the pursuit of greater and greater technological capability. By comparison, efforts to harmonize and integrate such advances with existing human capabilities are virtually non-existent. Micro-ergonomic manipulations facilitate individual interaction with particular machine systems and hence, at one level, act to ease local dysfunction. However, it is clear that such changes are, in and of themselves, insufficient and potentially impotent with respect to organizations. Our failures are marked by the increasingly spectacular breakdown of systems of larger scale and energy. Yet we have already constructed systems that cannot be shut down and systems that must not fail. Our inability to recognize these progressive symptoms may be reprehensible, but our failure to act upon them will certainly prove fatal. Hence, human factors in all its forms is not merely a cosmetic appendage to system design and operation, nor is it an advertising or marketing ploy to boost sales and acceptability. It is the decisive factor in our future ability to explore our environment and one key to collective survival.

The Problem of Representation

I have identified a number of problems. One of the more important, but least tractable is the question of representation. Essentially, the question of representation is a "how" question. How can any entity hold, in a finite storage space, information about potentially infinite combinations of instantaneous conditions and combine these with all previously stored information and expectation about future conditions? Further recognition that the storage medium itself is in a constant state of flux has proved an additional stumbling block for theoreticians. However, concerning human capabilities we have two important clues. Representation is necessary for successful behavior, and complete representation is probably beyond any limited entity. But we are both limited and successful entities, and therefore, there is a solution to the problem.

Theoretical answers to this question range from the postulation of a large storage size, or economies based on storage (e.g., plans, scripts, schemae), to more recently, postulation of a degree of information intrinsic to the environmental display. In this respect, science itself is the search for pattern in chaos and "the distinction of the possible from the actual." As suggested above, recent ecological approaches to human perception seek to reduce the dependency on internal representation and computation by fixing not simply order, but information as an environmental property.

In construction of displays for a remote sensing system, like air-traffic control, the design criteria have emanated from historical analog representations, emerging mainly from the industrial revolution, which are backed by the transformational basis of early experimental psychology and more recently the information processing paradigm. Early human factors research focused on issues such as compatibility in controls and displays, anchoring effects, and scale effects, each of which amended but never questioned the basic fundamental design. More recent displays use such contemporary technology as electronic representation but still persist largely with alpha-numeric coding for information content.

It was originally conceived that the process of sensing itself, particularly for vision, was one of information assemblage, based principally on the transformation of physical energy. Little wonder that displays founded on such a principle simply extended the concept of synthesis and expected the observer to perform the correct interpretation of the physical signals provided. However, the introduction of transformation, and particularly the use of metaphor, allows progressively greater interpretational latitude. This increase in latitude is benign in some realms such as art, where the artist may actually seek such breadth. However, in more veridical pursuits such as air-traffic control, displays are meant to represent a single unambiguous status, and thus increased latitude only increases potential misinterpretation.

In perception at humanscale, interpretation is largely unambiguous. Although we may be deceived by visual illusions, they are largely synthetic and rarely occur in our "natural" visual world. Partly as a result of this effect, the new thrust toward an "ecological" psychology refers to this effect as "direct perception." The comparative approach in human factors is the notion of direct displays. These are not simply displays that rely more on visual icons, and the desktop metaphor, as developed in the Apple® Macintosh™ interface. Rather, it is a move toward the application of ecological principles to systems in general. However, this is a global specification and does not indicate exactly how such displays might be constructed (Flach, 1989).

The Problem of Intention

If representation is a "how" question, then intention is a "why" question. What are the basic goals toward which action is directed. In reality, the question of intention or motivation for action is constrained by what can be conceived and, by inference, is an extension of what is perceived. However, due in part to reference to cause and effect, the notion of intention is often considered to precede actions.

Intrinsic to an argument of intention is the notion of control, "the acceptance of responsibility for the process and the product which results from intended actions." While we have substituted mechanical sources of energy for our own muscular efforts, and have gained in efficiency proportionally, the human has to date been the primary source of intentionality. Typically, automation is perceived as a substitute for the human in the process, but the prime source of control for intention, and the evaluation of product, is still thought to reside in the human observer.

Unfortunately, in numerous complex operations, particularly those without clear deterministic goals, the notion of intention is as much linked with process as with product. The worry of automation is then two-fold. First and foremost, it disenfranchises the human operator. Second, it becomes problematic as to whether the human can regain direct control of the operation of any system which has turned toward automation as an answer to demand.

While Hopkin (1991) examines ranges of these options, typically several such options become progressively infeasible in real-world operations. Removal of control brings numerous problems. First, lack of autonomy, "going along for the ride" as pilots put it, is a high source of stress. Indeed, the combination of low control and high demand is the most destructive working condition for the human operator (Hancock & Warm, 1989). Also, removal of control without removal of responsibility is both a dangerous and unfair eventuality. With progressive automation, not only do operators begin to lose skills, after some automated design iterations under automation, their skills become obsolete anyway.

Hence, we stand at a point of major decision. The original intention is the safe and efficient passage of aircraft in an air space. That is the predetermined goal of the system. The individual operator is then instrumental in achieving that aim, but does not alter the fundamental intentionality of the system.

Why then is there a problem in full automation? We still retain a fundamental distrust in automation. Although in suspecting that breakdown is still possible, we do need to consider that humans also breakdown and fail. The extension of this concern is what happens to subsume automated functions when they fail. The redundant belief is that humans can recover such failures. From a different perspective, it should be noted that we still underestimate the complexity and capability of human operators, particularly in solving non-algorithmic problems. However, what we have failed to do effectively is to place these unique resources in their most beneficial circumstance in a systems sense. The overemphasis on speed of response has appeared to make humans redundant before their time. What has not been examined in sufficient detail are the intrinsic rewards of work in a complex environment, and how to integrate such rewards into system function.

In aircraft control, the problem of chronic underload and uncontrolled oscillations in load is one that is assuming increasing importance. Perhaps the ultimate example of underload comes from the now classic work of Norman Mackworth on vigilance. Having observed subjects who had scored a series of target "misses," Mackworth entered the experimental environment only to find the subjects asleep. Mackworth is rightly credited with the discovery of many of the central factors in vigilance and sustained attention, but it is this apparently mundane finding that he failed to emphasize as the key characteristic of vigilance. No other task combines the elements of boredom and demand in such a manner as to tax the exposed performer (Hancock & Warm, 1989). With little extrinsic or intrinsic motivation, there is little surprise that one appropriate strategy in such conditions is sleep. (Parenthetically, Mackworth's subjects held Naval ratings and risked punitive action for their failure to comply, which is indicative of how seductive sleep is in such conditions. We might speculate on the number of aircraft lost when all flight deck personnel were asleep on a long haul, early in the morning, following extra duties all-round.)

The central point is that the inability to integrate an understanding of such human characteristics is clearly punished at some point in time. Perhaps the basic faculty of any individual is self-determined intention. In the name of efficiency, we have removed this capacity from the operator, and for the sake of efficiency have given them displays that remove them from direct experience of the objects they are required to control. This is a recipe for disaster.

Systems and Goals

From a theoretical perspective, the only essential goal of a system is self-perpetuation. However, for the purposes of the present argument I shall deal with a subset of this generality and frame my discussion around current and near future goals.

Goals may be defined as desired future states of the system. The way in which system goals can be achieved is critically dependent upon the structure of that system. An example can be seen in strategies to control complex systems such as aircraft themselves. Two general strategies are rule-based and knowledge-based response. In rule-based response, look-up tables describe the scenario and appropriate responses are recommended. This approach works well when routine demands occur. However, under novel, and particularly emergency situations, such preset responses are not contained in the series of rules, and potential disaster can occur. In essence, not all states of a complex interactive system can be specified before operation. Knowledge-based action requires operators to "know" the system they are controlling. The expectation is of superior performance in situations for which rules are unavailable, such as unforeseen emergencies.

The increasing size and rapidity of complex systems presents constant challenges for each strategy. The supposed answer lies in the process of automation. Many of the specifics of automation of air-traffic control have been identified by Hopkin (1991), who examines the broad sweep of potential forms of interaction.

Increasing flexibility of systems make them more adaptive to uncertainties but may make them somewhat less efficient under steady state operation. Of course, the aim of design is to produce both flexible and efficient systems.

Rigid structures, such as a strict hierarchy, contain clear advantages and limitations. Among the advantages is speed of response to repetitive deterministic input which requires stereotypical output. The principal limitation is the reliance on the single source of command for setting goals and ordering the activity of subsidiary components. In *The Art of War*, Sun Tzu, pointed out the importance of neutralizing an opposing general, as without orders from the single commander, a rigid hierarchy is essentially powerless to respond. This failure is compounded because in a strict hierarchy there is a unidirectional flow of information from the top down throughout the hierarchy.

This particular weakness is counteracted in a heterarchical organization by designing information flow which is purposely bi-directional, in that both top-down and bottom-up flow becomes possible. As the flow of information becomes less constrained, response becomes less stereotypical and the number of potential paths that can be generated toward a goal state proliferate. The trade-off between flexibility of response and the rapidity of reaction is one that recurs throughout the present argument concerning the application of automation to air traffic control.

A third mode of organization, the holarchy, has the character that any node of the system may assume command, depending upon the nature of the momentary task demands. For those familiar with theories on brain organization, it is clear that these various forms have been sequentially proposed as the manner in which perception-action systems are controlled. Amendments to structure may divide such a framework into a collection of such holarchies where specific demands are met by flexible action.

Again the problem of temporal integration resurfaces, as it does, for example, in the operation of current generation parallel processing (connection) machines. Solutions to the problem of conscious, sentient action have been proposed, based upon this integration of response from essentially unconscious component elements (Minsky, 1985). A consequence is the hope of conscious action from multi-processor machines. As a parenthetical note, we should recognize that the degrees of explanatory freedom expand with each of these sequential conceptions, and so the potential to explain resultant outcomes increases. The elegance of each theoretical construct is thereby weakened. Only careful examination and imposition of constraints can divide the actual from the possible in such conceptualizations.

Potential and Probable Solutions

The potential solutions to the numerous questions raised in the above brief examinations of a number of issues are as wide as imagination can make them. However, the probable solutions are far more constrained. For example, probable solutions must be framed in terms of an evolution of the present system (a particularly pertinent constraint). Further, probable solutions must be cost effective, and again the practical motivation of cost will dominate concern when compared to human issues such as operator contribution and satisfaction, although it is clear that the latter elements are critical to the operation of a safe and efficient system.

Despite the range of solutions identified by Hopkin (1991), two radical forms of solution suggest themselves. First, fly less. The demand is driven by an ever increasing need for air transportation. Yet we are clearly aware of the global effect of ever increasing demand for technology in a ecosystem of limited resources (Moray, 1990). Perhaps, with superior methods of electronic communication, international and national travel for business can be curtailed. Second, alternative routing for freight traffic to less crowded locations could restrict the need for ever more technology in air traffic control.

Such questions are directed right to the heart of what is expected out of life. However, it is an alternative that must receive serious consideration. The method of interaction that I have advocated in previous works (Hancock & Chignell, 1987; Hancock & Chignell, 1988), is one that seeks mutually adaptive capabilities, mediated principally through intelligent interfaces (Chignell & Hancock, 1988; Hancock & Chignell, 1988). The principles governing such mutual adaptation and their instantiation are contained in the referenced material.

Summary and Reprise

Air traffic control, like many other process control sequences, is the subject at the physical level of convergent evolution. With prototypical workstations emerging and interfaces based on metaphor, such as the desktop, becoming so common, it becomes progressively more difficult to distinguish between the work stations of a nuclear power plant operator, a secretary, and an air traffic controller. Unfortunately, these forms of work environment have begun to become as sterile as the production lines of early manufacturing. The difference, however, is that prolonged repetitive actions were manifest and clearly designed to be "mindless" on behalf of the worker. Today, we couch such interfaces with high technology and use terms such as "decision-aiding" to cover the progressive divorce of the operator from the work. But more insidiously, we have taken not only physical action but self-determination from tasks in which operators have traditionally prided themselves on their work skills. Added to this is the complexity of remote representation and the intrinsic use of metaphor, so that the next generation of work stations will become not only harder to distinguish from each other, but harder to distinguish from popular video games.

As automation increases, operator workload profile begins to emphasize periods of prolonged underload into which are injected transients of untenable overload. Perhaps, if individual loading is seen as a major criterion, air traffic flow might be assigned as a pilot problem. Or, if we must have ground-based operations, we might look forward to air traffic controllers working as operators of remotely piloted vehicles. Each of these eventualities seems only a distant possibiliy. However, as with many contemporary process control operations being forced to face similar increases in demands, the response of gentle progress as usual is insufficient. As noted in *Time*, July 18, 1988:

> The central question is whether technology may be pushing the fallible humans who operate it beyond their ability to make wise judgments instantly on the basis of what, with even the most sophisticated systems, will often be ambiguous information.

Our purpose is to answer this question, and a global appeal to greater automation will not serve.

Addendum

In the course of the Institute many fundamental assumptions were purposely examined and assessed in the light of the changes that automation might make. One principle that received little attention was the question of object (aircraft) density. The assumption being that a progessive increase in density leads in some systematic fashion to an increased probablity of collision. This is essentially a thermodynamic statement in which object collisions increase with object velocity, where the later can be directly equated with density. However, in air traffic control we have more than Maxwell's demon, that is a sentient controller of objects. We also have intentionality and control within the objects thenselves. In automobile traffic there is no individualized dynamic controller (there is general control such as traffic bulletins) and yet even with rudimentary user rules and almost no exclusion from driving for most individuals (at least in the United States) remarkably few accidents occur (on a *relative* scale) for the density of traffic present. Thus, one critical research question is the role of human and machine intentionality in the process of "sky-packing" or progressive increase in aircraft numbers in limited air-space. Althought I have tried on a global basis to address this issue in the present work, I believe the fundamental assumption unnderlying "sky-packing" is a topic worthy of much greater investigation.

References

Bronowski, J. (1978). *The common sense of science.* Cambridge, MA.: Harvard University Press.

Chignell, M.H., & Hancock, P.A. (1989). Intelligent interfaces. In: M. Helander (Ed.). *Handbook of human-computer interaction.* Amsterdam: North-Holland.

Eiseley, L. (1973). *The man who saw through time.* New York: Scribner.

Fjermedal, G. (1986). *The tommorrow makers.* Tempus: Washington.

Flach, J.M. (1989). The ecology of human-machine systems. Technical Report, Engineering Psychology Research Report, University of Illinois, EPRL-89-12.

Hancock, P.A., & Chignell, M.H. (1987). Functional synchronization: A theory for human-computer interaction. In: *Proceedings of the Second Conference on Human-Computer Interaction,* Honolulu, HI, August.

Hancock, P.A., & Chignell, M.H. (1988a). Mental workload dynamics in adaptive inter-face design. *IEEE Transactions on Systems, Man, and Cybernetics, 18,* 647-658.

Hancock, P.A., & Chignell. M.H. (Eds.). (1988b). *Intelligent interfaces: Theory, research, and design.* Amsterdam: North-Holland.

Hancock, P.A., & Chignell. M.H. (1990). A human factors theory. Manuscript in preparation.

Hancock, P.A., & Meshkati, N. (1988), (Eds.) *Human mental workload.* North-Holland: Amsterdam.

Hancock, P.A., & Newell, K.M. (1984). The speed accuracy trade-off in space-time. In: H. Heuer, U. Kleinbeck, & K.H. Schmidt (Eds.). *Motor behavior: Programming, control, and acquisition.* Berlin: Springer-Verlag.

Hancock, P.A., & Warm, J.S. (1989). A dynamic model of stress and sustained attention. *Human Factors, 31,* 519-537.

Hopkin, V.D. (1988). Air traffic control. In: E.L. Wiener & D.C. Nagel (Eds.). *Human factors in aviation,* 639-663. New York: Academic Press.

Hopkin, V.D. (1991). Keynote address. In: J. A. Wise, V. D. Hopkin, & M. L. Smith (Eds.) *Automation and Systems Issues in Air Traffic Control.* Berlin: Springer-Verlag.

Mackworth, N.H. (1961). Researches on the measurement of human performance. In: H.W. Sinaiko (Ed.), *Selected papers on human factors in the design and use of control systems.* 174-331, New York: Dover. Reprinted from: Medical Research Council Special Report Series No. 268. London: HM Stationery Office, (1950).

McCorduck, P. (1979). *Machines who think.* New York: Freeman.

McPhee, J. (1989). *The control of nature.* New York: Farrar, Straus, Giroux.

Minkowski, H. (1908). Space and time. Address delivered at the 80th Assembly of German Natural Scientists and Physicians, Cologne, Germany, 21 September.

Minsky, M. (1985). *The society of mind.* New York: Simon and Schuster.

Moray, N. (1990). Application of human factors to identified world problems. Plenary Address to the Fifth Mid-Central Ergonomics/Human-Factors Conference, Dayton, OH, May.

Morrison, P., Morrison, P., Eames, C., & Eames, R. (1982). *Powers of ten: About the relative size of things in the universe*. New York: ScientificAmerican Library.
Parasuraman, R., & Davies, D.R. (1976). Decision theory analysis of response latencies in vigilance. *Journal of Experimental Psychology, 2*, 578-590.
Porter, R. (1988) (Ed.). *Man masters nature*. New York: Braziller.
Rifkin, J. (1988). *Time wars*. New York: Bantam.
Schroots, J.J.F., & Birren, J.E. (1990). Concepts of time and aging in science. In: J.E. Birren & K.W. Schaie (Eds.). *Handbook of the psychology of aging*. New York: Academic Press.
Sharit, J. (1988). Issues in modeling supervisory control in flexible manufacturing systems. In: W. Karwowski, H.R. Parsaei, & M.R. Wilhelm, (Eds.). *Ergonomics of hybrid automated systems*. Amsterdam: Elsevier.
Shaw, R., & Kinsella-Shaw, J. (1988). Ecological mechanics: A physical geometry for intentional constraints. *Human Movement Science, 7*, 155-200.
Stroud, J.M. (1955). The fine structure of psychological time. In: H.Quastler (Ed.). *Information theory in psychology*, Illinois Free Press.
Tattersall, A. J. (1990). Effects os system induced stress on total system performance. In: J. A. Wise, V. D. Hopkin, & M. L. Smith (Eds.) *Automation and Systems Issues in Air Traffic Control*. Berlin: Springer-Verlag.
Tzu, S. (1983). *The art of war*. New York: Dell. (Original Publication, c. 500 B.C.).
Warm, J.S. (1984). *Sustained attention in human performance*. New York: Wiley and Sons.
Yates, F.E. (1988). The dynamics of aging and time: How physical action implies social action. In: J.E. Birren & V.L. Bengtson (Eds.). *Emerging theories of aging*. New York: Springer.

Human Factors in System Design

Paul O. Day

Plessey Research, Roke Manor, Romsey, Hampshire, England

Introduction

It is clear that, as Europe moves into the last decade of the century, its Air Traffic Control system is in need of major updating and re-organizing, if it is not to become the limiting factor in the projected growth in air traffic across the continent (Jackson, 1990). That a major investment in equipment and training will be required is now widely recognized, but there are important technological and systems issues to be resolved. One of the most important relates to the role of humans in ATC and the extent to which certain human roles can, and indeed should, be automated.

In ATC the principle of safe operation has, quite properly, been the overriding factor at all times, and that in turn relates closely to the traveling public's perception of safe system design and operation. While technologists are confident in their ability to quantify the failure modes of even fairly complex systems — to determine on a statistical basis how often they may fail, and then build in appropriate back-ups to cope with a wide range of eventualities — it is often felt that the flexibility and adaptability of the human in the system is not only an important safeguard against the unexpected, but also a means of preventing an automated system doing something fundamentally unreasonable or inappropriate. Thus, it is felt that humans have the "common sense" and inherent reasonableness that machines still lack, and may be relied upon to exercise it most of the time. The resolution of the fundamental dilemma of what the human should do within the system, and what the (remainder of the) system should do for itself, has always exercised Human Factors engineers, to the extent that we have a jargon term for the process of deciding those roles — "allocation of function."

The difficulties of allocation of function become especially prominent in the redesign or updating of systems where there is some prospect of automating roles or duties previously undertaken by a human operator. Our first reaction is to design for a "fall-back mode," whereby the automated function can be re-allocated to the human operator if the system fails in its new duties. We then fear, quite properly, for the performance of the out-of-practice human performing this duty on rare occasions, and at short notice. Should we ensure that he maintains his performance through frequent practice? Should we periodically test her by suddenly thrusting these duties upon her, with no notice? Should we simply re-allocate the function to humans, and abandon the attempt at automation altogether?

It is all too easy to tie ourselves in knots with these and other questions, if we allow ourselves to focus too closely on the incremental design of systems, and the process of automation in minuscule steps. What is clearly required is the exercise of a proper design methodology, looking at the design of the system in its entirety — from initial requirements specification through to implementation and beyond — and including undertaking the

NATO ASI Series, Vol. F 73
Automation and Systems Issues in Air Traffic Control
Edited by John A. Wise et al.

allocation of function according to the relative abilities and limitations of the human and the other, electronic, elements of the system.

Structured Design Methodologies

Since it is often easiest to identify Human Factors problems in the design of a system at the lowest level of the human-machine interface (the so-called "knobs and dials" level), this level has received a perhaps disproportionate amount of attention over the years. While it is of course important to get the design right at this level, it is merely one of the levels at which the design of a system must be appropriate.

In this context, it is useful to view a user interface in terms of a structured hierarchy of levels, each concerned with user-system interaction at a different level of abstraction. This approach draws upon the concept inherent in the well-known ISO Open Systems Interconnection (OSI) reference model. In the Human Factors context, the layered approach to interface design has emerged over the last decade, and finds its roots in the Command Language Grammar of Moran (1981), the 4-level Design Model of Foley and Van Dam (1982), and Nielsen's Virtual Protocol Model (1986). The most recent work in this area (of which the author is aware) is being conducted within the European RACE (Research into Advanced Communications in Europe) program and while generic in nature, is being applied to the design of advanced, multi-media user interfaces for systems that will utilize the proposed European Integrated Broadband Communications Network (Byerley, 1990). In this context, the user interface reference model identifies four layers, termed respectively the Goal, Application Concept, Interaction, and I/O Device layers (See Figure 1 below).

While this work on user interface reference models undoubtedly points the way ahead, it is by no means as well-developed as the seven-layer OSI model of the telecommunications world, which is now the accepted medium of exchange whereby communications engineers around the world can discuss systems concepts. In the Human Factors arena, our use of these reference models is still at the level of research, and they are not yet applicable to the procurement of major, software-intensive systems at an acceptable level of technological and financial risk. If we are to serve the needs of those procuring major ATC systems during this decade, it is appropriate to examine the interface between the discipline of Human Factors and the structured, top-down design approaches now quite widely used in the software world.

Structured analysis and design methods (SADMs), to use the general term — impose discipline in the software design process, outlining the steps to be followed, the notations to be adhered to, and the techniques to be used. They have proved invaluable in recent years in assisting those involved in the software development process, and have helped the evolution of that profession from an essentially craft-based activity to an engineering discipline in its own right; we use the term "software engineering" to highlight that new status. From the human factors viewpoint, however, SADMs still have some shortcomings, as Anderson (1988) has pointed out:

- They typically provide no support at all to the crucial allocation of function process, tending instead to lead the designer to automate whatever can be automated, leaving the remainder to the human.
- They give very limited help in specifying the interface to the user, assuming

instead that the designer's common sense will take care of that part of the design process.
- They do not adequately recognize the iterative nature of the design process.
- They do not support user involvement in the design process, the value of prototyping, or the need for usability testing.

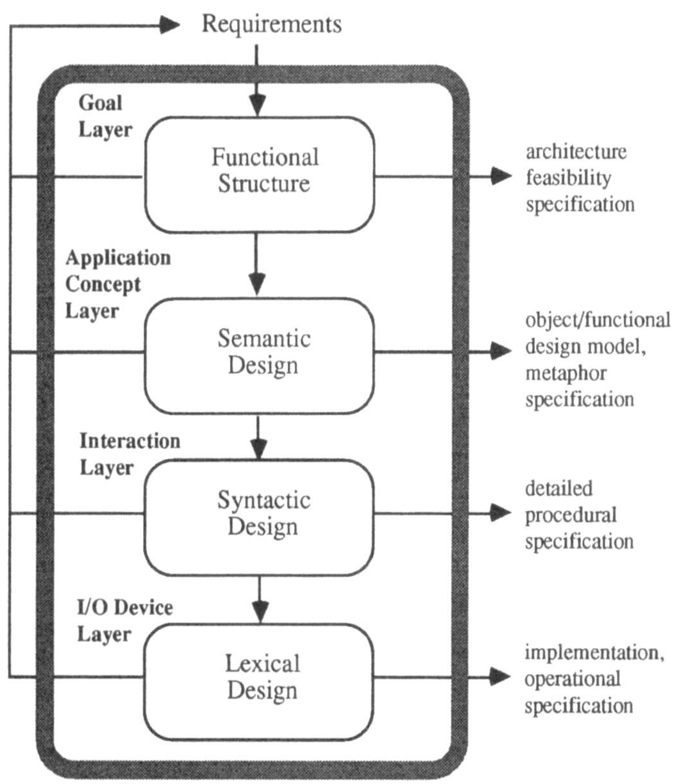

Figure 1. User-interface reference model(project Guidance)

These are important criticisms, and some Human Factors researchers are seeking ways to better match the needs of human factors to the major SADMs. Anderson (1988) concerns himself with SSADM (structured systems analysis and design method), the software design methodology required to be used by British Government contractors for (non-embedded) systems development. He specifies the additional Human Factors activities that should occur to supplement each of the design stages specified in the SSADM method. In a similar vein, Blyth and Hakiel (1989) concern themselves with RTSASD (real time system analysis and system design), another of the software design methodologies widely used in the software world. They describe how a user interface design methodology might integrate with this structured design method, and in a subsequent paper (Hakiel and Blyth, 1989), outline how modifications to RTSASD could lead to a better structured, more explicit allocation of function between the human and the remainder of the system. Walsh et al (1989), and also Carver (1988) attend to another widely used system design methodology, JSD (Jackson System Development), and describe how the JSD notation could be used as a means of both describing tasks and specifying the User Interface of a system.

Given the increasing software content of large electronic systems, and the need to use proven structured design methods in order to have any real prospect of completing the software elements on time and within budget, it is likely that a SADM approach will be adopted for the development of future ATC systems. It will then be necessary for the Human Factors specialist to fit in with whichever of the structured design methodologies are adopted for the design of the system. To that extent, a good deal more research is required, not only to put the human factors community into a better position to work with and through these design methodologies, but also to extend the software-based tools that already exist to support these structured design methodologies, so that they support human factors activities in the design process. Thus, human factors as a discipline must adapt itself to the methods used in the industrial and commercial world, rather than simply advocating a "user-centred" design approach and hoping that the world will eventually come around to the view that everything revolves around the user. The reality, of course, is that life is a little more complex than that.

The Role of Experimentation

Much of the research emphasis in the human factors world in recent years has lain in the area of human-computer interaction (HCI). This has been driven in part by the enormous advances that have occurred in integrated circuit technology, allowing greater resources in terms of processing power and memory to be devoted to the user interface of ever more available and powerful computers. This in turn has created the need to design usable interfaces, and significant advances have been made both in the USA and Europe in this area. The relative infancy of human-computer interaction as a distinct area of human factors engineering has, however, led to a considerable emphasis within the human factors research community on theoretical approaches to user interface design, and that in turn has led to the development and use of a complete lexicon of human-computer interaction jargon that is relatively inaccessible to non-human-computer interaction specialists. It is perhaps ironic that those who specialize in the design of user interfaces have largely isolated themselves from those system designers who would use the results of human-computer interaction research, if only they could understand them.

The origins of human factors engineering lie in the work of psychologists and physiologists who, faced with practical design problems, were keen to determine through objective means where the solutions lay. This legacy of experimentation is of great value to the practical human factors engineer, who knows the value of a well-designed, properly-controlled experiment not only in resolving uncertainties that lie beyond the reach of the established science of the time, but also in providing hard and credible results in a form that can be accepted by the most skeptical of engineering design teams.

In the context of future ATC systems design, then, despite the very proper emphasis that must occur on the top-down, structured, well-scheduled, ordered system and software development approach (crucial if we are to have any chance of controlling system builds of this complexity), there must also be a place for human factors experimentation at various phases during the system development process. It will be important for human factors specialists to have available to them experimental facilities of sufficient versatility and flexibility that uncertainties identified at any stage in the design process can be addressed experimentally, and unequivocal answers derived, without causing undue delay to the overall system development process. Such a facility will need to be available from the earliest phases of the procurement cycle, when arguably the most important human factors decisions, those concerned with allocation of function, and the degree of system automation, are taken (note that allocation of function decisions do not cease at that stage, although many important decisions are taken then (see e.g., Holman & Turley, 1988)). It must be stressed that this simulation facility need not solve, in computational terms, the problems of automation, but need only represent to the experimental subjects the functionality of the automated elements of the system. Through this approach we gain an understanding of how the limitations of the user affect the total system design, bearing in mind that our ability to "improve" the user (through selection and training) is far more limited than our ability to improve other elements of the system.

Human Factors Standards

Design standards have an important part to play in the development of future ATC systems, and it seems appropriate that any relevant human factors standards should be brought to bear. The recently published U.K. Ministry of Defence standard DEF STAN 00-25 (HMSO, 1989) has some useful information on specific user interface design features, and also, within Part 12, guidance on how to apply human factors data, methods, and techniques during the various stages of system design. The U.S. Department of Defense has applied human factors standards for many years, with Military Standard 1472D being the current version of its major standard. For the last five years, the U.S. Army has also had a program called Manprint, which aims to enhance human performance and reliability in the operation, maintenance and use of manned systems. This approach has now been adopted by the U.K. Ministry of Defence, initially for British Army equipment procurement, in a program (also called Manprint) which will "integrate the six domains of manpower, personnel, training, human engineering, system safety, and health hazard assessment" (M0D, 1990). It appears to be a move away from the "user-centred" approach advocated in some quarters in recent years, to a recognition that, if a system is designed properly, engineering and technology, environment, and people all interact and come together to form an integrated whole.

In practice, a Manprint Joint Working Group is set up for each project, and generates a Manprint Management Plan which allows human factors concerns to be identified and trade-offs to be recorded. The group can advise the operational requirements and project management branches of the Ministry of Defence, and also briefs the relevant senior staff at the various project reviews. By being a formal part of the total procurement process, the working Group ensures that human factors aspects receive proper attention throughout the project. They ensure that appropriate considerations are included in requirements documents, requests for proposals, and hence contracts. While it must be noted that this system has not been operational for very long in the UK, and so its effectiveness cannot yet be judged objectively, it would appear to be a major step forward, and might even serve (with suitable modifications) as a model for the procurement of future ATC systems.

Design Validation - Issues of Usability

The term "usability" has come into use in recent years, and while it is preferable to its predecessor "user-friendly," it must be admitted that there is still no generally agreed definition of usability. Briggs (1987) identified four main usability criteria by which systems might be judged:

- Increased efficiency
- Increased user satisfaction
- Reduced training needs
- Increased user understanding

These seem intuitively reasonable, although they may not all lend themselves to objective measurement. Shackel (1986) also proposed a four-part definition. This encompassed:

- "Effectiveness," which could be expressed in terms of time to perform a task, against particular error criteria
- "Learnability," which could be expressed in terms of time, schedules, etc., to learn a task to a particular level of performance
- "Flexibility," which related task performance to specified adverse variations in the environment
- "Attitude," which related to the acceptability of the system to users, as measured by standard subjective assessment techniques (questionnaires, rating scales, etc).

He then illustrated, for a sample task, how a usability requirement might be drawn up and the system, as finally developed, measured against it. In doing so, he endorsed sentiments expressed earlier by Gould and Lewis (1983), deriving from their work at the IBM Watson Research Center. Gould and Lewis referred to empirical measurement, "actual behavioural measurements of learnability and usability and conducting these experimental and empirical studies very early in, and throughout, the development process." In truth, it must be admitted that it would be difficult to adopt an approach of procuring a next-generation ATC system against a full usability specification, in that the techniques probably do not yet exist in most instances to derive reasonable usability requirements, except in the grossest of terms; even if they did, most contractors would not know how to set about designing a system to meet them. Nevertheless, the attraction of such an approach is that it transports the measuring of success in system design away from the realms of unstructured subjective assessment, to the realms of objective measurement. This approach has the additional advantage that it would get over the problem faced by human factors engineers the world over, which is that if they do their job well, the user fails to notice. In general, it is failure rather than success in user interface design that attracts attention.

The issue remains, however, that human factors engineering may ultimately need more than demonstrable adherence to a process to prove its worth in design, and the approach so often adopted — of pointing out the blunders of others who failed to adopt human factors principles in the design process, and consequently ended up with a sub-standard system — has never won many adherents to the human factors cause. Clearly, usability research and the objective measurement of usability is a key area for future research.

Conclusions

The large investment in ATC systems that will occur over this decade poses some significant challenges for the discipline of human factors. It is appropriate that human factors specialists should play a major part throughout the full design cycle, rather than (as hitherto) merely concerning themselves with the fine detail of the physical user interface, the "knobs and dials" (important as those are). Their greatest chance of doing so is by working through and with the structured design methodologies pioneered by the software specialists, modifying and improving those methodologies as the needs of human factors engineering dictate, and also taking the necessary steps to modify the supporting tools that make these methodologies so effective. In addition, this whole approach should be complemented by a rediscovery of the experimental method, the bed-rock on which the entire discipline of human factors lies, deriving as it does from experimental psychology.

For their part, the procuring authorities should take note of the steps that the military, both in the UK and in the United States, take to ensure that the systems they procure are properly designed from the user's viewpoint. The time may be ripe for them to seek to achieve from interested professionals a consensus view of the state of human factors, the role it may play in future ATC development, and what issues should be addressed (be they issues of research, standardization, co-ordination, or whatever) in the short and medium term, to advance the status and uptake of human factors engineering in this crucial area.

References

Anderson, J.M. (1988) "The integration of HCI principles in structured system design methods." In *Conference Proceedings Milcomp '88, Military Computers, Graphics and Software, 27th-29th September, 1988*, London.

Blyth, R.C. & Hakiel, S.R. (1989). "A user interface design methodology and the implications for structured systems design methods." In *Proceedings of the IEE Third International Conference on Command, Control, Communications and Management Information Systems, Bournemouth , UK*, 2-4 May 1989.

Briggs, P. (1987). "Usability assessment for the office: methodological choices and their implications." In Frese, M., Ulich, E., & Dzida, W. (Eds), *Psychological issues in the work place*. Elsevier Science Publishers B.V. (North-Holland), 1987.

Byerley, P. (1990) RACE Project R1067 GUIDANCE (Generic Usability Information for the Design of Advanced Network Communications in Europe). Position Paper. Applied Psychology Group, Standard Elektrik Lorenz AG, Pforzheim, West Germany.

Carver, M.K. (1988) "Practical experience of specifying the human-computer interface using JSD." In Oborne, D. (Ed), *Contemporary Ergonomics 1988, Proceedings of the Ergonomics Society's 1988 Annual Conference, Manchester, England, 11-15 April 1988*. London: Taylor & Francis.
Foley, J.D. & Van Dam, M.A. (1982). *Fundamentals of Interactive Computer Graphics*. Addison-Wesley, pp 217-222.
Gould, J.D. & Lewis, C. (1983). "Designing for usability: Key principles and what designers think". *Communications of the ACM, 28*, 3, 300-311.
Hakiel, S.R. & Blyth, R.C. (1989) "Keeping the Human in Context." In Lovesy, E.J. (Ed), "Contemporary Ergonomics 1990," *Proceedings of the Ergonomics Society's 1990 Annual Conference, Leeds, England, 3-6 April 1990*. London: Taylor & Francis.
HMSO (1989). "Human Factors for Designers of Equipment," Defense Standard 00-25. (Published in 12 parts).
Holman, R. & Turley, N. (1988). "Methods for integrating Human Factors into the early development of Surface Naval Command Systems." Document Ref ARE TM (AXC) 88001, March 1988, Admiralty Research Establishment, Cosham, Hampshire, UK.
Jackson, N. (1990) "A European planning strategy for air traffic to the year 2010". In *Proceedings of The Third European Aerospace Conference, London 22-24 May 1990*. Pub. Royal Aeronautical Society, Hamilton Place, London.
MoD (1990). "MANPRINT." Information leaflet A174, issued by ASD(5) Manprint, Ministry of Defence, Whitehall, London SW1A 2HB
Moran, T.P. (1981). "The Command Language Grammar - a representation of the user interface of interactive computer systems." *Int. Jnl. Man-Machine Studies, 15*, 1, 3-50.
Nielsen, J. (1986). "A virtual protocol model for computer-human interaction." Int. Jnl. Man-Machine Studies, 24, 3, 301-312
Shackel, B. (1986). "Ergonomics in Design for Usability." In Harrison, M.D. & Monk, A.F. (Eds), "People and Computers: Designing for Usability," *Proceedings of the Second Conference of the British Computer Society Human Computer Interaction Specialist Group, University of York, 23-26 September, 1986*. Cambridge: Cambridge University Press.
Walsh, P., Lim, K.Y., Long, J.B., & Carver, M.K. (1989) "Integrating Human Factors with system development," in Heaton, N. & Sinclair, M. (Eds), *Designing End-User Interfaces*. Oxford: Pergamon Infotech.

Automated Systems: The Human Factor

Daniel J. Garland

Embry-Riddle Aeronautical University, Daytona Beach, Florida, USA

Introduction

The need for a safe and expeditious air traffic system, along with recent and rapid developments in advanced computer-based technology, has been associated with remarkable changes in airway systems operations and in the system decision-making process. A key issue in air traffic systems concerns which human functions should be automated to maximize system performance and efficiency, without jeopardizing system safety. These decisions are quite simple for a few selected functions where human performance is superior, or in some instances inferior, to machines. The decisions become more critical in considering those functions where human and machine performance differences are subtle. These latter circumstances present the greatest challenge to the human factors professionals and system designers because of the potential human factors consequences (Adams, 1989).

Advanced automated systems require changes in the nature of the human factors questions to be asked and ultimately answered. If the human is to be a significant and contributing factor in the system (effectively "in the loop"), it is critical to optimize the human and machine interface in order to keep the human aware of system operations (and, conversely, to keep the machine aware of human operations). The advancement and sophistication of modern technology affords increased automation of functions, while concurrently requiring considerable ingenuity to prevent increased human workload and opportunity for errors by the new automated system. Recent technological innovations, if judiciously chosen and implemented, may allow for remarkable and welcomed benefits. However it must be emphasized that there may also be less beneficial human factors consequences. Awareness and prevention of potential human factors consequences should ultimately influence the decision-making process in automated systems design. If human factors implications are not considered, potential benefits of progressive automation may be unrealized.

The complexity inherent in automating airway systems functions raises important issues — issues that must be addressed as advanced technological developments provide the opportunity to automate practically every aspect of the airway system. O'Hare and Roscoe (1990) note two major areas where questions concerning application of automation remain unanswered. The first concerns immediate human-machine interactions. Human factors engineers and the aviation community are particularly concerned with consequences of potential computer failure. What safeguards can be developed and implemented to ensure accurate information input into the system? The second area is concerned with long-term human factors consequences of progressive automation. If the human is relegated to a passive systems monitor instead of a direct controller, what impact will this have on professional skills? Will the human gradually be taken "out of the loop," consequently negatively impacting skills?

NATO ASI Series, Vol. F 73
Automation and Systems Issues in Air Traffic Control
Edited by John A. Wise et al.

The purpose of this paper is twofold. First, there will be a cursory examination of a few undisputable differences in the capabilities of the human and the machine (Adams, 1989). While research on the relative capabilities of the human and machine is prolific (Price, 1985), an understanding of the optimal human-machine interface in advanced automated systems is still quite elusive. Secondly, several issues and concerns dealing with potential human factors consequences of automation will be presented. The issues raised need to be considered as advanced technology enjoys increasingly widespread implementation. While study of the relative capabilities of the human and machine and the automated systems issues are not new, this information is intended as commentary on the necessity of a scientifically-based design philosophy for human-centered automation.

Superior Human Capabilities

The Cognitive Advantage

Computers are indefatigable machines that obediently follow the rules of logic and serial processing. However, an appropriate response to various situations does not always afford such processing. Human decision-making processes are uniquely qualified to consider and weigh not only logical analyses, but also the importance of a myriad of relevant variables in making any decision. In normal operations, humans must consider information that may not be readily available to the computer (e.g., weather, air traffic, aircraft problems, company needs and expectations). During abnormal operations, the human is ultimately responsible for solutions to unforeseen problems. While automated systems would be a valuable aid in the problem-solving process, human experience and judgment are critical and imperative for successful problem resolution (Graeber, 1988).

The question of whether a computer can think (or rather,whether a machine can replace human function with indistinguishable results) (Turing, 1950), has received increasing attention with advanced technological innovations (Adams, 1989). While expert systems have been built to incorporate relevant human expert knowledge into a data base to parallel human expert knowledge, problems have been commonplace (Adams, 1989; Basden, 1983; Hopkin, 1984). It is fair to conclude that humans have a definite cognitive advantage over machines. However, a note of qualification is warranted. Wickens' (1984) research "... suggests a fairly pessimistic view of human performance in decision making and diagnosis" and that "... performance probably lies somewhere between chance and optimal, shifting further from the optimum as complexity and stress increase." Given this rather modest view of human decision-making abilities and continued technological advancements, the human decision-making advantage may be waning.

Auditory and Visual Pattern Recognition

Human auditory and visual pattern recognition is so efficient, accurate, automatic, and effortless that drawing comparisons with machine pattern recognition is tenuous at best. The incredible human ability to store and retrieve pattern information forms the foundation for higher-order perceptual and cognitive processes. Artificial intelligence research has provided some rather impressive developments in machine pattern recognition (Adams, 1989; Winston, 1984). Future knowledge-based computer systems may bring automated speech, as well as auditory and visual synthesis and recognition devices that rival human pattern recognition. However, despite impressive advancements in machine pattern recognition, these abilities are minimal in comparison to human capabilities.

Learning

The ability to learn, to adjust and adapt to new events and experiences that are expected or unexpected, is a remarkable characteristic of humans. While the field of artificial intelligence has made considerable advancements in computer learning (Cohen & Feigenbaum, 1982), machines are far below humans in the ability to change as a result of experience.

Memory

It has become commonplace to present computer memory and information processing as analogous in functioning to human memory and information processing. The computer and human comparison, while appealing, is quite superficial. Computer memory capacity can be specified exactly, unlike human memory, which is largely unlimited. In addition, human memory is multi-dimensional (Adams, 1989), allowing for memory on various dimensions (e.g., semantic, episodic, procedural, motor, visual, auditory). This multi-dimensional aspect of memory affords humans the unique residual ability of "common sense" (Waltz, 1982), which the field of artificial intelligence is having considerable trouble addressing.

Superior Machine Capabilities

Expeditious Processing Operations

The execution of processing or "mental" operations is remarkably fast for a computer, in comparison to human information processing. Human reaction and processing time is an eternity compared to those of modern advanced computer systems.

Advanced Sensory Capabilities

While the human sensory system possesses incredible capabilities, as indicative of established sensory absolute thresholds, it falls far below certain sensory capabilities of machine systems, such as the remarkable detection capabilities of radar and sonar. Machines are extremely useful in detecting colorless and odorless gases, radiation, and numerous electromagnetic energies (Adams, 1989).

Performance Stability and Consistency

The indefatigable nature of machines provides a distinct performance advantage over humans. Humans are unable to maintain constant performance due to limitations on endurance, fatigue, motivation, attention, and circadian rhythmicity. The variability of human performance is so large, it is a legitimate human factors concern.

Workload

As the number of information sources increases, particularly across sensory modalities, the human must decide on priorities. Humans must adopt different processing strategies to deal with labor-intensive and time-intensive situations. Invariably, human performance is degraded when faced with too many sources of information too quickly. Conversely, machines are not as easily affected. Machine design can anticipate and control for both labor-intensive and time-intensive situations.

Summary

Advanced automated systems present system designers with the fundamental problem of establishing a design philosophy and associated rules for the allocation of functions for the human and the machine (Price, 1985). Obviously, the strengths and limitations of the human and the machine must be considered in such design decisions. Adams (1989) provides a more substantial examination of the relative capabilities of the human and machine.

Human Factors Issues in Automation

Recent history has seen an unfortunate tendency for systems designers with little human factors training to believe that devices should control systems. In this section, I attempt to refute this fashionable practice by briefly presenting several current human factors concerns with automation.

The Human as Systems Monitor

Casting the human as a systems supervisor and monitor has been fashionable for three decades. Human factors professionals and systems designers have long recognized that functions in the lower levels of the systems loop should be automated, leaving the uniquely human higher-order skills, such as planning and decision-making, under direct human control. While casting the human in this supervisory and monitoring role is intuitively appealing, three decades of human factors research investigating the human as systems monitor (Howland & Wiener, 1963; Mackie, 1977; Sheridan & Johannsen, 1976; Warm, 1984; Wiener & Curry, 1980) have consistently identified the human as an unreliable monitor. The instability of human cognitive and perceptual vigilance in an automated control mode has been a primary factor. Contrary to the research, humans have consistently been cast in the self-defeating role of monitor of automated functions. Consequently, the human has gradually been taken "out of the loop," which has led to potential degradation of supervisory, monitoring, and emergency intervention skills.

Advanced automated systems have forced the need to re-examine the fundamental question of allocation of system functions (Price, 1985). Failure to do so will inevitably lead to recalcitrant human factors problems.

Deterioration of Skills

Another consequence of progressive automation is the potential for deterioration of professional skills. As automation increases, opportunities for direct human control decrease, resulting in a potential for reduced human proficiency. Research into this issue, particularly aviation research, is minimal. However, a study by Mengelkoch, Adams, and Gainer (1971) on the forgetting of previously learned flying skills does provide some provocative findings. Two groups of pilots performed a number of maneuvers during either five or ten training sessions in an aircraft simulator. The training period was followed by four months of no flight training. Subsequent performance testing indicated the greatest decline in skills due to the four month "down time" was in procedures. This decline in procedural skills presents obvious concerns, particularly in emergency situations.

The aviation community is justifiably concerned about how quickly and effectively the human can make the transition from automatic operation to manual operation ("re-enter the loop"). What impact does passively monitoring an automated system have on manual and procedural skills? Can the human maintain a passive monitoring or supervisory role in the automated system without depreciation of professional skills? Unfortunately, human factors research concerning this issue is minimal.

Workload

The introduction of advanced automated systems makes the measurement of human workload critically important because of the consequences of low-workload and high-workload phases. Workload is difficult to define precisely, and possibly more difficult to quantify. Various behavioral and physiological measures have proven unreliable, suggesting workload is essentially cognitive in nature. This cognitive component is reflected in the tendency for air traffic controllers to adopt different strategies for various situations (e.g., for dealing with different numbers of aircraft) (Sperandio, 1971).

Objective measures of workload would be beneficial in assessing the impact of introducing new system features, such as the automation of various functions in an existing system. Circumstances of system operations where the human is confronted with labor- and time-intensive situations naturally affords some level of function automation.

A consideration of workload and automation of functions presents the concern that, while automation may reduce workload problems in certain situations, there is the potential for exacerbated problems in others. While the goal of progressive automation is to maximize system safety and efficiency by reducing human workload, there is justifiable concern for increased human boredom, complacency, loss of situational awareness, and overreliance on and misuse of automated systems.

Ironically, automating various functions in order to reduce human workload during high-workload phases may contribute to boredom and complacency during low-workload phases. In addition, further workload reductions contribute to degradation of professional skills by curtailing skill utilization opportunities. This issue presents the need for some level of situation-based performance aids designed to optimize human performance during low- and high-workload phases (Hopkin, 1988).

Perceived Loss of Control

Advanced automated systems present an interesting and critical dilemma for the human: who is in charge, the human or the machine (Wiener, 1985; 1988)? With increasingly sophisticated and autonomous systems, there is obvious concern for perceived loss of system control, based in the realization that the machine may independently make decisions and initiate action without human input or consent. This concern, coupled with unnecessary workload reduction contributes to the larger problem of humans possessing a strong sense of being "out of the loop." The decisions on the allocation of authority to humans and machines may significantly impact human self-esteem, intrinsic motivation, and job satisfaction.

Psychosocial Effects of Automation

In considering the aforementioned issues, it is quite clear that advanced automation presents important consequences for several psychosocial factors, including job satisfaction, selection, and training (O'Hare & Roscoe, 1990; Boehm-Davis, Curry, Wiener, & Harrison, 1983). The nature of these consequences centers around the altered role of the human within the increasingly automated system. Several rather broad questions warrant attention. What steps can be taken to maximize job satisfaction, to keep the human as a direct and controlling contributor within the loop, and to ensure optimal emergency intervention performance? How should selection criteria and training methods be modified to better "fit" the human into the advanced automated system? Should training focus on developing skills and abilities for use in complex computer systems? Each of these questions becomes more critical as technology enjoys increased implementation.

Conclusion

For the past 20 to 30 years, there has existed an unwritten philosophy of automation which has espoused the assumption that maximum available automation is always appropriate. This philosophy has been based, in part, on the availability of increasingly sophisticated and advanced technological innovations, the need to reduce human workload, and the need for increased safety of flight. While automated systems have provided substantial benefits, several human factors consequences have arisen and incidents/accidents have occurred, forcing human factors professionals and the aviation community to re-examine automation practices. There is increasing awareness of the lack of a scientifically-based philosophy of automation (Air Transport Association of America, 1989). This philosophy must be based in an understanding of the relative capabilities of the human in the system, and the circumstances under which automation should assist and augment the capabilities of the human. Such a comprehensive, scientifically-based design philosophy for human-centered automation must be developed in order to avoid inevitable "one step forward and two steps backward progression.

References

Adams, J.A. (1989). *Human factors engineering.* New York: Macmillan Publishing Company.

Air Transport Association of America. (1989). *National plan to enhance aviation safety through human factors improvements.* Washington, DC: Author.

Basden, A. (1983). On the application of expert systems. *International Journal of Man-Machine Studies, 19,* 461-477.

Boehm-Davis, D. A., Curry, R. E., Wiener, E. L., & Harrison,R. L. (1983). Human factors of flight-deck automation: Report on a NASA-Industry workshop. *Ergonomics, 26,* 953-961.

Cohen, P. R., & Feigenbaum, E. A. (Eds.). (1982). *The handbook of artificial intelligence Volume 3.* Los Altos,CA: William Kaufmann.

Graeber, .R.C. (1988). Aircrew fatigue and circadian rhythmicity. In E. L. Wiener and D. C. Nagel (Eds.) *Human factors in aviation.* (pp 305-344). San Diego, CA: Academic Press.

Hopkin, V. D. (1984). Some human factors implications of expert systems. *Behavior and Information Technology, 3,* 79-83.

Hopkin, V. D. (1988). Air traffic control. In E. L. Wiener and D. C. Nagel (Eds.*), Human factors in aviation* (pp. 639-663). San Diego, CA: Academic Press.

Howland, D., & Wiener, E. L. (1963). The system monitor. In D. Buckner and J. J. McGrath (Eds.), *Vigilance: A symposium* (pp. 217-226). New York: McGraw-Hill.

Mackie, R. R. (Ed.). (1977). *Vigilance: Theory, operational performance, and physiological correlates.* New York: Plenum.

Mengelkoch, R. F., Adams, J. A., & Gainer, C. A. (1971). The forgetting of previously learned flying skills. *Human Factors, 13,* 397-405.

O'Hare, D., & Roscoe, S. (1990). *Flightdeck performance: The human factor.* Ames, IA: Iowa State University Press.

Price, H. E. (1985). The allocation of functions in systems. *Human Factors, 27,* 33-45.

Sheridan, T. B., & Johannsen, G. (Eds.). (1976). *Monitoring and supervisory control.* New York: Plenum.

Sperandio, J. C. (1971). Variation of operator's strategies and regulating effects on workload. *Ergonomics, 14,* 571-577.

Turing, A. M. (1950). I. Computing machinery and intelligence. *Mind, 59,* 433-460.

Waltz, D. L. (1982). Artificial intelligence. *Scientific American,* October.

Warm, J. (Ed.). (1984). *Sustained attention in human performance.* London: Wiley.

Wickens, C. D. (1984). *Engineering psychology and human performance.* Glenview, IL: Scott, Foresman and Company.

Wiener, E. L. (1985). Beyond the sterile cockpit. *Human Factors, 27,* 75-90.

Wiener, E. L. (1988). Cockpit automation. In E. L. Wiener and D. C. Nagel (Eds.), *Human factors in aviation* (pp. 433-461). San Diego, CA: Academic Press.

Wiener, E. L., & Curry, R. E. (1980). Flight-deck automation: Promises and problems. *Ergonomics, 23,* 955-1011.

Winston, P. H. (1984). *Artificial intelligence. Second Edition.* Reading, MA: Addison-Wesley.

Modelling Control Tasks in Complex Systems

Morten Lind

Institute of Automatic Control Systems, Technical University of Denmark, Lyngby, Denmark

Introduction

The development of control systems for industrial plants have traditionally been based on mathematical models of the causal dynamic interrelationships between plant parameters (e.g., by using knowledge of the physical properties of system components). These models are then used to synthesize control algorithms on the basis of optimization techniques. However, for many important industrial plant control problems, the design problem is not only to cope with the complexity of causal dynamic interactions, but also to cope with the complexity of the end or goal-means structure of the system. Knowledge about plant goal-means structure is important in diagnosis of fault and in planning of compensatory actions (i.e., for strategic plant resource management). Furthermore, it is also important for design of human machine interfaces, as goal-means structures can be used for management, interpretation, and presentation of plant information for the operator. Models of goals-means structures are therefore important for design of integrated supervision and control systems.

I have developed a modelling technique called Multilevel Flow Modelling (MFM) which can be used to model a system in terms of its goal-means structures (Lind, 1990b). An MFM model represents the plant in terms of a set of interrelated mass, energy, and information flow structures on several levels of abstraction. The levels are interdependent and form goal-means structures. An important aspect of MFM modelling is the modelling of control functions, which is done in terms of information flow structures. This paper shows how the concepts used for modelling information flow can be used as a generic framework for the representation of the functions of both the control and the supervisory systems. As MFM provides a unified conceptual framework, it is applicable for integrated system design.

This paper also describes the basic ideas of MFM and discusses how it can be used for modelling control tasks. Even though MFM has been developed for modelling material and energy production systems, the basic ideas are general and would also be applicable for modelling air traffic control systems.

Multilevel Flow Modelling

The purpose of Multilevel Flow Modelling is to model a system as an artifact (i.e., as a man-made purposeful system). In order to do this, MFM is based on concepts of goals and functions which traditionally are not used in the natural sciences but are used in the human

NATO ASI Series, Vol. F 73
Automation and Systems Issues in Air Traffic Control
Edited by John A. Wise et al.

sciences (sociology, cognitive science, and psychology). In this way MFM provides a link between two worlds of knowledge. This unusual combination is very important for solving systems engineering design problems where the two worlds meet. An important area of application is in the planning of human-machine interfaces and building intelligent control systems, because knowledge about system goals and functions is very effective for diagnosing malfunctions and for planning of control actions. The meeting of the two worlds is here in the acquisition of knowledge. Much common sense engineering knowledge of systems can be formulated by concepts of goals and functions. The understanding of the concepts of goals and functions is therefore an important one for using Multilevel Flow Modelling (Lind, 1990b).

Another important aspect of Multilevel Flow Modelling is the representation of a system by multiple descriptions on different levels of abstraction. This aspect is important for coping with complex systems. Multiple representations occur both because a system is described on many whole/part levels (levels of decomposition) and because of the context dependency of functional descriptions. Functional descriptions are context dependent because they relate to system goals. This context sensitivity is addressed specifically in MFM. In this way MFM departs also from modelling based on physical laws, which is (believed to be) independent of context.

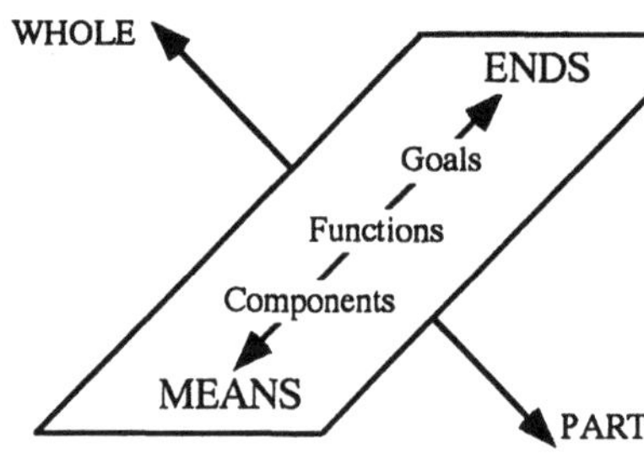

Figure 1. Dimensions of representation in MFM.

The basic dimensions of Multilevel Flow Modelling are depicted in figure 1. The figure illustrates the description of a system along two directions. A system is described in terms of goals, functions and the physical components (along the means/end axis). At the same time each of these descriptions can be given on different levels of decomposition along the whole/part axis.

Flow Concepts

Another distinctive feature of the MFM technique is the use of a set of primitive function concepts related to the processing of flows of material, energy, and information. Using these concepts, it is possible to represent knowledge of a process plant in a way which, at the same time, copes with system complexity (by using abstraction principles) and facilitates the formulation of design knowledge and high level heuristic knowledge for solving problems of diagnosis, planning, and design. The use of a rather small set of basic concepts provides in addition, a certain cognitive economy.

Multilevel Flow Modelling is described in detail in Lind (1990b). Appendix A shows the basic concepts and their graphic representation.

MFM in Control Systems Design

Multilevel Flow Models can be used for control system design. This aspect is best discussed from a system development perspective. The design of control systems for industrial plants involve several stages, which can be described in terms of figure 2 shown below. This is a simplified model depicting the development of control systems in terms of three phases.

The first phase comprises a systems analysis aiming at the identification of goals functions, and physical resources of the system to be controlled. The second phase includes the design of control system algorithms and plans of control sequences. The third phase comprises the implementation of the designed algorithms in software and hardware.

Most existing control system design techniques relate to the second phase of development of algorithms, but they are based on a prior, informal, overall analysis of the control problem to be solved. MFM models are intended to be a tool for this overall analysis where system control tasks are defined (i.e.,before decisions are made on how to implement the control functions). This type of analysis is also needed when coping with major system disturbances or in situations which require reevaluation of system goals. In fact, models such as MFM which support the first phase of control systems development can be used in intelligent control systems for plant resource management.

Resource management and control tasks in complex systems can be categorized along the two dimensions as shown below. Along the vertical axis I have distinguished between three decision making levels. The levels of operational, tactical, and strategic decision making correspond to three levels of uncertainty in systems control. And along the horizontal axis I have related the tasks to the three main system goal categories.

The level of the operational tasks represents the level where fixed control algorithms can be applied and in systems where detailed models are available for design of control algorithms. These tasks comprise all control tasks handled by the conventional regulation and optimizing control systems.

On the tactical level, we find the tasks related to change of operating mode involving change in functional structures typical to start-up and shut-down and to change of production (sequential control).

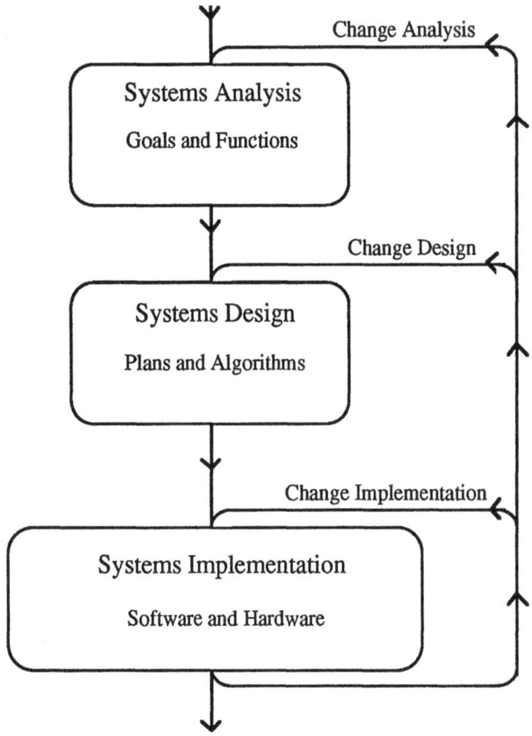

Figure 2. Phases in control systems development.

Decision Level	Goal		
	Safety	Production	Economy
Stategic	Diagnosis, goal evaluation and planning		
Tactical	Shut down & interlock	Sequence control	
Operational	Regulation and control		

Figure 3. Relations between system goals, decision level, and resource management tasks.

On the strategic level, we find tasks involving problem-solving activities. These levels relate to major system upsets and include diagnosis, planning, and goal evaluation. Problem solving on this level requires system representations (such as MFM) which can support reasoning over goals and resources.

Models for Control Systems Development

There is a long tradition within the engineering sciences to base modelling of systems on laws of physics and chemistry. These models represent the behavior of the system in terms of basic physical mechanisms. They are useful for accurate predictions of system behavior and for optimization, but are not adequate for coping with design problems which include strategic or tactical decisions about the management of system resources. If conventional models should be used for coping with more large scale decision problems, the models tend to be very complex and unmanageable because the level of detail introduces many unknown factors. In such cases, more qualitative models of relations between goals and resources, such as MFM, seem to be more adequate.

MFM can be used to formulate some of the control system design strategies implemented by practicing control engineers. The concepts used by control engineers in strategic considerations are very important to identify. because they represent the way the plant has been functionally organized. These functions will be reflected in the choice of control system structure. A convenient set of function concepts to choose as a basis for the MFM technique must accordingly reflect, in a natural and convenient way, the overall plant strategies, for resource management and control. By providing concepts to represent these strategies we can communicate design intentions to the operator and provide an understanding which supports proper reasoning about the system.

System Concepts of MFM

The purpose of Multilevel Flow Modelling is to represent complex systems as artifacts (i.e., as human-made and purposeful entities) Concepts of goals and functions are important for understanding the nature of artificial systems, and are therefore basic to MFM. MFM is, in this way, different from modelling techniques representing systems on the basis of knowledge of physical mechanisms in terms of laws of physics and chemistry or causal relationships between system states or events. The focus is accordingly on representation the systems "intentions," not the causality. Models of causality does not represent intentional aspects of systems.

Blauberg (1977) provides a comprehensive discussion of the philosophical and methodological problems associated with systems theory. Of particular interest in the present context, is the discussion of the systems approach originally formulated by Churchmann (1968) and systems analysis as an application of the systems approach to management and control problems. Concepts of goals and functions play an important role in the systems approach.

I will use Blauberg's analysis as a vehicle for the definition of my systems concepts, since Multilevel Flow Modelling can be considered a formalized representation of the result of a system analysis. In this way, I also relate MFM to established systems engineering traditions.

The Systems Approach

The purpose of a systems analysis is to provide a basis for decision making and for planning of actions (i.e., for design, resource management, and control). These actions may be modifications to system elements or to their interrelations, or it may be a change in their states.

But analysis is more than a preparation for synthesis or design. The way the system is perceived by the analyst must take into account or tacitly assume possible means of action and procedures or strategies for solving the design problem.

The description of systems analysis given by Blauberg (op.cit.) reflects this close relation between analysis and synthesis. Blauberg describes the result of a systems analysis as including the following logical elements, comprising basic categories:

1) a goal (objective) or a set of goals

2) alternative means that can be used to reach the goal(s) and are sets of system elements or strategies

3) resource expenditures on the system including a quantification of the amount of resources spend in achieving goals (time, material, energy, knowledge, personnel etc.)

4) a mathematical model or logical model — i.e., a representation of relations between the goals, alternative means, environment, and requirements imposed on resources

5) criteria by which the preferred alternatives are selected

These concepts define the terms by means of which the results of a systems analysis task are formulated. This formulation implicitly assumes the existence of an agent faced with a design problem. The agent has a goal the system should achieve, and the logical model (the strategies and criteria which are identified as a result of the analysis), provides the knowledge used by the agent for synthesis of possible courses of action. The agent is assumed to perceive his problem as a problem of resource management.

The systems approach represents, accordingly, a special point of view in modelling, in the sense that the value or utility of system elements for achieving some identified goals is put into focus. There is accordingly a contrast to systems analyses based on physical modelling (i.e., based on physical and chemical laws). In physical modelling, concepts of value and utility have been systematically eliminated in order to reduce the influence of subjective or contextual factors. But as only mechanisms are described, these models are not able to capture information which is important for coping with the complexity of many design problems.

Categories of Means

Note that different categories of means for goal achievement are mentioned in the description of the results of systems analysis above. First of all, the model describes the means (physical or more abstract) which are available for goal achievement in the system under investigation (the object for the agent's actions).

But the definition also includes the means available to the agent in selecting between different courses of action. These means include, in addition to the model (knowledge about the system object), a set of strategies for generating alternative solutions to the design problem and a set of decision criteria to choose among the alternatives. The definition includes also means of control to ensure that the actions planned actually achieve the goals as intended.

Strategies describe how to achieve goals and criteria to select among alternatives. The strategies represent the accumulation or generalization of previous experience of the agent, or knowledge provided by the agent who designed the system to be managed. The designer has provided certain functional or physical means the agent can use to achieve his goals.

Accordingly, a systems analysis involves much more than a physical description of the system which is the object of the agent's design problem. Intentions (goals) of the agent and design knowledge (the strategies and criteria) are also important.

An Example

In order to illustrate the concepts of goals and functions I have chosen a simple system: a water circulation circuit of a central heating system. The system shown below depicts how the physical parts (components) of the central heating system provide different functions and how these functions contribute to the achievement of system goals.

The circulation system provides several functions at the same time and these functions each contribute to one or several goals. This is a general feature of most systems, even relatively "simple" systems such as the central heating system. The nature of these mappings between functions and goals is important for the understanding of complex systems and therefore also to the task of a systems manager. Multilevel Flow Modelling provides formalized concepts to describe these mappings.

Flow Structures for the Central Heating System

Now extend the example above to comprise the whole central heating system and to show how we can describe functions of the system in terms of mass and energy flow structures. The system is shown below. As mentioned in Lind (1990b), it is not meaningful to describe system functions without the identification of system goals. Goals and functions go hand in hand. I therefore define the goals of the system first.

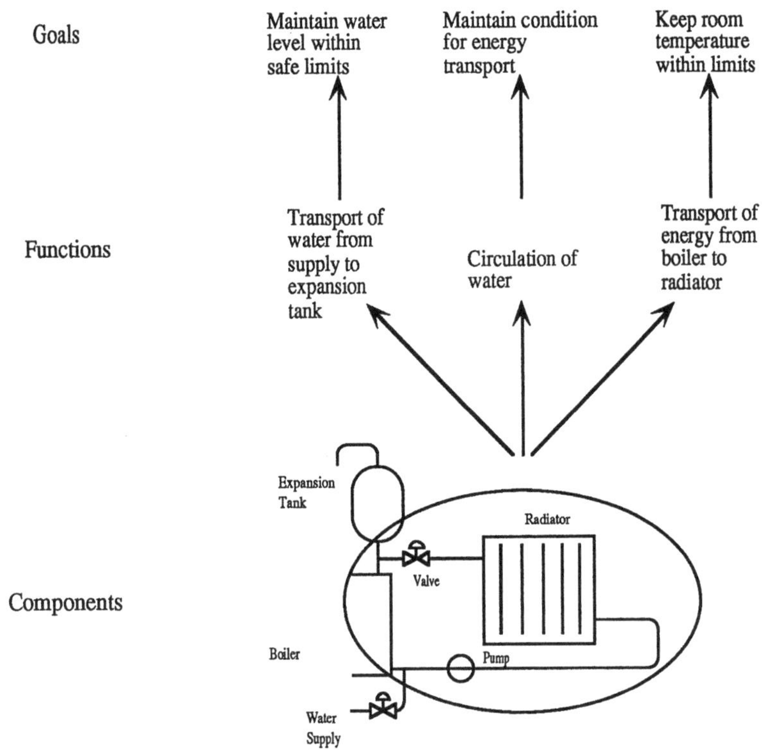

Figure 4. Goals and functions of a circulation system.

I have selected the following goals as the basis for my example:

Goals:
G0: Maintain the room temperature within limits
G1: Provide circulation of water
G2: Provide lubrication to the pump
G3: Maintain water inventory in the expansion tank

For each of the goals G0, G1, G2, and G3, I have shown the functions which are provided by the system for their achievement. These functions comprise the energy production functions, the water circulation functions, the water supply functions and the pump lubrication functions. Each function is described in terms of a set of interrelated flow functions (i.e., flow structures).

Each flow structure represents a functional view of the system,and for each of these views, a corresponding selection of physical components. These sets of components are shown in the figure below. Now, let us discuss each function separately and relate them to the goals and the sets of physical components.

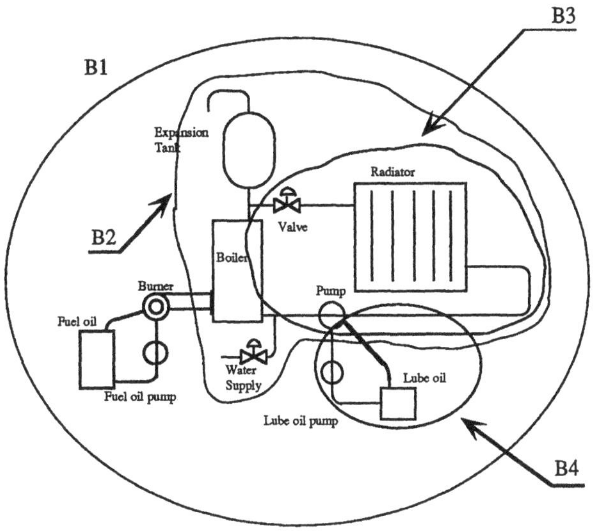

Figure 5. The central heating system example.

The Energy Production Functions

These functions are modeled in terms of an energy flow structure; the components realizing these functions are indicated by the boundary B1 in the physical diagram. The energy flow structure describes the functions provided by the system in order to achieve goal G0. Note that all components are involved. The energy source f1 describes the function of the burner system (including the fuel oil supply). The energy transport f2 connected to the source describes the function of the air/gas path in the boiler. The energy storage f3 describes the heat accumulation function of the boiler, the circulation circuit,and the radiator. The energy transport f4 describes the function of the circulation system. The energy sink f5 describes the function of the environment of the central heating system (the room).

The Water Circulation Functions

These functions are modeled in terms of a mass flow structure; the components realizing these functions are indicated by the boundary B3 in the physical diagram. The mass flow structure describes the functions provided by the system in order to achieve goal G3. In this case, only some of the components of the system are involved. The mass transport f7 describes the function of the pump and the piping, and the balance f6 describes the function of the radiator.

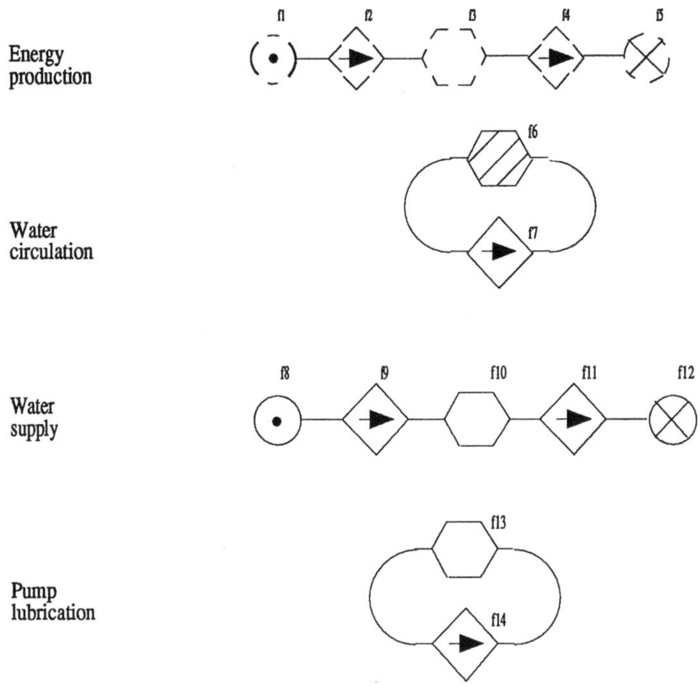

Figure. 6. Flow structures for the central heating system.

The Water Supply Functions

These functions are modeled in terms of a mass flow structure; the components realizing these functions are indicated by the boundary B2 in the physical diagram. The mass flow structure describes the functions provided by the system in order to achieve goal G4. In this case, only some of the components of the system are involved. The mass source f8 describes the function of the environment of the system (the water works). The mass transport f9 describes the function of the circulation circuit in the context of achieving G3. The mass storage f10 describes the mass accumulation capability of the piping, the boiler and the expansion tank. The mass transport f11 describes the function of the outlet piping on the expansion tank, and the sink f12 describes the ability of the environment to receive surplus mass.

The Pump Lubrication Functions

These functions are modeled in terms of a mass flow structure; the components realizing these functions are indicated by the boundary B4 shown in the physical diagram. The mass flow structure describes the functions provided by the system in order to achieve goal G2. As before only some of the components of the system are involved. The mass storage f13 describes the function of the oil tank, and the mass transport f14 describes the function of the oil pump and the piping.

The plant functions modeled above in terms of mass and energy flow structures can now be combined into a multilevel flow model. Each functional level comprises a level of abstraction in the modelling of the system. Three levels can be interrelated by means of means/end and conditioning relations (see appendix). The resulting MFM model of the central heating system is seen below. We can explain the model by following the vertical connections by condition and achieve relations.

Explanation:
The energy transport function f4 will only be available provided that there is circulation of water in the circulation circuit. It is therefore related by a condition relation with G1 which expresses that condition.
But G1 is the goal of the water circulation functions. G1 is accordingly related with the water circulation flow structure.
The mass transport f7 will only be available if there is sufficient water in the system and if the pump is lubricated. These two conditions are shown in the model in terms of G2 and G3.
But G2 is a goal for the lubrication function and G3 is a goal for the water supply function of the system. We get, accordingly, the condition and the achieve relations as shown in the figure.

The condition and achieve relations interrelate the levels of the model and represent knowledge about the plant which is efficient in diagnosing faults and in planning control actions. Applications of MFM for diagnosis are reported in Lind (1982, 1988b, 1989, and 1990a).

Modelling of Decision Agents

Decision agents in a system introduce some special modelling problems in MFM. Decision agents would typically be manual or automated control and supervisory systems. These agents would obviously be information processors, as their purpose is to observe system states, evaluate system goals, and to coordinate and execute of actions on the system in order to achieve selected goals. Furthermore, most systems comprise several agents which communicate and cooperate in the management of system resources. In the following. we will discuss problems in modelling the function of these agents.

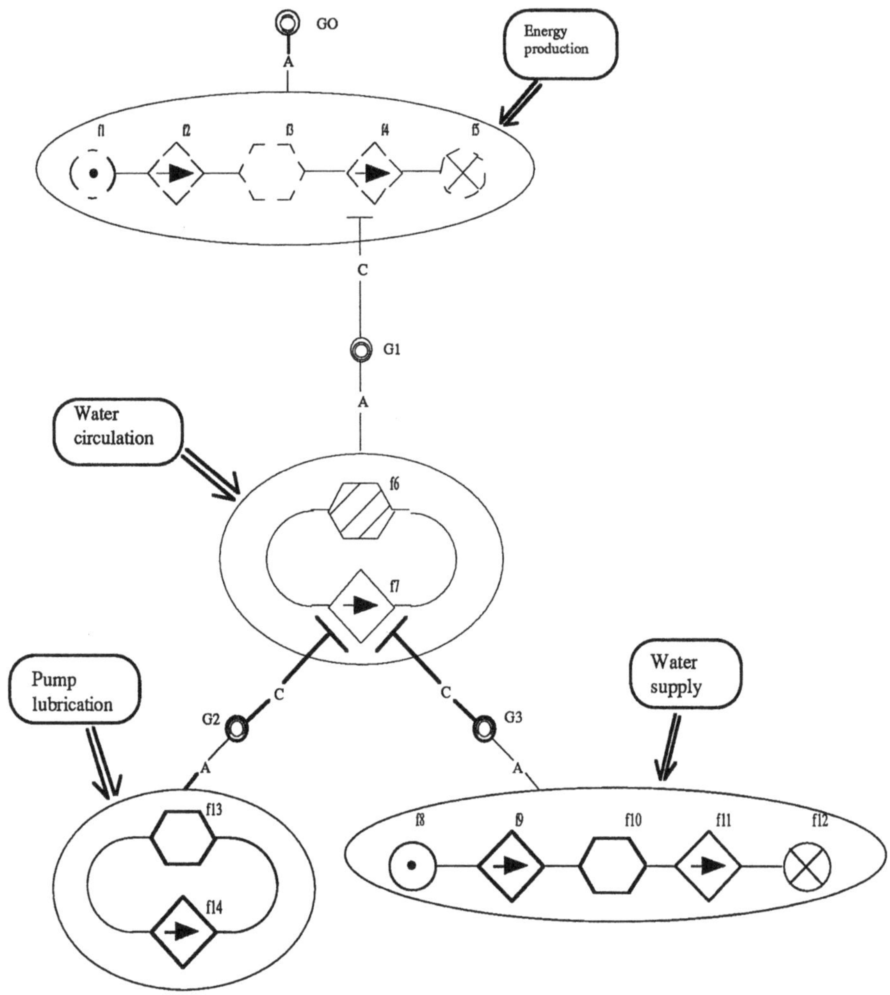

Figure 7. Multilevel Flow Model of the Central Heating System.

Control Systems

The functions of control systems can be implemented in a variety of ways. They can be based on mechanical principles (pneumatic or hydraulic), they can be based on analog electrical systems or on programmable digital computers using sophisticated software maybe based on knowledge based techniques. Control tasks can also be performed manually by humans.

Even though many types of implementations can be applied, these systems perform the same basic functions, since their purpose is to maintain specified conditions or constraints in the environment. Explicit modelling of these functions will, accordingly, clarify the functional structure of a complex system.

A control system analyzes system states, and plans and executes actions in the system. In a systems analysis, we would describe a control system as a means for goal achievement. But, since the decision problem solved by the control system includes an analysis (state identification) and a synthesis (planning of control actions) phase, the control system is an agent solving a design problem. Using the concepts of the systems approach, this design problem can be described in terms of the goals, system resources, strategies, criteria, and a model relating those elements. But in the case of modelling the agent, the resources would be information processing resources such as measured plant information, models of the system to be controlled, and means of manipulation of plant systems or components. The goal would be the requirements on control performance.

This means that, in systems with decision agents, we need to use the concepts of systems analysis for representational purposes. The structure of systems analysis comprises the basis for modelling of the agents. The understanding of the structure and conceptual categories of systems analysis become a pertinent problem for the use of MFM.

Control Systems in MFM

The representation of a control function in an MFM model is illustrated below in terms of an example from our central heating system. We have taken the water supply function, with its associated goal G3 and assumed that we have provided a control function f15 for the achievement of G3. As the control function is a resource which is provided for goal achievement, it is related with the goal G3 and the mass flow structure in terms of a special achieve relation called an achieve-by-control relation (see appendix A or Lind, 1990b). Note that the model does not tell how the control function is achieved; it only shows the structural relationship between a goal and a set of resources. The modelling of how this control function is achieved is the topic of the next section.

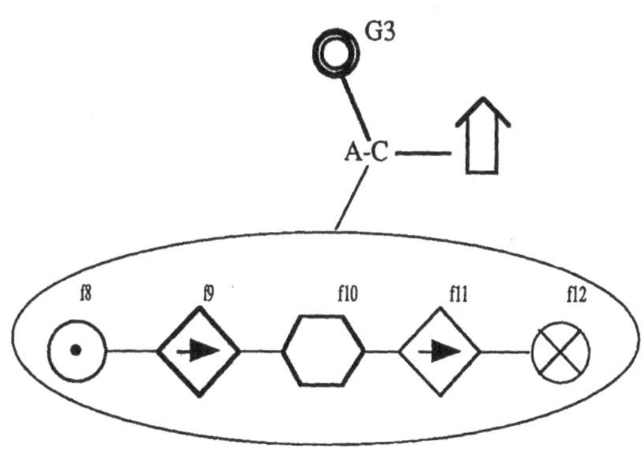

Figure 8. The water supply function with a control f15.

Modelling Information Processing Resources

The control function f15 in Figure.8 provides only an abstract representation of the role of the control system in achieving a goal G3. The means required in order to achieve control are described in terms of information processing resources. I use here a model of decision making such as that developed by Rasmussen (1986). The model is shown below in Figure 9, and before I describe its use in MFM, I will explain the elements of the model.

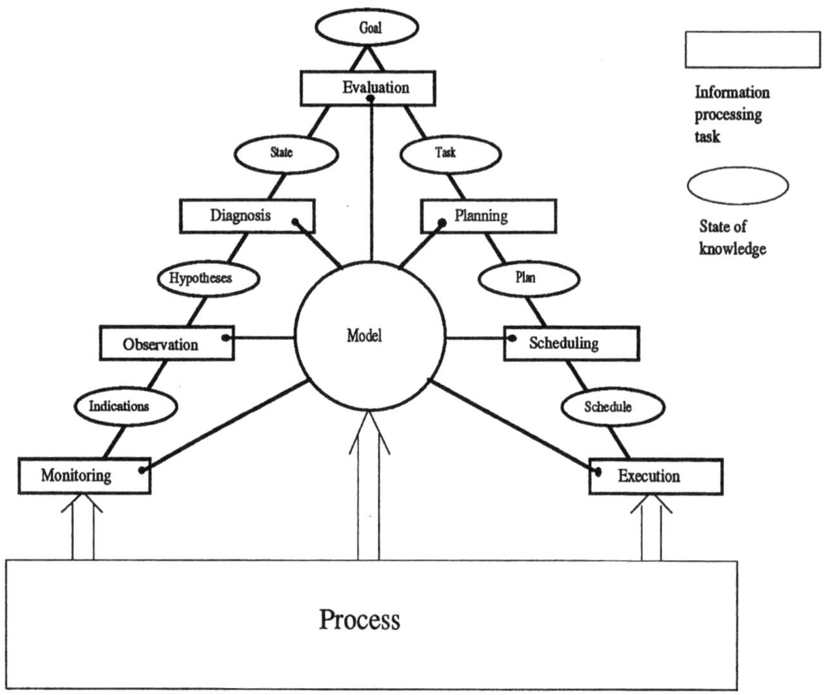

Figure 9. The decision model (Rasmussen, 1986).

The Decision Model

The model describes the information processing tasks which are involved in systems control. They are interconnected via states of knowledge which represent the information about the system as it evolves during the process of gathering measured plant data, the interpretation of the data, the evaluation of plant operational goals, and the subsequent planning and execution of control actions. As mentioned in Rasmussen (1986), several paths can be followed in the model, corresponding to different shortcuts of the information processes. I will not deal with these aspects of the model here as they have been analyzed in (Lind, 1984). In the present context I will consider each of the information processing tasks (and the states of knowledge) simply as resources for achieving control.

I have, in comparison with Rasmussen (1986), extended the model to include the process controlled and the knowledge (symbolized by the circle in the center) required for the different information processing tasks. The double arrow directed from the process to the model indicates a relation of representation. The model is a representation of the process under control and the model is a resource for the decision maker. We can consider the decision model as a representation of the information processing functions required for control; it is accordingly connected to the control function in Figure 8 by an achievement relation.

Modelling Control Tasks

As mentioned in Lind (1984), the decision model can be used in three basically different ways, in three different contexts. This is illustrated in Figure 10. Three levels of the decision model are shown, representing three of the control tasks involved in controlling the plant process R1. As an example, the plant process could be the water filling process in the central heating system (as modeled in fig 8), or it could be any other of the processes in the system described in Figure 7. Let us now examine the three control tasks.

Let us first consider the decision model representing information processing resources R2 involved in the control of plant resources R1 (e.g., the water filling process). The goal for this decision task is to achieve G3 (the requirement to the water level) (see Figure 8). The model M1 is a representation of the plant processes under control, and is used for diagnosing the plant state and for planning compensatory actions. The model M1 could be a representation of the mass flow structure in Figure 6 or it could a partial model thereof.

The information processing resources R2 are controlled by another decision maker (shown on top of the box named R2). The goal G4 of the decision maker is to ensure that control is accomplished (i.e., to ensure that the goal G3 is met with the satisfaction of control criteria). The goal G4 is accordingly a requirement on the use of the information processing resources R2. The decision maker manages the different alternatives for processing the information available about R1, and would accordingly have a model (M2) describing alternative processing strategies. The different shortcuts in the decision model described by Rasmussen would here correspond to different alternative information processing task sequences.

I have also shown how one of the processing tasks in R2 can be decomposed into a set of information processing resources (R3) and how a decision maker using these resources could accomplish a goal (G5). I have taken the diagnosis task as an example, but the same decomposition can be done with any of the tasks in R2 (resulting in different goals). In the case of the diagnosis task under consideration, G5 would be the diagnostic goal (e.g. to identify the state of every plant function). The processing resources R3 are decompositions of the diagnosis task and states of knowledge from R2. The model M3 would be a representation of the interactions between the tasks and states of knowledge in R3. M3 would be the basis for the diagnostic strategy implemented by the decision maker.

It is seen that we are able to model three different levels of control, including plant control, supervisory functions (diagnosis), and the management of information processing resources.

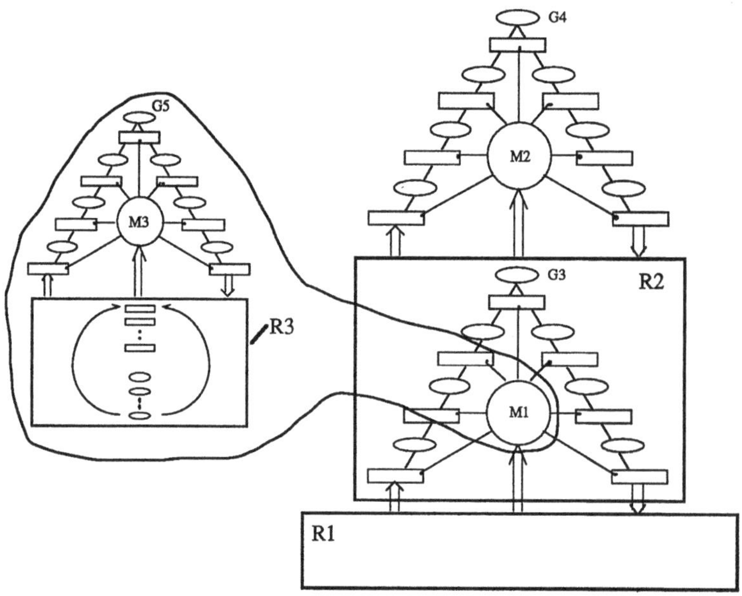

Figure 10. The use of the decision model in three different contexts.

Implementation of the Modelling Framework

The modelling concepts described above have been implemented in a preliminary version in a modelling system called ABSTRACTIONS developed by the author (Lind, 1990c). The implementation supports structured implementations of diagnostic systems. A system is under development which use a MFM model of plant as the basis for the reasoning. The diagnostic strategies (not the implementation) are described in Lind (1988b).

Conclusions

This paper has described concepts for modelling control tasks in complex systems. The concepts provide extensions of the MFM modelling technique (Lind, 1990a) and comprise a unified framework for the design of integrated control and supervisory systems. Even though the modelling technique has been developed for process plant applications, the basic concepts are so general that they could be applied to other domains, such as air traffic control, as well.

References

Blauberg, I.V. et. al. (1977). Systems Theory - Philosophical and Methodological Problems. Progress, USSR.

Churchmann, C. W. (1968). *The Systems Approach.* New York: Dell Publishing.

Lind, M.. (1982). Multilevel Flow Modelling of Process Plant for Diagnosis and Control. Ris¢-M-2357. Ris¢ National Laboratory. Denmark.

Lind, M.. (1984). "Decision Models and the Design of Knowledge Based Systems" .In Hol-nagel, E., Manzini, G. and Woods, D.(eds). *Human Decision Making in Process Environment* (proceedings from NATO ASI meeting in San Miniato, Italy.)

Lind, M.. (1988b). Diagnosis using Multilevel Flow Models.ESPRIT Project 96 (Expert System Builder), July.

Lind, M. et.al. (1989). "Human-Machine Interface for Diagnosis based on Multilevel Flow Modelling". Presented at: 2nd. European Meeting on Cognitive Science Approaches to Process Control, Siena, Italy, October 24-28.

Lind, M.. (1990b). Representing Goals and Functions of Complex Systems-An Introduction to Multilevel Flow Modelling.Institute of Automatic Control Systems, July.

Lind, M.. (1990c). ABSTRACTIONS- Description of Classes and their use. Report for the SIP project. Institute of Automatic Control Systems, July.

Lind, M. (1990e). "Representations and Abstractions for Interface Design Using Multilevel Flow Modelling". In G. Weir and J. Alty (Eds). *Human-Computer Interaction and Complex Systems.* London, Academic Press .

Rasmussen, J. (1986). *Information Processing and Human-Machine Interaction.* New York: North Holland.

APPENDIX A: Symbols used for graphical representation of MFM concepts.

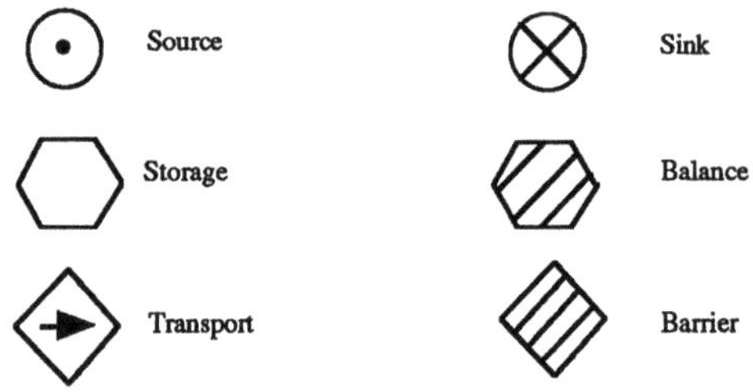

MASS FLOW FUNCTIONS

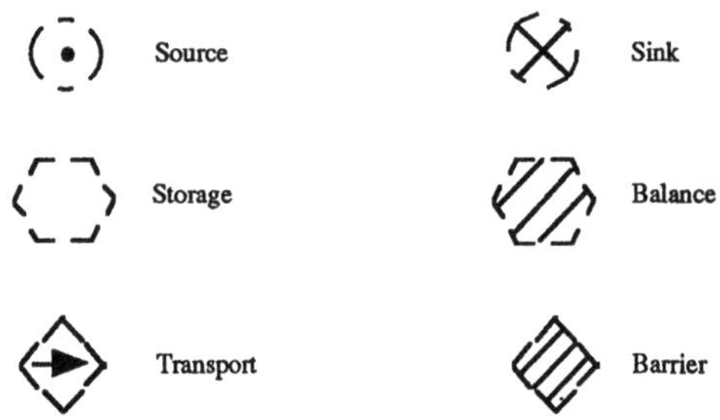

ENERGY FLOW FUNCTIONS

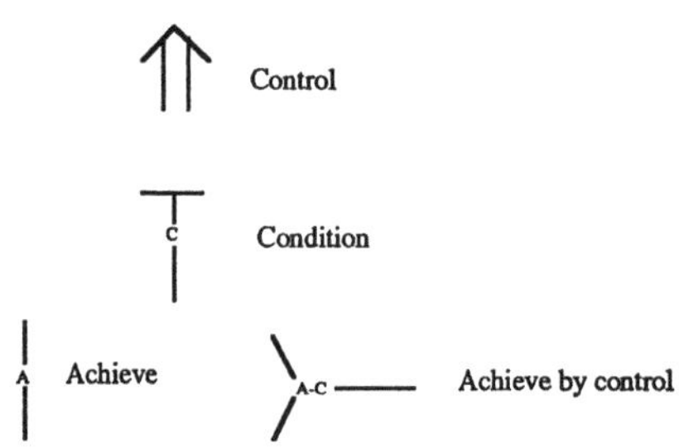

Mental Models in Air Traffic Control

Richard H. Mogford

Hickling Corporation, Ottawa, Ontario, Canada

Introduction

Air traffic control (ATC) is one of a number of situations in which humans must use remote control. Since it is impossible for the air traffic controller to directly observe and communicate with aircraft, electronic devices must be employed to collect and represent the required information and to relay the operator's decisions.

As display technologies continue to improve, an increasing amount of creative latitude is possible in the construction of the human-machine interface at air traffic control workstations. It is important that design be driven by an understanding of the cognitive dynamics inherent in the user-system interaction, rather than by technological innovation alone. Also, because special skills are required of the operators of ATC systems, study of the cognitive demands of the tasks involved will assist with selection and training.

Wickens (1984) observed three levels of representation in such a system. These are (a) the physical system, (b) the operator's internal representation of the system, and (c) the interface between (a) and (b). That is, "the display surface upon which changes in the system are presented to 'update' the operator's internal model and to form the basis for control action and decision" (Wickens, 1984, p. 179). He stressed that training is a method of creating the operator's internal model of a system.

A more detailed concept of the human-machine relationship is by Norman (1986), who distinguished between the Design Model, the System Image, and the User's Model. The Design Model is the planner's conceptual model of the intended system, taking into account the user's background and abilities. The domain, processes, and equipment to be controlled are also undoubtedly included in the designer's considerations.

The operator's mental model of the system is called the User's Model. However, it is not derived directly from the Design Model, but from the System Model, the physical expression of the hardware and software decided upon by the designer. Norman (1986) stated that the goal in system design is for the User's Image (gained through interaction with the system) to be compatible with the Design Model. He suggested that the System Image must be explicit, intelligible, and consistent.

According to Wickens (1984) and Norman (1986), the operator's internal representation (or User's Model) of a system is a key element in understanding the human-machine interaction. It is expected that the improvement of our inferences about the air traffic controller's internal representations will assist with the design of the human-machine interface and with the selection and education of operators.

NATO ASI Series, Vol. F 73
Automation and Systems Issues in Air Traffic Control
Edited by John A. Wise et al.

Definitions of the Mental Model

A mental model is a hypothetical construct which refers to the operator's learning and concepts about a system and what it is supposed to do. Wickens (1984) noted that the internal model helps guide and explain the operator's information gathering strategies. It is also used to design the necessary control inputs to achieve future goals, is the basis of the operator's expectancies about co-varying events, and so is useful in the detection of problems. Another statement of this idea is:

> Mental models are the mechanisms whereby humans are able to generate descriptions of system purpose and form, explanations of system functioning and observed system states, and predictions of future system states. (Rouse & Morris, 1986, p. 351)

Norman (1986) described this concept in the following:

> Mental models seem a pervasive property of humans. I believe that people form internal, mental models of themselves and of the things and people with whom they interact. These models provide predictive and explanatory power for understanding the interaction. Mental models evolve naturally through interaction with the world and with the particular system under consideration. These models are highly affected by the nature of the interaction, coupled with the person's prior knowledge and understanding. The models are neither complete nor accurate, but nonetheless they function to guide much human behavior. (p. 46)

In the case of air traffic control, the controller's training, coupled with the information available from the radar screen and other displays, supports the construction of a mental model of airspace, the objects within it, and the rules which govern their dynamics. The fact that some kind of internal model exists is demonstrated by the controller's ability to project the future positions and speeds of aircraft in order to anticipate possible conflicts. A controller is able to continue to do this for a period of time even if deprived of the radar display (Whitfield, 1979).

The Characteristics of the Mental Model

It is tempting to assume that the controller's model is simply a visualization of aircraft, navigational markers, terrain, and other pictorial elements of the system. However, there may be other aspects of this concept which are not visually represented.

Rassmussen (1979) described mental models as having different levels of abstraction, including physical form, physical function, functional structure, abstract function, and functional meaning. Johnson-Laird (1983) distinguished between physical and conceptual mental models. These comments suggest that there may be a number of mental models of a system, ranging from the "concrete" (an analog, visual image) through various forms of increasing abstraction depicting functional relationships and operating rules, to a level subsuming the overall meaning and purpose of the system (Wilson & Rutherford, 1990). Although some type of model based on a visual image is probably involved in many air traffic control tasks, it may not be sufficient in itself for all aspects of the work. Other kinds of conceptual, verbal, and numerical information must also be present and be applied as needed.

It should be noted that the definition of the mental model is not fully agreed upon. Wilson and Rutherford (1990) commented that "We have shown the several different interpretations of the concept and its utility to be a weakness, which militates against the widespread use of mental models in systems design" (p. 629). When this concept is applied to the study of a particular domain, a different set of findings will emerge, depending upon which aspect of the internal model is being emphasized.

Theory and Method in Research on Mental Models

Rouse and Morris (1986) discussed some of the issues regarding research on mental models. They emphasized that a clear definition of terms is important in order to distinguish a mental model from knowledge about things in general. Accordingly, the themes of describing, explaining, and predicting the behavior of a system or device were stressed. In an attempt to guide investigative work, Rouse and Morris (1983) suggested that internal models of various types be defined by two dimensions, the nature of model manipulation and the level of behavioral discretion. Their concept is diagrammed in Figure 1.

The "nature of model manipulation" axis refers to the degree of one's awareness of the presence of an underlying mental model. The "level of behavioral discretion" axis describes the degree to which behavior is open to choice or is dictated by a task.

Inferential (or indirect) methods of studying the mental model are useful when the operator is assumed to have little discretion. This is because the constraints of the system or domain restrict behavior and make it feasible to infer mental operations from task descriptions and direct comparisons of input and output. Verbal (or direct) methods are effective when the domain involves higher levels of explicit manipulation, because cognitive activities are more accessible to introspection. It can be seen from this chart that situations represented in the upper left of the space in Figure 1 will be difficult to study, using any of these approaches, because of the low level of self-awareness and high degree of choice or variability.

Where would the activity of air traffic control lie in this space? As a task, it is similar to supervisory control, in that there is considerable monitoring of variables with the goal of error detection. However, there is also an ongoing requirement for intervention, which introduces elements of manual control. Therefore, for the experienced controller, model manipulation may only be partially conscious, with a range of levels of behavioral discretion, depending upon whether the operator is passively monitoring or actively controlling. According to the preceding guidelines, it should be possible to study this area using both inferential and verbal methods. However, there will be limitations on the effectiveness of both techniques.

Rouse and Morris (1986) discussed the usefulness of discovering the forms, structures, and parameters of mental models. When researching the mental representations of the air traffic controller, it will be important to discover the form the model takes (sensory code, spatial dimensionality, etc.), its structure or organization, and its parameters (number of items included, capacity, accuracy, etc.). These authors also noted that "experts' mental models are not just more elaborate or accurate; evidence suggests that they are fundamentally different from novices' models" (Rouse & Morris, 1986, p. 356).

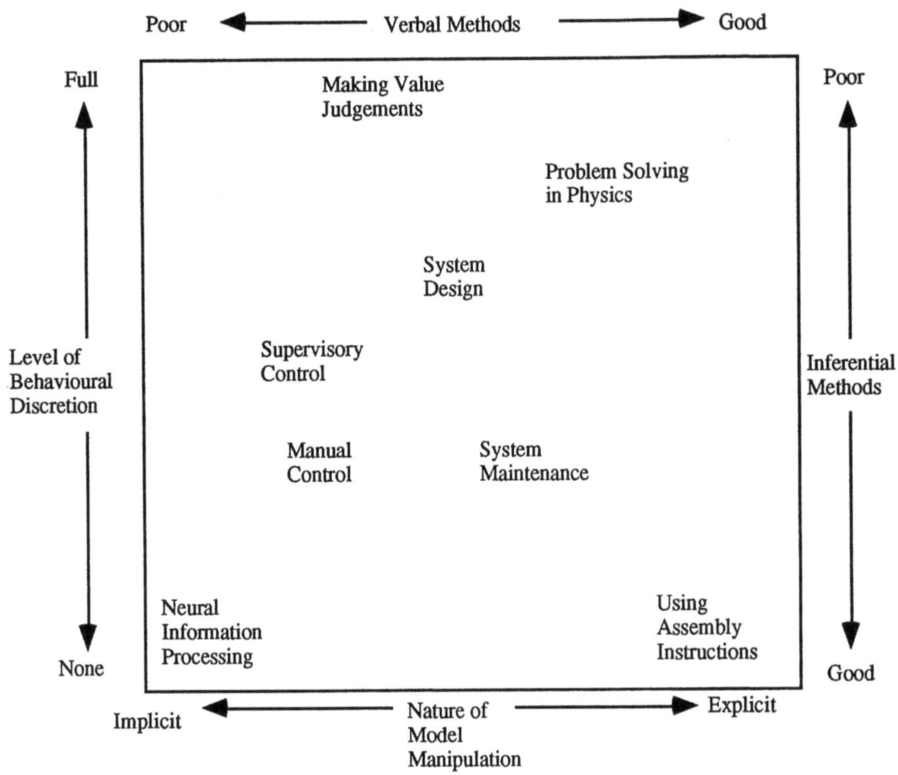

Figure 1. Distinctions among domains of mental models. Adapted from "On looking into the black box: Prospects and limits in the search for mental models" by W. B. Rouse and N. M. Morris, 1986, *Psychological Bulletin, 100*, p. 354.

Previous Research on Mental Models in ATC

Research on the air traffic controller's mental model (controllers call it the "picture") has been limited. Whitfield (1979) and Whitfield and Jackson (1982) elicited verbal reports from experienced Canadian controllers and suggested that such studies could assist with interface design, training, and selection. "Thus a better understanding of the picture should yield information on controller capacity based on the underlying psychological processes, rather than only on system measures" (Whitfield, 1979, p. 19).

In Whitfield's (1979) study, experienced controllers seemed to comprehend the picture in larger units, and three different kinds of memory seemed to be involved: a static model of the air traffic route structures, separation guidelines, and procedures; a dynamic model including the changing information about particular aircraft (climb or descent rates, turns, etc.); and working memory for the current position and status of each aircraft. It was also noted that most subjects (especially controllers with non-radar experience) reported the picture to be a three-dimensional, geographical representation which assisted with the understanding of the air traffic control situation. The identification of problems was also associated with the picture. It was sometimes described as a plan against which aircraft behavior was compared and which helped in the detection of conflicts. Other interesting research which will be reviewed briefly below includes work by Lafon-Milon (1981); Lapin (1982); Landis, Silver, Jones, and Messick (1967); Means, Mumaw, Roth, Schlager, McWilliams, Gagné, Rice, Rosenthal, and Heon (1988); Bisseret (1970); and Moray (1980).

In spite of these efforts, the question of whether the controller's picture is an internal spatial representation in three dimensions, as suggested by Whitfield (1979,; whether it is only two-dimensional, or even lacking in any visuospatial qualities, remains open. However, it appears that the controller's mental model possesses various kinds of information which are reliant on different types of memory. Maps, flight plans, aircraft performance information, separation standards, and procedures are learned through training and experience and stored in memory. However, as Moray (1980) and others have suggested, this part of the controller's mental model, once learned, may be largely unconscious.

The term "picture" probably refers to the more conscious parts of the internal model, such as could be maintained in short term memory. If it is true that the unconscious aspects of the model control sampling and organization, then the influence of this underlying learning may be reflected in overt operator activities. For example, Means et al. (1988) observed a traffic drawing task and discovered that controllers segregated aircraft into groups. The unconscious rules guiding this patterning behavior were inferred from an analysis of these responses.

Applying Mental Model Research to ATC Systems

The concept of the mental model provides a convenient way of thinking about the operator's cognitive processes. However, apart from being an interesting, if at times ambiguous, vehicle for theorizing about complex systems, can it be applied to assist with practical problems? How can the mental model be objectively measured? Is there any evidence that the air traffic controller's ability to form a picture is an important skill?

If we could look inside the air traffic controller's mind and examine the picture, the resulting knowledge could assist with the selection and training of controllers. It is reasonable to expect that ATC candidates have varying amounts of aptitude for developing and maintaining a mental model. If this could be demonstrated, along with evidence that this skill is related to the quality of on-the-job-performance, it would be important to develop screening tests to determine whether ATC candidates possess this ability. It would also be useful to measure how well students are able to build a picture as they progress through ATC training. In addition, methods might be developed to actively foster this skill.

Significant effort is expended during the design of advanced ATC workstations to provide the information the operator needs in the form most useful for task execution. In addition, it is desirable to configure the hardware and software so that it is "transparent" or does not obscure or confuse the user's understanding of the domain (in this case, aircraft in airspace) which must be monitored and acted upon. Vincente (1988) presented a convincing argument that the effectiveness of a user interface can be measured by how well it supports the operator's mental model. If techniques can be developed to assess the quality of the model, they would allow the comparison of various candidate systems on the basis of how well each supplies the required information, allows the necessary activities, and offers the fewest impediments to the formation of the controller's mental model.

The above suggestions, while intriguing, presuppose the existence of methods to measure the air traffic controller's picture. Rouse and Morris' (1986) work suggests that the air traffic controller's model should be accessible using both inferential and verbal methods. As mentioned above, Whitfield (1979) employed verbal reports from controllers. Other researchers have used a variety of techniques, two of which will be discussed further.

Lapin (1982) and Landis, Silver, Jones, and Messick (1967) applied multidimensional scaling (MDS) to explore the air traffic controller's conceptual space. MDS is a mathematical procedure which assists in the exploration of data by spatially representing the similarities between objects. Subjects are asked to make comparison judgments between pairs of stimuli (such as aircraft on a radar display). MDS treats the matrix of interobject comparisons as analogous to a set of interstimulus distances, and plots the stimuli in a Euclidean "stimulus space." The dimensions of this space correspond to the common stimulus attributes the subject has chosen to compare one object with another (such as altitude or speed). A related technique, cluster analysis, employs the same kind of data but displays the stimuli in groups based on the subject's similarity judgments.

Lapin (1982) used MDS to investigate the spatial characteristics of the controller's picture. Two groups, "highly qualified" and "satisfactory" ATCs, were tested. The MDS results showed that the skilled group's stimulus space was three-dimensional, while the less skilled group's space had only two dimensions. Landis, Silver, Jones, and Messick (1967) successfully used MDS to look at the similarities between various air traffic control problems. They discovered a number of common themes controllers used when comparing different situations they might encounter. The complexity of the judgments increased with the skill level of the controllers. More recently, Murphy, Reaux, Stewart, Coleman, & Kelly (1989) suggested that MDS be applied to capture the mental picture of current and future air traffic situations, and to explore knowledge of ATC rules and operating procedures.

Another approach which has been successfully employed is based on the operator's recall of radar screen information. This was the technique suggested by Vincent (1988) for studying interface transparency (in process control) and was used by Means et al. (1988), Bisseret (1970), and Moray (1980) with regard to air traffic control. Means et al. (1988) found that controllers recalled 86% of the aircraft at the end of a dynamic simulation exercise, and that certain target characteristics (such as position), were remembered more frequently. Bisseret (1970) found that highly qualified controllers had better recall for aircraft than average controllers or trainees; altitude and position information were recalled most reliably. The author has just completed experiments with similar results (Mogford, 1990).

The preceding has demonstrated that objective tools exist for exploring some facets of the operator's mental model of a complex system. However, this discussion has not yet revealed any empirical proof that the air traffic controller's ability to create a picture has any relevance to the job of controlling air traffic.

Whitfield's (1979) interviews with controllers resulted in conclusions that the picture helps the controller maintain an understanding of the situation, acts as a "supplementary display," enhances the feeling of being in control, and replaces some information lost during equipment failure. A number of the research studies reviewed above have also shown that skilled controllers (as compared to less skilled controllers or trainees) conceive of the domain differently and have better memory recall for aircraft.

Recent research by the author has gone a step further toward demonstrating that the quality of the picture is related to air traffic control expertise. In a study involving Canadian air traffic controller trainees, it was found that their ability to correctly recall aircraft information from a radar simulation was significantly related to how well they eventually scored in final simulation tests (Mogford, 1990). This result, combined with other related research, suggests that the ability to form a good picture of air traffic is operationally important. This work also reveals that a realistic, dynamic simulation of the controller's tasks provides an effective environment for evaluating the mental model.

An air traffic controller's ability to accurately remember the large quantity of information represented in the picture is, as Vincente (1988) has emphasized, probably related to its meaningfulness. Meaning evolves as a user gains knowledge of a system. However, not every ATC student appears to be able to apply this knowledge to develop a good mental model, and this has been shown to undermine the capability to control traffic to some degree. These individual differences suggest that a measure of how well an applicant for ATC training can form a picture should be considered for inclusion in future screening test batteries. A trainee's recall of the picture could be a useful method of assessment.

MDS, cluster analysis, and recall of the picture could be employed to investigate specific computer interface design questions. These techniques could assist with such problems as menu design, screen organization, color coding, and computer aids.

In conclusion, the concept of the mental model in the context of air traffic control is not only a useful theoretical construct, it also has practical applicability. Although the techniques discussed above are still in an early stage of development, there are promising indications that these tools can help us measure some aspects of the controller's mental model or picture. There is also evidence that the quality of a controller's picture is related to operational skill level. Scaling techniques such as MDS and cluster analysis, as well as memory recall of aircraft information, have potential for exploration of the mental model. On a practical level, they may be used in new selection testing procedures and applied to the assessment of human-computer interfaces.

As air traffic control systems continue to become more automated, the effects of moving the user to ever higher levels of supervisory control is of considerable concern. Being out of touch with detailed aircraft information and restricted to monitoring automatic systems, how will the controller of the future retain a good picture? What sort of mental model will be important for someone using this kind of system? Research has not yet answered these questions. However, it is hoped that further explorations of the air traffic controller's mental model will result in the theory and tools with which these problems can be addressed.

References

Bisseret, A. (1970). Mémoire opérationelle et structure du travail. *Bulletin de Psychologie, 24*, 280-294.

Johnson-Laird, P. N. (1983). *Mental Models.* Cambridge, England: Cambridge University Press.

Lafon-Milon, M. T. (1981). Représentation mentale de la separation verticale au cours du diagnostic dans le controle aerien: III Représentation des états futurs (Report No. CO 8107 R66). Rocquencourt, France: INRIA.

Landis, D., Silver, C. A., Jones, J. M., & Messick, S. (1967). Level of proficiency and multidimensional viewpoints about problem similarity. *Journal of Applied Psychology, 51*, 216-222.
Lapan, Y. A. (1985). Spatial representation and the activity of the air traffic controller. Vestnik Moskovskogo Universiteta Seriya 14 *Psikhologiya, 14*, 68-70.
Means, B., Mumaw, R., Roth, C., Schalger, M., McWilliams, E., Gagne, E., Rice, V., Rosenthal, D., & Heon, S. (1988). ATC training analysis study: Design of the next-generation ATC training system (OPM Work Order No. 342-036). Richmond, VA: HumRRO International.
Mogford, R. H. (1990). Explorations of the Air Traffic Controller's Mental Model. Doctoral thesis in preparation, Carleton University, Ottawa, ON.
Moray, N. (1980). Human information processing and supervisory control (Report No. AD-A092 840). Cambridge, MA: MIT.
Murphy, E. D., Reaux, R. A., Stewart, L. J., Coleman, W. D., & Harwood, K. (1989). Modeling air traffic controller performance in highly automated environments. *Proceedings of the Human Factors Society* (pp. 47-51). Santa Monica, CA: Human Factors Society.
Norman, D. A. (1986). Cognitive engineering. In D. A. Norman and S. W. Draper (Eds.), *User Centered System Design* (pp. 31-61). Hillsdale, NJ: Lawrence Erlbaum.
Rasmussen, J. (1979). On the structure of knowledge: A morphology of mental models in a man-machine system context (Report No. RISO-M-2192). Roskilde, Denmark: Riso National Laboratory.
Rouse, W. B., & Morris, N. M. (1986). On looking into the black box: Prospects and limits in the search for mental models. *Psychological Bulletin, 100*, 349-363.
Saxe, J. G. (1955). The blind men and the elephant. In W. P. Percival & J. G. S. Brash (Eds.). *Poems to Enjoy* (pp. 76-77). Toronto: Thomas Nelson & Sons.
Vincente, K. J. (1988). Adapting the De Groot memory recall paradigm to evaluate interface transparency. (Report No. RISO-M-2691). Roskilde, Denmark: Riso National Laboratory
Whitfield, D. (1979). A preliminary study of the air traffic controller's picture. *CATCA Journal, 11*, 19-28.
Whitfield, D., & Jackson, A. (1982). The air traffic controller's 'picture' as an example of a mental model. In G. Johannsen & J. E. Rijnsdorp (Eds.), *Analysis, Design, and Evaluation of Man-Machine Systems* (pp. 45-52). Duesseldorf, W. Germany: International Federation of Automatic Control.
Wilson, J. R., & Rutherford, A. (1990). Mental Models: Theory and application in human factors. *Human Factors, 31*, 617-634.
Wickens, C. D. (1984). *Engineering Psychology and Human Performance*. Columbus, OH: Charles E. Merrill.

The Controller in Human Engineering

Michael Tonner

Canadian Air Traffic Control Association, Ottawa, Ontario, Canada

Introduction

Throughout most of the world, but most especially in the industrialized nations, major changes are underway to modernize and update existing air traffic control systems. Not since the advent of the use of radar for air traffic control purposes following World War II have there been so many profound changes underway or on the near horizon. New high speed, large capacity computers, enhanced automation, and improved radar and communication capabilities offer us a chance to make a great step forward.

Much of this change is being driven by the burgeoning growth in air traffic on a global basis. Crowded skies, congested airports, and shortages of air traffic controllers have literally become major international problems. Increased productivity, efficiency, reliability, and safety are the demands of governments and agencies concerned with aviation directed to their planners and the supporting industry.

Can we do more with fewer air traffic controllers? Can we increase the controller's productivity without jeopardizing safety? Can technology permit decreased separation standards so that more airplanes can safely operate in less airspace? Can we produce an air traffic control system that can reduce the chance of human error thus enhancing aviation safety? Can we build and operate a system that will reduce or hold down current costs? The list could go on, but I hope this and other similar forums can provide both answers and guidance for all concerned with building and operating national and international air traffic control services.

My role, as I perceive it, is to provide input and insight from my experience as an operating air traffic controller; to share with you my thoughts, concerns; and knowledge of operational requirements that may aid in the design, development or refinement of new or improved ATC equipment or systems.

I joined air traffic control in Canada and qualified as an airport controller (tower) in 1963, followed by an instrument flight rules (IFR) rating for both en-route and oceanic control in 1965. My operating experience has been at the Area Control Centre (ACC) in Gander, Newfoundland. I hold a pilot's licence and have a number of years of service as an officer for the Canadian Air Traffic Control Association (CATCA). At the national level, I have co-chaired the Transport Canada-CATCA Joint Technical Committee and currently am serving as the national Vice President, Technical of CATCA.

Both my training in computers and automated systems along with my technical role for the Association have provided me with opportunities to identify both the weaknesses and strengths of the existing Canadian Automated Joint En-route and Terminal System (JETS), and to provide input into the Radar Modernization Project (RAMP), which is now in the

NATO ASI Series, Vol. F 73
Automation and Systems Issues in Air Traffic Control
Edited by John A. Wise et al.

delivery phase at a cost of approximately $750 million (Canadian), and the Canadian Automated Air Traffic Control System (CAATS) contract ($350 million Canadian), which has just been let to Hughes.

Having said this, let me turn my attention back to my stated purpose of sharing with you my thoughts and concerns as an air traffic controller. The nature of the job requires a capability to assess complex traffic situations and make rapid decisions that will keep aircraft moving both safely and expeditiously. The consequence of error can be enormous and, coupled with the environment in which the air traffic controller works, makes this a highly stressful occupation. In spite of considerable efforts to find effective screening processes, most administrations continue to experience high attrition rates, particularly in early training stages. In Canada, following successful entry into the national training institute, more than 50% fail to qualify as licensed controllers. In fact, for those in the direct entry stream for IFR, the success rate, on average, is less than 25%.

There is a great sense of job satisfaction that feeds back at the end of each shift. You know if you have done a good job, and the challenge for tomorrow is "Can I do better or at least as well?" The ability to see dozens of airliners responding to your commands can be a very rewarding experience. As we move to the application of sophisticated systems, both on the ground and in the cockpit of those aircraft, is this sense of being in-charge going to be diminished for either controller or pilot or both? As the human portion of the system becomes less, will the job satisfaction alter or change? Are the types of controllers and pilots we have today going to mesh with the coming technological changes? Can we continue the human decision making process in a meaningful way, as more and more tasks become automated?

Is this the future?

Let me give you a scenario that gives me and, I expect, my pilot colleagues some concern when we look to the future and what technology can/might do to us.

In the *Aviation Week and Space Technology* issue of April 23, 1990, an article appeared entitled "Boeing Condor Raises UAV Performance Levels." The Condor, a twin engined unmanned aerial vehicle (UAV) made of light weight composite materials and building on new and existing technologies operated aloft for 2.5 days at altitudes up to 65,000 feet. Some 60,000 lines of code are used to accomplish navigation, flight control, and vehicle management functions that permit, if desired, autonomous operations based on the pre-planned mission and desired flight profile. Navigation is accomplished through inertial navigation, but later vehicles will also be able to use the Global Positioning System. In the landing and departure phases, a microwave landing system (MLS) is used for guidance. The flight operated, in part, in air traffic control areas and was accompanied by a chase plane in the traffic pattern. This UAV has a wing span (200 feet) greater than that of a Boeing 747 and a 20,000 lb. gross take off weight.

Given that software programs now exist to predict traffic confliction and provide solutions, it seems highly probable that as we enter the 21st century we will see the black boxes in unmanned aircraft communicating and exchanging flight data with black boxes on the ground and receiving in return, from unmanned (at least by air traffic controllers) ATC centres required instructions and clearances. Where or how will the future pilot and controller fit in? If we are going down this path, then I hope this scenario is evolutionary and one in which the human-machine interface is carefully thought out. Both controllers and pilots are looking for safe, soft landings, should this be the future.

What do we want to know?

Now let me come back to today and the more immediate problems we face in the next decade. Let me start by asking questions for which I (and most other controllers) don't have the answers.

- How best can computer-software programs assist air traffic controllers?
- Can controllers (humans) really be relied on to back up computers?
- Can controllers cope with the increased efficiencies arising from the application of technology?
- If all routine tasks are eliminated, can controllers cope with an endless succession of "executive decisions"? At the moment, some routine tasks give the controller a chance to catch his or her breath.
- Can controllers stay ahead of the machine?
- How can we make the transition from basically human to automated or computerized systems?
- Does the body of expertise such as is represented here today recognize the disparity between the existing air traffic control systems around the world? In some areas , they are still struggling to obtain basic radio equipment, while elsewhere , clearances are being delivered via data links requiring no voice communication.
- Do the human factors people appreciate the differing working conditions, skill, and knowledge levels found in ATC, such as between an airport controller and a radar controller in an en-route area control centre, or the different skill sets between a sophisticated Terminal Control Unit Controller and one operating as a procedural controller without radar and limited communications?
- What new types of training or qualifications are called for by the new technologies or capabilities?
- How do we do effective re-training? Can we effectively train "old dogs" to do "new tricks"?
- How do we compensate pilots and controllers for a diminished sense of command and control that contributes to their sense of worth and job satisfaction?

- Do we have a grasp of the appropriateness of duty times for controllers in the new work environment?

Do studies cover these points?

Now let me suggest some things you should be aware of when evaluating the human factors (ergonomics) of the human/computer interface. We must assure that specialists such as yourself fully understand:

- Operational requirements

- Environment in which the equipment will be used

- Impact on the controllers of "doing it differently"

- Provision of appropriate training/counseling to get us from today to tomorrow

- Use of simulation to the greatest possible extent to confirm hypotheses of increased efficiencies and/or capabilities, as well as the logical placement of tools and assists required by controllers

- Full testing of mockups of work stations and associated controls, displays, alerting functions, colours, etc. — initially in the lab and then ideally in the operational environment or, as the minimum, in a realistic simulator

- Testing/evaluation should be more than "sit down and give us your opinion," testing must be long enough in duration to overcome resistance to change or initial acceptance "just because it is new and has lots of bells and whistles"

- Let's make sure that what is delivered is driven by legitimate controller requirements and not by companies anxious to sell new technology or unneeded capabilities

- There must be a total human factors assessment in the development of systems, equipment, and workstations

- A chair that is really a 24 hour chair, one that will accommodate to rotation movement, good back support, good range of easy adjustments (perhaps with memory lock)

- An arrangement of display information in a logical sequence, easily read

- Visual cueing that draws a controller's attention to a required task

- Combinations of visual and audio cues to priorities

- Head sets that are light, effective, and adjustable — or perhaps better still — built into the wings of the chair at ear level

- Radar or video displays that can stand normal ambient lighting

- Radar displays that don't wash out in the tower cab in daylight

- Keyboards and function keys that are designed for human interface, set up (functions) for ease of use and to reflect priority or frequency of use and compatible to either hand use

- Fail safe systems

- Room environments that are healthful and contribute to the job controllers are doing

- Work stations that offer ease of reconfiguration that can be collapsed to one person operation or expanded to team levels, as traffic and circumstances dictate

- Simulations that reflect and portray the actual operation

- Expert systems that reduce clerical or secondary duties and aid in the control decisions controllers have to make

- Assurances and safeguards that the electronic world and visual displays won't harm a person, or an unborn child

- Will low-level radiation problems require limitations on duty times?

- Will constant use of visual displays affect controllers' eyesight?

How do we do it?

The prime source for information and feedback on the development of new automated air traffic control systems and equipment is the person working with today's developing systems —the air traffic controller. We are the experts in our field as you are in yours. Together, we possess the knowledge and ability to design the best possible system for the future air traffic control system. Separately, we will not succeed! We must guard against the temptation to each "do our own thing." It is vital to the success of the endeavour that each plays his or her part from project inception through testing, simulation, and installation. These controllers must be active, line controllers, responsible for separating aircraft in today's system.

Considering the size of today's system and the high investment required, second chances will be hard to come by. While modifications and enhancements will undoubtedly follow, there will likely be no major changes in the immediate future. Let us work together, as a team, and get it right the first time!

Finally, and again, you need to understand controllers' requirements, how they work, their priorities, and their interaction with the users. You need to work with us and we with you to build a safe, effective, and efficient air traffic control operation.

Acceptance of New Systems by Air Traffic Controllers

Egbert Seidel

Air Traffic Control Consultant, Dreieich, Germany

Introduction

Air traffic control is a most desirable and exciting career which affords responsibility, stress, flexibility, and experience in a high technology system. Controllers must be selected, educated, trained, and licensed before they become familiar with the daily workload and sovereign in making quick decisions.

In air traffic control, new procedures and equipment cause frequent change in the daily working circumstances for the controller. Though relatively small, these changes make a difference. New subsystems, like weather-information displays, and new technical systems, like radar displays, interrupt the controllers' routine and render the habits of years invalid overnight. New equipment and systems—because of their complexity, size, task, and performance—demand an indoctrination phase, not only for technical reasons but also for human ones. Controllers must be given opportunity to accept advantages or reject possible disadvantages in handling and operating new technology. Besides habitual experience, they need to gain confidence in the technical systems, and gaining confidence takes time.

The Controller Generations

The first generation of air traffic controllers was trained as procedural controllers, radar not being available for civil purposes yet. The staff gained experience in applying standard separation for the piston engined aircraft at that time. The change from procedural to radar controller was relatively complicated and took time.

The second generation was trained as procedural controllers as well, but many of them were committed to becoming approach (radar) controllers from the very beginning. The change from analog (primary) radar to synthetic display met reluctance from some controllers, since the targets' turns could not be observed as quickly and precisely as on the previous, familiar displays. Then the appearance of radar simulators in the air traffic control schools altered the training completely. From this moment on, radar controllers came out of training fully qualified for radar. But, worldwide, they were all used to the conventional control strip.

NATO ASI Series, Vol. F 73
Automation and Systems Issues in Air Traffic Control
Edited by John A. Wise et al.

Stripless System

The paper control strip is a typical example for a tool of great importance to the air traffic controller, having remained in the system for decades. Besides radar information, current flight plans are the most important data in air traffic control. In the old days, they were handwritten on different colored paper strips, the color itself denoting some information (direction of flight, in- or outbound, etc.) automatically. Important information was marked in red pencil, some data in bigger size than less important ones. The pencil was the proved, accepted "input device" for the data assistant — as it still is for controllers — to mark quickly all flight progress data according to their clearance. (The "check mark" is still an issue in the evaluations performed today on electronic displays.)

At the very beginning of the computer era, computerized flight-strip processing (with all the advantages of the network-system) reduced flight-plan coordination by telephone. Different-colored strip-holders now replaced the different-colored paper strips. For more than ten years, several national air traffic control administrations have been evaluating various concepts for replacing the printed control-strips by electronic data display. The system of the future shall be a system without traditional paper control strips.

Experimental Centers

Most of the European countries established their own experimental and test centers for air traffic control to evaluate suggestions, ideas, and concepts. Different kinds of simulators are used to simulate reality as closely as possible. These simulations are performed in cooperation with experienced, active controllers, to gain realistic results and encourage the acceptance of new systems by their colleagues. These experimental centers are equipped with the most modern technology available at their construction (for example, color raster scan high resolution displays and various input devices).

The Federal Republic of Germany established the experimental workposition simulator, at the test center of the Federal Administration of Air Navigation Services, as a suitable means to evaluate new subsystems before their implementation into operational deployment. In 1988, a simulation was carried out to test a new data input device with new presentations of flightplan and flight movement data. The goal is partly automated coordination between electronic data displays and touch input devices.

In 1989, experimental workposition simulator began real-time simulations of a stripless ATC system. These are continuing at two sectors of the Upper Area Control Center at Karlsruhe, to verify results previously achieved in control sectors of the lower airspace. To date, the acceptance of the investigated concept is predominantly positive. However, a number of relevant aspects remain to be investigated in detail. These are the big steps to use electronic data displays in replacing the control strip in medium term development (in Germany, one sub-system has already been installed; the second one is due to be installed in autumn 1990; both are supposed to support the replacement of paper control strips in the control towers and approach control offices).

Computer Oriented Metering, Planning and Advisory System (COMPAS)

COMPAS is a computer-supported planning system for inbound traffic to encourage the air traffic controller in sequencing aircraft approaching a high density airport. It is designed as an advisory system without restricting the controller's domain of decision and responsibility. COMPAS was developed by the German Aerospace Research Establishment at the Institute for Flight Guidance, in cooperation with the German Federal Administration of Air Navigation Services (BFS).

In 1988, the system was investigated with respect to its applicability in air traffic control systems, based on a simulated situation of the Frankfurt Regional Control Center. Besides measurable operational improvements (reduction of coordination effort, increased use of runway capacity, smoothing of flight paths), the acceptance of COMPAS by the BFS air traffic control staff taking part in this simulation must be emphasized. Since September 1989, BFS has been using COMPAS in daily operation at its Regional Control Center in Frankfurt.

Departure Coordination System (DEPCOS)

Another electronic system is DEPCOS (Departure Coordination System), a system for advanced flightplan data processing in air traffic control. With DEPCOS, the paper control strips in the control tower and at the departure controllers' position will be replaced by a separate computer-system with an input device for all departing aircraft. DEPCOS is due for operation in September 1990. Before actual operation, an intensive training period is planned for each tower and approach controller, as well as for each flight data assistant. This training will be performed in a separate mock-up in the tower building, during parallel installation of the DEPCOS system in the tower-cabin.

Conclusion

For more than 25 years, extensive planning and evaluation of future air traffic control systems has been in progress. First-generation computer systems and radar technology had their impact of automation on ATC systems (Hopkin, 1967). The increasingly-rapid development of several generations of computer systems along with more complex, sophisticated radar systems and their display possibilities, have further influenced air traffic control systems and their administrations. The air traffic control system is a human-machine system which places primary reliance on the human component; the human operator is often the limiting factor in system output (Dailey, 1969). The acceptance of new systems by the individuals has to be promoted by:

- Motivation
- Volunteers
- Participation of a large number of controllers in evaluation
- Assistance by experienced air traffic controllers in research and development.

References

Dailey, J. T. (1969). “Management factors in reducing ATCS stress.” *The Controller: IFATCA Journal of Air Traffic Control.* (March).

Hopkin, V. D. (1967). “Human factors engineering problems in air traffic control tasks.” *The Controller: IFATCA Journal of Air Traffic Control.* (October).

Effects of the System on the Human: Stress and Workload

Stress and Workload Management in Air Traffic Control

Andrew J. Tattersall,[1] Eric W. Farmer,[2] and Andrew J. Belyavin[2]

[1]School of Psychology, University of Wales, College of Cardiff, Cardiff, UK
[2]RAF Institute of Aviation Medicine, Farnborough, Hants, UK

Introduction

There has been much interest in recent years in the subject of workload and stress among air traffic controllers. Although some of the problems are fairly long-standing, many are becoming increasingly significant because of increasing traffic loads, the changing nature of this traffic, and technological advances in equipment. It is therefore becoming extremely important to understand the nature of the complex demands of this work and the management of those demands by controllers. Also, and perhaps most important in this type of work, we must determine whether and how the quality of system performance is affected.

The research to be outlined briefly in this paper is concerned with the relationships between work demands, well-being, and performance, along with the task factors and personal factors which influence those relationships. A main focus of the paper is on the techniques and methods currently available for evaluating the workload and stress experienced by people at work. As will become apparent, a multi-measurement approach is likely to be necessary in order to understand fully the performance of the whole system.

The popular conception, influenced by much media coverage especially during the summer months, is that increased air traffic places demands on controllers that are reaching almost intolerable levels. However, a number of reviews (e.g., Melton, 1982) have not found much support for this. Clearly the task is to determine what effects these demands actually have, on both individual and system performance, as air traffic safety may be at risk. Achieving this may depend on the development of effective techniques for the measurement of the processes underlying the relationship between work demands and individual responses to stress.

The Nature of Workload and Occupational Stress

Over the last few decades there has been a fundamental change in the nature of jobs, from work which is primarily physical to more cognitively oriented work. Physical workload is comparatively easier to assess than mental workload (e.g., Rohmert, 1987). This is partly due to the difficulty in defining mental workload and the availability of straightforward tests and measures.

NATO ASI Series, Vol. F 73
Automation and Systems Issues in Air Traffic Control
Edited by John A. Wise et al.

What is workload anyway? Having too many things to do? Too little time to carry out the task? Too much to remember? Too much responsibility?

In spite of the difficulties, it is important to develop sensitive and accurate measures of work demands, because if task demands exceed operator resources, then errors may be made (the costs of which may be great), and physiological and psychological stress may be experienced by controllers.

But what is stress?

The term was first used in a systematic manner by Selye (1956), following his work with animals and the finding of a generalized response to a variety of harmful physical stimuli. This response-based theory is, however, not particularly appropriate for explaining human stress, as it assumes that the response to a stressor will always take the same course. Current theories of human stress focus instead on transactional models (e.g., Cox, 1978; Harrison, 1978; Lazarus, 1966), in which appraisal and coping are central features. The crucial point is that stress is not simply a dimension of the physical or social environment, in terms of demands or events, nor simply the responses associated with the consequences of stress.

Stress occurs when there is some imbalance between the perceived demands on an individual and the resources available for dealing with those demands (e.g., Appley and Trumbull, 1986). The effects of environmental changes are moderated by individual interpretive and evaluative processes.

This notion of a mismatch is seen in models of occupational stress such as the person-environment fit model, which illustrates that a mismatch between work demands and opportunities, and individual needs and resources, is associated with job strain (Caplan, Cobb, French, Harrison & Pinneau, 1975; Harrison, 1978). Factors which are important in such a model include the nature, intensity, duration, and predictability of "stressors" and the availability and selection of individual resources such as styles of behaviour in response to an event, personality, previous experience of similar events, perception of control, and social support. Potential stressors include heavy work demands, role ambiguity, role conflict, job loss, shiftwork, noise, pollution, poor relations between workers and supervisors, and lack of participation in decision-making and control over work processes (e.g., Caplan et al., 1975; Cooper and Marshall, 1976; Smith, 1987).

Response to Stress and Mediating Factors

What of the effects on an individual of such a mismatch between demands and resources?

A mismatch between desired and perceived states of the environment may lead to changes in affective state, as well as, physiological or behavioural responses. Reported effects of occupational stress include increased blood pressure, coronary heart disease, respiratory symptoms, ulcers, lowered job satisfaction, depression, job related tension, excessive drinking and smoking, and absenteeism (e.g., Broadbent, 1985; Cooper and Marshall, 1976; Cox, 1985; Karasek, 1981; Parkes, 1987).

Studies of air traffic controllers are equivocal in their findings of health changes as a consequence of the work. Cobb and Rose (1973), for example, reported that the incidence of hypertension was four times higher in a sample of over 4,000 American air traffic controllers than in a comparison group of over 8,000 other aviation workers. The incidence of diabetes and ulcers was also higher in the controller group. Rose, Jenkins and Hurst (1978) carried out a three-year study of controllers, taking repeated blood pressure measures, and confirmed that hypertension was the most frequent chronic illness among controllers (and occurred more often than in comparison groups). However, a number of other studies reviewed by Smith (1980) and Melton (1982) failed to find clear evidence of health problems associated with air traffic control work.

A number of the above mentioned outcomes are physical effects, and it seems that it is not possible to gain a complete understanding of the relationship between demanding situations and changes in well-being and health, without investigating changes in biological and biochemical body state, both in the short- and the long-term. There is a need to consider both the physiological and the affective state (in terms of mood and well-being), and cognitive activity (as implicated in performance). As mentioned earlier, however, the outcomes are not common to all individuals and situations. Among factors are being increasingly recognized as having an important influence on the relationship between stressful experiences and health are personal control, level of responsibility, and the way in which an individual copes with a particular situation.

The importance of control over a work situation has been demonstrated in numerous studies over the past two decades, and this concept is receiving a great deal of attention in the recent literature (Sauter, Hurrell and Cooper, 1989; Steptoe and Appels, 1989).

Karasek (1979, 1981), in large scale social surveys of US and Swedish workers, was able to distinguish between demand and control (job decision latitude). Demand may be described in terms of the number of calls for action and the rate of work; control, in terms of discretion over how problems of work can be met. What is crucial is the interaction of these two factors. Strain symptoms, such as physical ill-health and reduced psychological well-being, were more often found in jobs having both high demand and low discretion, whereas these symptoms were not as prevalent in jobs with other combinations of the two factors.

The extent to which a stressor may be modified by direct action seems to be the key factor here. It is not only the execution of control but simply the knowledge that control is available that is effective (Glass and Singer, 1972). The desire for control in situations in which opportunities for control have been diminished has been reported by Maule and Hockey (1988), in a study of a chemical plant process. Operators were observed to make unscheduled and voluntary manual interventions during an automated process, even though they involved a great deal of risk and concentrated effort.

This last example highlights one of the possible dangers of automation. As observed by (Danaher, 1980, p544), "The introduction of automation and other efforts to reduce controller and pilot workload and to augment their performance have not always reduced human errors in the ATC system." Operators often report being disengaged from the task, and there are problems of boredom and distraction when the skilled components of tasks are removed. The "out-of-the-loop" role that may be forced upon an operator in a highly automated system may thus impair performance, as well as increasing frustration and anxiety (Bainbridge, 1987; Wickens, 1984a; Wiener, 1985).

Automation, as others in this volume have discussed, has the potential to reduce the workload on controllers, but care must be taken to evaluate the outcomes in terms of changes in well-being and performance. The introduction of new technology could actually increase the demands placed on controllers and reduce the control that controllers have over the system as a whole.

The responsibility that work incurs has had little research attention until recently. One attempt to assess the role of cost responsibility is that of Martin and Wall (1989), who studied operators of computer-based manufacturing technology. They found increased stress levels only in those jobs which had both high attentional demand and high cost responsibility.

Furthermore, Warr (1987) has suggested that attentional demand and responsibility are the main determinants of psychological and physiological well-being.

Within the air traffic control domain, Cobb and Rose (1973) found that air traffic controllers suffered from a higher incidence of hypertension and peptic ulcers when working in high density traffic areas. To use Warr's (1987) terms, there seems to be a joint effect of high attention required and high levels of reponsibility as measured by the cost of an error.

The manner in which an individual deals with, or reacts to, the working environment is acknowledged to be an extremely important mediator, or possible moderator, in the work stress process. Lazarus (1976; Lazarus and Folkman, 1984) proposed three processes relating to the perception of, and response to, stress. These are primary appraisal (perceiving threat), secondary appraisal (considering potential responses), and coping (process of executing response). He further distinguished problem-focussed and emotion-focussed coping (concern with task goals or one's own state).

Coping may not always be positive, however (Schonpflug and Battmann, 1988). There may be negative instead of beneficial consequences, in that stress states may be aggravated or prolonged. In addition to the obvious consequences of the suppression or avoidance of demands (and various types of maladaptive coping such as drinking and smoking), active coping, involving increased effort over long periods, may also deplete available resources. An active coper may accept fatigue, for example, as a short-term cost in order to avoid the build up of bigger problems. Therefore, the effects of work on well-being depend upon the way in which individuals manage, or regulate, their internal and external resources.

Workload Regulation

The concern of much recent work in this area is with human performance problems in complex industrial and medical settings, of which process control, air traffic control and medical monitoring are prime examples. These complex jobs often require the operator to monitor large numbers of sources of information in order to detect certain system states, and perhaps to diagnose, rectify and make decisions about certain occurrences. One of the major problems for designers and managers of these systems is that it is difficult to estimate the load of these cognitive tasks on the operator, and therefore when a breakdown in skills might occur. It might be argued that workload does not matter if the task is successfully accomplished.

Certainly, the safety record in air traffic control is very impressive, but what happens when the load on controllers is increased? A critical instance may only arise after a change in working procedure or with an additional concurrent task or possibly as a result of the cumulative effect of sustained and high task demands over a long period. It is important to be able to assess the workload experienced when task demands reach the limits of, or exceed, cognitive resources, in which case operators may experience physiological stress or make errors.

It is not possible to infer the load on an operator simply by measuring task performance or objective task demands. Operators may compensate for any increase in task demands by increasing the amount of effort invested in the task. Therefore, observed performance levels may remain constant while the operator experiences increased workload. Conversely, a reduction in the level of performance may result either because operators cannot maintain the level of effort expenditure required, or because they lower their criteria for adequate performance. This notion of resource-limited processes as opposed to data-limited processes, which are constrained by the quality of information rather than by increases in effort, is described in more detail by Norman and Bobrow (1975) and later developments of resource theory (Navon and Gopher, 1979; Wickens, 1984b).

The implication is that operators are able to exercise some control over some aspects of their task, such as work scheduling, planning, and how much effort is put into a task at any one time. It would seem that air traffic controllers may have less scope for this sort of control, as the task is, to a certain extent, paced or determined by the flow of traffic. However, the notion of effort regulation involving active task involvement is nevertheless an important aspect of successful performance.

An explicit model of the regulatory process of state control developed by Hockey (1986a; Hockey and Hamilton, 1983; Hockey, Briner, Tattersall and Wiethoff, 1989; Hockey and Tattersall, 1989), illustrates how trade-offs between performance efficiency and individual state are logically possible. The main components are a state monitor and an executive process. Well practised tasks have a target state in memory. The state monitor checks for discrepancies between target and actual state and feeds information to the executive. When there is such a discrepancy, different outcomes are possible and the model distinguishes indirect and direct active modes of action from a passive one. For example, trying harder is a direct method of reducing the difference between target and actual states. An indirect method might be to accept a lower target rate or level, resulting perhaps in a slower rate of work than that previously required.

It can be seen that active coping is highly effective but effortful, resulting in high costs. These costs may manifest themselves as changes in affective state (mood) and/or physiological activity. Passive coping is low cost but also has low effectiveness. Failures may be found in workload regulation, however, due to fatigue or suboptimal internal states, or as a result of particular environmental conditions such as noise. Prolonged active management of the resources required to meet task demands may lead ultimately to a deterioration in performance, but there may also be implications for short-term well-being and longer-term health. The experience of workload is thus not likely to depend simply on task load, but on the interaction of task demands, how these demands are dealt with by an operator, and the level of performance achieved. Task demands are important but are mediated by effort and the priority placed on the particular tasks.

There is not a simple relationship between workload and performance, which makes the prediction of errors and a breakdown in skills extremely difficult. Studies of supervisory monitoring and control in complex systems show that open-loop strategies are seen to be more skilled and efficient (Rasmussen and Rouse, 1981; Sheridan and Johannsen, 1976). The open-loop mode involves greater planning and broader understanding of the system as a whole. However, Umbers (1979) has reported that even experienced operators resort to the closed-loop mode when under high levels of workload or when unfamiliar situations occur. The important implication is that those operating in the closed-loop mode may only apply an action once a fault or emergency occurs, rather than being able to predict impending catastrophic events. Air traffic controllers were found by Sperandio (1978) to vary their strategies according to task demand. Similarly, in a study of process control, Bainbridge (1974) found that, when under pressure, subjects used quicker, less accurate methods of finding data values.

Other studies of complex task performance highlight the transactional nature of workload. Tattersall and Hockey (1990), for example, identified different activities in a simulated flight engineer task which resulted in different cardiovascular costs and subjective ratings of effort and concern. Hockey and Wiethoff (1989) carried out a longitudinal study of junior doctors into patterns of adaptation to changes in workload. When looking at the effects of low and high workload days on the patterns of mood changes, they found a dissociation of fatigue and tension depending upon the coping strategies employed. They also found that particular patterns of endocrine activity were associated with different aspects of performance following extended work shifts. Earlier, Frankenhaeuser (1979) had demonstrated that changes in catecholamine and cortisol excretion are associated with increased work demands. The patterning of these changes appears to reflect active management of work and opportunity for control over work, but the exact relationships are not yet clear. Cortisol appears to be influenced by distress and lack of control. Adrenaline and noradrenaline are probably associated with mental and physical effort investment (Frankenhaeuser, 1979; Frankenhaeuser and Johansson, 1986).

Three particular domains may show the effects of different stressors. These are performance, subjective state (in terms of indices of mood or well-being), and bodily state (measures of cardiovascular or endocrine activation for example). Many different measures have been developed, but their use in particular situations is likely to depend mainly on their sensitivity to changes in demand, their ability to distinguish different kinds of demand, and their suitability or relevance to that situation. Urine analysis allows measures of longer-term changes in state through assessment of cortisol and catecholamine concentration. Sustained stress states show increased levels of these hormones. Blood or saliva samples may provide shorter-term measures of fluctuations in state.

Other physiological measures are related to cardiac function. A number of studies now suggest that the power in the heart rate variability (HRV) spectrum (0.07-0.14 Hz) is related to the level of mental effort invested in a task by an individual (Tattersall and Hockey, 1990; Aasman, Mulder and Mulder, 1987; Mulder, 1980; Vicente, Thornton and Moray, 1987). Other measures include pupil dilation and brain function. The advantage of measurement of cardiac function is that it can be continuously and independently measured without intrusion.

Subjective measures are also relatively easy to employ. It is intuitively attractive to simply ask controllers or other workers about the levels of demands they are experiencing. Brief questionnaires or inventories can assess mood (e.g., Mackay, Cox, Burrows and Lazzerini, 1978; Warr, 1989) or workload (e.g., Hart and Staveland, 1989; Reid, Eggemeier and Shingledecker, 1982). Some of the most widely used and reliable measures are of short- and long-term well-being and health. Importantly, significant relationships have been found between specific physiological and subjective responses, such as cortisol and subjective distress (Frankenhaeuser and Johansson, 1986) and adrenaline and effort (Hockey and Wiethoff, 1988).

See Gopher and Donchin (1986), Hockey (1986b), O'Donnell and Eggemeier (1986), and Hancock and Meshkati (1988) for more thorough reviews of these techniques.

Studies of Stress and Workload in Air Traffic Control

Two studies now to be described form part of an investigation of stress in Air Traffic Control Officers employed by National Air Traffic Services (NATS). They were commissioned by the Civil Aviation Authority and conducted by the Royal Air Force Institute of Medicine and the

MRC/ESRC Social and Applied Psychology Unit at the University of Sheffield. The aims were to identify sources and effects of stress in air traffic control work, to investigate differences in stress experienced by controllers and other occupational groups, and to determine the role of individual differences such as personality and use of coping strategies in the process of dealing with work demands.

The first study was a survey carried out among the population of air traffic controllers in the UK (Farmer, Belyavin, Berry, Tattersall and Hockey, 1990). The aims were to identify sources of stress in air traffic control, to assess the impact of these factors on health and well-being, and to evaluate the role of mediating factors such as coping strategies and personality variables.

Three main sections were included in the survey. The first section contained the Occupational Stress Indicator (Cooper, Sloan & Williams, 1988), which comprises a series of questionnaires relating to general sources of occupational stress, individual characteristics such as locus of control, coping strategies and the effects of stress on the individual and the organization. The second questionnaire was devised to complement the other information collected by requesting information specifically related to air traffic control work. This included biographical information such as sex, age, domestic circumstances, air traffic control unit, and job functions. It also addressed potential sources of stress in air traffic control, including items concerning duty hours, equipment, workload, ergonomics, and other factors relevant to air traffic control duties. Further questions concerned sleep quality during periods of night work, other shifts and leave; the use of sleep aids such as medication or relaxation techniques; and smoking, drinking, the use of medication, and exercise. Finally, workload and mood were assessed. Respondents rated workload during the last shift they had worked, using the NASA Task Load Index (TLX)(Hart & Staveland, 1989). A 20-point scale was used for self assessment on six dimensions of workload: mental demand, physical demand, temporal demand, performance, effort, and frustration. Respondents were also asked to rank order these dimensions in terms of their importance to the task. Mood during the last shift was rated, using an adjective checklist comprising 20 items (Warr, 1989) relating to anxiety, depression and fatigue. The third section contained the 16PF (e.g., Cattell, Eber and Tatsuoka, 1970), which was used to assess the role of individual differences in personality in air traffic control stress.

These questionnaires and rating scales were sent to 1122 operational NATS controllers in the UK, with 618 completed questionnaires returned (55.1% response rate). Controllers from 25 different units in the UK were represented in the sample. The respondents had a mean age of 40.5 years and a mean level of operational service of 17.2 years. The mean time at their present unit was 10.4 years, and length of time in present job grade was 9.8 years.

A necessarily brief overview of the results of the survey will be given here. Concerning individual characteristics, the controllers sampled rated themselves as intelligent, rather reserved, and free-thinking. The overall level of trait anxiety in the sample was similar to that of the general population, although the controllers exhibited a greater than average level of behavioural control. Type A scores (indicating competitiveness, aggression, ambition, and time consciousness) were lower than in several of the comparison occupational groups.

Of the general sources of stress, organisational structure and climate caused the most concern to controllers. This factor includes items such as lack of opportunities for consultation, low staff morale, staff shortages, and lack of training. Controllers appeared to perceive their work activities to be more open to external control than other occupational groups. As discussed earlier, the perception of control is an important factor in occupational stress. A principal components analysis of the more specific air traffic control-related sources of stress revealed three factors which accounted for the major sources of variation. These are related to equipment (e.g., reliability and layout), workload, and external factors (e.g., housing problems, public comment, etc). The workload factor was made up of items concerning duty hours, traffic levels, shifts, extraneous traffic, and other aspects of demands on controllers. The major sources of stress were associated with peak traffic loads, extraneous traffic, length of time at peak load, and unbroken duty periods with difficulties in finding extra help.

Multiple regression techniques were then used to produce models of stress in air traffic control. Some of the major results are given here. Self rated workload was related to the stress created by workload and external factors rather than factors relating to equipment. Sleep quality was directly influenced by workload, but the major effect of workload here was to affect adversely the perception of organisational structure and climate, which in turn influenced the change in sleep quality. Satisfaction with the job itself was related to workload, but was more strongly associated with the intermediate effect of workload on the stress of the managerial role. Overall job satisfaction was negatively associated with the stress of work facilities and workload, both directly and through the effects of these variables on the perception of climate.

Mental health was affected adversely by workload, and physical health declined with the stress created both by workload and by external factors. Subjective ratings of mood showed that depression was increased by workload stress, with perception of climate an intermediate factor; fatigue was increased by workload stress; and state anxiety was elevated by the stress of both workload and external factors, directly and by increasing the stress of the job itself.

The results of this study have a number of implications for the effective management of air traffic control operations. Although the results do not imply that air traffic control is an inherently stressful occupation, there are considerable levels of occupational stress among the controllers sampled. A number of problems could be alleviated by relatively minor changes to work organization and design. For example, shift work can be a problem. The length of duty periods (maximum two hours at present in the UK) was of considerable concern at busy units, as was the length of some shifts. Shift patterns in general were not rated as particularly stressful, but the problems that were mentioned primarily concerned psychologically unsound patterns which violated, for example, the principle that shifts should delay (morning/evening/night) rather than advance (Folkard and Monk, 1985).

Job satisfaction was found to be generally low among the controllers sampled, although they were satisfied with the job tasks themselves. The organisational structure and climate were perceived very unfavourably, indicating problems with management structure and prospects for promotion. One possible way of improving the sense of personal control is perhaps through organisational changes such as increased participation in decision making. Other design issues, particularly relating to workstations and radio and telephone communications, can be identified; problems specific to individual units can be solved.

There are also implications for selection, in terms of individual placement in specific jobs, and training, which may include training in stress management, and effective planning and scheduling of work. Knowledge, skills and effective effort regulation may play a moderating role in meeting work demands. If it is not possible to remove all potential sources of stress from the work, then it is certainly possible to develop potentially efficient strategies for coping with them.

This study is typical of many cross-sectional studies of occupational stress and health, and is a useful way to gain background information about possible sources of stress, and the coping strategies and behavioural styles which might be generally effective in moderating job strain. However, it is also important to investigate the time course of the effects and exposure to different job demands. Some may be chronic effects, others acute.

The interest of the second study is in the trade-off between task demands and the costs of regulatory activity: how, for example, sustained effort expenditure may result in impaired performance, or whether misplaced effort, which does not result in improved performance, results in fatigue or other changes in mood state. It is a more focused investigation of the work of 65 controllers. Data were gathered on two occasions (during busy and quiet periods in the year), in four sites varying in traffic load. Cognitive performance measures were taken at the beginning and the end of each shift, through tests of verbal reasoning and paced visual monitoring, together with subjective ratings of workload, effort and affective state. Physiological state regulation has been assessed, using analyses of ECG data (heart rate and heart rate variability) and catecholamine and cortisol excretion.

One major difficulty in systems such as air traffic control is in measuring overall performance, both of the individual and the system. The changes may be small and infrequent. Minor slips may not even be detected. Therefore, simulated tasks are used which are thought to be sensitive to fatigue and demands of this type of work. There may be relationships between overall levels of workload and change in performance at these tasks over the day. A range of tasks have been developed to measure vigilance, visual search performance, memory, and tracking performance. The aim is to provide measures of the state of various cognitive and perceptual abilities under different conditions. Different aspects of air traffic control work may produce changes in specific tasks (for example in vigilance performance) after prolonged periods of monitoring.

The analyses are not yet fully completed, but the description of the study gives a good example of the techniques currently in use for assessment of work demands and response to those demands by controllers. Further studies, both controlled laboratory-based studies and field-based studies, are necessary to refine the techniques, but it should be possible to develop a model of stress and workload which will allow the accurate prediction of states or situations in which a breakdown in skills might occur. Such an example is "losing the picture," as air traffic controllers describe not being able to remember accurately all relevant information about current aircraft.

Conclusions

It has been impossible in this short paper to summarise all the data from the studies described. However, the main aim of the paper was to give an overview of the current methods available for assessing work demands and management of those demands by air traffic controllers.

The primary aim of work such as this is to understand the relationships between work demands and strain through the analysis of individual differences in cognitive control processes. The application of these processes may again depend on individual differences in long term styles and short term strategies, but may be used to prevent decrements in work performance under difficult conditions. However, control may attract costs as well as benefits in terms of changes in affective state and physiological activity. Thus, strain associated with work demands may only be measurable in the form of tradeoffs between performance and other domains of individual activity. The methodological implications are to identify patterns of adjustment to workload variations using repeated measures and longitudinal sampling of performance, as well as subjective and physiological measures.

In summary, multi-level measurement methods are necessary to investigate and understand individual differences in workload regulation in relation to varying task demands. Possible enhancements to workload and stress management at both individual and organizational levels may only then be suitably effective.

References

Aasman, J., Mulder, G., & Mulder, L. J. M. (1987). Operator effort and the measurement of heart-rate variability. *Human Factors, 29*, 161-170.

Appley, M. H., & Trumbull, R. (1986). *Dynamics of Stress*. New York: Plenum.

Bainbridge, L. (1974). Analysis of verbal protocols from a process control task. In E. Edwards & F. P. Lees (eds.), *The Human Operator in Process Control*. London: Taylor and Francis.

Bainbridge, L. (1987). Ironies of automation. In J. Rasmussen, K. Duncan, & J. Leplat (Eds.), *New Technology and Human Error*. London: John Wiley & Sons.

Broadbent, D. E. (1985). The clinical impact of job design. *British Journal of Clinical Psychology, 24*, 33-44.

Caplan, R. D., Cobb, S., French, J. R. P., Harrison, R. V., & Pinneau, S. R. (1975). Job demands and worker health. NIOSH Research Report. Washington, DC: USHEW, US Government Printing Office.

Cattell, R. B., Eber, H. W., & Tatsuoka, M. M. (1970). *The 16PF Test*. Champaign, Illinois: IPAT.

Cobb, S., & Rose, R. M. (1973). Hypertension, peptic ulcer, and diabetes in air traffic controllers. *Journal of the American Medical Association, 224*, 489-492.

Cooper, C. L., & Marshall, J. (1976). Occupational sources of stress: A review of the literature relating to coronary heart disease and mental ill-health. *Journal of Occupational Psychology, 49*, 11-28.

Cooper, C. L., Sloan, S. J., & Williams, S. (1988). *Occupational Stress Indicator*. Windsor: NFER-Nelson.

Cox, T. (1978). *Stress*. London: Macmillan.

Cox, T. (1985). Repetetive work: Occupational stress and health. In C. L. Cooper & M. Smith (Eds.), *Blue collar stress*. Chichester: Wiley.

Danaher, J. W. (1980). Human Error in ATC System Operations *Human Factors, 22*, 535-545.

Farmer, E. W., Belyavin, A. J., Berry, A., Tattersall, A. J., & Hockey, G. R. J. (1990). Stress in Air Traffic Control I: Survey of NATS Controllers. RAF Institute of Aviation Medicine Report No. 689.

Folkard, S., & Monk, T. H. (1985). *Hours of Work: Temporal Factors in Work Scheduling*. Chichester: Wiley.

Frankenhaeuser, M. (1979). Psychoneuroendocrine approaches to the study of emotion as related to stress and coping. In H. E. Howe, & R. A. Dienstbier (Eds.), *Nebraska Symposium on Motivation*. University of Nebraska Press.

Frankenhaeuser, M., & Johansson, G. (1986). Stress at work: psychobiological and psychosocial aspects. *International Review of Applied Psychology, 35*, 287-299.

Glass, D. C., & Singer, J. E. (1972). *Urban Stress*. New York: Academic Press.

Gopher, D., & Donchin, E. (1986). Workload - An examination of the concept. In K. R. Boff, L. Kaufman & J. P. Thomas (Eds.), *Handbook of Perception and Human Performance Volume II: Cognitive Processes and Performance*. New York: John Wiley & Sons.

Hancock, P. A., & Meshkati, N. (1988). *Human Mental Workload*. Amsterdam: Elsevier.

Harrison, R. V. (1978). Person-environment fit and job stress. In C. Cooper & R. Payne (Eds.), *Stress at work*. New York: Wiley.
Hart, S. G., & Staveland, L. E. (1989). Development of a NASA TLX (Task Load Index): Results of empirical and theoretical research. In P. Hancock & N. Meshkati (eds.), *Human Mental Workload*. Amsterdam: Elsevier.
Hockey, G. R. J. (1986a). A state control theory of adaptation and individual differences in stress management. In G. R. J. Hockey, A. W. K. Gaillard, & M. G. H. Coles (Eds.), *Energetics and Human Information Processing*. Dordrecht: Nijhoff.
Hockey, G. R. J. (1986b). Changes in operator efficiency as a function of environmental stress, fatigue, and circadian rhythms. In K. R. Boff, L. Kaufman & J. P. Thomas (Eds.), *Handbook of Perception and Human Performance Volume II: Cognitive Processes and Performance*. New York: John Wiley & Sons.
Hockey, G. R. J., Briner, R. B., Tattersall, A. J., & Wiethoff, M. (1989). Assessing the impact of computer workload on operator stress: the role of system controllability. *Ergonomics, 32*, 1401-1418.
Hockey, G. R. J., & Hamilton, P. (1983). The cognitive patterning of stress states. In G. R. J. Hockey (Ed.), *Stess and fatigue in human performance*. Chichester, Wiley.
Hockey, G. R. J., & Tattersall, A. J. (1989). The maintenance of vigilance in automated monitoring. In A. Coblentz (Ed.), *Vigilance and Performance in Automatized Systems*. Dordrecht: Kluwer.
Hockey, G. R. J., & Wiethoff, M. (1990). Assessing the patterns of adjustment to the demands of work. In A. Oliverio & S. Pugliso-Allegra (Eds.), *The Psychobiology of Stress*. Dordrecht: Kluwer.
Karasek, R. A. (1979). Job demands, job decision latitude, and mental strain: Implications for job redesign. *Administrative Science Quarterly, 24*, 285-308.
Karasek, R. A. (1981). Job design latitude, job design, and coronary heart disease. In G. Salvendy & M. J. Smith (Eds.), *Machine Pacing and Occupational Stress*. London: Taylor & Francis.
Lazarus, R. S. (1966). *Psychological Stress and the Coping Process*. New York: McGraw-Hill.
Lazarus, R. S. (1976). *Patterns of Adjustment*. New York: McGraw-Hill.
Lazarus, R. S., & Folkman, S. (1984). *Stress, Appraisal and Coping*. New York: Springer.
Mackay, C., Cox, T., Burrows, G., & Lazzerini, T. (1978). An inventory for the measurement of self-reported stress and arousal. *British Journal of Clinical Psychology, 17*, 283-284.
Martin, R., & Wall, T. D. (1989). Attentional demand and cost responsibility as sources of stress in shopfloor jobs. *Academy of Management Journal, 32*, 69-86.
Maule, A. J., & Hockey, G. R. J. (1988). The role of unscheduled manual interventions in process control. Proceedings of 7th European Conference on Decision Making and Manual Control, Paris.
Melton, C. E. (1982). Physiological stress in air traffic controllers: A review. Washington, DC: Federal Aviation Administration Report No. FAA-AM-82-17.
Mulder, G. (1980). The Heart of Mental Effort. University of Groningen, The Netherlands.
Navon, D., & Gopher, D. (1979). On the economy of the human processing system. *Psychological Review, 86*, 214-255.
Norman, D. A., & Bobrow, D. G. (1975). On data-limited and resource-limited processes. *Cognitive Psychology, 7*, 44-64.
O'Donnell, R. D., & Eggemeier, F. T. (1986). Workload assessment methodology, In K. R. Boff, L. Kaufman & J. P. Thomas (Eds.), *Handbook of Perception and Human Performance Volume II: Cognitive Processes & Performance*. New York: John Wiley & Sons.
Parkes, K. R. (1987). Relative weight, smoking and mental health as predictors of sickness and absence from work. *Journal of Applied Psychology, 72*, 275-286.
Rasmussen, J., & Rouse, W. B. (Eds.) (1981). *Human Detection and Diagnosis of System Failures*. New York: Plenum.

Reid, G. B., Eggemeier, F. T., & Shingledecker, C. A. (1982). Subjective workload assessment technique. Proceedings of the AIAA Workshop on flight testing to identify pilot workload and pilot dynamics. Edwards Air Force Base, CA, pp. 281-288.

Rohmert, W. (1987). Physiological and psychological work load measurement and analysis. In G. Salvendy (Ed.), *Handbook of Human Factors*. New York: John Wiley & Sons.

Rose, R. M., Jenkins, C. D., & Hurst, M. W. (1978). Air traffic controller health change study. Washington, DC: Federal Aviation Administration Report No. FAA-AM-78-39.

Sauter, S. L., Hurrell, J. L., & Cooper, C. L. (Eds.) (1989). *Job Control and Worker Health*. Chichester: John Wiley.

Schonpflug, W., & Battman, W. (1988). The costs and benefits of coping. In S. Fisher & J. Reason (Eds.), *Handbook of Life Stress, Cognition and Health*. Chichester: John Wiley.

Selye (1956). *The Stress of Life*. New York: McGraw-Hill.

Sheridan, T. B., & Johannsen, G. (Eds.) (1976). Monitoring and Supervisory Control. New York: Plenum.

Smith, M. J. (1987). Occupational Stress. In G. Salvendy (Ed.), *Handbook of Human Factors*. New York: John Wiley & Sons.

Smith, R. C. (1980). Stress, Anxiety and the Air Traffic Control Specialist. Washington, DC: Federal Aviation Administration Report No. FAA-AM-80-14.

Sperandio, J. (1978). The regulation of working methods as a function of workload among air traffic controllers. *Ergonomics, 21*, 195-202.

Steptoe, A., & Appels, A. (Eds.) (1989). *Stress, Personal Control and Health*. Chichester: John Wiley.

Tattersall, A. J., & Hockey, G. R. J. (1990). The assessment of workload in a complex monitoring and fault diagnosis task. In D. Brogan (Ed.), *Visual Search*. London: Taylor & Francis.

Umbers, I. G. (1979). Models of the process operator. *International Journal of Man-Machine Studies, 11*, 263-284.

Vicente, K. J., Thornton, D. C., & Moray, N. (1987). Spectral analysis of sinus arrhythmia: A measure of mental effort. *Human Factors, 29*, 171-182.

Warr, P. B. (1987). *Work, Unemployment and Mental Health*. Oxford: Oxford University Press.

Warr, P. B. (1989). The measurement of well-being and other aspects of mental health. MRC/ESRC Social and Applied Psychology Unit Memo Mo. 1035. University of Sheffield.

Wickens, C. D. (1984a). *Engineering Psychology and Human Performance*. Columbus, OH: Merrill.

Wickens, C. D. (1984b). Processing resources in attention. In R. Parasuraman & D. R. Davies (Eds.), *Varieties of Attention*. New York: Academic Press.

Wiener, E. L. (1985). Beyond the sterile cockpit. *Human Factors, 27*, 75-90.

Shiftwork and Circadian Variations of Vigilance and Performance

Giovanni Costa

Institute of Occupational Medicine, University of Verona, Policlinic Hospital, Verona, Italy

Introduction

Circadian Rhythms of Body Functions

Man is a day-light creature, having associated his own state of wakefulness and activity with the day-light period and, consequently, his rest and sleep state with the night-dark period, with a regular rhythm over the 24-hour span. This conditioning derives from the evolutionary adaptation of the human species to the physical environment and from the social habits of each individual. In relation to this periodic alternation, all the body functions (metabolic, respiratory, digestive, cardio-circulatory, renal, nervous) show a circadian ("circa diem" = about 24 hours) rhythm, with higher functional levels during the "ergotropic" phase (light-wake-activity) and lower levels during the "trophotropic" phase (dark-sleep-rest). This circadian rhythm is sustained by an endogenous oscillator or "body clock," localized in the brain at the level of the suprachiasmatic nuclei, and modulated by external influences of such socio-environmental "synchronizers" as light period, timing of sleep, activity pattern, and meals. Under normal conditions of the sleep/wake cycle, the internal and external factors are in constant relation, determining a clear circadian variation of the psycho-physiological state (Minors and Waterhouse, 1986).

Circadian Rhythms of Vigilance and Performance

The body's rhythmic structure plays an important role in influencing human performance and work efficiency: apart from the factors relevant to the state of health and motivation, findings show a circadian rhythm also of alertness and efficiency of many psycho-motor functions.

Daily variations in human performance efficiency have been recognized for a long time (von Bechterew, 1893; Kraepelin, 1893; cited in Rutenfranz and Colquhoun, 1979). Kleitman and Jackson (1950) successively found variations of performance efficiency strictly in conjunction with the oral temperature circadian rhythms; similar results have been reported by other authors, showing that both performance and temperature were high during the day (with peak levels in the afternoon) and low during the night. Those studies stressed the occurrence of these variations on a number of tasks, such as: signal detection, card sorting, letter cancellation, time estimation, simple reaction times, calculation, and simulating flight (Colquhoun, 1971).

NATO ASI Series, Vol. F 73
Automation and Systems Issues in Air Traffic Control
Edited by John A. Wise et al.

Although it is possible that, in such experiments, some confounding effects can derive either from sleep interruption or sleep deprivation, nevertheless the circadian rhythm appears sufficiently well detectable (Rutenfranz and Colquhoun, 1979). Moreover, in most recent years, well-controlled laboratory experiments have shown that performance can vary during the day in different ways, according to different types of mental activity involved. (For example, simple perceptual motor tasks show an improvement during the day, while tasks requiring a high short-term memory load worsen; performance of tasks involving verbal reasoning and mental arithmetic shows a more alternate trend). The fluctuations vary from 10% to 30% (Folkard and Monk, 1979; Klein and Wegmann, 1979; Folkard, 1990).
However, many other factors can influence the circadian variation both in terms of phase and amplitude. These factors include:

a) Arousal: the relation between arousal and performance is postulated to take the form of an inverted U-shaped curve (Blake, 1971). The circadian variability of performance efficiency has been found to increase as much as the arousal is under or above the "optimal" levels (Colquhoun, 1971). Otherwise, according to other authors (Hockey and Hamilton, 1983), such variations can be related to different arousal states;
b) Workload: the heaviness or the complexity of the task increases the circadian fluctuations;
c) Motivation: a higher motivation decreases the circadian fluctuation of performance, while lack of interest and boredom have the opposite effect (Alluisi and Chiles, 1967);
d) Sleep deprivation: the longer the previous waking time, the higher the performance oscillation; short sleep periods or naps reduce the fluctuation, although a temporary grogginess might appear if the person is awakened during the "deep" sleep phase (Gillberg, 1985);
e) Working time / Sustained operation : the performance of prolonged operations shows a deeper decrease when work starts at night rather than during the day (Klein and Wegmann, 1979).
f) Personality and behavioral characteristics: the circadian pattern of "morning types" makes them more active and alert during the first part of the day and less so during the evening and the night, while the opposite behavior is seen in the "evening types." Introverted personalities appear to behave as morning types, while extroverts are more likely to behave as evening types: the differences are more evident when these characteristics are associated with neurotic instability (Colquhoun and Folkard, 1978).

Interferences of Shiftwork on Vigilance and Performance

Psychophysiological Problems Connected to Shiftwork

Shiftwork, in particular nightwork, is a stressful condition which interferes with life rhythms, forcing people to change life schedules (sleep, eating, family and social contacts) and to carry on doing jobs at times when body fitness is impaired (Reinberg et al., 1981; Folkard and Monk, 1985; Costa et al., 1990).

Nightwork, in particular, compels the worker to invert his normal "activity-rest" cycle, forcing him to adjust his body function to the night activity period. Such "adjustment" entails a partial phase shift of the body functions' circadian rhythms (in advance or in delay, according to the shift rotation). This increases with the number of successive night-shifts, but seldom reaches a complete inversion, particularly in the case of the more common shift schedules, e.g., week rotation. In fact, the human body is exposed to continuous stress in the attempt to adjust as quickly as possible to the new working hours, while at the same time being invariably frustrated by the continuous "change-over." In many cases, a flattening of the curve (Knauth et al., 1978 and 1981) or a delinking ("desynchronization") of the different functions (Reinberg et al., 1988), with different speeds and directions of re-adjustment, can be observed: this testifies to the strain of the organism under such a stressful condition.

On shifts (particularly on night shifts), the main function altered is sleep, which undergoes a shortening and a lessening of quality in relation to the different work shifts. As a matter of fact, shorter sleep is recorded during the morning shift period, thanks to the early-morning wake. On the other hand, during night shifts, sleep is reduced due to the inversion of the normal "sleep/wake" cycle, leading to difficulty in falling asleep while body temperature is rising. Besides, it is very difficult to sleep properly in unfavorable environments, such as over-crowded, built-up areas; proximity to main roads; noises made by electric household appliances; crying and screaming children; ringing door-bells and telephones (Knauth and Rutenfranz, 1975). Moreover, sleep quality is also impaired (Akerstedt, 1985). Day-sleep reduces the "deep-sleep" component, which is important for recovering from physical fatigue. On the other hand, a significant reduction of the late-stage "REM sleep" (that related to dreams) can be recorded when the morning shift starts too early.

Disturbing the rhythmic structure and losing sleep play an important role in influencing health and work capacity. Many studies have reported falls in performance efficiency at night, stronger feeling of fatigue, and more mistakes, all of which can lead to injury to people and damage to things (see next paragraph). Moreover, such a condition, in the long run, can not only give rise to permanent and severe sleep disturbances, but also cause problems for the nervous system. Chronic fatigue, changes in behavior patterns, and persistent anxiety or depression are usually observed. They often require medical treatment with administration of psychotropic drugs, which can further interfere and worsen the performance efficiency. Psychotropic drugs are, in fact, well known substances affecting both behavior and psychomotor performance, causing changes in attention, vigilance, learning, discrimination, intellectual capacity, and emotional response. This is not only the case with tranquilizers and hypnotics (e.g., benzodiazepines and barbiturates), but also with more common drugs such as alcohol and tobacco. Shiftworkers generally consume more coffee, tea, and alcoholic beverages, as well as tobacco, to maintain alertness on night shift; on the other hand, they often take sleeping pills to fall asleep more quickly after the night shift: this is a dangerous "vicious circle."

Performance Efficiency on Shift and Nightwork

In real working conditions, a decrease of work performance during the night is frequent, showing a fluctuation which can exceed 100%. This leads to errors or accidents by switchboard operators (Browne, 1949), gas-work log-keepers (Bjerner et al., 1955), car drivers (Prokop et al., 1955), spinners (Wojtczac-Jarosowa and Pawlowska-Skyba, 1967), ship workers (Mann et al., 1972), train-drivers (Hildebrandt et al., 1974), truck drivers (Harris, 1977), and nurses (Folkard et al., 1978). It is also common to see a less pronounced "post-lunch dip," which is not necessarily related to the meal taken but seems determined by a 12-hour rhythm of alertness and performance superimposed on the more evident 24-hour rhythm (Hildebrandt et al., 1974).

Other studies support the assumption that the human sleep/wake cycle can be better described by ultradian (less than 24 hours) rhythms with a period of about four hours (Zulley and Bailer, 1988). Moreover, Murrel (1971) suggests that a high level of performance cannot be maintained for more than two consecutive hours, and that a fixed ten-minute break per hour actually increases productivity. Besides, Lavie (1985) indicated the possible existence of a 90-minute biological "working unit."

A curious phenomenon, described among such night workers as naval officers and printers (Rudolp, 1946ab), nurses (Folkard et al., 1984) and air traffic controllers (Folkard and Condon, 1987), is "night shift paralysis." According to Rudolp's description (1946b), it is a sustained immobility of the voluntary muscles during consciousness: the subject can see and hear and is aware of the inability to move. Such attacks have an insidious and sudden onset, and can last from a few seconds to a few minutes. The incidence of these attacks has been recorded in 6% to 12% of the subjects, peaks around 5 a.m., and is directly related to the number of successive night shifts worked and sleep deprivation.

These are very important aspects to consider, particularly when consistently high performance levels are a must because public health is at stake (e.g,. air traffic control). New technologies, which increase cognitive tasks and require more alertness and vigilance, are often more vulnerable to errors than manual work activities. It may be a coincidence, but it is worth mentioning that the two nuclear-reactor accidents, at Three Mile Island (1979) and Chernobyl (1986), and the Bophal chemical plant disaster (1984), all started during the night hours (at 0400, 0125, and 0057, respectively): in all situations human error has been claimed as an important factor.

However, the studies concerning work accidents among shiftworkers are quite controversial: some investigations have reported more accidents on night shifts, others found more on day shifts, and still others reported less frequent but more serious events on night shifts (Carter and Corlett, 1978). Besides the interference of many other variables, the different findings can be explained by, on one hand, the various work sectors and situations examined (whether prone to minor or major accident risk, firm size, specific jobs); and on the other hand, by taking into account that "conditions are rarely, if ever, the same by day and by night" (Colquhoun 1976). In fact, the reduction in psycho-physical performance at night is not necessarily associated with a higher incidence of accidents, as many factors linked with work environment and organization can vary (i.e., reduction of the work load, night-interruption of higher-risk jobs, slowing-down of work-pace, reduction of extraordinary maintenance to the minimum, added automation). On the other hand, many studies have pointed out that the peak accident hours during day shifts are around 10-11 a.m. and 3-4 p.m. when, though the performance curve reaches its best levels, work activity is also at its peak. However, as concerns the morning shift, some studies on bus and train drivers showed that higher frequencies of errors and accidents are associated with earlier starting hours (Hildebrandt et al., 1975; Pokorny et al., 1981).

Finally, we should also consider the high variability of short- and long-term adjustment to shift and night work among people (Monk and Folkard, 1985). This can be influenced by several personal factors, including age, work experience, personality, sleep needs, and behavior patterns. As already mentioned, the circadian phase position of the biological rhythms, and their adjustment to the night shift period, seem to be related to some personality and behavioural characteristics. For example, Colquhoun and Folkard (1978) found that "neurotic extrovert" subjects show the greatest degree of adjustment of their body temperature rhythm to phase-shift; while Nachreiner (1975) pointed out that more introverted and emotionally unstable subjects are less tolerant of nightwork. The difference in circadian phase position between "morning" and "evening" types is also believed to influence their different adaptability to night-work: the "evening" types are more able to cope with night shifts because of their normally delayed circadian phase position (Ostberg, 1973; Breithaupt et al., 1978; Hildebrandt and Stratman, 1979). Flexibility of sleeping habits and ability to overcome drowsiness may also be important factors in determining a better adjustment and tolerance to night work (Folkard et al., 1979; Costa et al., 1989a).

Some Recommendations to Improve Alertness, Performance Efficiency, and Health on Shift and Night Work

Recent studies and research have promoted some design recommendations for shiftwork systems aimed at avoiding or reducing dangerous effects on the fitness, well-being, and efficiency of shiftworkers (Knauth and Rutenfranz, 1982). They can be summarized in the following points:

a) Adopt rapidly-rotated shift systems, instead of the weekly (or longer) rotation, in order to cause less disturbance of the normal circadian rhythm of the body functions;
b) Reduce the number of consecutive nightshifts as much as possible (one or two at most) and have the day-rest after the nightshift period. This prevents accumulation of sleep deficit and fatigue, and allows quicker recovery;
c) Start the morning shift later (i.e., 0700 or later), in order to allow a normal sleep length and save the "paradoxic sleep";
d) Prefer the forward rotation (morning-afternoon-night) to the backward one (N-A-M), to allow a longer rest interval between shifts. The forward parallels the "natural" tendency of body functions to lengthen the circadian rhythm over 24 hours when in "free-running" (without external synchronizers) conditions (Wever, 1985);
e) Reduce the length of the nightshift according to the physical and mental work-load connected to the job. This can become a necessity in work requiring high levels of vigilance and performance where public safety is at stake (for example, among fire fighters, train and aircraft drivers, air traffic controllers, workers in nuclear and petrochemical plants). Andlauer et al. (1982), after the Three Mile Island accident, proposed doubling up the night shift with two teams and providing satisfactory rest facilities for the off-duty team so that "no operator should work more than 4 or 4.5 hours total in the night shift"; and
f) Make the shift rotation as regular as possible, so the shift cycle will not be too long and will include some free weekends (this allows a better organization of personal, family, and social life).

Some other arrangements should include:

g) Incorporate nap periods into the nightshift. The possibility of having a short sleep during the night has been found to have favorable effects on performance (Angiboust et Gouars, 1972; Gillberg, 1985; Rogers et al., 1989), physiological adjustment (Minors and Waterhouse, 1981; Matsumoto et al., 1982) and tolerance to night work (Kogi, 1982);

h) Arrange a sufficiently long pause (45-60 min) for the meal during the workshift, and provide hot meals.

Individuals may also adopt some personal strategies of re-arranging their sleep and eating:

1) Keep to a tight sleeping schedule while on shiftwork and, as much as possible, avoid disturbances (arrange the bedroom so that it is as silent and dark as possible; use ear plugs; make some arrangements with family members and neighbours);
2) Do not use sleeping pills, save in exceptional cases and only under medical control;
3) Try to maintain usual meal times, which can act as a good synchronizer of body functions, as much as possible;
4) Eat light meals not later than two hours before going to sleep; avoid caffeinated drinks and alcohol; do some relaxation before going to sleep (light exercise, reading, watching TV, listening to music);
5) During the nightshift, have the main meal preferably before 0100, thereafter taking only light snacks with carbohydrates and soft drinks (fruit juices, milk).

In some particular cases, re-scheduling has been proposed to improve performance efficiency. For example, after the Three Mile Island accident, special shift schedules utilizing the synchronizing effect of mealtimes were suggested for the nuclear power industry. In fact, establishing scheduled eating habits and using meal times as a time index can speed the adjustment of circadian rhythms to phase-shifts, maintaining a satisfactory level of alertness and performance (Ehret, 1981).

Finally, it is worth mentioning that careful selection of workers for shift and nightwork and periodical health checks are very important measures for improving work efficiency, safety, and health (Rutenfranz, 1982). Taking into account the different personal factors influencing short- and long-term adjustment and tolerance to nightwork, it may be sensible to select or assign (when possible) primarily those persons whose psycho-physical state, personality, behavioural characteristics, health situation, and living conditions suit them to nightwork. According to the research carried out in this field in recent years, personality tests, standardized questionnaires, rating scales, and self-recordings may be helpful in this respect (see, for example, Horne and Ostberg, 1976; Folkard et al., 1979; Reinberg et al., 1979; Torsvall and Akerstedt, 1980; Moog, 1981). Such screening can enhance the reliability of the medical examination made prior to starting shiftwork in evaluating "fitness for work," and in predicting the successive adjustment and tolerance to shift and nightwork (Monk and Folkard, 1985; Wynne et al., 1986; Moog, 1987; Vidaceck et al., 1987; Bohle and Tilley, 1989; Costa et al., 1989a).

Work Load and Performance of Air Traffic Controllers on Shiftwork

General Considerations

The job of the air traffic controller (ATC) is generally considered one of the most stressing, thanks to the high levels of attention, vigilance, and responsibility it requires (Crump, 1979; Stammers, 1978). The basic task is to convert a traffic demand into a traffic flow (Ratcliffe and Gent, 1974) which needs continuous adjustments according to differing circumstances. Therefore, the task requires high attention and surveillance of various instruments (radio, telephone, radar displays, etc.), and interaction with different people (pilots, assistants). It is quite evident that the ATC's cognitive and operational processes vary not only according to the number of aircraft under control, but also with the number and complexity of problems to be solved (Karlbeek, 1976; Melton et al., 1978; Hurst and Rose, 1978). Furthermore, if the work-load were a simple linear function of the number of aircraft under control, the ATC would rapidly become crammed. Actually, they alter their strategy, progressively reorganizing their system of processing flight information by changes in operating methods (in particular: cognitive processes, verbal exchange, mnemonic activity, co-ordination with the assistants, anticipation and solution of the developing problems) (Sperandio, 1978).

This very complex task requires the contemporary activity of all the mental faculties at their best level. Furthermore, the same levels of vigilance and attention are required at all hours of the day, irrespective of the work load: if the performance of the ATC can be impaired at certain hours of the day by an excessive work load, it can also be impaired during the night by lowered mental and physical efficiency, even with a lower work load. In fact, in many cases, the nocturnal drop in body efficiency can be further increased by external understimulation, becoming a dangerous factor in case of emergency.

This observation points out the importance of correct analysis and measurement of the ATC's job and work load (Hopkin, 1980), arranging work hours and shift schedules to maintain the highest possible performance at unusual hours.

Some Indications Coming from an Italian Study

A recent study on stress and performance in a group of ATCs, working in three shifts at the Regional Air Control Center of Rome, investigated some aspects connected to these problems (Costa et al., 1989b).

We examined 20 ATCs selected from a population of 350, aged between 32 and 59 (mean 40) and with a length of service between 12 and 35 years, during their normal 6-day shift cycle (which was organized on "fast rotation": 1st day: Afternoon [1300-2000]; 2nd day: Morning [0700-1300]; 3rd day: Night [2000-0700]; 4th day: rest; 5th day: off; 6th day: off). We recorded the number of aircraft under control per hour as an index of work load (stress) and both subjective (fatigue, mood, physical fitness, pace rating) and objective (heart rate, hormones excretion, visual reaction time, critical flicker fusion frequency, oral temperature) measures as indices of strain and performance.

The data recorded showed that the air traffic, and consequently the work load, varied greatly during the day, with remarkable differences between day and night both in the overall group (Figure 1) and per individual (Table 1); nevertheless, the heart rates of the group showed quite a constant trend on all the three shifts, irrespective of the work load. The same pattern appeared in the controllers' excretion of vanillyl mandelic acid (V.M.A.), a metabolic product of the hormones related to stress activation (catecholamines). V.M.A. remained high during both the night and the day shifts (Table 2), regardless of the reduced work load. This indicated highly active body functions, as controllers maintained high levels of awakeness and alertness in spite of external understimulation. The subjects' mood and physical fitness decreased similarly, while fatigue increased, on all three shifts, but particularly on night shift (Table 2). As concerns the tests of performance, both Flicker Fusion Frequency and Visual Reaction Time registered statistically insignificant differences between the beginning and end of each shift, but showed a normal circadian pattern (with the best efficiency in the afternoon). Also the circadian rhythm of oral temperature, while showing a slight modification of the noctural depression during the night shift, determined by state of awakeness and activity, was not altered in its normal circadian phase (Figure 2).

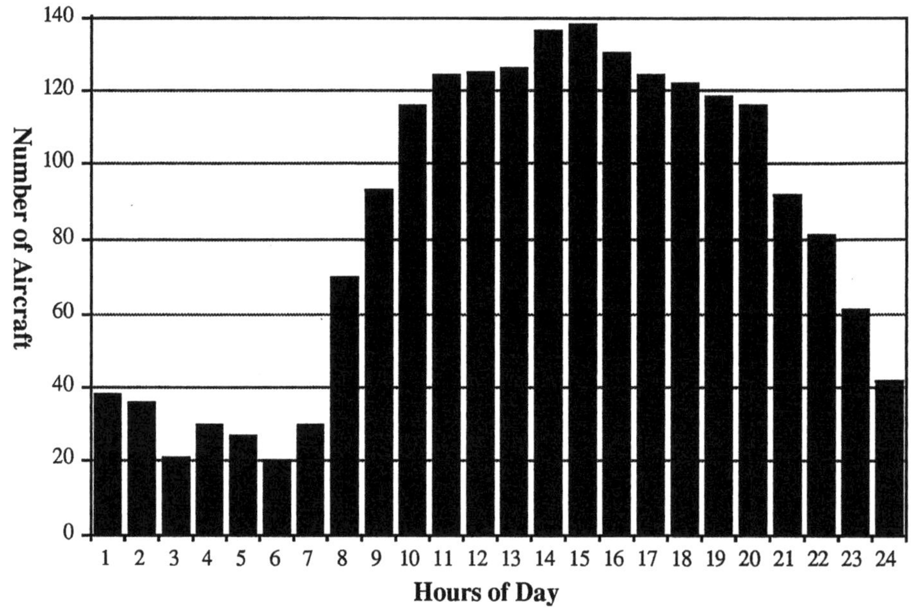

Figure 1: Hourly distribution of the air traffic during the day, in the Rome Regional Center (1986: annual average).

ATCs on the night shift experienced increased arousal and body activation equal to their dayshift peers, mainly due to the maintenance of awakeness. Nor were their performance levels impaired during the night, probably because of the good work organization and compensative measures adopted. In fact, the arrangement of the shift schedule on fast rotation, with two days off after the single night shift, did not alter significantly their normal circadian synchronization. Moreover, during the night shift, the ATCs didn't work all the 11 hours (from 2000 to 0700), but worked together until 2300, after which half of them rested in a room next to the control-room until 0300, then relieved their colleagues.

For a further indicator of performance efficiency and alertness, we reviewed records of near-misses for the last two years. Fifteen near-misses had occurred during this period (twelve between May and September), with nine happening during peak traffic hours (between 1130 and 1600). Only one occurred at night (0438). We conclude most of them happened due to poor attention on part of the ATC in evaluating altitude, speed, or distance between aircraft; four resulted from misunderstandings in communication between ATC, assistants, and pilots; and one derived from radar failure.

Table 1. Mean values (standard deviation and range) of work-load (nr.aircraft, pace rating) and strain (heart rate) indices in ATCs' shifts.

	MORNING	AFTERNOON	NIGHT
Nr. AIRCRAFTS/hour	9.2 ±4.7 (12 - 28)	0.2 ±4.5 (6 - 24)	9.4 ±4.0 (4 - 16)
PACE RATING	5.7±0.6 (4.5 - 6.5)	5.5±1.5 (3.5 - 8.0)	3.5 ±1.7 (2.0 - 6.5)
HEART RATE/min	83.3±5.1 (60 - 125)	87.2 ±7.1 (64 - 122)	82.1±5.0 (56 - 118)

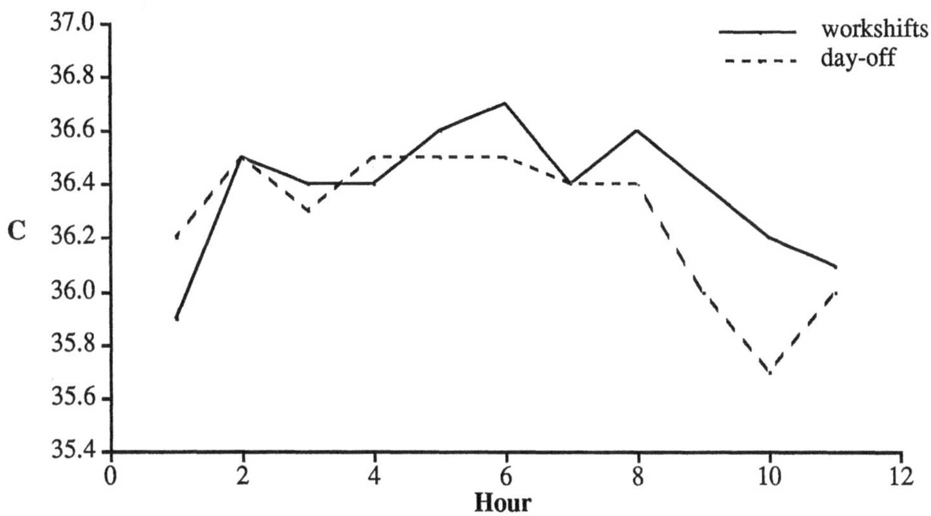

Figure 2. Circadian variation of oral temperature during the three workshifts (taken as on day) and day off.

Table 2: Tests of performance, self ratings of mood and fatigue, and Vanillyl Mandelic Acid excretion at the beginning and at the end of the three shifts (mean and standard deviation)

	BEFORE 0700	AFTER 1300	BEFORE 1300	AFTER 2000	BEFORE 2000	AFTER 0700
REACTION TIME	226.1	225.5	223.1	228.7	210.5	223.0
(msec)	22.7	23.5	19.9	20.9	22.7	21.8
FLICKER FUSION FREQ	38.5	38.3	37.2	37.3	39.0	39.2
(Hz)	2.5	2.2	2.8	2.6	2.1	2.0
MOOD	5.1	** 4.1	4.1	4.1	4.6	3.9
(1 min - 7 max)	1.0	1.0	1.0	1.2	1.1	1.0
PHYSICAL FITNESS	4.5	* 3.5	4.1	* 3.3	4.4	* 3.4
(1 min - 7 max)	1.3	1.4	1.6	1.3	1.1	1.2
STRONG-WEAK	3.2	3.5	3.5	3.7	3.2	** 4.2
(1 min - 7 max)	0.8	0.8	0.8	0.9	0.9	1.0
RELAXED-TENSE	2.9	** 3.7	3.5	3.9	3.2	3.6
(1 min - 7 max)	0.9	1.0	1.4	1.1	1.0	1.1
REFRESHED-TIRED	3.3	** 4.3	3.6	** 4.8	3.3	** 4.7
(1 min - 7 max)	0.9	1.1	1.3	1.5	1.2	0.9
VIGOROUS-EXHAUSTED	3.2	3.8	3.5	* 4.2	3.3	** 4.1
(1 min - 7 max)	1.0	0.9	1.2	1.1	1.0	0.9
AWAKE-SLEEPY	2.7	2.9	2.7	3.3	2.6	** 4.2
(1 min - 7 max)	1.1	0.9	1.4	1.1	1.1	1.4
OVERALL FATIGUE	15.2	** 18.2	16.8	** 19.9	15.6	** 21.2
(5 min - 35 max)	3.6	3.3	4.5	4.2	4.7	3.7
V.M.A. excretion	5.09	4.91	4.67	5.06	4.97	5.02
(mg/ml)	0.98	1.03	0.90	0.96	0.90	0.93

* "p" (probability) < 0.05
** "p" (probability) < 0.01

References

Akerstedt, T. (1985). "Adjustment of physiological circadian rhythms and the sleep-wake cycle to shiftwork." In S. Folkard and T.H. Monk (Eds.) *Hours of Work. Temporal Factors in Work Scheduling*. Chichester: John Wiley & Sons, pp 185-197.

Alluisi, E.A., and Chiles, W.D. (1967). "Sustained performance, work rest scheduling and diurnal rhythms in man." *Acta Psychologica, 27*, pp 436-442.

Andlauer, P., Rutenfranz, J., Kogi, K., Thierry, H., Vieux, N., and Duverneuil, G. (1982). "Organization of night shifts in industries where public safety is at stake." *Int. Arch. Occup. Environ. Health, 49*, pp 353-355.

Angiboust, R., and Gouars, M. (1972). "Tentative dévaluation de lé fficacité operationelle du personnel de l'aeronautique militaire au cours de veilles nocturnes." In W.P. Colquhoun (Ed.) *Aspects of Human Efficiency*. London: The English Universities Press Ltd, pp 151-170.

Bjerner, B., Holm, A., and Swensson, A. (1955). "Diurnal variation in mental performance - A study of three-shift workers." *Brit. J. Industr. Med., 12*, pp 103-110.

Blake, M.J.F. (1971). "Temperament and time of day." In W.P. Colquhoun (Ed.) *Biological rhythms and Human Performance*. London: Academic Press, pp 109-148.

Bohle, P., and Tilley, A.J. (1989) "The impact of night work on psychological well-being." *Ergonomics, 32*, 9, pp 1089-1099.

Breithaupt, H., Hildebrandt, G., Dohre, D., Josch, R., Sieber, U., and Werner, M. (1978). "Tolerance to shift of sleep as related to the individual's circadian phase position." Ergonomics, 21, pp 767-774.

Browne, R.C. (1949). "The day and night performance of teleprinter switchboard operators." *Occup. Psychol., 23*, pp 1-6.

Carter, F.A., and Corlett, E.N. (1978). *Shift Work and Accidents*. Dublin: European Foundation for the Improvement of Living and Working Conditions.

Colquhoun, W.P. (1971). "Circadian variations in mental efficiency." In Colquhoun, W.P. (Ed.) *Biological Rhythms and Human Performance*. London: Academic Press, pp 39-107.

Colquhoun, W.P. (1976). "Accidents, injuries and shiftwork." In P.G. Rentos and R.D. Shepard (Eds.) *Shift Work and Health*. Washington: HEW Publ. No. (NIOSH) 76-203, pp 160-175.

Colquhoun, W.P., and Folkard, S. (1978). "Personality differences in body temperature rhythm, and their relation to its adjustment to night work." *Ergonomics, 21*, pp 811-817.

Costa, G., Cesana, G.C., Kogi, K., and Wedderburn, A. (Eds.) (1990). *Shiftwork: Health, Sleep and Performance*. Frankfurt a.M.: Verlag Peter Lang Gmbh.

Costa, G., Lievore, F., Casaletti, G., Gaffuri, E., and Folkard, S. (1989a). "Circadian characteristics influencing interindividual differences in tolerance and adjustment to shiftwork." *Ergonomics, 32*, 4, pp 373-385.

Costa, G., Pinchera, E., Pistilli, M., Battisti, S., and Munafo', E. (1989b). "Stress and performance of air traffic controllers." *Ergonomia, 12*, 1, pp 93-100.

Crump, J.H. (1979). "Review of stress in air traffic control: its measurement and effects." *Aviat. Space Environ. Med., 50*/3, pp 243-248.

Erhet, C.F. (1981). "New approaches to chronohygiene for the shift worker in the nuclear power industry." In A. Reinberg, N. Vieux and P. Andlauer (Eds.) *Night and shift Work: Biological and Social Aspects*. Oxford: Pergamon Press, pp 263-270.

Folkard, S. (1990). "Circadian performance rhythms: some practical and theoretical implications." *Phil. Trans. R. Soc. Lond.*, B 327, pp 543-553.

Folkard, S., and Condon, R. (1987). "Night shift paralysis in air traffic control officers." *Ergonomics, 30*, 9, pp 1353-1363.

Folkard, S., Condon, R., and Herbert, M. (1984). "Night shift paralysis." *Experientia, 40*, pp 510-512.

Folkard, S., and Monk, T.H. (1979). "Shiftwork and performance." *Human Factors, 21*(4), pp 483-892.

Folkard, S., and Monk, T.H. (Eds.) (1985). *Hours of Work. Temporal Factors in Work Scheduling*. Chichester:John Wiley & Sons.
Folkard, S., Monk, T.H., and Lobban, M.C. (1978). "Short and long-term adjustment of circadian rhythm in permanent night nurses." *Ergonomics, 21*, pp 785-799.
Folkard, S., Monk, T.H., and Lobban, M.C. (1979). "Towards a predictive test of adjustment to shift work." *Ergonomics, 22*, 79-91.
Gillberg, M. (1985). "Effects of naps on performance." In S. Folkard and T.H. Monk (Eds.) Hours of Work. Temporal Factors in Work Scheduling. Chichester: John Wiley & Sons, pp 77-86.
Harris, W. (1977). "Fatigue, circadian rhythms, and truck accidents." In R. Mackie (Ed.) *Vigilance. Theory, Operational Performance, and Physiological Correlates*. New York and London: Plenum Press, pp 133-146.
Hildebrandt, G., Rohmert, W., and Rutenfranz, J. (1975). "The influence of fatigue and rest period on the circadian variation of error frequency of shift workers (engine drivers)." In P. Colquhoun, S. Folkard, P. Knauth and J. Rutenfranz (Eds.) *Experimental Study of Shiftwork*. Opladen: Westdeutscher Verlag, pp 174-187.
Hildebrandt, G., and Stratmann, I. (1979). "Circadian system response to nightwork in relation to the circadian phase position." Int. Arch. Occup. Environ. Health, 43, pp 73-83.
Hildebrandt, G., Rohmert, W., and Rutenfranz, J. (1974). "12 & 24 rhythms in error frequency of locomotive drivers and the influence of tiredness." *Int. J. Chronobiol., 2*, pp 175-180.
Hockey, R., and Hamilton, P. (1983). "The cognitive patterning of stress states". In R. Hockey (Ed.) *Stress and Fatigue in Human Performance*. Chichester: Wiley & Sons, pp 331-362
Hopkin, V.D. (1980). "The measurement of the air traffic controller." *Human Factors, 22*, 5, pp 547-560.
Horne, J.A., and Ostberg, O. (1976). "A self assessment questionnaire to determine morningness-eveningness." *Int. J. Chronobiol., 4*, pp 97-110.
Hurst, M.W., and Rose, R.M. (1978). "Objective workload and behavioural response in airport radar control rooms." *Ergonomics, 21*, 7, pp 559-565.
Kalsbeek, J.W.H. (1976). "Some aspects of stress measurements in air traffic control officers at Schipol Airport." Symposium on Stresses of the Air Traffic Control Officers, University of Manchester, Department of Postgraduate Medical Studies, pp. 39-42, (cited by Crump, 1979).
Klein, E.K., and Wegmann, H.M. (1979). "Circadian rhythms of human performance and resistance: operational aspects." In Sleep, Wakefulness and Circadian Rhythm. AGARD Lectures Series No. 105, pp 2.1-2.17.
Kleitman, N., and Jackson, D.P. (1950). "Body temperature and performance under different routines". *J. Appl. Physiol., 3*, pp 309-328.
Knauth, P., and Rutenfranz, J. (1975). "The effects of noise on the sleep of night-workers." In W.P. Colquhoun, S. Folkard, P. Knauth and J. Rutenfranz (Eds.) *Experimental Studies of Shiftwork*. Opladen: Westdeutscher Verlag, pp 57-65.
Knauth, P., and Rutenfranz, J. (1981). "Duration of sleep related to the type of shift work." In A. Reinberg, N. Vieux and P. Andlauer (Eds.): *Night and Shift Work: Biological and Social Aspects*. Oxford: Pergamon Press, pp 161-168.
Knauth, P., and Rutenfranz, J. (1982). "Development of criteria for the design of shiftwork systems". *J. Human Ergol., 11*, Suppl., pp 337-367.
Knauth, P., Rutenfranz, J., Herrmann, G., and Poppel, S.J. (1978). "Re-entreinment of body temperature in experimental shift work studies." *Ergonomics, 21*, pp 775-783.
Kogi, K. (1982). "Sleep problems in night and shift work." *J. Human Ergol., 11*, Suppl., pp 217-231
Lavie, P. (1985). "Ultradian Cycles in Wakefulness. Possible implications for work-rest schedules." In S. Folkard and T.H. Monk (Eds.) Hours of Work. *Temporal Factors in Work Scheduling*. Chichester: John Wiley & Sons, pp 97-106.
Mann, H., Rutenfranz, J., and Aschoff, J. (1972). "Untersuchungen zur Tagesperiodik der Reaktionzeit bei Nachtarbeit. I. Die Phasenlage des positiven Scheitelwertes und Einflusse des Schlaft auf die Schwingungsbreite." *Int. Arch. Arbeitsmed., 29*, pp 159-174.

Matsumoto, K., Matsui, T., Kawamori, M., and Kogi, K. (1982) "Effects of nighttime naps on sleep patterns of shiftworkers." *J. Human Ergol., 11*, Suppl., pp 279-289.
Melton, C.E., Smith, R.C., and McKenzie, J.M. (1978). "Stress in air traffic personnel: low density towers and flight service stations." *Aviat. Space Environ. Med., 49*/10, pp 724-728.
Minors, D.S., and Waterhouse, J.M. (1981). "Anchor sleep as a synchronizer of rhythms on abnormal routines." In L.C. Johnson, D.I. Tepas, W.P. Colquhoun and M.J. Colligan (Eds) *Biological Rhythms, Sleep and Shift Work. Advances in sleep research. Vol. 7.* New York: Spectrum, pp 399-414.
Minors, D.S., and Waterhouse, J.M. (1986). "Circadian rhythms and their mechanisms." *Experientia, 42*, 1, pp 1-13.
Monk, T.H., and Folkard, S. (1985). "Individual differences in shiftwork adjustment." In S. Folkard and T.H. Monk (Eds.) *Hours of Work. Temporal Factors in Work Scheduling.* Chichester: John Wiley and Sons, pp 227-237.
Moog, R. (1981). "Morning-Evening types and shiftwork." In A. Reinberg, N. Vieux and P. Andlauer (Eds.) *Night and Shift Work: Biological and Social Aspects.* Oxford: Pergamon Press, pp 481-488.
Moog, R. (1987). "Optimization of shiftwork: physiological contribution." *Ergonomics, 30*, pp 1249-1259.
Murrel, H.F.H. (1971). "Industrial work rhythms." In W.P. Colquhoun (Ed.) *Biological Rhythms and Human Performance.* London: Academic Press, pp 241-272.
Nachreiner, F. (1975). "Role perceptions, job satisfaction and attitudes towards shiftwork of workers in different shift systems as related to situational and personal factors." In W.P. Colquhoun, S. Folkard, P. Knauth and J. Rutenfranz (Eds.) *Experimental Studies of Shiftwork.* Opladen: Westdeutscher Verlag, pp 232-243.
Ostberg, O. (1973). "Circadian rhythms of food intake and oral temperature in "morning" and "evening" groups of individuals." *Ergonomics, 16*, pp 203-209.
Pokorny, M., Blom, D., and Van Leeuwen, P. (1981). "Analysis of traffic accident data (from bus drivers). An alternative approach (I)." In A. Reinberg, N. Vieux and P. Andlauer (Eds.) *Night and Shift Work: Biological and Social Aspects.* Oxford: Pergamon Press, pp 271-278.
Prokop, O., and Prokop, L. (1955). "Ermudung und Einschlafen am Steuer". *Dtsch. Z. gerichtl. Med., 44*, pp 343-355.
Ratcliffe, S., and Gent, H. (1974). "The quantitative description of a traffic control process." *Journal of Navigation, 27*, pp 317-322.
Reinberg, A., Motohashi, Y., Bourdeleau, P., Andlauer, P., Levi, F., and Bicakova-Rocher, A. (1988). "Alteration of period and amplitude of circadian rhythms in shift workers." *Eur. J. Appl. Physiol., 57*, pp 15-25.
Reinberg, A., Vieux, N., and Andlauer, P. (Eds.) (1981). *Night and Shift Work: Biological and Social Aspects.* Oxford: Pergamon Press.
Reinberg, A., Vieux, N., Chaumont, A-J., Laporte, A., Smolensky, M., Nicolai, A., Abulker, C., and Dupont, J. (1979). "Aims and conditions of shift work studies." *Chronobiologia, 6*, Suppl.1, pp 7-23.
Rogers, A.S., Spencer, M.B., Stone, B.M., and Nicholson, A.N. (1989). "The influence of a 1h nap on performance overnight." *Ergonomics, 32*, 10, pp 1193-1205.
Rudolp, G.deM. (1946a). "Psychological aspects of a conscious temporary generalized paralysis." *J. Mental Sciences, 92*, pp 814-816.
Rudolp, G.deM. (1946b). "Night-nurse paralysis: a temporary tonic motor paralysis." *Bristol Med.-Chir. J., 43*, pp 132-135.
Rutenfranz, J. (1982). "Occupational health measures for night and shiftworkers." *J. Human Ergol., 11*, Suppl., pp 67-86.
Rutenfranz, J. and Colquhoun, W.P. (1979). "Circadian rhythms in human performance." *Scand. J. Work Environ. Health, 5*, pp 167-177.
Sperandio, J.C. (1978). "The regulation of working methods as a function of work-load among air traffic controllers." *Ergonomics, 21*, 3, pp 195-202.

Stammers, R.B. (1978). "Human factors in airfield air traffic control." *Ergonomics, 21*, 6, pp 483-488.

Torsvall, L., and Akerstedt, T. (1980). "A diurnal type scale." *Scand. J. Work Environ. Health, 6*, pp 283-290.

Vidacek, S., Kaliterna, L., and Radosevic-Vidacek, B. (1987). "Predictive validity of individual differences measures for health problems. Preliminary results." In A. Oginski, J. Pokorski and J. Rutenfranz (Eds.) *Contemporary Advances in Shiftwork Research.* Krakow: Medical Academy, pp 99-106.

Wever, R.A. (1985). "Man in temporal isolation: basic principles of the circadian system." In S. Folkard and T.H.Monk (Eds.) Hours of Work. *Temporal Factors in Work Scheduling*. Chichester: John Wiley and Sons, pp 15-28.

Wojtczac-Jaroszowa, J., and Pawlowska-Skyba, K. (1967). "Night and shift work. I:Circadian variations in work." *Medycyna Pracy, 18*, 1. (Cited by Wojtczac-Jaroszowa, J. "Physiological and psychological aspects of night and shift work". Washington, D.C.: U.S. Department of Health, Education and Welfare. Publication No.(NIOSH) 78-113.

Wynne, R.F., Ryan, G.M., and Cullen, J.H. (1986). "Adjustment to shiftwork and it's prediction: results from a longitudinal study." In M. Haider, M. Koller and R. Cervinka (Eds.) *Night and Shiftwork: Longterm Effects and Their Prevention.* Frankfurt a.M.:Verlag Peter Lang Gmbh, pp 101-108.

Zulley, J., and Bailer, J. (1988). "Polyphasic sleep/wake patterns and their significance to vigilance." In J.P. Leonard (Ed.) *Vigilance: Methods, Models, and Regulations*. Frankfurt a.M.: Peter Lang Gmbh, pp 167-180.

Operator Workload as a Limiting Factor in Complex Systems

P. G. A. M. Jorna

National Aerospace Laboratory (NLR), Amsterdam, The Netherlands

Introduction

There are a growing number of activities in which the human operator is assigned duties that involve mental processing of information. Such processing can be experienced by the operator as being exhausting or even stressful, and such feelings were at the origin of the concept of "mental workload." It is used as a moderator variable to explain why applications of modern technology sometimes cause such unexpected variabilities in human performance as accidents, errors, or increases in the number of health problems.

This paper will start with a summary of the more general human operator problems observed in operating complex systems so far. Issues concerning the complexity of definition and measurement of workload will be discussed in some detail. The purpose is to provide background information on the origin of particular limitations in human performance, and to focus on possible ways for managing workload. That information is subsequently used in a discussion concerning workload problems in air traffic control.

High Tech and Human Resources

The introduction of micro electronics and computers dramatically influenced the type of tasks that humans had to perform, with respect to both their structure and their content. A general expectation was that routine chores would be reduced and productivity increased, as humans were now able to spend more time on creative planning, decision making, etc. The contention that tasks would also become easier to perform has proven to be a wrong assumption. On the contrary, complexity was generally increased, and when the mental component of jobs gained importance, complaints about job conditions and workload also increased. Mental workload and stress experiences are apparently powerful enough to create health problems for operators.

Besides these technology induced shifts from the physical towards the mental aspects of work, other reasons for workload increments can also be identified. One is that the capabilities of a system will also be expanded. This phenomenon creates new situations to which humans have to adapt and master new strategies for handling the increased complexities. The trend of "stretching" system capability is a "natural" one if put in an economic perspective where growth in system output is the primary concern, with the risk of

NATO ASI Series, Vol. F 73
Automation and Systems Issues in Air Traffic Control
Edited by John A. Wise et al.

overloading the operator. Only if serious health problems occur, or if accident rates boom and economic or safety standards are surpassed, is attention allocated to these human factor related problems. As a consequence, history shows that technological evolution can and will cause an increase in human factor related problems. The sequences for such a process are illustrated in table 1.

Table 1. Impact of technology and initial human factor problems

1. New technology application
2. Increased system capability
3. More complex information processing (workload)
4. More extreme working conditions (risks)
5. Taxing human resources (altered strategies, complacency)
6. Emerging accidents, errors, training failures, operator shortages

Typical examples in this respect are problems encountered with adapting pilots to low level flying with new generations of high performance military jets, or safety problems that occurred with control over the operation of nuclear power plants. One solution tried so far was rigorous automation in order to reduce the total amount of work, while at the same time reducing the impact of the fallible human operator on the system. Unexpected technical or procedural failures are, however, always possible, and such events still force the human to adopt the role of a "flesh and blood" back-up. In order to perform that function of "manual reversion" effectively, the operator needs awareness of system status, knowledge about strategies, and control over means for handling emergencies. Being uninformed or unprepared, while maintaining the highest standards for safety, could again lead to subjective feelings of being mentally taxed or frustrated. Typically, task demands were not reduced effectively, but simply shifted to other types of duties like monitoring. In these more passive system roles, human operators proved to be unstable in their performance levels. Complaints about workload did not cease, but their content was changed towards task characteristics like boredom and underload.

A lesson to be learned is that, in designing new tasks for humans, provisions should be made to accomplish something like an appropriate "workload balance" that ensures both adequate performance levels as well as sufficient human endurance. This is, however, easier said than done, as workload is not easily accessible to measurement.

Workload as a Fuzzy Concept

Typically, the concept of workload seems poorly defined, and it generates much confusion amongst practitioners as well as researchers. The reasons for that confusion are matters of definition.

The concept of workload is used to describe several things:

- It describes a certain characteristic of a task without any specific reference to the performer (i.e., it is defined as a stimulus property like task complexity).
- It is also used to index the performance efficiency of a particular individual (i.e., it is defined as some human response). Learning and practice serve to reduce workload, as performance is improved without an apparent change in the physical or objective appearance of the task.
- Workload is used as a composite of all imaginable aspects that contribute to the "burden" of work. Both physical and mental factors contribute. Composite mental load has many origins like the task (responsible for "mental task load"), or the environment (which generates "emotional load" by being dangerous.) Other psychological factors like "time pressure" or "responsibility" can impose additional difficulties by their requirement to perform under emotionally taxing or stressful circumstances.

The confusion about terminology is very common. The most common conception of workload, as used by system designers is something like "task difficulty" or "complexity." Such a notion simply replaces one ill defined concept for the other, and the disguise is not very helpful. Workload clearly is a multi-faceted and multi-dimensional concept, but one common denominator can be identified in most descriptions; that is the concept of "human costs" involved in achieving particular levels of performance under specific conditions. Performance will suffer if the demands exceed capability. The idea of such "discrepancies" is also prevalent in contemporary theories about "Stress" (Appley & Trumbull, 1986). Mental load and stress are argued to represent related concepts on a continuum, rather than separate issues. The conceptual difference is that mental load is not necessarily stressful by definition, provided that positive emotions are involved and the operator still perceives control over the situation. It is not accidental that the assessment techniques for both of them are essentially the same.

Workload Measurement

Well known techniques that were reviewed many, many times are the easy to use subjective ratings, the more cumbersome physiological indices, and the complex to define performance measures (including dual-task assessments of spare capacity). It is evident that the evaluation of workload in a certain person-task loop, and the use of specific assessment methods, are not only a matter of task analysis or convenience, but also require some basis in human performance theory. In the literature, two conceptional models of human information processing prevail.

The first model is essentially a single channel model with a limited capacity for information processing (Kantowitz 1987; Kantowiz & Casper, 1989). A consistent workload concept can be defined, in this case, by referring to the inverse of the spare capacity that is left during processing of task information. More difficult tasks will require more capacity, so there is less to spare, etc. The single channel model enables a global or overall assessment of workload, and the concept is often used implicitly in applied research. Secondary tasks are essentially based on this model and subjective scales like the often used Bedford scale (Hancock 1989) actually require the subject to report that level of spare capacity.

The second, so called multiple resource model (Wickens, 1984) describes the human information processing system as consisting of several separate processing resources or capacities. The implications for workload concepts are quite clear: workload is multidimensional, and each resource can now be loaded independently. Measurement techniques should be designed to be "diagnostic" for the different processing resources. Task structure and task distribution are now very critical for timesharing, and a proper design can distribute workload across several capacities, instead of overloading one. The implication is that there will not be a single best "secondary task method" for assessing spare capacity. The technique can, however, remain valid by designing several sets of tasks, each with its own demands on a specific type of information processing. A detailed discussion about this model and its implications for task design is beyond the scope of this paper. For now, it is sufficient to realize that the discussion about workload dynamics are not necessarily restricted to one channel.

Indices for workload or stress is known to provide contradictory results if they are compared directly with performance measures. This "dissociation" can be understood by considering the strategy of the operator. Increasing the demands of a task will not always decrease the level of performance, as long as the operator is able and willing to compensate for it by trying harder (i.e., investing more effort). Performance will not be related to workload until beyond the point where the maximal available capacity is exceeded. That maximum will be different for each individual, depending on ability, experience, level of fatigue, etc. Performance and costs relations are therefore dynamic and complex, as they are under strategical control of the operator. Workload measurement is concerned with patterns between many indices (Derrick, 1988), and no single best measure can be obtained.

An illustration of a particular trade off is provided by a study by Frankenhaeuser and Johansson (1976), in which subjects had to perform a cognitive task under two levels of demands as created by increasing the number of conflicts in the available data. The results of this manipulation on performance scores and workload indices are depicted in Figure 1. Clearly, performance was hardly affected, while self-reported distress and adrenaline levels increased dramatically.

When designing complex systems, performance requirements most often represent the primary design driver in meeting economical demands or other external requirements. An increased task load will not necessarily reduce human performance directly, but it will reduce spare capacity first. The endurance of the operator can also be impaired by cumulative fatigue. In the case of air traffic control, safety issues should prevail. Adequate workload management is essential to obtain that goal.

Workload Management

The concept of workload is related to limits in human information processing. The impact of workload on a system design will therefore depend on the level of "discrepancy" that exists with human capabilities. Such discrepancies can be manipulated at different levels. The first one is provided by organizational policies and control options over hiring and training people for the job at hand, with the purpose of fitting human capabilities to its requirements. The second option for control is provided at the level of the individual operator. Aspects of this "human control level" represent strategical control over timing of task demands, strategies for task execution and the regulation of personal effort that has to be invested in the process.

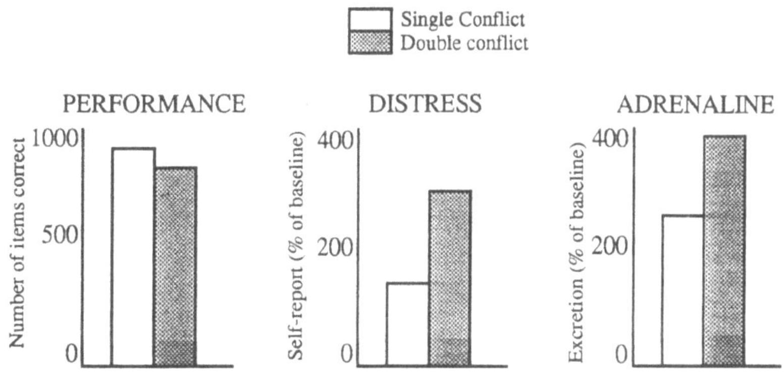

Figure 1. Effects of an increase in task demands, from "single" to "double" conflicts in stimuli, on performance and corresponding "human costs" indices like self reported distress and the level of adrenaline excretion. Based on Frankenhaeuser & Johansson, 1976.

Organizational control Jobs with the reputation of being "demanding" ones often share the characteristic that no performance whatsoever is to be expected without appreciable levels of practice and training. This is true for piloting an aircraft as well as for air traffic control. Another characteristic of such jobs is that training is not completed successfully by every trainee. In general, these observations imply that the system is approaching some of the limits in human information processing capabilities, and, consequently, special measures are taken to select humans with a high potential for the jobs as assessed by tailored selection procedures. Training should further develop that potential into the necessary skills. No guarantee for proficiency is, however, obtained immediately, as in many of those jobs additional "experience" is mandatory to perform at a comfortable level of workload.

Workload as a "human costs" factor is so to say, part of a "performance circle" that defines a particular balance between task demands, basic human capacities, and the operator competence required to perform up to a prescribed standard. The interrelationships are illustrated in figure 2.

A change in system demand can have an impact on selection and training requirements. The need to handle more aircraft in air traffic control could be achieved theoretically by extending the training of the controllers, but it is unlikely that sufficient numbers of those superior operators could be delivered to fulfill the operational needs. Clearly, in this case, tasks are in need of a re-design. Predicting the consequences of re-design for training and selection procedures is important and extensive prototyping, as well as simulations resembling operational conditions, are required to prevent flaws in system design. These could easily lead to the occurrence of new types of errors, and often they do. Such errors can be quite subtle. Most of you will be familiar with the use of simple (?) copier machines and the resulting frustration of not getting what you wanted. One obscure cause is the combined use of command selection keys with an entry buffer that will hold or remember a series of commands. These will be executed sequentially after finishing the instruction at hand. No indications are, however, given about the size of that buffer. As a result, some part of the instructions to the machine will be lost, since they were not recorded, and performance will be impaired unexpectedly. It is important to stress that a new design should also be tested with naive subjects who receive training to operate the system during the experimental phase. System designers themselves are clearly not good subjects. They conceived the design and consequently adjust their expectations to its limits. Experienced "old system users" are also tricky subjects, as it may be expected that they will import at least some of their old and familiar strategies to the new situation, which could produce negative transfer of training. Other strategies could be better, and maximally exploit the potential of the new system.

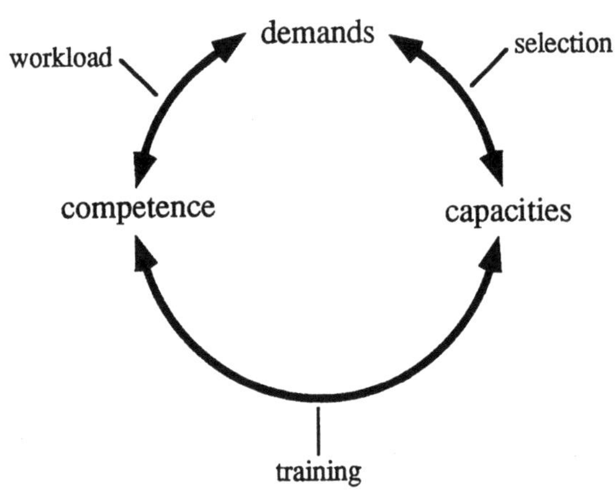

Figure 2. Relation between task demands, basic human capacities, and operator competence in meeting task requirements. The higher the competence level of the operator, by either training or selected aptitude, the less the experienced workload.

Operator control In many jobs operator control is severely limited and timing of tasking is paced by either machines, as in a factory, or by procedures based on economical requirements. Many task designs force the human operator to adjust to imposed variations in rate of input. Human information processing resources are however limited in their capacity, and these limits can be exacerbated by a sub-optimal energetic state of their supporting physiological systems (Sanders, 1983; Hockey et al., 1986). A simplified model of human information processing is depicted in figure 3.

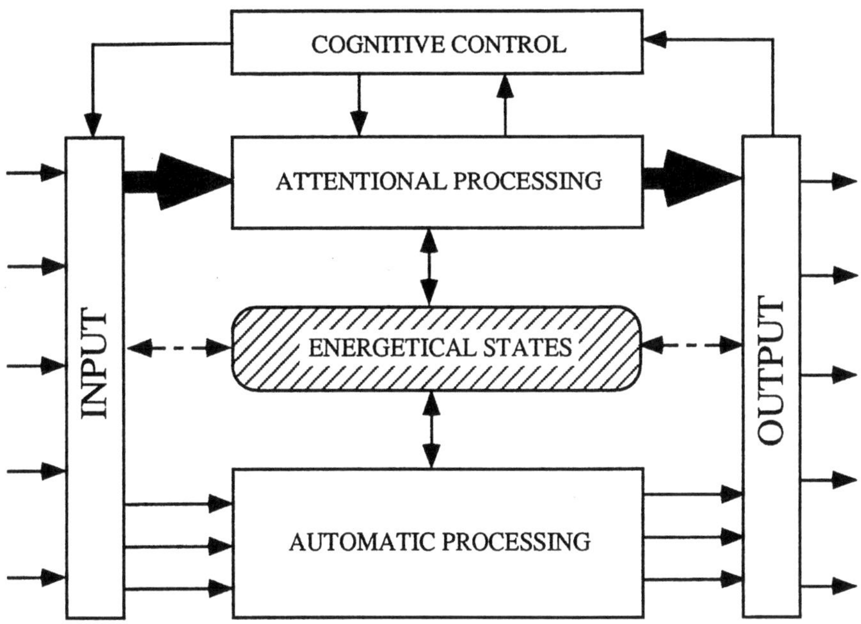

Figure 3. Simplified model of human information processing with two modes: an attentional and automatic one. Supervisory cognitive control compares intended with actual performance and assesses level of control. It is able to adjust input handling strategies or compensation of load by increasing effort, which will alter the state of the energetical mechanisms. Extraneous factors like drugs, sleep deprivation , etc., can also affect the efficiency of the "energetics." The experience of "losing control" will trigger strong emotional responses that lead to highly activated states and possible changes in performance strategy.

The existence of different modes for the processing of information is well established and can be identified in many theories about human performance, as will be illustrated. The attentional or conscious mode is associated with serial, time consuming, laborious analytical processing in working memory and is severely limited in capacity. It is used for problem solving in novel situations and in task conditions that are characterized by continuous variation in, or even inconsistent relations between, input and output (a condition known as "varied S-R mapping") (Shiffrin & Schneider, 1977). The automatic mode is used to process information fast, in parallel, and with little effort. This mode can only be acquired after extensive practice in consistent and predictable conditions. The limitations of both modes are clearly different, and they are also associated with different types of error. "Mistakes" are made during active attentional processing, while "slips" are representative of the execution of routine types of duties and, as such, are related to inattention or absentmindedness (Reason, 1987; Stager, 1991).

These modes of processing typically are accompanied by different energetical states. A general construct in psychology around the 1960's, was that of "Arousal." Early human performance theories postulated a direct relation between level of arousal and the quality of information processing. Such a uni-dimensional concept proved to be far too global. A direct manipulation of energetical state by depriving subjects of sleep hardly ever induces a consistent performance decrement. Subjects are apparently able to compensate, if motivated to do so. This possibility for active human compensation is very common and is used by truck drivers on long duration drives. Boring or non-stimulating tasks are far more difficult to perform under such conditions (Hockey & Hamilton, 1983). Drug studies performed in the laboratory revealed that substances that were supposed to act as energizers (e.g., amphetamine) also did not improve all aspects of performance. They only produced very selective effects on the speed of motor activity — i.e., the execution of responses — instead of speeding up the information processing required to select that response (Frowein, 1981). The theoretical construct of a single energetical dimension was clearly too simple, and it also assumed a passive operator.

Several types of energetical mechanisms were found to be relevant for human performance. Pribram and McGuiness (1975) identified three separate, although interacting, energetical systems in the neurological structure of the brain. The appeal of these energetical concepts is their apparent functional relation with the stages in human information processing (Sanders, 1983).

They refer to:

- Arousal, as only reflecting the immediate responses to input stimulation, which represents a phasic reaction
- Activation, as a more preparatory and tonic readiness to respond
- Effort, as a coordinating mechanism linked to reasoning and active attentional control, which is able to compensate for sub-optimal states of activation or arousal.

A relevant consideration, with respect to workload now, is the ability of the information intake and response activation functions to operate adequately without the involvement of compensating effort. Such "automatic" activation is supposed to occur with well-practiced tasks in familiar situations and is accompanied by a specific and, for that task, optimal pattern of energetical states (Hockey, 1986). Familiar manifestations of activation are "the butterflies" caused by anxiety or risks of failure. Boring tasks such as continuous monitoring without any action, also disturb energetical patterns (but now by their lack of stimulation which results in de-activation and being non-prepared). If external requirements such as safety regulations, do force the operator to stay alert, a conflict is created. In that case, considerable effort will be needed to adjust the level of activation and stay attentive. Humans strongly dislike such conditions, and they rate boredom and fatigue consistently as being very stressful, even more stressful as those conditions that will overload the operator with information (Macky, 1987).

When given the opportunity of control over their own task load, most operators will select to be busy. In non-stimulating circumstances, it is tempting to seek for information or something to do, a strategy known as "hunting for traffic" in the context of air traffic control (Weston, 1976). Hunting for non-task related information introduces the risk of distraction from the task to be performed. Pilots and air traffic controllers who were asked to decide on a suitable display for splitting workload amongst them both selected the display that provided the highest level of "perceived control" (Kantowitz & Casper, 1989).

In conclusion, it is essential to provide the operator with adequate control over task demand, with respect to both timing and the strategies or skills available for its execution. Preferably, such control should include the manipulation of rate of input. Demanding and stimulating working conditions will normally be accompanied by an "activated" energetical state that is optimal for attentional processing of information under cognitive control. Performance decrements under overload conditions are therefore predicted to be caused by limitations in individual processing capacity like the capability for problem solving, etc. Errors will occur if accuracy of performance is traded for speed in order to meet the imposed demand. Underload conditions are normally not expected to occur, provided that strategical control over task load is available. External pacing of tasks, or automating without such opportunity for control, will only leave the strategy of compensating effort to manage the energetical systems. Performance decrements, in that case, are predicted to be related to inattention, induced fatigue, and distraction by seeking other activities.

Workload in Air Traffic Control

The general view is that workload within the ATC system has risen dramatically in the past few years. Further growth is predicted in the near future, and a doubling of the traffic volume is not excluded. Traffic growth has been substantial, but traffic complexity has also increased. A corresponding increase in the number of controllers will not solve these problems, since it is impossible to create new sectors endlessly. Such a strategy would provide other problems caused by the handover of aircraft, etc. A solution discussed earlier was to increase training by such an extent that many more aircraft could be handled by one single controller. This strategy will probably not deliver enough controllers to do the job. Moreover, performance duration is expected to be restricted enormously, as the endurance of such operators would also be very limited. Even with present systems and highly skilled controllers, problems of endurance and cumulative effects of mental load could be identified or were suspected (Cobb & Rose, 1977; Finkelman & Kirschner, 1980).

The task of a controller is to develop a mental image of the aircraft in the sector and to consider the relevant data concerning the intentions of those aircraft. Controllers must update this image continuously, while at the same time communicating with pilots to instruct them how to proceed. Typically, this type of task will require the controller to operate in the "attentional or controlled mode" of information processing (Shiffrin & Schneider, 1977), as described earlier. This mode is highly limited in capacity, and processing in working memory is restricted to about seven separate pieces or "chunks" of information. Limitations of controllers are expected to have a similar range, and guiding a total of four to six aircraft does not pose any serious problems (Finkelman & Kirschner, 1980). A continuous refreshment of the information is needed, however, in order to maintain the content of working memory active. The mental work of the controller in handling and managing memory can be divided into several components with operational relevance. These have been named "mental interfaces," and their images concern aspects like the current picture of traffic, the projected picture of traffic based on intentions of aircraft and flight plan information, and the load of the total sector (Murphy et al., 1989).

The peril for the controller is the occurrence of some form of interference with that updating, as it will cause the controller to "lose the picture." Such interference can, as an example, be created by excessive requirements for communication. Controllers report that, when around 50 to 60% of their time has to be devoted to talking with pilots and processing their responses, updating memory will be out of pace and the mental image gets disturbed. Aids that serve to support memory functions are therefore welcomed by controllers, and flight strips are often used in that way.

One of the first symptoms of controller overload predicted on the basis of the required mode of information processing, is an intentional reduction of possibilities for distraction and interference, or the "get lost" syndrome. The next symptom is a reduced service to the pilots. When someone observes controllers, it is rapidly noticed that they take pride in providing ample service to pilots, in the form of additional information about traffic situation, possible straight-in approaches, etc. Generally, this kind of information will increase the pilots' situational awareness and contributes to flight safety. More aircraft mean less time available for issuing such information to each one of them. Additionally, more frequent application of the existing separation limits will also be in need. The net effect is a reduction in the available slack or spare capacity for solving problems, while at the same time creating more critical circumstances. Performance decrements for the memory function will create errors during the processing of information or even forgetting things that were in the cue for execution. These types of overload errors have indeed been reported in safety investigations (Rosenthal & Mellone, 1989).

Recent safety studies with Canadian data on incidents and occurrences in air traffic control failed to show a consistent relation with workload levels, as determined by a "pace factor" that is defined by both traffic volume and complexity. The Canadian Aviation Safety Board reported the finding of this paradox of ATC, as occurrences often happened during periods of light, non-complex traffic. One would apparently expect more errors during times of overloading the controller. One comment on this investigation concerns the use of that "pace factor." Stated in the report is that "Controllers do not have control over their own workload." The pace of the task is typically externally controlled. However true this observation may seem, it is not entirely correct. Consider the options available to the controller described in the human information processing model of Figure 3. If feedback of the traffic situation indicates a high pace of incoming aircraft, the strategy available is to divert them into holding areas. Such "parking" can be argued to represent active control by the operator of the working environment. The clear effect of a controller overload is, in this case, translated by the controller into traffic delays (and happily enough, not into too many forgotten or uncontrolled aircraft). Controllers are considered to be professionals, and they will bias their decision making to exert "caution" (Bisseret, 1981).

The major portion of errors and incidents were reported to occur during low workload, or more precisely, low task or traffic densities. Errors, in that case, were typically of the kind representing "not doing things," like inattention or failure to coordinate. These types of errors closely agree and stipulate the difficulty that human operators have in regulating their energetical states appropriately under non-stimulating conditions. Maintaining preparation or alertness under such conditions does require extensive effort. This will create the ATC paradox of high levels of human workload, while traffic densities and low complexity erroneously suggest conditions of low system workload.

Conclusions

Workload will remain a confusing concept for the simple reason of its complexity. For that reason, I could have decided to change the title of this paper to: "Operator workload as a complex factor in systems." Recent air traffic control studies identified several problems and limitations in system performance that can be traced directly to problems with workload of the operator. The first one is the occurrence of traffic delays. Controllers are (naturally) biased to be cautious, and they will prevent the occurrence of "overload errors" by putting the aircraft into holding areas. Next to planning, this is one of the few available options available for actively controlling the density of traffic. Underload errors will provide the next limitation for complex systems. Automation, as an example, can result in moments of boredom or underload, and this will occur rather quickly, especially with highly competent operators. Dissociations between demand and competence will alter performance strategies and cause excessive fatigue when alertness has to be maintained and specific types of error will occur, such as distractions from the task at hand.

Well motivated operators will try to keep their "workload balance" by all means, such as hunting for traffic or chasing other, non-task relevant, information. The impact of low levels of stimulation on efficiency and frustration of the human information processing system are generally underestimated.

An opportunity to pay respect to this issue is to exploit more fully the generic ability and motivation of the human operator to be busy. This will necessitate the adoption of a new design philosophy aimed at matching human *capabilities* instead of human *limitations*. A first step of accomplishing such design is to test and evaluate prototypes on their "intuitive value," by testing their compatibility with basic human information processing resources by using non- or newly trained subjects as additional experts for system review. No such comparisons are available at this moment. Performing them is of the highest priority.

References

Appley, M.H., and Trumbull, R.(Eds). (1986). *Dynamics of stress*. New York: Plenum.

Bisseret, A. (1981). Application of signal detection theory to decision making in supervisory control: The effect of the operator's experience. *Ergonomics, 24*, 81-94.

Canadian Aviation Safety Board. (1990). Report on a special investigation into air traffic control services in Canada. Report no. 90-SP001. ISBN 0-662-17693-6.

Cobb, S., & Rose, R.M., (1977). Hypertension, peptic ulcer and diabetes in air-traffic controllers. *Journal of the American Medical Association, 224*, 489-492.

Derrick, W.L. (1988). Dimensions of operator workload. *Human Factors, 30*, 95-110.

Finkelman, J.M. and Kirschner, C. (1980). Information-processing interpretation of air traffic control stress. *Human Factors, 22*, 561-567.

Frankenhaeuser, M. and Johansson, G.(1976). Task demand as reflected in catecholamine excretion and heart rate. *Journal of Human Stress, 2*, 15-23.

Frowein, H. W. (1981). Selective drug effects on information processing. PhD Thesis. Institute for Perception TNO. Soesterberg The Netherlands.

Gopher, D. and Donchin, E. (1986). Workload- an examination of the concept. In: K.Boff, L. Kauffman and J.P. Thomas (Eds) *Handbook of perception and human performance, Vol 2*, Wiley: New York.

Hancock, P. A. (1989). *Human Factors Psychology*. North Holland.
Hockey, G. R. J. (1983). *Stress and fatigue in human performance*. Wiley: Chichester.
Hockey, G. R. J. (1986). Changes in operator efficiency as a function of environmental stress, fatigue and circadian rhythms. In: K. Boff, L. Kauffman and J.P. Thomas (Eds) *Handbook of perception and human performance, Vol 2*, Wiley: New York.
Hockey, G. J., Coles, G. H. and Gaillard, A.W. K. (1986). Energetical issues in research on human information processing. In: *Energetics and Human Information Processing* by Hockey, Gaillard & Coles (Eds). NATO ASI series no 31, Nijhoff publishers.
Kantowitz, B. H. (1987). Mental workload. In: *Human Factors Psychology* by Hancock (ed). North Holland publ.company.
Kantowitz, B. H. and Casper, P.A., (1989). Human workload in aviation. In: *Human Factors in Aviation* by Wiener & Nagel (eds). Academic press.
Mackie, R. R. (1987). Vigilance research-Are we ready for counter measures? *Human Factors, 29*, 707-723.
Manzey, D. (1989). Workload assessment by secondary tasks and the multidimensionality of human information processing resources. In: *Proceedings of the Fifth International Symposium on Aviation Psychology*. Columbus, Ohio.
Murphy, E. D., Reaux, R. A., Stewart, L. J., Coleman, W. D., & Bruce, K. (1989). Where's the workload in air traffic control? In: *Proceedings of the Fifth International Symposium on Aviation Psychology* Columbus ,Ohio.
Pribram, K. H. & McGuinness, D. (1975). Arousal, activation and effort in the control of attention. *Psychological Review, 82*, 116-149.
Reason, J. (1987). A framework for classifying errors. In : J.Rasmussen, K.Duncan, and J.Leplat (Eds), *New technology and human error*, 22-30, Chichester: Wiley.
Rosenthal. L. J., & Mellone, V. J. (1989). Human factors in ATC operations: Anticipatory clearances. In: proceedings of the *Proceedings of the Fifth International Symposium on Aviation Psychology*. Columbus ,Ohio.
Sanders, A. F. (1983). Towards a model of stress and human performance. *Acta psychology, 53*, 61-97.
Shiffrin, R. M and Schneider, W. (1977). Controlled and automatic human information processing II: Perceptual learning, automatic attending and a general theory. *Psychological Review, 84*, 127-190.
Stager, P. M. (1991). Error models for operating irregularities:implications for automation. In: J. A. Wise, V. D. Hopkin, & M. L. Smith (Eds.) *Automation and Systems Issues in Air Traffic Control*. Berlin: Springer-Verlag.
Weston, R. C. W. (1976). Human factors and air traffic control. In: *Pilot error, The Human Factors*, R. Hurst & L. R. Hurst (Eds). Granada.
Wickens, C.D. (1984). *Engineering psychology and Human Performance*. Merrill.

Attention, Workload, and Automation

Barry P. Goettl

Department of Psychology, Clemson University, Clemson, South Carolina, USA

Introduction

Discussion of the effects of automation in air traffic control requires an understanding of automation, the air traffic control environment, and the human operator. Without an understanding of what automation can and cannot do, it is impossible to know how it will affect the work environment. Likewise, without a clear understanding of the tasks, jobs, and goals of air traffic control, one can neither determine whether automation is appropriate nor comprehend the effects that automation will have on the work. It also requires an understanding of human abilities and limitations, preferences and biases, perceptions and cognitions. But without understanding automation and the air traffic control task, an expert in human performance can, at best, provide only vague predictions and generic warnings about how the human operator will respond to automation in air traffic control.

This paper is not intended to provide a comprehensive review of automation, or the air traffic control task, or even human performance. This paper represents my own attempts to reach a better understanding of these topics and the issues involved in automation in air traffic control. It is hoped that my own attempts to understand these issues will provide the reader with a better appreciation of them.

What is Automation?

Automation can be defined as a machine performing some operation by itself. This definition may be accurate but does not describe the diversity of automation. It only takes a little thought to realize that there are many different levels of automation and many different kinds of mechanisms, devices, and operations that have been labeled automatic. Different levels of automation have different effects on the job being automated. Thus, it is important to understand the different levels of automation.

Butera and Thurman (1984) describe two factors important in understanding how automation affects work: the relationship between automation and job tasks, and the complexity of automation. First, they point out that automation does not directly affect the jobs but rather the specific tasks that comprise them. As a consequence, the effect of automation on the work environment can be very complex, with some tasks eliminated, other tasks modified, and still others created.

NATO ASI Series, Vol. F 73
Automation and Systems Issues in Air Traffic Control
Edited by John A. Wise et al.

Second, Butera and Thurman (1984) identify four different types of automation.

Type 1 automation (i.e., electronic mechanization) consists of computer aided machines that perform specific well-defined tasks (e.g., packaging, assembly). These types of machines require constant human intervention to load and unload materials and to perform other needed operations.

Type 2 automation, used in conditions of low uncertainty, is a complex form of automation involving the automation of an entire plant. Humans in these systems primarily perform monitoring tasks but are often held responsible for any problems that arise.

Type 3 automation is used in situations having a moderate to high degree of uncertainty and requiring little manual support. Adjustments are needed due to disturbances in the input, throughput, and output processes. Human operators are involved at a higher level of intervention involving the control and adjustment of automated equipment.

Type 4 automation is used in situations of moderate uncertainty but in which manual work is high. In type 4 automation, production is highly automated but manual intervention is required for "...refitting, setting up, and change overs..." (Butera & Thurman, 1984, p. 110) as a result of varying production requirements. Human intervention here seems to be primarily making adjustments to the machines themselves rather than the automated process.

Each of the categories of automation has different implications for the work environment. For example, Type 2 automation may eliminate the need for skilled workers. Automation may change the requisite skills of the worker as tasks are performed in new ways (type 3 and type 4). It is also possible for humans to be assigned tasks and jobs for which they are not particularly well suited, such as monitoring automated processes (Mackworth, 1948). In applying automation to air traffic control, it is important to understand that: (1) only certain types of automation will be appropriate for aiding air traffic control, and (2) different levels of automation have characteristic effects on the work environment.

Automation and Air Traffic Control

It is apparent that air traffic control has seen dramatic changes over the years as a direct result of technological advances. In the early days, controllers used only flight strips and radio communications to control aircraft. Later, radar was developed and radar scopes were incorporated into air traffic control. However, the first radar scopes provided little information about the aircraft in the airspace. Experienced controllers tell of the days when the only way to distinguish one aircraft from another was to request one aircraft to change course and observe which "blip" on the scope changed course. Computers have been incorporated into air traffic control for presenting identification labels and flight information on scopes, for transmitting and receiving flight strips from other control centers, and for displaying updated weather conditions and forecasts. More recently computers have been developed for vector and approach path calculations, problem resolution, collision avoidance, and aircraft sequencing.

It is debatable whether all of these technological advances in ATC are examples of automation, and which type of automation they represent. Early advances may best be described as Type 1 automation: electronic mechanization. Later technologies seem to fit into Type 3 automation and perhaps even Type 2 automation. Since human operators have a sense of responsibility that computers do not, it seems unlikely that a fully automated system (Type

2) will be in service in the near future. Nevertheless, as technology moves from the piecemeal electronic mechanization of Type 1 automation into Type 3, where the operators are moved to a higher level of control, designers must become more aware of the human's role in the system and the consequences of automation. As the operator is moved to a higher level, will the job be as satisfying? Will the operator understand what the computer is doing at any point in time? Will the operator be left with tasks that are difficult or boring? Will the operator's workload really be reduced? These are empirical questions that human factors specialists and human performance researchers have begun to address.

Automation and Human Performance

Wickens (1984) has outlined the major benefits and drawbacks of automation. He describes three categories of purposes of automation which, in general, deal with overcoming certain limitations in human performance. Automation may be used for tasks that humans cannot perform at all, perform poorly, or perform with limitations. In addition, several researchers have pointed out the economic advantages of automation (Hopkin, 1970; Weiner & Curry, 1981; Wickens, 1984). Drawbacks described by Wickens include: (1) increased monitoring, (2) out-of-the-loop familiarity, (3) false sense of security, (4) proliferation of components, (5) higher- level operator error, and (6) loss of cooperation between human operators. The first three drawbacks presented here pertain to changes in workload and attention and will be the focus of the remainder of this paper.

Workload

Some types of automation relieve the operator of workload. For example, the incorporation of radar scopes relieved controllers the burden of remembering how many aircraft were in the airspace, their positions, headings, identification numbers, etc. Another example of how automation may reduce workload is the ground proximity warning signal (GPWS) (Danaher, 1980). The GPWS monitors several pieces of information that the pilot normally would have to monitor and integrates them into a single warning signal.

Unfortunately, not all types of automation lead to a reduction in workload. Some types of automation, such as Butera & Thurman's (1984) type 2, may increase the number of monitoring tasks assigned to human operators. Automation may also place humans in the role of "supervisory controllers" (Sheridan & Johannsen, 1976) or "supervisor-sentinels" (Pew, 1969 cited in Wickens 1984). In these types of systems, the operator may have less to do but more to monitor, and is also responsible for the system and any problems that might develop. Early research on vigilance has documented the relatively poor performance of humans in monitoring tasks (Mackworth, 1948). Vigilance declines after 30 minutes.

Although technological feasibility, not to mention user acceptance, may not permit full scale type 2 automation in air traffic control for some time, it is still important to consider the extent of monitoring tasks in air traffic control. To date, many of the advances in air traffic control have been in the areas of providing more information to controllers, and these advances have eased some of the workload problems. However, care must be taken to avoid overwhelming the controller with too much information. There is a push to increase the capacity of the air traffic control system, and automation may be required to meet these demands (Hopkin, 1970). However, consideration must be given to how automation is used and how the human's role will change.

Attention

Human limitations in attention have implications for automation. Limitations in vigilance have already been mentioned. In addition, research has shown that humans take longer and are less accurate in detecting system failures when they are monitoring the system than when they are actively controlling the system (Wickens & Kessel, 1980; Ephrath & Young, 1981). This observation has been attributed to out-of-the-loop familiarity. Operators become unfamiliar with the current state of the plant when they sit back and monitor rather than control the operations. Wickens (1984) suggested that humans need to interact with the system in order to maintain an accurate perception of the state of the system. Interaction with the system serves to update their representation of the system. This phenomenon and analysis should serve to caution officials against over-automation of the air traffic control environment.

Another effect of automation that is related to attention is a false sense of security. That is, the operators may come to believe that a fallible system is infallible and absolutely safe. If a system is perceived as infallible, operators may become complacent and less diligent while monitoring the system. If an automated system is performing some operation or calculation, the operator with a false sense of security may neglect to evaluate the solution. Errors may go undetected and cause serious problems. A false sense of security may even influence the diagnostic process, once a problem is detected, and actually lead to an incorrect diagnosis. For example, at Three Mile Island, operators believed that a relief valve had closed when it had not. They developed and acted on an incorrect diagnosis of the problem, and the actions they took aggravated the situation (Perrow, 1984).

Air traffic control is an environment in which many things can go wrong: from pilots misunderstanding or even ignoring instructions, to mechanical failures, to changes in weather conditions. It is often critical for operators to have correct information to detect dangerous situations. If automated systems provide incorrect information that is thought to be correct, dangerous situations may develop before the human can detect and diagnose the problem. Measures should be taken to avoid controller complacency.

Summary

As demands on air traffic control are increased and technology improves, more and more of the air traffic control environment will become automated. It is critical to consider the effects of automation on human workload and attention. This paper has discussed the different types of automation and some of the human performance issues involved. Human factors issues must become an integral part of the planning of future air traffic control systems, if automation is to play a larger role in the control of aircraft.

References

Butera, F. & Thurman, J. E. (1984). *Automation and Work Design.* New York: Elsevier Science Publishing.

Danaher, J. W. (1980). Human error in ATC systems operations. *Human Factors, 22,* 535-545.

Ephrath, A. R., & Young, L. R. (1981). Monitoring vs. man-in-the-loop detection of aircraft control failures. In J. Rasmussen & W. B. Rouse (Eds.), *Human Detection and Diagnosis of System Failures.* New York: Plenum Press.

Hopkin, V. D. (1970). *Human factors in the ground control of aircraft.* NATO AGARDograph No. 142, AGARD-AG-142-70.

Mackworth, N. H. (1948). The breakdown of vigilance during prolonged visual search. *Quarterly Journal of Experimental Psychology, 1,* 5-61

Perrow, C. (1984). *Normal Accidents: Living with High Risk Technologies.* New York: Basic Books.

Pew, R. W. (1969). Comments on 'promotion of man': Challenges in socio-technical systems; designs for the individual operator. In E. O. Attinger (Ed.),*Global Systems Dynamics.* Munich/New York: Karger.

Sheridan, T. B. & Johannsen, G. (1976). *Monitoring Behavior and Supervisory Control.* New York: Plenum Press.

Weiner, E. L., & Curry, R. E. (1980). Flight deck automation: Promises and problems. *Ergonomics, 23,* 995-1012.

Wickens, C. D. (1984). *Engineering Psychology and Human Performance.* Columbus, OH: Charles Merrill Publishing.

Wickens, C. D., & Kessel, C. (1980). The processing resource demands of failure detection in dynamic systems. *Journal of Experimental Psychology: Human Perception and Performance, 6,* 564-577.

Approaches to Error and Failure

ICAO and Civil/Military Coordination

John N. Bradbury

International Civil Aviation Organization (ICAO), Montreal, Quebec, Canada

Introduction

Several circumstances in recent years, including the destruction of Korean Airlines Flight KAL 007 (September 1983) and the shootdown of Iran Air Airbus A300 on flight IR 655 (July 1988), have resulted in a number of actions by the International Civil Aviation Organization (ICAO) to improve ICAO Standards and Recommended Practices (SARPs) in order to help prevent recurrence of such tragic incidents.

Present SARPs pertinent to activities potentially hazardous to civil aircraft are located in Annex 2 - Rules of the Air; Annex 6 - Aircraft Operations (Parts I and II); Annex 10 - Aeronautical Telecommunications (Volumes I and II); Annex 11 - Air Traffic Services and Annex 15 - Aeronautical Information Service.

The key provisions in Annex 11 have their origin mainly in amendments adopted by the Council on 4 January 1975 and 4 March 1981 as a result of studies by the Air Navigation Commission and consultation with States and international organizations. The purpose of the earlier study was to develop provisions aimed at eliminating or reducing the need for interception of civil aircraft and to reduce the risks associated with interceptions undertaken as a last resort. The purpose of the latter study was to identify measures to improve the coordination between authorities engaged in potentially hazardous activities and appropriate air traffic services and aeronautical information services.

Annex 11, paragraph 2.15 requires air traffic services authorities to establish and maintain close co-operation with military authorities responsible for activities that may affect the flights of civil aircraft, and arrangements made to permit prompt exchange of information relevant to the safe and expeditious conduct of civil aircraft flights. Paragraphs 2.15.2 and 2.16 of Annex 11 specify requirements for coordination of the arrangements for activities potentially hazardous to civil aircraft with the appropriate air traffic services authorities. The objectives of such coordination and recommendations for methods of application are set out in paragraph 2.16. The required coordination shall be effected early enough to permit timely promulgation of information regarding the activities in accordance with the provisions of Annex 15.

The most recent ICAO study regarding SARPs and provisions pertinent to activities potentially hazardous to civil aircraft, completed in January 1989 following consultation with States and international organizations, indicated that ICAO SARPs are adequate in respect of the actions required to be taken in the event that activities potentially hazardous to civil aircraft are planned and conducted. The study concluded that, if the current provisions are properly implemented and applied by all concerned, they should be capable of ensuring the safety of civil aircraft operations. The study recognized, however, the general nature of the

NATO ASI Series, Vol. F 73
Automation and Systems Issues in Air Traffic Control
Edited by John A. Wise et al.

SARPs concerned does not provide for specific steps and detailed actions to be taken with respect to every situation which may arise. As a result, on 9 February 1989, the Air Navigation Commission (ANC), at the eighth meeting of its 120th Session, agreed that the relevant ICAO SARPs should be amplified as necessary and presented as guidance material.

The guidance material, which is advisory in nature, was subsequently developed by the Secretariat with the assistance of an air navigation study group consisting of both civil air traffic services and military experts from seven Contracting States and three international organizations. The material is available in ICAO Doc 9554-AN/932 - Manual Concerning Safety Measures Relating to Military Activities Potentially Hazardous to Civil Aircraft Operations, First Edition, 1990.

The Problem

Civil and military aviation share the same airspace and have learned to co-exist in a symbolic relationship which essentially responds to its own set of rules. As a result, the coordination of civil and military aviation has been, and continues to be, one of the biggest dilemmas facing States, their air traffic services, and military authorities.

There are a number of factors which affect the objectives of safe and efficient use of airspace, some of which are not directly related to aviation interests. Geographical, economic, and political influences have significant impact on the air traffic control (ATC) environment in which civil and military aircraft operate. Many, if not most, of the present ATC systems accommodating the mix of civil and military aircraft have evolved over the years and have adapted to changing military philosophy and advancing technology the best way they could.

The center of the issue is that civil aviation interests differ from those of the military, and the two are often in direct conflict with each other. These divergent interests, which are often incompatible, create considerable pressure on the air traffic services (ATS) provided by States which are required to produce a service and airspace structure accommodating the needs and requirements of both. As a result of individual national aviation interests and security requirements, the ATC systems are usually not homogeneous organizations.

Airspace remains a finite resource, in the sense that it is not renewable, nor can it grow larger or be expanded. Therefore. better and more innovative ways need to be realized in order to utilize this resource more efficiently.

Part of the Solution

The ICAO Manual Concerning Safety Measures Related to Military Activities Potentially Hazardous to Civil Aircraft Operations (Doc 9554) provides guidance to States on the matter and is considered an important part of the solution regarding problems associated with civil/military coordination. Major parts of the new ICAO guidance material are presented below.

In order to set the stage for improved civil/military coordination, it is important to be aware of and understand certain principles embedded in the Convention on International Civil Aviation (Chicago, 1944), the Annexes thereto, and certain policies and practices adopted by the ICAO Assembly.

Article 3(a) of the Convention on International Civil Aviation states that "this Convention shall be applicable only to civil aircraft, and shall not be applicable to state aircraft." However, Article 3(d) of the Convention states that "the Contracting States undertake, when issuing regulations for their state aircraft, that they will have due regard for the safety of navigation of civil aircraft". Article 3(b) states that "aircraft used in military, customs, and police services shall be deemed to be state aircraft."

The ICAO Assembly has developed a "consolidated statement of continuing policies and associated practices related specifically to air navigation,"which is updated at each session of the Assembly and was last set forth in Assembly Resolution A27-10 (Doc 9551). Appendix P to this Resolution reads as follows:

16. Coordination of civil and military air traffic

WHEREAS the airspace as well as many facilities and services should be used in common by civil aviation and military aviation;

WHEREAS Article 3(d) of the Convention requires that the Contracting States undertake, when issuing regulations for their state aircraft, that they have due regard for the safety of navigation of civil aircraft;

WHEREAS satisfactory solutions to the problem of common use of airspace have not been evolved in all areas; and

WHEREAS although full integration of the control of civil and military air traffic may be regarded as the ultimate goal, improvement in coordination in many States offers, at the present time, an immediate approach towards resolution of existing difficulties;

THE ASSEMBLY RESOLVES that:

1. the common use by civil and military aviation of airspace and of certain facilities and services shall be arranged so as to ensure the safety, regularity and efficiency of international civil air traffic; and

2. the regulations and procedures established by Contracting States to govern the operation of their state aircraft over the high seas shall ensure that these operations do not compromise the safety, regularity and efficiency of international civil air traffic and that, to the extent practicable, these operations comply with the rules of the air in Annex 2.

Associated Practices

1. Contracting States should as necessary initiate or improve the coordination between their civil and military air traffic services to implement the policy in Resolving Clause 1 above.

2. The Council should ensure that the matter of civil and military coordination in the use of airspaces is included, when appropriate, in the agenda of divisional and regional meetings.

3. When establishing the regulations and procedures mentioned in Resolving Clause 2 the State concerned should co-ordinate the matter with all States responsible for the provision of air traffic services over the high seas in the area in question.

The International Standards and Recommended Practices in Annex 11 to the Convention, Chapter 2, 2.15, and 2.16, contain provisions for coordination between military authorities and air traffic services and coordination of activities potentially hazardous to civil aircraft. These provisions specify that air traffic services authorities shall establish and maintain close cooperation with military authorities responsible for activities that may affect flights of civil aircraft. The provisions also prescribe that the arrangements for activities potentially hazardous to civil aircraft shall be coordinated with the appropriate air traffic services authorities and that the objective of this coordination shall be to achieve the best arrangements which will avoid hazards to civil aircraft and minimize interference with the normal operations of such aircraft.

Coordination between Military Authorities and ATS Authorities

Coordination between the responsible military authorities and the appropriate ATS authorities is essential to the safety of civil aircraft operations whenever activities potentially hazardous to such operations are planned and conducted by any military units. The "appropriate ATS authority" is, by definition, "the relevant authority designated by the State responsible for providing air traffic services in the airspace concerned." This coordination is necessary regardless of whether the activities take place over the territories of States, over the high seas, or over territories of undetermined sovereignty, and whether the military and ATS authorities belong to the same or different States.

In the event that a sudden outbreak of armed hostilities or any other factors preclude this normal coordination process, appropriate State and ATS authorities, civil aircraft operators and pilots-in-command of aircraft must assess the situation, based on the information available, and plan their actions so as not to jeopardize safety.

ATS authorities should be alert to military operations in areas of conflict, or the potential for armed conflict, when such operations include a potential for hazardous activity, and react accordingly.

If the potentially, hazardous activities are planned to take place on a regular or continuing basis, a coordinating group should be given the task of ensuring that the operational needs of all parties concerned are adequately coordinated. This group should consist of representatives of the military organization(s) concerned, the appropriate ATS authorit(y/ies) and the operators of civil aircraft.

Coordination with regard to activities potentially hazardous to civil aircraft operations over the high seas should be effected, even if the States whose military organization and ATS authorities are concerned find themselves temporarily in diplomatic disagreement. If direct coordination with the appropriate ATS authorities via aeronautical or diplomatic channels is not possible, the coordination should be effected with the assistance of the appropriate regional office of ICAO or the ATS authorities of another State.

Coordination of activities potentially hazardous to civil aircraft should be effected with all air traffic services authorities responsible for providing services in the airspace concerned.

The State(s) of the military organization(s) planning the potentially hazardous activities should initiate the coordination process. When the military organization involved is located in a State other than that responsible for the provision of air traffic services in the airspace affected, the initial coordination should be effected through the ATS authority responsible for the airspace over the State where the organization is located or via other agreed channels. For example, a naval force of State A, operating in the FIR of (friendly) State B, plans a potentially hazardous activity in the FIR of State C and States A and B agree through prior arrangements, the ATS authority of State B may co-ordinate the potentially hazardous activity directly with the ATS authority of State C. The ATS authority will be able to provide information and assistance in achieving coordination with all appropriate ATS authorities and ATS units and to give advice as to the impact which the planned activity is likely to have on civil aircraft operations in the area.

The first step in the normal coordination process is the transmission or delivery of a message to the appropriate ATS authority or authorities, containing details of the planned activity. This message should describe the nature of the activity; the geographical area(s) affected, including its horizontal and vertical dimension(s); the proposed date(s), time(s), and duration(s) of the activity; any special safety measures to be taken, if necessary; and the means and methods of coordination between participating military units and ATS units concerned, including use of radio communications.

This first step should be taken early enough to permit a dialogue regarding the planned activity and the detailed arrangements, bearing in mind the need to avoid hazards and to minimize interference with all airspace users.

The timing of the first step should also allow for promulgation of information regarding the activity, following the coordination dialogue, in sufficient time to reach flight crews well before the start of the activities. The advance notice required by the International Standards in Annex 15 to the Convention - Aeronautical Information Services, Chapter 5, is at least 7 days in the case of activation of established danger, restricted, or prohibited areas, and in the case of activities requiring temporary airspace reservations other than for emergency operations. However, a greater lead-time is recommended by the Annex regarding the establishment and/or withdrawal of, or premeditated significant changes in temporary danger, restricted and prohibited areas. Greater lead times are also recommended when significant navigational hazards, military exercises and mass movements of aircraft are involved.

The objective of the normal coordination process should be to reach agreement on:

- the selection of location(s) or area(s), time(s), and duration(s), so as to avoid closure or realignment of established ATS routes, blocking of the most economic flight levels, or delays of scheduled aircraft operation, unless no other options exist

- the smallest possible size of the airspace designated for the conduct of the activity consistent with the attainment of its goals

- any special safety measures which need to be taken by the unit(s) conducting the activity, the ATS unit(s) concerned, or civil aircraft operating in the vicinity of the area

- the coordination needed between the ATS authority or unit and the military organization or unit(s) during the conduct of the activity

- the means and method of effecting
- Exchanges of information regarding the start(s) and stop(s) of the activity
- Exchanges of information regarding the identity of civil aircraft, when necessary
- Coordination of special safety measures, including alerting and search and rescue services
- Coordination in the event that civil aircraft emergencies or other unforeseen circumstances require discontinuation of the activities or parts thereof

In the event that agreement is not reached on satisfactory arrangements, the ATS authorities should not withhold the promulgation of the essential information to flight crews, but should take the necessary steps to ensure that the safety of civil aircraft is not jeopardized, including any necessary re-routing of aircraft to avoid the area of activity. If the arrangements are considered completely unsatisfactory from the point of view of civil aviation, a report to that effect should be forwarded by the ATS authorities promulgating the essential information to the appropriate military authority for action. If the military organization belongs to another State, the report should be sent through the civil aviation authority of that State, and a copy should be forwarded to the appropriate regional office of ICAO.

The most suitable communication means for achieving coordination are those which can ensure rapid and reliable exchanges between the authorities concerned regarding the arrangements for activities potentially hazardous to civil aircraft.

The aeronautical fixed service (AFS) is available for this purpose. The AFS comprises all types and systems of point-to-point communications in the aeronautical telecommunication services. The most extensive and widely used part of the AFS is the aeronautical fixed telecommunication network (AFTN), which is a network of teletypewriter circuits that links all international airports, air traffic services units, and notice to airmen (NOTAM) offices, and most national aviation administrations throughout the world. The AFTN is the principal means for the international exchange of aeronautical printed communications and also meets national aeronautical requirements. Messages exchanged between government civil aviation authorities in the AFS are classified as aeronautical administrative messages.

General procedures applicable to the use of the AFS, including the AFTN, are set forth in Volume II of Annex 10 - Aeronautical Telecommunications. Abbreviations and NOTAM codes approved for use in aeronautical messages are contained in ICAO Doc 8400, Procedures for Air Navigation Services, ABC - ICAO Abbreviations and Codes, while 4-letter location indicators for use in addresses are specified in ICAO Doc 7910, Location Indicators, and 3-letter addressee indicators are contained in ICAO Doc 8585, Designators for Aircraft Operating Agencies, Aeronautical Authorities and Services.

When a requirement exists for coordination between a military organization in one State and the ATS authority in another State, the AFTN can be used. The military organization may gain access to this system through the ATS authority of its own State, based on local arrangements.

The use of military liaison officers and, where appropriate, civil liaison officers would do much to smooth the process of coordination and to ensure necessary follow-up action. In the case of extensive military operations, invitations from the planners to civil aviation administrations to participate in planning conferences at an early stage have been found fruitful, and exchange visits by operations personnel are also useful.

Promulgation of Information

The International Standards in Annex 15, Chapter 5 specify that a NOTAM shall be originated and issued by States regarding the presence of hazards which affect air navigation (including obstacles, military exercises, displays, races, and major parachuting events outside promulgated sites).

The responsibility for initiating action regarding the promulgation of information on activities potentially hazardous to civil aircraft rests with the appropriate ATS authority of the provider State, following the coordination process described above. The information to be promulgated is forwarded to the Aeronautical Information Service (AIS) of the same State, which has the responsibility for originating and issuing the requisite NOTAM. The task of issuing the NOTAM for international distribution is undertaken by the International NOTAM Office (NOF) of the State, which is part of the AIS.

The Standards in Annex 15 prescribe that at least seven days'advance notice shall be given of the activation of established danger, restricted or prohibited areas, and of activities requiring temporary airspace restrictions other than for emergency operations. Notice of any subsequent cancellation of the activities or any reduction of the hours of activity or the dimensions of the airspace should be given as soon as possible, preferably at least 24 hours in advance.

As regards temporary danger, restricted and prohibited areas,and navigational hazards, military exercises, and mass movements of aircraft, Annex 15 recommends that the regulated system (AIRAC - Aeronautical Information Regulation and Control) should be used to promulgate information. The AIRAC system is based on a series of common effective dates at intervals of 28 days. It is aimed at the issuance of a NOTAM at least 42 days in advance of the effective date, with the objective of reaching the recipients at least 28 days prior to the effective date.

A NOTAM is given either Class I distribution, normally via the AFTN, or Class II distribution, normally via airmail. Specifications regarding the format of a NOTAM Class I are contained in Appendix 5 to Annex 15. Distribution of a NOTAM Class I is predetermined according to agreements between International NOTAM Offices, based on the needs for pre-flight planning. Information regarding the current exchanges of Class I NOTAM between International NOTAM Offices is contained in Part 2 of ICAO Doc 7383, Aeronautical Information Services provided by States.

Promulgation of information regarding activities potentially hazardous to civil aircraft should not be withheld in the event that agreement cannot be reached with the originator on the details to be promulgated. Any disagreement should be the subject of separate remonstrations toward the organization planning the activities.

It is important that the exchange of NOTAM regarding potentially hazardous activities continue even in the case where two States find themselves temporarily in diplomatic disagreement. It must be borne in mind that the information is required not only for the airlines of the two States but also for international operators flying routes through the airspace affected by the activities. These operators will, in all probability, have no part in the disagreement, and there can be no justification for penalizing them by denying them information which is essential for the safety of their operations. In such cases, the Aeronautical Information Services (AIS) should be treated in a similar manner as the meteorological services. Therefore, preservation of their existing channels of communication should be guarded in the light of their international obligations.

In the event that a State is unable, for some reason, to comply fully with the provisions of Annex 15 and the ICAO regional air navigation plan regarding promulgation and distribution of information concerning potentially hazardous activities, it should seek assistance from adjacent States or the appropriate regional office(s) of ICAO.

Coordination between Military Units and ATS Units

An essential part of the coordination between military and ATS authorities at the planning stage is to seek agreement on the coordination needed between the military unit(s) conducting potentially hazardous activities and the appropriate ATS unit(s), as well as on the means and methods for effecting the coordination.

The International Standards in Annex 11, Chapter 6 specify the need for facilities to permit communications by direct speech and, when a written record is required, by printed communications between area control centers, flight information centers, approach control offices or aerodrome control towers, and appropriate military units providing a service within their respective areas of responsibility. The phrase "appropriate military units" in this context is taken to mean any permanently established military units controlling activities that may affect flights of civil aircraft. It is recommended, however, that the same requirement be applied also to temporarily established or mobile military units conducting potentially hazardous activities.

Ideally, the communication facilities should permit direct voice communication between the officer in charge of the conduct of the military activities and the air traffic control unit(s) responsible for the provision of air traffic services in the airspace affected. To this end, agreement should be sought on the use of military or civil land-line, microwave. or satellite-relayed telephone circuits or radiotelephone channels, individually or in combination. The use of ATS air-ground radiotelephone channels for such coordination should be avoided, except as a last resort, as it may interfere with the communications between pilots and air traffic controllers.

The need for "printed" communications between military units and ATS units, such as teletype, telex, or facsimile, depends on the nature of the potentially hazardous activity and the resultant need for exchanges of information in printed form. When the military unit(s) require(s) a constant flow of information regarding the planning and actual flight progress of civil flights in the area of activity, consideration should be given to the use of the AFTN both for that purpose and, with the approval of the appropriate civil aviation authority, for arranging for a special link to that network. If delays encountered on the AFTN are unacceptable, discrete teletype, telex, or facsimile connections should be established.

In cases where the military activity results in temporary route or altitude restrictions for civil aircraft, the controlling military unit should inform the appropriate ATS unit when the activity begins and ends, as well as when any temporary interruptions occur which would permit civil aircraft to be routed through the area of the activity. The availability of direct voice communications would also permit the ATS unit to request that the activity be discontinued completely or partly, as necessary, in the event of an accidental or emergency incursion of a civil aircraft in the area of activity.

Normally, civil aircraft should not operate in an area of hazardous military activity; however, in the event that civil aircraft are permitted to operate through, or in the immediate vicinity of, an area of potentially hazardous military activity, the safety of the civil aircraft may depend on positive identification by the military units. In such circumstances, arrangements should be made in all cases to provide the responsible military units with advance information regarding regularly, or seasonally scheduled flights, as well as non-scheduled commercial flights and general aviation flights, through the area.

The information on planned flights should include:

- aircraft identification to be used in communications (i.e., flight number and aircraft registration marks)
- aircraft type
- point and estimated time of departure
- route(s)
- flight level(s)
- destination and estimated time of arrival
- individual SSR code, if assigned in advance

Any changes to the foregoing information should be forwarded as soon as they become known.

Information on the actual progress of flights should include:

- actual take-off time, or last reported position, time, and level
- next reporting point and estimated time
- SSR code

Warnings to civil aircraft regarding hazards and the need to take evasive action should normally be co-ordinated with, and issued only by, the appropriate ATS unit, unless other coordination procedures have been established.

If an extreme emergency is deemed to exist, the military unit may attempt to transmit a direct warning to the aircraft on the VHF emergency channel 121.5 MHz. It must be realized, however, that not all aircraft will always have the capability to maintain a continuous listening watch on 121.5 MHz in addition to the ATS channel(s) (unless the carriage and monitoring of the emergency frequency has been made a mandatory provision for operations in the area by requiring this in a NOTAM) and that, in the case of aircraft having this capability, other communication tasks may result in interruptions of the listening watch on the emergency frequency.

It must also be realized that, unless the warning contains the correct call sign of the aircraft, the SSR Mode A code, or identification of the aircraft clearly and unambiguously by reference to its position, and unless standard aviation phraseology and a common international language is used, the warning may not be understood by the civil aircraft. A warning transmitted by the appropriate ATS unit on the normal ATS frequency is therefore more likely to be heard, understood, and acted upon.

As indicated above, clarity in communication between military units and ATS units or, in emergencies, between military units and civil aircraft, is essential to ensure proper understanding and avoid potentially disastrous results. Some examples of phraseologies which may be suitable for use in air-ground communications are given in Appendix A. Additional phraseologies may, of course, be selected for use by agreement between specific military and ATS units to suit the circumstances associated with particular military activities.

In the event that the demand for routine or ad hoc coordination exceeds the capacity of the officer-in-charge of the conduct of the military activity and/or the duty air traffic controller or supervisor, the designation of separate liaison officers might be necessary. The use of separate liaison officers may also be desirable in other cases. In the case of large-scale activities covering a wide area, the establishment of a special coordination cell in the area control center concerned may be the optimum solution. Such a cell should then comprise both military and civil liaison personnel, and may be given the tasks of correlating and disseminating critical flight plan and flight progress data, including SSR transponder codes, co-ordinating the use of routes and flight levels, and assisting in resolving identification problems and coping with emergencies.

Familiarization of Personnel with the Area of Activity

In order that due regard will be given to the safe and efficient operation of civil aircraft, States should ensure that military authorities responsible for planning and conducting activities potentially hazardous to such aircraft are fully informed, and conversant with, the following in respect of the area of activity:

- the type(s) of civil aircraft operations
- the ATS airspace organization and responsible ATS unit(s)
- ATS routes and their dimensions
- relevant regulations and special rules, including airspace restrictions

During the coordination process described earlier, reference should therefore be made to the relevant ICAO regional Air Navigation Plan(s) (ANP) and States' aeronautical information publications and related documents and charts, which contain the information mentioned above. If necessary, a special briefing regarding the civil aviation infrastructure should be arranged with the assistance of ATS specialists from the State(s) concerned.

In the event that civil aircraft are permitted to operate through an area of military activity, military staff conducting the activity should also be fully informed of, and familiar with:

- means and methods of identifying civil aircraft
- means and method(s) of coordination with the ATS unit(s)
- terminology and phraseologies for use in communications with ATS units or, as a last resort, with civil aircraft

The foregoing information should be provided by the appropriate ATS unit, unless it is known that the military staff concerned has already been properly briefed.

In the event of large-scale, complex, or unusual military activities, a special briefing should be arranged for civil ATS staff concerning the nature and scope of the activities, potential risks, necessary airspace reservations, and means and methods of coordination with participating military units. The objective of such briefings should be to make civil ATS staff familiar with all aspects which will, or may, affect the flights of civil aircraft and the air traffic services to be provided during the military activities.

Identification of Civil Aircraft

The safety of civil aircraft operations may, in certain circumstances, depend on the capability of military units to secure positive identification of civil aircraft. Special arrangements may therefore need to be made to ensure or enhance that capability.

The identification process is based on the correlation by military units of information derived from several sources, which include:

- air traffic services units
- airline flight schedules
- filed flight plans and related messages
- departure messages and flight progress reports
- electronic emissions from aircraft, including emissions from airborne weather radar and radio altimeters, and SSR responses
- visual observations

Advance information on regularly or seasonally scheduled civil flights can be derived from the airline flight schedules published in the ABC World Airways Guide or the Official Airline Guide (OAG), or from repetitive flight plans filed with, and stored by, the ATS units concerned. If weekly or monthly lists of planned regular flights are prepared by the ATS units concerned for their own purpose, copies of such lists should be forwarded to the military units planning to conduct potentially hazardous activities. If no lists are being prepared as a matter of routine, the specific needs of the military units in terms of data and format should be ascertained and agreement reached on the best way of meeting the needs without imposing an excessive workload on the ATS units.

The provision of filed flight plan information to military units on a routine basis may not always be necessary, if weekly or monthly lists of planned regular flights are forwarded routinely. However, arrangements must be made so that information on any additional flights and significantly late departures of regular flights will be provided on an ad hoc basis.

In this context it should be noted that the scheduled departure time in the ABC World Airways Guide or the Official Airline Guide and in filed flight plans is the scheduled or estimated departure time from the airport terminal, the so-called "off-block" time, (i.e., start taxi time) and not the estimated take-off time.

Information on the actual progress of flights through the area of activity may be obtained by the military unit in one of the following ways, or a combination thereof:

- electronic means, such as re-transmission of radar pictures from the ATS unit concerned

- monitoring of the appropriate civil air traffic services frequency/ies

- forwarding of departure times and position reports from the ATS unit concerned

The methods outlined above would ensure immediate access to the required information and would keep the routine workload on the ATS unit concerned to a minimum, while keeping the communications link between the two units free for urgent ad hoc exchanges. The monitoring of appropriate civil AIS frequencies will, of course, require that the military unit be provided with VHF equipment and HF equipment where appropriate, and that military personnel be assigned specifically to the task of monitoring the civil ATS air-ground frequency/ies and be familiar with standard aeronautical position reporting procedures and the names, name codes, and locations of designated reporting points. The forwarding of departure times and position reports from the AIS unit concerned can lead to unacceptable delays if it is selected as the only or primary method, and it may also require the employment of additional personnel in the ATS unit in order to be effective. In any case, it is essential that information be requested from the ATS unit concerned on an ad hoc basis, if any doubts regarding the identity of a flight remain in spite of radar monitoring and/or monitoring of reports transmitted on the air-ground frequency/ies.

Identification of civil aircraft by primary surveillance radar alone presupposes correlation of a radar return with the identity of an aircraft whose flight plan and/or present position is known from other sources (i.e., those described above). Identification by means of secondary surveillance radar (SSR) is easier if the SSR code allotment plan for the area of activity and the SSR codes assigned to individual flights are known.

The SSR code allotment plan for a given area may be found in the Aeronautical Information Publication (AIP) of the State(s) concerned, or it may be obtained, on request, from the relevant ICAO regional office. The SSR code allotment plan will indicate the SSR code groups in Mode A/3 allotted to individual States for assignment to international or domestic flights, or the code groups allotted to groups of States for assignment to international transit flights. The discrete SSR codes assigned to individual aircraft in flight can be obtained, on request, from the responsible ATS unit(s). The current Procedures for Air Navigation Services, OPS - Aircraft Operations (ICAO Doc 8168, Volume I, Part VIII) require civil aircraft equipped with a serviceable SSR transponder to operate the transponder at all times during flight and to transpond on Mode A Code 2000, unless otherwise directed by ATS units or prescribed by regional air navigation agreement. The latter can be ascertained from the relevant ICAO regional office.

An additional means of identifying a particular radar response as that of a civil aircraft is to request a given aircraft to squawk IDENT (i.e., to activate the special position identification (SPI) feature of the transponder). Such a request should be made, when required, through the responsible ATS unit, as any direct transmission from a military unit could interfere with normal air traffic control communications and lead to confusion if the aircraft is within ATS radar coverage.

Identification of the "electronic signature" of civil aircraft will be enhanced by requiring all aircraft operating in a given area to carry serviceable SSR transponders and requiring aircraft equipped with weather radar and radio altimeters to operate these continuously during flight. However, as long as the carriage of such equipment is not mandatory world-wide, reliance by a military unit on the absence of an "electronic signature" could lead to misidentification of civil aircraft and the inherent hazards.

If circumstances permit visual observations of aircraft, identification as civil aircraft is possible with reference to the nationality and registration markings under the wings and on the fuselage. The airline logo and distinctive colors on the tail plane and/or other parts of commercial aircraft are additional identification features. The use of full cockpit and cabin lights and illumination of logo lights, if possible, by civil aircraft should also facilitate identification.

Warnings and Navigational Assistance

In the event that a military unit observes a civil aircraft is entering, or is about to enter, a designated prohibited, restricted, or danger area, or any other area of activity which constitutes potential hazards, a warning to the aircraft should be issued through the responsible ATS unit. The warning should include advice on the change of heading required to leave, or circumvent, the area.

If the military unit is unable to contact the responsible ATS unit immediately, and the situation is deemed to be a genuine emergency, an appropriate warning to the aircraft may be transmitted on the VHF emergency channel 121.5 MHz. If the identity of the aircraft is not known, it is important that the warning includes the SSR code, if observed, and describes the position of the aircraft in a form meaningful to the pilot (e.g. by reference to an ATS route and/or the direction and distance from an airport or an aeronautical radio navigation aid, an established waypoint, or reporting point). Examples of phraseologies for use in such circumstances are given in Appendix A.

In the case where an unauthorized aircraft is observed visually to be flying in, or about to enter, a prohibited, restricted, or danger area, the following visual signal is prescribed by the International Standards in Annex 2, Appendix A to indicate that the aircraft is to take such remedial action as is necessary:

By day and night: A series of projectiles discharged from the ground at intervals of 10 seconds, each showing, on bursting, red and green lights or stars.

The importance of co-ordinating with the responsible ATS unit(s), whenever possible, the issuance of any warnings and advice to civil aircraft regarding changes of flight path should be emphasized in any briefings or instructions given by military authorities to their units, since uncoordinated warnings and associated navigational advice, when followed, may result in a potential risk of collision with other aircraft in the area.

Air Traffic Restrictions

The objective of the coordination between the military authorities planning activities potentially hazardous to civil aircraft and the responsible ATS authorities is to reach agreement on the best arrangements to avoid hazards to civil aircraft and minimize interference with the normal operations of civil aircraft. Ideally, this means the selection of locations outside promulgated ATS routes and controlled airspace for the conduct of the potentially hazardous activities.

If the selection of such locations is not possible due to the nature and scope of the planned activities, temporary restrictions imposed on civil air traffic should be kept to a minimum, through close coordination between the military units and the ATS units.

Whenever possible, a flight level should be designated, at or above which civil aircraft may continue to operate normally without any hazard. In areas where the majority of civil aircraft would be in the en-route phase of flight, this flight level should ideally be at or below the lowest cruising levels normally used.

If the temporary closure of certain ATS routes is unavoidable, agreement should be sought by ATS authorities with the State(s) concerned on the temporary use of promulgated alternative routes bypassing the area of activity or, if no convenient promulgated alternative routes exist, on the establishment of temporary routes.

Note: If the routes concerned are in airspace over the high seas or of undetermined sovereignty, the provision of Annex 11, 2.1.2 will apply.

It may be necessary to make special arrangements for aircraft descending into, or climbing out of, airports in the vicinity of the area of activity. If the nature of the military activity permits, approaches and departures may have to be restricted to a particular direction; otherwise, temporary discontinuation of the activity should be considered, to permit normal approach and departure maneuvers.

Special Measures in the Event of Armed Conflict or the Potential for Armed Conflict

In the event of armed conflict or the potential for armed conflict, the Convention on International Civil Aviation does not affect the freedom of action of any Contracting State affected, whether as belligerent or as neutral States (Article 89 of the Convention). Nevertheless, the need for close coordination between civil and military authorities and units is even more critical. ATS authorities should ensure that civil aviation is warned of the conflict and directed away from the area. This point should form a part of the normal contingency planning for such situations.

The responsibility for initiating the coordination process rests with the States whose military forces are engaged in the conflict. The responsibility for instituting special measures to ensure the safety of international civil aircraft operations remains with the States responsible for providing air traffic services in the airspace affected by the conflict, even in cases where coordination is not initiated or completed.

Based on the information available, the State responsible for providing air traffic services should identify the geographical area of the conflict, assess the hazards or potential hazards to international civil aircraft operations, and determine whether such operations in or through the area of conflict should be avoided or may be continued under specified conditions. An international NOTAM containing the necessary information, advice, and safety measures to be taken should then be issued and subsequently updated in the light of developments. All those concerned with initiating and issuing of NOTAM should be aware of the provisions governing the duration of the published NOTAM. Annex 15, 5.3.1.2 states that a NOTAM given Class I distribution shall be superseded by a NOTAM given Class II distribution when the duration of the circumstances notified is likely to exceed three months, or the NOTAM has remained in force for three months. Examples of such NOTAMs are given in Appendix B.

If the necessary information is not forthcoming from the States whose military authorities are engaged in the armed conflict, the State responsible for providing air traffic services should ascertain the nature and scope of the hazards or potential hazards from other sources, such as aircraft operators, the International Air Transport Association (IATA) and the International Federation of Airline Pilots' Associations (IFALPA), adjacent States, or in some cases the relevant ICAO regional office, in order to take the action outlined above.

The safety measures required to be taken will depend on the assessment by the State responsible for providing air traffic services of the nature and scope of the hazards, or potential hazards, to civil aircraft and the decision as to whether flight operations by civil aircraft through the area can be considered without risk.

If civil aircraft operations through the area are allowed, immediate attention should be given by the States concerned to special arrangements regarding coordination between military units and ATS units, briefings of personnel, identification of civil aircraft by military units, issuance of warnings and navigational advice, and air traffic restrictions. It is also essential that flight crews are reminded to be particularly vigilant when operating in, or transiting, the area of armed conflict.

In planning the conduct of operations through areas of armed conflict or the potential for armed conflict, operators should give due regard to the availability and serviceability of aircraft equipment needed to facilitate identification of the aircraft by military units and to permit guarding of the appropriate frequencies.

Safety measures prescribed by the State responsible for providing air traffic services may include, for example, the following requirements applicable to all aircraft operating in the area:

- continuous monitoring of the emergency frequency 121.5 MHz

- carriage and continuous operation of SSR transponder with Mode C capability

- carriage and continuous operation of weather radar
- continuous display of aircraft exterior lighting and cabin lighting, and illumination of logo lights, if possible
- presence on the flight deck of transport aircraft of a full flight crew complement, augmented by additional personnel as required by the situation

The assistance of the ICAO regional office(s) concerned and the ICAO Headquarters in Montreal may be available on request, in the event that a State experiences difficulties in defining an area of armed conflict; assessing the impact upon civil aircraft operations; and/or developing, promulgating, and implementing the requisite safety measures. Assistance may also be provided by ICAO at the request of IATA or IFALPA.

ICAO may assist in the development, coordination, and implementation of necessary safety measures in the event that the State(s) responsible for the provision of air traffic services in an area of armed conflict cannot, for some reason, adequately discharge the responsibility. The specific nature and scope of such action will depend upon the particular circumstances involved. In such circumstances, ICAO will work in close coordination with the responsible State, with other provider and user States concerned, and with IATA and IFALPA.

Summary

The process of civil/military coordination is an essential part of harmonizing the civil air traffic control system with national security requirements. The way in which this process is brought into effect is directly related to the safety, regularity, and efficiency of international civil aviation. ICAO, through the principles embedded in the Convention on International Civil Aviation and the Standards and Recommendations contained in the Annexes thereto (Annexes 2, 6, 10, 11 and 15), as well as certain policies and continuing practices adopted by the ICAO Assembly, provides the framework for civil/military coordination to be successfully conducted. The newly-developed ICAO Manual Concerning Safety Measures Relating to Military Activities Potentially Hazardous to Civil Aircraft Operations (Doc 9554-AN/932), which is advisory in nature, provides further guidance to States and civil ATS and military authorities in amplification of present ICAO Standards and Recommendations.

References

ICAO (1990a). Designators for Aircraft Operating Agencies, Aeronautical Authorities and Services. (Doc 8585). 77th edition.

ICAO (1990b). Location Indicators. (Doc 7910). 61st edition.
ICAO (1990c). Manual concerning Interception of Civil Aircraft. (Doc 9433). 2nd edition.
ICAO (1990d). Manual on Safety Measures Relating to Military Activities Potentially Hazardous to Civil Aircraft Operations. (Doc 9554). 1st edition.
ICAO (1989a). Aeronautical Information Services Provided by States. (Doc 7383). 73rd edition.
ICAO (1989b). Air Navigation Plans. Africa-Indian Ocean Region. (Doc 7474). 26th edition.
ICAO (1989c). Assembly. Resolutions. 27th Session, Montreal, 19 September - 6 October 1989. (Doc 9551).
ICAO (1989d). Procedures for Air Navigation Services. ABC - ICAO Abbreviations and Codes. (Doc 8400). 4th edition.
ICAO (1987a). Annex 11 - Air Traffic Services. 8th edition.
ICAO (1987b). Annex 15 - Aeronautical Information Services. 7th edition.
ICAO (1986a). Annex 2 - Rules of the Air. 8th edition.
ICAO (1986b). Annex 6 - Operation of Aircraft. Part III - International Operations - Helicopters. 1st edition.
ICAO (1986c). Assembly Resolutions in Force (as of 10 October 1986). (Doc 9509).
ICAO (1986d). Procedures for Air Navigation Services. OPS - Aircraft Operations. (Doc 8168) Volume I - Flight Procedures. 3rd edition.
ICAO (1985a). Air Navigation Plans. Caribbean and South American Regions. (Doc 8733). 13th edition.
ICAO (1985b). Air Navigation Plans. European Region. (Doc 7754). 23rd edition.
ICAO (1985c). Air Navigation Plans. Middle East and Asia Regions. (Doc 8700). 14th edition.
ICAO (1985d). Annex 10 - Aeronautical Telecommunications. Volume I (Part I - Equipment and Systems; Part II - Radio Frequencies). 4th edition. Volume II (Communication Procedures including those with PANS status). 4th edition.
ICAO (1985e). Procedures for Air Navigation Services. RAC - Rules of the Air and Air Traffic Services. (Doc 4444). 12th edition.
ICAO (1984). Air Navigation Plans. North Atlantic, North American and Pacific Regions (Doc 8755). 12th edition.
ICAO (1983a). Annex 6 - Operation of Aircraft. Part I - International Commercial Air Transport - Aeroplanes. 4th edition.
ICAO (1983b). Annex 6 - Operation of Aircraft. Part II - International General Aviation - Aeroplanes. 3rd edition.
ICAO (1980). Convention of International Civil Aviation. (Doc 7300). 6th edition.

APPENDIX A

EXAMPLES OF AIR-GROUND TRANSMISSIONS ON 121.5 MHz

1. Identification of civil aircraft

Military unit: AIRCRAFT POSITION (position[10]) SQUAWKING MODE A CODE (Code) [ESTIMATED] ALTITUDE (level) AND GROUND SPEED (ground speed) THIS IS (call sign of military unit) PLEASE IDENTIFY YOURSELF AND STATE YOUR INTENTIONS.

Aircraft:(call sign of military unit) THIS IS (call sign of aircraft) POSITION (position*) SCHEDULED (or CHARTER or PRIVATE) FLIGHT FROM (point of departure) TO (destination) FLIGHT LEVEL (level) [CLIMBING TO or DESCENDING TO] (level).

Military unit:(Call sign of aircraft) THIS IS (call sign of military unit) ROGER OUT.

2. Navigational warning

Military unit: (Call sign of aircraft) THIS IS (call sign of military unit) YOU ARE APPROACHING DANGER AREA (designation) (or AN AREA OF POTENTIAL HAZARDS) TURN RIGHT (or LEFT) TO HEADING (figure) TO AVOID CONFLICT.

Aircraft: (Call sign of military unit) THIS IS (call sign of aircraft) WILCO TURNING RIGHT (or LEFT) TO HEADING (figure).

NOTE: It is emphasized that advice to civil aircraft regarding changes to flight path should be coordinated with the responsible ATC unit(s) since uncoordinated advice may result in a potential risk of collision with other aircraft in the area.

1) (distance) NM (direction) OF (navigation aid or reporting point or airport) [ON AIRWAY (designation) or ON ATS ROUTE (designation)]
Example: 25 NM WEST OF WILLY ON AIRWAY A97

2) [(distance) NM] ABEAM (navigation aid or reporting point)
Example: ABEAM NILLY ON AIRWAY R54

3) (distance) NM (direction) OF (name of well-known geographical location, such as town, city, island or mountain)
Example: 4 NM SOUTH OF DAISY ISLAND

4)(latitude) NORTH or SOUTH (longitude) EAST or WEST
Example: 1630 NORTH 14245 EAST.

---

10. Position shall be expressed as follows, in order of preference. Use ofgeographical co-ordinates should be limited to cases where no other moresuitable references are available.

APPENDIX B

EXAMPLES OF NOTAM CLASS I REGARDING MILITARY ACTIVITIES POTENTIALLY HAZARDOUS TO CIVIL AIRCRAFT

1. GG BCDEYNYX CDEFYNYX DEFGYNYX ...
   091532 EJKLYNYX
   A707 NOTAMN
   A) EJKL FIR B) WIE c) UFN APRX DUR
   D) NAVAL FORCES INCLUDING AIRCRAFT CARRIERS WILL BE OPERATING IN THE AREA (describe area with reference to latitude and longitude). IN ORDER TO FACILITATE CO-ORDINATION AND ENSURE SAFETY, CIVIL AIRCRAFT OPERATING BELOW FL ... IN THIS AREA ARE REQUIRED TO OPERATE THEIR SSR TRANSPONDER, WEATHER RADAR AND RADIO ALTIMETER CONTINUOUSLY. CIVIL AIRCRAFT ARE ALSO REQUIRED TO MAINTAIN Continuous LISTENING WATCH ON THE EMERGENCY FREQUENCY 121.5 MHZ AND TO RESPOND TO REQUESTS FOR IDENTIFICATION BY GIVING AIRCRAFT IDENTIFICATION, TYPE OF AIRCRAFT, ALTITUDE, POINTS OF DEPARTURE AND DESTINATION AND TYPE OF FLIGHT (IFR or VFR). AIRCRAFT UNABLE TO COMPLY ARE REQUESTED TO AVOID THE AREA OR TO NOTIFY EJKL ACC.

2. GG DCBAYNYX ACCOYNYX BADCYNYX ...
   171814 CBADYNYX
   A747 NOTAMN
   A) CBAD FIR B) WIE C UFN APRX DUR
   D) PARAMILITARY FORCES REPORTED OPERATING IN AREA (describe area with reference to latitude and longitude). CIVIL AIRCRAFT ARE REQUESTED TO MAINTAIN AT LEAST FL ... WHILE TRANSITING THE AREA IN ORDER TO AVOID A POTENTIAL THREAT (describe threat).

Error Models for Operating Irregularities: Implications for Automation

Paul Stager

Department of Psychology, York University, Toronto, Ontario, Canada

Introduction

With progressive automation in air traffic control (ATC) (Hopkin, 1987, 1989a, 1989b, 1991; Stein, 1987), there is a particular need to recognize that the most effective design solutions in the automation process will be sensitive to our growing knowledge of human error. An indiscriminate application of automation technology carries with it the potential to contribute its own sources of system error. Ultimately, however, automation is intended to reduce the processing demands on the controller and to reduce the potential for human error. A better understanding of the current factors underlying the human behaviors associated with system errors will contribute directly to improved design decisions in the automation process.

In the present paper, the terms *system error* and *operating irregularity* will be used interchangeably. An ATC system error is defined as an operational error that results either in less than the required separation minima or in a hazardous situation. In part, this paper reviews what we have observed about system errors and makes a preliminary attempt to summarize available data. Much of the error data has been categorized within *taxonomies* which tend to describe errors in a language that is directly tied to controller tasks.

In part, also, this paper is intended as an introduction to cognitive *framework models* of human error that represent an alternative approach to structuring what we know about human error in operational systems. Framework models outline the principal characteristics of human information processing, focussing on cognitive control mechanisms. In addition, the paper will review selected comments on the relevance of *mental models* in the design of the computer-human interface and in understanding behavior in complex operational environments. However, the primary focus of the paper will be to review the attempts that have been made to develop generic framework models to account for human error and to examine what implications these models have for the implementation of automation technology in air traffic control. Some of the human factors issues that are seen to be related to the implementation of automation will also be discussed.

NATO ASI Series, Vol. F 73
Automation and Systems Issues in Air Traffic Control
Edited by John A. Wise et al.

Analysis of system errors

How can error models be applied in ATC?

There are at least three potential applications of error models in ATC systems that can be considered in the present context of automation and systems issues.

The first is in the design of the controller interface. There is an interest in avoiding system designs which promote error at the operator interface, for example, although Rasmussen and Vicente (1989) have suggested that attempting to remove the potential for human error through training programs or optimal interface design is not an efficient approach.

> ...Errors tend to reflect either systematic interference between various [cognitive control structures]..., or the effects of the adaptive mechanisms involved in learning....Reliable human-system interaction will be achieved by designing interfaces which tend to minimize the potential for control interference and support recovery from errors. (p. 517)

The second potential application of error models is in system evaluation which is undertaken as an inherent part of system development. Rasmussen and Vicente (1989) viewed error in this context as an *independent variable*. What are the predicted effects of error on system performance? Conversely, human error can be used as a *dependent variable* and as a measure or index of system performance. The questions then become, "What are the error measures to be used?" and "What levels of error can be tolerated?"

The third potential application of error models is in the assessment of changes within existing systems. For example, what are the effects on controller performance of changes in operating procedures, in selection policies (see for example, Eißfeldt, 1991), in training programs, or in equipment modification? Human error is sometimes an effective index of the impact of changes in other system parameters.

What do we know about errors in ATC?

Perhaps the clearest conclusion to be drawn from the results of the error analysis completed by Stager and Hameluck (1988) was that the error category system that is currently in use in Canada did not enable either the situational factors or the controller states uniquely associated with the different errors to be determined. The question then was how to proceed in the development of a revised category system or error taxonomy in which a distinction between direct causal activities and underlying or contributing variables could be maintained.

Stager and Hameluck (1988) reported that in their analysis of 301 Fact Finding Board reports, between January 1985 to May 1987, the occurrence of incidents was not related directly to rated workload. Operating irregularities were associated with conditions of moderate or low workload, normal complexity, and intermediate pace levels. (Pace level is a rating based on traffic volume and traffic complexity for traffic operating under instrument flight rules.)

The most frequently cited primary categories of error in the incident reports were *attention*, *judgment*, and *communication*. Within these three categories, non-recognition of conflict, inattention, deviation from required operating procedures, failure to coordinate, and poor judgment were the most frequently cited errors. None of the primary categories was cited more frequently than another within the different levels of workload, complexity, or pace rating. The likelihood of a category being cited was independent of the traffic volume and incident rate. Allowing for the fact that more than one cause could be assigned by a review board to the same incident, both attention and judgment errors were cited as the cause in over 60% of the cases examined.

A recent report by the Canadian Aviation Safety Board (1990) stated that at least half of all system errors have "...causal or contributory factors which are directly attributable to breakdowns in the information transfer process — usually in oral communications" (p. 6). In the Board's analysis of 217 incidents selected at random from a pool of 437 occurrences, 60% of the system errors were accounted for by planning errors, judgment errors, and lapses in attention. Nearly 75% of the inattention errors occurred under relatively low workload and low complexity situations.

Kinney, Spahn, and Amato (1977) earlier analyzed data from the U.S. Federal Aviation Administration's (FAA) data base called the System Effectiveness Information System (SEIS). Their analysis was limited by the absence of normative data, but the investigation indicated that more than 90% of the recorded system errors were attributed directly to factors within the broad categories of *attention*, *judgment*, and *communications*. Characteristics such as traffic load, traffic peaks and complexities, and average controller experience were not related to system performance.

In developing classification systems and error models, the conditions that are either conducive or not conducive to the desired behaviors (Wojcik,1989) will follow from the analysis. However, the question of what language is used will depend on whether the taxonomy or model building is being driven by operational data or by a more generic schema. For example, Empson's (1987, 1991) terminology of controller errors (e.g., *discrimination failures* , *program assembly failures*, *test failures*, *sub-routine failures*, and *storage failures*) is the language of human cognition. However, Morrison and Wright (1989) chose to retain the terminology of ATC, grouping controller errors within two broad concepts: control or communications. In this work, control errors included the sub-groups of *monitoring*, *timeliness*, and *coordination*. (Coordination refers to control activities that affect aircraft traversing jurisdictional boundaries of airspace, including the transfer of information between controllers.) Communications errors included *clearance composition*, *phraseology and delivery technique*, and *read/hearback errors*. Again, systems-specific language is used to describe the environmental factors predisposing controller performance errors: *numbers of aircraft*, *frequency congestion*, and *combined position/sector operations* were the factors most often cited in the reports.

Rosenthal and Mellone (1989), in their investigation of anticipatory clearances (e.g., fast sequence clearances to expedite traffic flows in high volume situations), reported that controller performance deficiencies were often a function of *memory* (e.g., forgetting about other cleared traffic, forgetting to issue a planned clearance) or *judgment and planning* (e.g., not expecting the unexpected, miscalculating landing speed). Grayson and Billings (1981) found that the most frequently occurring communications problems were inaccuracies in content (stemming from errors in *judgment and planning*) and *forgetting* to originate or transmit a message.

The concepts of attention, judgment, communication, and memory are frequently associated with system errors. What might their relationship be in a conceptual schema that distinguishes between sources of error and error itself?

Taxonomies for classifying errors.

Kinney et al. (1977) attempted to go one step beyond what were called system error elements (e.g., equipment, human, procedural, and traffic workload) to the more fundamental conditions or underlying factors that caused a system error. For example, the SEIS included the human element categories of "attention", "judgment", "communication", and "stress," Cause factors listed under "attention" included failure to maintain constant surveillance of the flight data display or traffic situation, failure to correctly perform an accurate and complete radar handoff, and loss of retention or forgetting. The underlying factors were considered to be either prevailing, such as job attitudes or work habits, or momentary, such as distractions and lapses of memory, but Kinney et al. do not make a further conceptual differentiation within this grouping.

An additional step toward a conceptual model is represented by Grayson's (1981) examination of data from the NASA Aviation Safety Reporting System (ASRS). In Grayson's analysis of ATC coordination failures, a distinction is again made between human errors and their predisposing factors. The variables that comprised each group are shown in Table 1. A review of Grayson's two groupings suggests that there is a primitive distinction being made between basic cognitive resources (if not actual cognitive processes) and their more contextual or situational determinants. The *predisposing factors* which Grayson lists, as well as the *underlying factors* of Kenny et al. (1977), are very similar to the *performance affecting factors* advocated by Rasmussen (1987a) (e.g., stressful work environments, affective situations, and inadequate instruction).

TABLE 1

Human factors associated with coordination failures (adapted from Grayson, 1981).

Human Errors	Predisposing Factors
Non-recall	Distraction
Technique error	Excessive workload
Perceptual error	Experience/ Training level
Failure to monitor	Complacency
Message inaccuracy	Airspace configuration
Misidentification	Procedural problem
	Automation mind-set (when system degraded)
	Equipment failure
	Interpersonal relationships

A more general and comprehensive general systems model for classifying errors has been proposed by Reason (1987a). Reason's classification system provides rather fine discriminations between the descriptions of particular error situations by means of only a limited number of categories on each of the dimensions of the taxonomy. However, Reason is more concerned with developing a conceptual approach to classifying human errors than with developing a particular taxonomy, and his approach to modelling cognitive control mechanisms is implicit in the proposed system. Reason has proposed that *primary error groupings* (PEGs), such as attention failures or unintended words or actions, can be derived from a matrix of *basic error tendencies* (BETs), on one dimension, and cognitive or *information processing domains* (IPDs), on the other. Resource limitations and cognitive models or schema, for example, are taken as basic error tendencies, and input selection or judgment processes as cognitive domains. Within each error grouping, predictable error forms (PEFs) are distinguished according to the situational factors (SFs) associated with their occurrence. Reason's matrix for classifying primary error groupings is shown in Figure 1.

Basic Error Tendencies

Ecological Constraints | Change Enhancement | Resource Limitations | Schema Properties | Strategies Heuristics

Processing Domains — *Primary Error Groupings*

Sensory Registration
Input Selection — Attentional Failures
Volatile Memory — Memory Lapses
Long-Term Memory — Inaccurate Recall
Recognition Processes — Misperceptions
Judgemental Processes — Errors of Judgement
Inferential Processes — Reasoning Errors
Action Control — Unintended words/actions

Figure 1. Matrix for classifying primary error groupings (adapted from Reason, 1987a).

In Reason's classification system, it is possible to make probabilistic statements both about the circumstances that will promote the occurrence of a predictable error form and about the likely form that it will take. The structure of the matrix acknowledges that

> ... BETs can manifest themselves somewhat differently in different cognitive domains...[and] that a further important source of variation is introduced by the kind of situation involved and the type of activity being carried out. (Reason, 1987a, p. 6)

Reason's approach is intended to reflect the fact that the basic error tendencies "... constitute the fundamental roots of most, if not all, of the systematic varieties of human error" (1987a, p. 6). For example, many predictable error forms can be traced to the limitations of human information processing and memory capacity. However, the efficiencies of information processing that humans do achieve are largely attributable to the internal schema and mental models that govern our interactions with the world about us (e.g., See the following section, Perspectives on mental models). The same schema and mental models also influence the strategies by which the limited processing resources are applied to problem solving requirements. It is primarily these three basic error tendencies (e.g., resource limitations, schema, and strategies) that determine the primary error groupings shown in Figure 1. (e.g., attentional failures, memory lapses, inaccurate recall, misperceptions, errors in judgment, reasoning errors, and unintended words or actions).

A current dominant perspective on resource limitations is represented by Wickens' (1984) multiple attentional resource model, which is illustrated in Figure 2. Given the demands placed on specific processing resources, the model provides a basis for predicting the area of performance likely to be degraded. In multiple task situations, tasks will interfere more if more resources have to be shared. When tasks demand separate rather than common resources on any of the three dimensions, time-sharing is likely to be more efficient. Wickens' multiple resource model was used by Murphy, Reaux, Stewart, Coleman, and Bruce (1989) in developing their very detailed model of controller workload. Parenthetically, it is likely that Reason's (1987a) classification scheme could be validated on the data that would be generated by the application of the Murphy et al. workload model to real-time control operations.

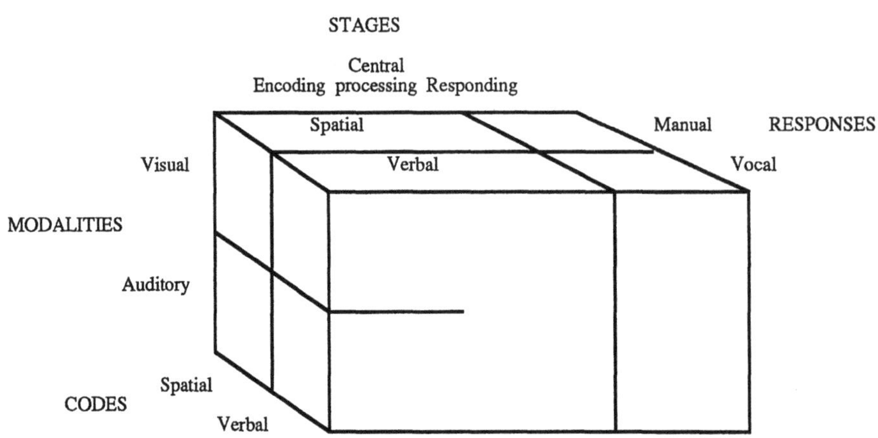

Figure 2. The structure of processing resources (adapted from Wickens, 1984).

Perspectives on mental models

Wilson and Rutherford's (1989) recent review of mental models focused on the relation between the operator's mental model and models of the operator. Norman (1983, 1986) earlier made the distinction among the *conceptual model* of the operational system, the *system image* or operation of the system as it is characterized by the displays (for example, at the operator's interface with the system), and the operator's *mental model* of the system. In addition, we have a requirement for a model of the operator, or what Wilson and Rutherford call a "*designer's conceptual model*" (p. 631) of the user. Edwards (1991) has provided an overview of how Artificial Intelligence could be applied to ATC through the incorporation of explicit models of the controller and of the system which would support intelligent dialogue. Explicit models embody the constraints and objectives of the controller as well as the system, and represent an effective means of introducing human factors considerations during system development.

An error model for operating irregularities in ATC constitutes a predictive model of the user. However, this same model will itself incorporate the operator's mental model as one of the determinants of human error. Reason's use of *schema properties* as one of the basic error tendencies, and Norman's (1981) proposal that mistakes result from misperceptions and inappropriate models, exemplify this approach.

One of the concerns in modelling human error is that the optimal design of the operator interface be compatible with the operator's mental model of the system characteristics (Hollnagel, 1988; Lind, 1988; Van der Veer, 1987) and that the nature of the displays match the nature and level of information processing at which the operator is working (Moray, 1988; Rasmussen, 1985; Rasmussen and Vicente, 1989). Some success has been achieved at deriving approximations of users' mental models through the use of hierarchical cluster analysis (e.g. Kellogg and Breen, 1987). Rasmussen and Vicente (1989) have also cautioned that operators may have various mental models at different levels of abstraction and that interference between mental models is to be avoided. Externalizing the effective mental models might support memory and facilitate the kind of reasoning required by the task (p. 524-525). They have also suggested that the interface should not only support the properties of human cognition but also "...demand or invite appropriate behaviors" (p. 531).

Murphy et al. (1989) have taken the approach that it is also possible to look beyond the controller's interface with the logical displays, to the *interface between the different mental models* with which the controller must work. Murphy et al. identify these mental models as the current mental picture, the projected mental picture, the sector load picture, and the decision-making criteria. It is their contention that a large portion of mental workload can be traced to these mental interfaces.

Framework models of human performance and error

Framework cognitive models (Reason, 1988a) "...outline, in deliberately general terms, the principal characteristics of the human-information processing system" (p. 36), focusing on the cognitive control mechanisms that determine human performance. These global performance models are more closely allied with cognitive scientists and cognitive engineers rather than with traditional cognitive psychology.

In thinking about the application of mental models, and particularly models of human error, to automation and system design, it is appropriate to include Rasmussen's (1986) suggestion that higher-level structural models of human mental activities will have broader application than detailed models of actual mental processes. Rasmussen's own framework model for the classification of behavior into skill-based, rule-based, and knowledge-based categories (abbreviated as SRK) has to be viewed as a high level functional model, rather than a psychological model of the operator's mental processes. However, Rasmussen's SRK model has

> ...provided system designers and engineers with a common language for talking about operators' mental processes; it has served as a unifying rubric for research and modelling in this area; and it has also effected a reorganization of the way we now perceive, and therefore classify, the human operator's role in complex systems. (Sanderson and Harwood, 1988, p. 21)

Rasmussen's SRK model.

The basic SRK model of Rasmussen is shown in Figure 3. Typical error mechanisms can be derived from the characteristics of the three types of behavior. For example, errors occur at the knowledge-based level when causal conditions and side effects are not considered or an inappropriate model is applied. At the rule-based level, errors occur when isolated items are forgotten, mistakes among alternatives are made, or recall is incorrect. Skill-based behavior is subject to error when the task signs trigger a related and habitual action. Additional information on the SRK model will be presented in other sections of this paper.

Reason's generic error-modelling system (GEMS)

Reason (1987b) has used Rasmussen's SRK model as a vehicle to integrate two areas of error research, slips and mistakes. His generic error-modelling system (GEMS) is intended to provide a framework for addressing these two more predictable varieties of human error.

A slip refers to an action in which the intention is correct but the intended action is not executed as planned. This type of error is distinguished from a mistake, in which there is a faulty understanding of the situation or in which an inappropriate assumption is made (e.g., a misperception).

Human activity is described (Reason, 1987b, p. 64; 1988a) as being governed by two modes of control: attentional and schematic. The attentional mode is associated with analytic information processing, working memory, and responding to change. The schematic or automatic mode can process familiar information rapidly in parallel without conscious involvement. Skill-based and rule-based behaviors involve schematic control. The attentional mode is associated with knowledge-based activity. Empson (1991) has cautioned that there may be a tendency in ATC systems to automate those functions which already involve schematic control and to leave the controller with those tasks that involve more difficult decision making under attentional control.

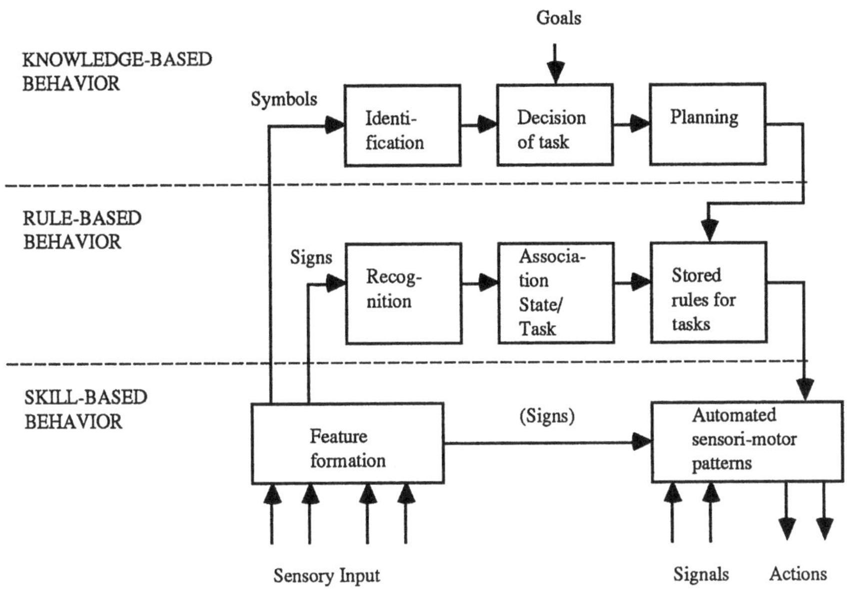

Figure 3. The SRK formulation of cognitive control (adapted from Rasmussen, 1987b).

The early work of Norman (1981) and Reason and Mycielska (1982) on slips and lapses has shown the dependency of slips on failures in attention, either in terms of inattention or overattentiveness (i.e., an attentional check at an inappropriate moment in the sequence). Slips were found to be associated with the automatic execution of well established routines, with distractions or preoccupation, and with absentmindedness in relatively familiar environments. Heckhausen and Beckmann (1990) have recently attempted to further extend the work of Reason and Mycielska, by differentiating those slips that can be associated either with the initiation, the deactivation (termination), or the implementation of an intended sequence of actions toward an intended goal.

The GEMS model yields three distinct error types: skill-based slips, rule-based mistakes, and knowledge-based mistakes. These three types are summarized in Table 2. As shown, errors at the skill-based and rule-based levels tend to take the form of *"strong-but-wrong."* In the case of the skill-based error, the guidance of action tends to be captured by the most active motor schema in the vicinity of the node in the absence of monitoring. Norman (1981) has described these as *capture errors*.

> Similarly, the most probable error at the rule-based level involves inappropriate matching of readily available but irrelevant patterns, [or] scripts....In both cases, the error forms are *already available* within the individual's stored repertoire or action sequences or rules. (Reason, 1987b, p. 70)

However, at the knowledge-based level, the mistakes are more difficult to predict because they depend on the interaction between the operator's reasoning and what is likely an incomplete or inaccurate mental model. One implication of these distinctions is that expertise can be defined in terms of the repertoire of appropriate subroutines that accrues with experience and which enables the expert to deal with a variety of contingencies.

Reason (1987b) has also described how "errors at the three levels of performance are further distinguished by the nature of the psychological and situational factors which combine to shape characteristic error forms" (p. 72). Some of the "error-shaping" factors apply at more than one level. The error-shaping factors are summarized in Table 3.

TABLE 2

Distinctions among the three basic error types (from Reason, 1987b).

FACTORS:	SKILL-BASED SLIPS	RULE-BASED MISTAKES	KNOWLEDGE-BASED MISTAKES
ACTIVITY	Routine actions	Problem solving	
CONTROL	Mainly automatic processors (Schemata)	(Rules)	Resource-limited conscious processes
FOCUS OF ATTENTION	On something other than present task	Directed at problem-related issues	
FORMS	Largely predictable "Strong-but-wrong' error forms (Actions)	(Rules)	Variable
DETECTION	Usually fairly rapid	Hard, and often only achieved with help from others	

As noted earlier, skill-based slips frequently take the form of *capture errors,* in which the strength of the associated action schemata is determined by the relative frequency of successful behaviors in the past. Strong-but-wrong sequences are likely to occur, therefore, in the absence of appropriate monitoring. In terms of predicted quantity of errors, Reason believes that the quantity of errors at the skill-based and rule-based levels will be small but that will usually take the strong-but-wrong form. However, at the knowledge-based level, a relatively large number of errors are likely and their forms will be difficult to predict. Given the nature of controller activities in the ATC environment, it would be expected that many of the error forms will be predictable and therefore open to efforts to reduce their occurrence and impact.

In another context, Reason (1987c) has observed that one of the inherent characteristics of human information processing is our ability to simplify complex informational tasks through the application of automatic (schematic) rather than conscious (attentional) modes of processing. "Human beings are strongly disposed to exploit the effortless, parallel and preprogrammed operations of highly specialized low-level processors and their associated heuristics (intuitive rules of thumb)" (p. 212). Thus the various predictable errors become the byproducts of human cognition. This is consistent with Rasmussen's (1985) observation that the analysis of human error has to be viewed as being a part of the study of normal human behavior in real world situations.

TABLE 3
Major error shaping factors for each type of error
(adapted from Reason, 1987b, and Baganara and Rizzo, 1989).

PERFORMANCE LEVEL ERRORS	ERROR-SHAPING FACTORS
SKILL-BASED SLIP	
Intention or correct action not carried out as planned	1. Recency and frequency of previous use 2. Environmental control signals 3. Shared schema properties 4. Concurrent plans
RULE-BASED MISTAKE	
Faulty application of well-known rules or procedures both in identifying a situation and/or in adopting a plan of action	1. Mind set[1] (It has always worked before) 2. Availability (First come, best preferred) 3. Matching bias (like relates to like) 4. Over-simplification 5. Over-confidence
KNOWLEDGE-BASED MISTAKE	
Selected plans do not fit with the problem to be solved	1. Bounded rationality: • Selectivity • Working memory overload • Out of sight, out of mind 2. Allocating either too little or too much time 3. Memory cueing/ reasoning by analogy 4. Matching bias revisited 5. Incomplete/ incorrect mental model

[1] Reason (1987c, 1988) has also described "frequency gambling," which refers to the use of the more recent and frequently successful option when requirements are incompletely specified.

Perspectives on automation

The nature of air traffic control is about to begin a fundamental evolution over the next decade. The implementation of systems such as the Advanced Automation System (AAS) in the United States (Benel, Dancey, Dehn, Gutmann, and Smith, 1989; Scardina, Ryberg, and

Gerstenfeld, 1989) and the Canadian Automated Air Traffic System (CAATS) in Canada (Stager, 1991) will significantly alter the characteristics of information processing.

One question that continues to receive focused attention in discussions on automation is "What constitutes an appropriate level of automation for ATC operations?" What level will provide the required system adaptability for conditions of uncertainty, and yet enable the efficiencies associated with steady-state operations to be realized? The resolution of this question will impact significantly on the qualitative nature of the controllers' work as well as the levels of workload and stress that will be encountered (Hancock, 1991; Hancock and Warm, 1989; Tattersall, 1991).

Sources of error in automation technology.

In his commentary on cognitive engineering,Woods (1988a) observed that we are continuing to create a larger world of cognitive tasks but also expanding "...the potential to assist and augment human cognitive activities in complex problem-solving worlds" (p. 115). One of most significant developments will be the increased capability in forward planning with extrapolated displays that will be available to an operator. Woods has referred to this development as "*conceptualization power*" (p. 124). At the same time, we have begun to see the hazards of over-reliance on automated systems and the concomitant failure to monitor system parameters when required. With the introduction of the new generation of "glass cockpits," there has been an increase in the number of reports to the ASRS concerning these aircraft and events related to the use or disuse of the automated systems. These reports are now tagged by the ASRS as "advanced technology" incidents (Lauber, 1987).

Wiener (1987) has cautioned that "automation essentially relocates and changes the nature and consequences of human error, rather than removing it..." (p. 179). For example, in the context of human error in the cockpit, Wiener has proposed the development of an "electronic cocoon" (Wiener, 1989; Wiener and Curry, 1980) which would provide a protective envelope for pilots. Forecasting algorithms would alert pilots before the electronic envelope was penetrated, and would aid the pilot in taking corrective action. The electronic cocoon concept represents an alternative to increased levels of cockpit automation. The application of the concept would ensure that the pilot continues to participate actively in the monitoring and control of the aircraft.

In the foreseeable future, the automation of systems will not be wholly complete, leaving a complex of automated, partially automated, and manual tasks for controllers in the aviation system and for operators in process control systems. Without careful task analyses and the assessment of controller/operator workload in these hybrid systems, the potential for human error remains unchanged and, in some instances, may be increased relative to the current operational environment. Ironically, automation technology can carry with it the potential to contribute its own sources of human error (Bainbridge, 1987).

In spite of our attempts to anticipate system function and operator requirements, the possibility remains that we will overlook what Reason has called "resident pathogens" (1987c, p. 211; 1988b). The resident pathogens could be design errors that contribute to minor failures in the system and which are analogous to the resident pathogens in the human body. As Reason indicates, they are usually checked by sensors or other functions within the system or detected and rectified by human operators. However, it remains possible for several of these small failures to combine in unforeseen ways and to result in larger and more critical failures (Wiener, 1989).

Mancini (1988) has underlined the human tendency to rely on the cognitive heuristics of "similarity matching" and "frequency gambling" (Reason, 1987c; 1988b) in unforeseen situations. However, problem situations do not have to be particularly complex to quickly exceed our ability to track unfolding events (Woods, 1988b). Incidents frequently follow a time-course, requiring the controller or operator to reassess the unfolding circumstances and to update the mental model. Initial assessments of the situation must be revised as the situation unfolds or these same perceptions can lead to erroneous interpretations with later developments. With time, the operator can become "decoupled from the actual state of the world" (p. 134).

One of the design features of the Canadian CAATS project (Stager, 1991) is that automatic conflict-resolution options would not be provided for radar traffic; however, this operational feature would be available for non-radar airspace. One of the consequences of this decision is that the controllers working in a radar environment can remain directly within the processing loop and retain a situational awareness. At the same time, the fully automated real-time flight data processing and distribution system will ensure that the information required for strategic decision making is readily available.

Error detection and error recovery.

Future system designs in general will have to be more error tolerant (Rasmussen, 1987a, 1987c; Rasmussen and Vicente, 1989; Rouse, 1985; Wiener, 1987). Rasmussen and Vicente (1989) have advocated an approach to interface design which attempts to increase the system's tolerance to errors by providing the operator with improved means of controlling their effects. Cognitive engineering has matured to the point where predictive or forecasting algorithms can facilitate planning and verification of potential responses, as well as providing an alerting function. However, attention has to be given to how the occurrence of errors can be made more conspicuous at the interface and how the relevant information for reversing errors is to be provided in a timely manner (Lewis and Norman, 1986; Rasmussen, 1985). Rizzo, Bagnara, and Visciola (1988) and Bagnara and Rizzo (1989), for example, have pointed to the current imbalance between the attention given to the cognitive analysis of sources of error and the attention given to error detection and recovery.

What, then, are the characteristics of error detection and what are the strategies that are utilized for correction or recovery (Lewis and Norman, 1986)? Rizzo et al. and Bagnara and Rizzo rely on a synthesis of the cognitive modelling represented by Rasmussen's (1986) SRK model, Reason's (1987b) generic error-modelling system, and Norman's (1981) categorization of slips and mistakes for their approach to these questions.

The detection of skill-based slips and rule-based mistakes will require an involvement of the attentional control processes. However, skill-based behavior is normally under schematic control and must compete with other demands for attentional resources (e.g., conscious monitoring of feedback). Rule-based mistakes are more likely to be detected earlier; the involvement of working memory means that cues about the expected outcome of the action will be present during the monitoring. In contrast, knowledge-based mistakes are difficult to detect; their detection depends on a revision of the mental model driving the behavior and on an acknowledgement that the current actions are inappropriate.

Bagnara and Rizzo (1989) have proposed three processes by which errors may be detected: forcing functions, external feedback, and inner feedback (p. 609). With a *forcing function*, the system does not permit the operator to continue with an action. With *external feedback*, the information that is eventually received is inconsistent with the expectations of the operator. In the case of *inner feedback*, a mismatch occurs between the cognitive extrapolation of expected outcomes and the information being received. Bagnara and Rizzo suggest that these events can be understood in terms of the notion of *cognitive distance* developed by Hutchins, Hollan, and Norman. Cognitive distance refers to the cognitive transformation that is required to resolve the perceived mismatch between expectations and the feedback. The speed and accuracy with which error recovery occurs depends on the nature of the transformation required.

The critical question for error detection and recovery is "What characteristics of the information displayed at the interface will enable the operator to construct a corrective action in an optimal manner?" Findler, Bickmore, and Cromp (1985) earlier described a dynamic causal modelling system for an ATC environment that could help to address this question. Their modelling system would enable controllers to monitor current and extrapolated displays in real time. With the addition of an extrapolated display, the controller can observe causes and effects at the same time. This capability could have a beneficial application in the investigation of display characteristics and error-recovery strategies.

Summary observations

This paper has provided a brief review of the current efforts to identify predictable forms of human error through the use of classification systems and generic framework models. The review included some of the human factors issues that link questions of human error and questions of automation. The potential for predicting and controlling human error in complex systems is evident in the more mature framework models of human performance. However, a significant effort is still required to validate these models in the context of air traffic control.

The more comprehensive taxonomies for classifying human error are most effective when they can be related implicitly to a more generic framework model. Circumscribed descriptive taxonomies perhaps appear more useful initially because the language in which they are couched is more directly tied to actual controller tasks. This relationship permits a simple mapping of controller activities onto the taxonomy. However, with technological change, error models or taxonomies that are tied to the elements of task analyses become inadequate (Rasmussen, 1987a). Under these circumstances, the advantages of classifying human error in terms of human cognition and the relevance of generic models for system design become clear.

Reason's classification of errors (1987a), his generic framework model (1987b), and Leplat's (1987) model of accident analysis have drawn on the intense effort over the past decade to understand human error in complex operational systems. Their work holds promise for developing a procedure for the classification and analysis of human errors in the context of air traffic control.

In the analysis of accidents, Leplat assumes that the production of errors occurs after a change or variation in some usual condition. Although his analysis focuses on retracing the antecedent conditions, Leplat's approach is more clearly characterized as a response or action analysis, given the difficulty of tracking subtle variations that are difficult to observe directly. These variations can include states of the operator as well as the environment. Stager, Hameluck, and Jubis (1989) have proposed a hybrid of the approaches to human error taken by Reason and Leplat, respectively, to classify and analyze system errors. The proposed hybrid approach would, like the original Leplat model, retrace events that eventually led to the overt system error. Identifying the convergence to the direct cause of the system error within Reason's framework for classifying errors would entail tracing the variables that affected the information processing mechanisms and, in turn, the processing failures that contributed to the inappropriate action. The analyses based on this hybrid approach should provide useful information concerning which actions or omissions tend to be associated with conditional outcomes. Subsequent analysis of the database might also provide information on what predictable error forms, within each primary error grouping (Reason, 1987a), are distinguished according to the situational factors associated with their occurrence.

Acknowledgements

The preparation of this paper was supported by Transport Canada through two contracts (No. AQ-0859 and No. T8080-9-5072/02-ZA). The views and opinions expressed are those of the author and not necessarily those of the Transport Canada.

References

Bagnara, S., and Rizzo, A. (1989). A methodology for the analysis of error processes in human-computer interaction. In M.J. Smith and G. Salvendy (Eds.) *Work with computers: Organizational, management, stress, and health aspects*, (pp. 605-612). Amsterdam: Elsevier.

Bainbridge, L. (1987). Ironies of automation. In J. Rasmussen, K. Duncan, and J. Leplat (Eds.), *New technology and human error*, (pp. 271-283). Chichester: Wiley.

Benel, R. A., Dancey, R. D., Dehn, J. D., Gutmann, J. C., and Smith, D. M. (1989). Advanced automation system design. *Proceedings of the IEEE*, *77*(11), 1653-1660.

Canadian Aviation Safety Board. (March 1990). *Report on a special investigation into air traffic control services in Canada.* Report No. 90-SP001. Supply and Services Canada Catalog No. TU4-5/1990E.

Edwards, J. L. (1991). Intelligent dialogue in air traffic control systems. In J. A. Wise, V. D. Hopkin, & M. L. Smith (Eds.) *Automation and Systems Issues in Air Traffic Control.* Berlin: Springer-Verlag.

Empson, J. (1987). Error auditing in air traffic control. In J. A. Wise and A. Debons (Eds.). *Information Systems: Failure Analysis. NATO ASI Series F: Computer and Systems Sciences, Vol. 32,* (pp. 191-198). Berlin: Springer-Verlag.

Empson, J. (1991). Cognitive failures in military air traffic control. In J. A. Wise, V. D. Hopkin, & M. L. Smith (Eds.) *Automation and Systems Issues in Air Traffic Control.* Berlin: Springer-Verlag.

Eißfeldt, H. (1991). Automation in ATC - How does it affect the selection of controllers? In J. A. Wise, V. D. Hopkin, & M. L. Smith (Eds.) *Automation and Systems Issues in Air Traffic Control. Berlin: Springer-Verlag.*

Findler, N. V., Bickmore, T. W., and Cromp, R. F. (1985). A general-purpose man-machine environment with special reference to air traffic control. *International Journal of Man-Machine Studies, 23*, 587-603.

Grayson, R. L. (1981). Information transfer in the surface component of the system: Coordination problems in air traffic control. In C. E. Billings and E. S. Cheaney (Eds.). Information transfer problems in the aviation system. *Information transfer problems in the aviation system..* Moffett Field, CA: National Aeronautics and Space Administration, NASA Technical Paper 1875.

Grayson, R. L., and Billings, C. E. (1981). Information transfer between air traffic control and aircraft: Communication problems in flight operations. In C. E. Billings and E. S. Cheaney (Eds.). Information transfer problems in the aviation system. *Information transfer problems in the aviation system..* Moffett Field, CA: National Aeronautics and Space Administration, NASA Technical Paper 1875.

Hancock, P. A. (1991). The aims of human factors and their application to issues in automation and air traffic control.In J. A. Wise, V. D. Hopkin, & M. L. Smith (Eds.) *Automation and Systems Issues in Air Traffic Control.* Berlin: Springer-Verlag.

Hancock, P. A., and Warm, J. S. (1989). A dynamic model of stress and sustained attention. *Human Factors, 31*, 519-537.

Heckhausen, H., and Beckmann, J. (1990). Intentional action and action slips. *Psychological Review, 97*(1), 36-48.

Hollnagel, E. (1988). Mental models and model mentality. In L. P. Goodstein, H. B. Andersen, and S. E. Olsen (Eds.), *Tasks, errors, and mental models* , (pp. 261-268) London: Taylor & Francis.

Hopkin, V. D. (1987). Human factors implications of progressive air traffic control automation. *Proceedings of the Fourth International Symposium on Aviation Psychology,* Columbus, OH, 179-185.

Hopkin, V. D. (1989a). Implications of automation on air traffic control. In R. S. Jensen (Ed.), *Aviation Psychology* (pp. 96-108). Brookfield, Vermont: Gower.

Hopkin, V. D. (1989b). Man-machine interface problems in designing air traffic control systems. *Proceedings of the IEEE, 77*(11), 1634-1643.

Hopkin, V. D. (1991). The impact of automation on air traffic control systems. In J. A. Wise, V. D. Hopkin, & M. L. Smith (Eds.) *Automation and Systems Issues in Air Traffic Control.* Berlin: Springer-Verlag.

Kellogg, W. A., and Breen, T. J. (1987). Evaluating user and system models: Applying scaling techniques to problems in human-computer interaction. In J. M. Carroll and P. P. Tanner (Eds.), *Human factors in computing systems - IV and graphics interface: Proceedings of CHI and GI '87 Conference* (pp. 303-308). New York: Association for Computing Machinery.

Kinney, G. C., Spahn, M. J., and Amato, R. A. (1977, December). *The human element in air traffic control: Observations and analyses of the performance of controllers and supervisors in providing ATC separation services.* McLean, VA: METREK Division of the MITRE Corp., MTR-7655.

Lauber, J. K. (1987). New technology, human performance and transportation system safety. *Proceedings of the Human Factors Society 31st Annual Meeting,* 15-20.

Leplat, J. (1987). Accidents and incidents production: methods of analysis. In J. Rasmussen, K. Duncan, and J. Leplat (Eds.), *New Technology and Human Error* (pp. 133-142). Chichester: Wiley.

Lewis, C., and Norman, D. A. (1986). Designing for error. In D. A. Norman and S. W. Draper (Eds.) *User-centered system design,* (pp. 411-432). Hillsdale, NJ: Erlbaum.

Lind, M. (1988). System concepts and the design of man-machine interfaces for supervisory control. In L. P. Goodstein, H. B. Andersen, and S. E. Olsen (Eds.), *Tasks, errors, and mental models* , (pp. 269-277). London: Taylor & Francis.

Mancini, G. (1988). Commentary: Models of the decision maker in unforeseen accidents. In E. Hollnagel, G. Mancini, and D. D. Woods (Eds.), *Cognitive engineering in complex worlds,* (pp. 177-185). London: Academic Press.

Moray, N. (1988). Prologue. In L. P. Goodstein, H. B. Andersen, and S. E. Olsen (Eds.), *Tasks, errors, and mental models*, (pp. 12-17). London: Taylor & Francis.
Morrison, R., and Wright, R. H. (1989). ATC control and communications problems: An overview of recent ASRS data. *Proceedings of the Fifth International Symposium on Aviation Psychology*, Columbus, OH.
Murphy, E. D., Reaux, R. A., Stewart, L.J., Coleman, W. D., and Bruce, K. (1989). Where's the workload in air traffic control? *Proceedings of the Fifth International Symposium on Aviation Psychology*, Columbus, OH.
Norman, D. A. (1981). Categorization of action slips. *Psychological Review*, *88*, 1-15.
Norman, D. A. (1983). Some observations on mental models. In D. Gentner and A. L. Stevens (Eds.), *Mental models* (pp. 7-14). Hillsdale, NJ: Erlbaum.
Norman, D. A. (1986). Cognitive engineering. In D.A. Norman and S. Draper (Eds.), *User-centered system design* (pp. 31-61). Hillsdale, NJ: Erlbaum.
Rasmussen, J. (1985). Trends in human reliability analysis. *Ergonomics*, *28*, 1185-1195.
Rasmussen, J. (1986). Information processing and human-machine interaction: An approach to cognitive engineering. Series Volume 12. North-Holland Series in System Science and Engineering. New York: North-Holland.
Rasmussen, J. (1987a). The definition of human error and a taxonomy for technical system design. In J. Rasmussen, K. Duncan, and J. Leplat (Eds.), *New technology and human error*, (pp. 23-30). Chichester: Wiley.
Rasmussen, J. (1987b). Cognitive control and human error mechanisms. In J. Rasmussen, K. Duncan, and J. Leplat (Eds.), *New technology and human error*, (pp. 53-61). Chichester: Wiley.
Rasmussen, J. (1987c). Reasons, causes, and human error. In J. Rasmussen, K. Duncan, and J. Leplat (Eds.), *New technology and human error*, (pp. 293-301). Chichester: Wiley.
Rasmussen, J., and Vicente, K. J. (1989). Coping with human errors through system design: implications for ecological interface design. *International Journal of Man-Machine Studies*, *23*, 517-534.
Reason, J. (1987a). A framework for classifying errors. In J. Rasmussen, K. Duncan, and J. Leplat (Eds.), *New technology and human error* (pp. 5-14). Chichester: Wiley.
Reason, J. (1987b). Generic error-modelling system (GEMS): A cognitive framework for locating common human error forms. In J. Rasmussen, K. Duncan, and J. Leplat (Eds.), *New technology and human error*, (pp. 63-83). Chichester: Wiley.
Reason, J. (1987c). An interactionist's view of system pathology. In J. A. Wise and A. Debons (Eds.). *Information Systems: Failure Analysis. NATO ASI Series F: Computer and Systems Sciences, Vol. 32*, (pp. 211-220). Berlin: Springer-Verlag.
Reason, J. (1988a). Framework models of human performance and error: a consumer guide. In L. P. Goodstein, H. B. Andersen, and S. E. Olsen (Eds.), *Tasks, errors, and mental models*, (pp. 35-49). London: Taylor & Francis.
Reason, J. (1988b). Cognitive aids in process environments: Prostheses or tools? In E. Hollnagel, G. Mancini, and D. D. Woods (Eds.), *Cognitive engineering in complex worlds*, (pp. 7-14). London: Academic Press.
Reason, J., and Mycielska, K. (1982). *Absent-minded? The psychology of mental lapses and everyday errors*. Englewood Cliffs, NJ: Prentice-Hall.
Rizzo, A., Bagnara, S., and Visciola, M. (1988). Human error detection processes. In E. Hollnagel, G. Mancini, and D. D. Woods (Eds.), *Cognitive engineering in complex worlds*, (pp. 99-114). London: Academic Press.
Rosenthal, L. J., and Mellone, V. J. (1989), Human factors in ATC operations: Anticipatory clearances. *Proceedings of the Fifth International Symposium on Aviation Psychology*, Columbus, OH.

Rouse, W. B. (1985). Optimal allocation of system development resources to reduce and/or tolerate human error. *IEEE Transactions on Systems, Man, and Cybernetics*, SMC-15(5), 620-630.

Sanderson, P., and Harwood, K. (1988). The skills, rules and knowledge classification: A discussion of its emergence and nature. In L. P. Goodstein, H. B. Andersen, and S. E. Olsen (Eds.), *Tasks, errors, and mental models* , (pp. 21-34). London: Taylor & Francis.

Scardina, J. A., Ryberg, P-Y., and Gerstenfeld, A. (1989). Future ATC automation aids based upon AI technology. *Proceedings of the IEEE*, *77*(11), 1625-1633.

Stager, P. (1991). The Canadian Automated Air Traffic System (CAATS): An overview. In J. A. Wise, V. D. Hopkin, & M. L. Smith (Eds.) *Automation and Systems Issues in Air Traffic Control.* Berlin: Springer-Verlag.

Stager, P., and Hameluck, D. (1988, February). *Factors associated with air traffic control operating irregularities: An analysis of Fact Finding Board reports.* Report prepared under Transport Canada Contract No. AP-0264 for the Air Traffic Services Branch, Evaluation Division, Ottawa, Canada.

Stager, P., Hameluck, D., and Jubis, R. (1989). Underlying factors in air traffic control incidents. *Proceedings of the Human Factors Society 33rd Annual Meeting*, 43-46.

Stein, E. S. (1987). Where will all our air traffic controllers be in the year 2001? *Proceedings of the Fourth International Symposium on Aviation Psychology*, Columbus, OH.

Tattersall, A. J. (1991). Effects of stress and workload on system performance in air traffic control. In J. A. Wise, V. D. Hopkin, & M. L. Smith (Eds.) *Automation and Systems Issues in Air Traffic Control.* Berlin: Springer-Verlag.

Van der Veer, G. C. Mental models and failures in human-machine systems. In J. A. Wise and A. Debons (Eds.). *Information Systems: Failure Analysis. NATO ASI Series F: Computer and Systems Sciences, Vol. 32,* (pp. 221-230). Berlin: Springer-Verlag.

Wickens, C. D. (1984). Processing resources in attention. In R. Parasuraman and R. Davies (Eds.), *Varieties of attention* (pp. 63-102). New York: Academic Press.

Wiener, E. L. (1987). Fallible humans and vulnerable systems: Lessons learned from aviation. In J. A. Wise and A. Debons (Eds.). *Information Systems: Failure Analysis. NATO ASI Series F: Computer and Systems Sciences, Vol. 32,* (pp. 163-181). Berlin: Springer-Verlag.

Wiener, E. L. (1989). Reflections on human error: Matters of life and death. *Proceedings of the Human Factors Society 33rd Annual Meeting*, 1-7.

Wiener, E. L., and Curry, R. E. (1980). Flight-deck automation: Promises and promises. *Ergonomics*, *23*, 995-1011.

Wilson, J. R., and Rutherford, A. (1989). Mental models: Theory and application in human factors. *Human Factors*, *31* (6), 617-634.

Wojcik, L. A. (1989, July). Probabilistic risk assessment and aviation system safety. *Flight Safety Digest.*, pp. 1- 5.

Woods, D. D. (1988a). Commentary: Cognitive engineering in complex and dynamic worlds. In E. Hollnagel, G. Mancini, and D. D. Woods (Eds.), *Cognitive engineering in complex worlds*, (pp. 115-129). London: Academic Press.

Woods, D. D. (1988b). Coping with complexity: The psychology of human behaviour in human systems. In L. P. Goodstein, H. B. Andersen, and S. E. Olsen (Eds.), *Tasks, errors, and mental models* , (pp. 128-148). London: Taylor & Francis.

Cognitive Failures in Military Air Traffic Control

Jacob Empson

Department of Psychology, University of Hull, North Humberside, UK

Introduction

As performance becomes skilled, less and less attention seems to be required to be devoted to it. As the Cambridge psychologist Sir Frederick Bartlett once perceptively remarked, "The skilled batsman has all the time in the world." Implicitly, the more skill is gained, the more behavior is delegated to the emission of well-drilled responses, and the more attention may be paid to the anticipation of events, and the grander aspects of the overall plan. Despite the enormous gains in performance by organizing motor behavior in this way, however, if there is a trade-off between attention being paid to the larger aspects of the plan at the expense of attending to the details of motor control, the scope for minor errors in execution remaining undetected may increase.

The study of cognitive failure, or absent-mindedness, was relatively neglected by psychologists after Freud's (1914) analysis of "the psychopathology of everyday life" until the last fifteen years or so, when it has found a role in the analysis of plans of action, and even of attentional processing. Reason (1977, 1979) explained many of the failures in skilled performance in terms of the inappropriate use of sub-routines which were open-loop — not routinely attended to as to outcome. This work was explicitly informed by the theoretical approach of Miller, Galanter, and Primbram (1960), who provided a novel unit for the analysis of behavior — the TOTE, or "Test-Operate-Test-Exit" unit. One of the most important ways in which this theory departed from the conventional behaviorist models of the time was the recognition of intention. The "tests" in the TOTE unit were to check any incongruity between the current state of affairs and a desired state of affairs. Furthermore, TOTE units could be subordinated to higher-level TOTE units, so that the "operate" part of the unit could be one which activated lower-order TOTE units. It was a theory which accounted for action in terms of intention and provided the basis for a grammar of actions. Some of these actions, subordinate to others, could also be assumed to be open-loop, and lacking comprehensive checks as to whether they had been satisfactorily completed.

Reason's analysis of errors, or, as he called them, "actions not as planned," was initially based on a diary study of 35 individuals who kept records of their lapses over a two week period. The error classification system which he developed (see below) accounted for 95% of the instances of error recorded. While all the errors logged were trivial, their categorization in this theoretical framework provides a system which is potentially extremely valuable. Accidents are commonly the consequences of errors, although very few errors result in accidents. The understanding of error should contribute to our understanding of the causes of accidents, and ultimately, to their prevention.

NATO ASI Series, Vol. F 73
Automation and Systems Issues in Air Traffic Control
Edited by John A. Wise et al.

Table 1. Error types*

A. Discrimination Failures
1. Perceptual Confusion
2. Functional Confusion
3. Spatial Confusion
4. Temporal Confusion

B. Program Assembly Failures
1. Spoonerisms
2. Confusions between current programs
3. Confusion between stored and ongoing programs

C. Test Failures
1. Stop-rule overshoots
2. Stop-rule undershoots
3. Branching error
4. Multiple side-tracking

D. Sub-Routine Failure
1. Insertions
2. Omissions
3. Misordering

E. Storage Failure
1. Forget previous action
2. Forget discrete items of the plan
3. Forget substance of the plan.

*Adapted from Reason (1979)

In the context of air traffic control, this classificatory system was used by Langan-Fox and Empson (1985) in their analysis of errors made by military air traffic controllers at an R.A.F. station in the U.K. Another system of categorization (Norman, 1981) was also tried out on these errors, but was found to be difficult to use, presumably because Norman's system explicitly disregarded slips of speech, and of course many of the air traffic controllers' errors were of this type.

The station involved was responsible for the last squadron in the U.K. of obsolescent (Lightning) fighter aircraft, and had few of the advantages of new technology. The working environment, while not ideal, was well tested, and all the controls and displays were relatively transparent in function. All eight controllers were observed both as Director controllers and Approach controllers, for a total of five hours each. Director controllers are responsible for directing returning aircraft to the airfield, before normally being passed on to the precision radar controllers for the last nine miles. Approach controllers provide a climb-out service for departing aircraft and a service on request to any other traffic in the 30 mile area around the airfield.

A diary study over a two week period, involving 15 controllers, elicited 13 self-reported errors. The systematic audit of errors by Langan-Fox (himself a trained air traffic controller), sitting next to each of eight senior controllers as they were working, elicited 131 errors over two short shifts (see Table 2)

It was found that the number of errors made per hour significantly increased with workload (number of aircraft on frequency) in Director controllers. Approach controllers showed no increase in errors with the number of aircraft on frequency. There was no evidence of any particular error type increasing more than others, although programming errors were significantly more numerous than the other two types (Langan-Fox and Empson, 1985; Empson, 1987a).

Table 2. Classification of errors

	Director	Approach	Total
Perceptual			
Discrimination Failures	19	13	32
Programming			
Program Assembly Failures	21	19	40
Sub-routine Failures	11	8	19
Memory			
Test Failures	10	6	16
Storage Failures	10	6	16
Uncategorizable	4	4	8
TOTAL	75	56	131

The results were interpreted in terms of the time-paced nature of the Director's job in directing returning aircraft, which had to be dealt with immediately as they called in. In addition, the Lightning jets involved, now all superseded, were invariably short of fuel when approaching their home base. Any delay caused by inefficient air traffic control management would at least lead to recriminations from the pilots involved (all known personally to the controllers), might result in re-direction of the aircraft to another station, or at worst could result in catastrophic fuel shortage (fortunately, never known to have occurred at this station). The Director's job was thus not only time-stressed but also intrinsically psychologically stressful.

While examples have been given elsewhere (Langan-Fox and Empson, 1985), it may be appropriate to give some others here to provide an idea of the nature of the cognitive failures made by our air traffic controllers.

Discrimination Failures (Example: Temporal Confusions)

Controller A in Director, with no aircraft on frequency, was monitoring Approach's frequencies while he was out of the room. Two Alpha Jets called on the Approach frequency and he responded, glancing at the D/F to see what the aircraft Q.D.M. had been, to aid identification. Just then Approach returned and resumed his seat.

"A"'told Approach of the two aircraft, and pointed to two close returns crossing the coast to the S.E., range 20 miles, saying "That's them." Approach took control.

Approach said to the Alpha Jets "confirm crossing the coast," and they confirmed, but in fact were N.E. near Spurn Point. The Q.D.M. that "A" had read had been from his own transmission. Subsequently, Approach developed considerable problems controlling the aircraft, looking as he did at the wrong returns.

Program Assembly Errors (Example: Spoonerisms)

Controller D in Director had four aircraft on recovery. 44 was at 11 miles finals, with information transfer outstanding. Two other aircraft had just been taken on frequency from a handover and had interrupted the rapid information transfer to and from 44, and other aircraft at 8 miles finals was just transferring to talkdown frequency. D passed the approach minima to 44 saying, "The minima procedure for this approach is.... What is your decision height?"

He should have said, "The procedure minima...etc."

Test Failures (Example: Multiple Side-tracking)

Controller G in Director had one aircraft on frequency for ILS. He retained the aircraft since he was not busy, monitoring its progress by looking at the PAR console on his left. Midland radar called on a landline, with Eastern radar patched through, to hand over two more aircraft. "G" was busy taking the details and identifying the aircraft when his ILS A/C reached three miles from touchdown. At this point "G" should have checked the landing gear and flaps of the ILS aircraft and obtained a clearance from local. He failed to do so. The supervisor stepped in and completed the ILS procedures.

"G" was reminded when a moment later, he finished his take-over call and looked back to see the ILS aircraft at one mile from touch-down.

Sub-routine Failures (Example: Insertions)

Controller E in Director had one aircraft on frequency and had just handed one to Local. He turned over the plaque of the aircraft that had just gone, to fill in the details, when there was a message telling him of a phone call for him at the desk behind. The aircraft had free-called at 5000', and therefore his intended action was to tick off the >3000 box. He did this and also ticked the <3000 box. He looked at it, puzzled, then put it down, told Approach to monitor his aircraft for a moment and went to the phone.

On his return, he was asked about the two ticks. He realized then what was wrong, and declared that he could not understand what it was that had puzzled him. Something was wrong and he could not see it.

Storage Failures (Example: Reverting to earlier plans)

Controller H in Director had a pair of aircraft downwind on recovery for GCA, over the mast (6 miles S.E.) heading south. The aircraft were high and "H" intended to take them to 13 miles South to lose height before turning in.

Coningsby called for traffic information on "H's" pair indicating their traffic 5 miles N.E. of Coningsby heading west in their GCA pattern. "H" coordinated action with Coningsby controller, saying, "I'll turn mine in at 10 or 11 miles, to keep clear of yours."

In the next minutes he let his aircraft run to 13 miles before giving instructions to turn. The turn circle took them to 15 miles. They remained clear, but separation eroded (in plan) to 3 miles.

Ergonomic Implications

Many of the errors in the Langan-Fox and Empson study seemed to be related to particular pieces of equipment, and shortcomings in human factors design were made evident in some of them. For instance, the landlines for incoming calls were common to both controllers, so that as a line called, both controllers were obliged to identify the relevant line and decide whether to answer it.

The direction-finding equipment was involved in two serious errors, both due to the controller misinterpreting the stored data derived from the machine's treatment of the controller's transmissions as if they were from an aircraft.

Errors were all either slips of speech or in handling equipment. It might be hypothesized that some categories of error would more commonly highlight ergonomic problems than others — for instance, perceptual errors might commonly involve a display.

The record of each error was therefore reconsidered in terms of whether any equipment was involved which was directly relevant to the failure (Empson, 1987b). As in the original study, analysis of error type was done in terms of three grand categories, mainly in order to ensure large enough numbers of examples in each category for statistical analysis:

• Perceptual errors:	discrimination failures
• Programming errors:	program assembly failures
	sub-routine failures
• Memory errors:	test failures
	storage failures

Chi square analysis revealed a statistically significant interaction between error type and device involvement. That is, while programming errors were associated with equipment on fewer than half the occasions, memory failures were associated with pieces of equipment almost twice as often than not, and perceptual failures were also more often than not associated with equipment. Table 3 gives the frequencies, with examples of all the items of equipment which were involved more than once with any one class of error.

Perceptual and mnemonic errors seem to be more diagnostic of difficulties with controls and displays than programming errors. So far as reducing such errors is concerned, the improvement of displays might be expected to reduce perceptual errors, and the improvement of mnemonic systems, and any ambiguity in choice of controls (such as buttons and knobs), to reduce memory failures.

These results suggest that error auditing may benefit from an analysis in terms of these three gross categories of Reason's taxonomy, and at a practical level, may have a potential use in identifying ergonomic failings in the working environment, as well as identifying performance that is less than ideal.

Table 3. Frequency and examples of errors associated with equipment.

Error Types	No Device Implicated	Device Implicated	Typical Items	
Perceptual	14	18	Landlines	6
			Screen knobs	2
			QDM (pos'n)	2
			Wind dials	2
			MTI knobs	2
Memory	12	20		
			Plaque	9
			Desktop	4
			Landlines	2
			R/T sel'n	2
Programming	42	17		
			Plaque	4
			Desktop	2
			Landlines	7
			R/T sel'n	2

Chi square (df=2) = 11.87, df 2, p<.01

Implications for Selection

In this study two controllers contributed over 20 errors each, while one individual made fewer than 10. With such small numbers of subjects, any meaningful study of individual differences is impracticable, but, for what it is worth, the error scores for all 8 controllers correlated significantly with Neuroticism, as measured by the E.P.I ($r = 0.83$, $p < .05$), and there was a large but statistically insignificant correlation with Extroversion ($r = 0.61$). Bartram and Dale (1982) , among others, have assessed the use of personality tests in selection for tasks involving high attentional load (in their case, pilot training), finding that the Neuroticism score of the E.P.I. was negatively related to success in training. While the correlation was low, some improvement in prediction of success would be achieved if it could be taken into account. A problem they noted with using inventories of this type is that of candidates "faking good," particularly in response to questions with a medical flavor, asking about "having dizzy turns," or getting "palpitations or thumping in your heart."

It would be useful indeed to have a psychological instrument which indicated which individuals were more liable to make errors than others. The concept of "accident proneness," first developed almost 60 years ago from psycho-analytic ideas of self-destructive tendencies, as the notion of a neurotic personality type liable to have accidents in any situation, has proved to be of little practical use. The recent development of the Cognitive Failures Questionnaire (CFQ) by Broadbent et al (1982) offered the possibility of approaching error and personality from a new direction. The questionnaire measures self-reported frequencies of failures in memory, perception, and motor function. Correlations between self-report ratings and those made about the same individuals by their spouses were positive and, while low, statistically significant (r = 0.32 for husbands and 0.36 for wives). As they pointed out, "the respondent's own view of his or her liability to cognitive failure does tend to be shared by somebody who had a good opportunity to judge."

Broadbent et al (1982) also assessed the relationship of CFQ scores with a measure of neuroticism (using the Eysenck Personality Questionnaire) and the Middlesex Hospital Questionnaire (MHQ), which is an inventory of relatively mild neurotic symptoms, providing a measure of psychological state, rather than of personality. Neuroticism was found to be positively correlated with CFQ scores among 221 student nurses (r = 0.28). Interestingly, while the CFQ did not correlate well with the MHQ scores of these students before they started clinical training, it did predict their MHQ scores when they were subsequently placed in a stressful working environment. That is, there is evidence that a high CFQ increases vulnerability to subsequent stress. There is also evidence (Fisher and Hood, 1987) that students suffering from home sickness have higher CFQ scores than those who do not. Self-reported lapses thus seem to be indicative of vulnerability to psychological stress. In a study of the relationship between CFQ scores and performance on tasks of visual search and of focused attention, Broadbent et al (1986) failed to find any straightforward relationship, although there was evidence that individuals with low cognitive failure scores showed different patterns of attention when anxious, compared with high scorers.

While a promising research instrument, the CFQ cannot in any way be regarded as a diagnostic test for identifying individuals likely to make more errors than others. There are three reasons for this, which Broadbent (1989) has suggested in the context of identifying error-prone individuals in the electrical generating industry. Firstly, being a self-report, the CFQ may be as sensitive to self-awareness or self-critical tendencies as to error rate itself. Secondly, a major difficulty of validating any test which might identify error-prone individuals is that the experienced error rate in many practical tasks is low. Thirdly, while it would clearly be wise not to employ people overtly showing signs of stress (and who might score highly on a self-report questionnaire such as the MHQ) on highly demanding and responsible tasks, one cannot rely on employees answering questions truthfully when it is clear that their job may depend on the nature of their answers, and when their level of insight into their own problems may be low.

In this study we have found relatively high error rates when these are monitored by a skilled observer, and very low rates when relying on self-report. (Ten times as many errors were recorded than by self-report, in half as many subjects, observed over one-seventh of the time.) Error auditing may therefore represent another approach to selection for error-free performers, as the problems of low error rate and of bias from self-reporting (either about cognitive failures or about psychological stress or neuroticism) do not apply. It is still unclear whether differences in observed error rates between individuals represent stable traits, or rather the influence of temporary factors. Broadbent's work on the CFQ has established that self-report scores are relatively stable over time, but further research is needed to establish whether this is also true of objectively audited error rates. If it is, the development of performance tests to assess freedom from cognitive failure (as distinct from skill) may help in the future selection of air traffic controllers.

Implications for Automation

The role of the operator in controlling complex systems has been described by Rasmussen (1980, 1982, 1986), in a model based originally on verbal protocols collected from operators engaged in electronic "trouble-shooting." Errors may be either slips (of the kind described by Reason) or the consequences of inappropriate intentions. The organization of behavior is described in terms of a hierarchy of skills, rules, and knowledge, with the expert operator moving appropriately between different levels, and the novice beginning with some knowledge but little skill and few rules, then gradually acquiring qualitatively different skills, rules, and knowledge of the system.

It is uncontroversial to assert that as skills develop they become increasingly undemanding of attention. Sequences of responses are chained in units, with their own syntax. An obvious example is the touch typist, phenomenally speedy when typing in his or her own language, but almost as slow as a novice when typing nonsense sequences or in a foreign language. The skilled driver has little difficulty in carrying on a conversation while driving, while the novice needs to devote all his or her attention to the task of controlling the car.

One approach has been to explain these phenomena in terms of the development of "automaticity" (Shiffrin and Schneider, 1977). An even more interesting approach, for the understanding of errors of action, has been one which treats the control of action as an integral part of the control of attentional processing (Norman and Shallice, 1986). In this, the allocation of attentional resource is normally achieved by "contention scheduling" — an automatic system containing simple rules as to the relative importance of competing habits. These habits, or schemata, may be skilled open-loop sub-routines (such as gear-changing in a car, or the typing of a well-rehearsed word or phrase) and while under contention scheduling control, the driver responds automatically to the demands of the road, the typist to the material being copied. The model also includes a "supervisory activating system," or SAS, capable of changing the criteria for contention scheduling and changing the direction of behavior.

For a full review of error models for "operating irregularities," including some which are less psychological and more specifically tied to controller tasks, the reader can do no better than to consult Stager (1991). So far as psychological models are concerned, in recent years there has been an encouraging convergence of thinking by psychological theorists addressing issues of attention, memory, and performance, with the common ground being that attentional resources may be limited and interchangeable within certain domains. Framework models of human performance and error such as these are perhaps the most appropriate for the understanding of the air traffic controller's task;in a recent review, Reason (1988) has summarized the common ground on automatic and controlled processing in a table (see Table 4).

It is suggested that devising a safe automated working environment for air traffic controllers should include careful consideration of the nature of the work being demanded of the operator in terms of models of information processing, in addition to engineering considerations. The effects of automation are complex but will undoubtedly alter the nature of work, and may well be to delegate the tasks which could have been done automatically by the operator to the machine, leaving the operator with the more difficult decision making "under attentional control." The result would then be a job that is more demanding, rather than less so. The importance of human factors in the design of such work will therefore be substantially increased, and it could be argued that error auditing should play an important role in the development of automated systems of air traffic control, and that their categorization in terms of perceptual, memory, and programming errors will provide a useful addition to the designer's intellectual toolkit.

Table 4. Contrasting the properties of attentional and automatic control (After Reason, 1988)

Attentional Control	Automatic Control
Serial	Parallel
Slow	Rapid
Laborious	Effortless
Analytic	Intuitive
Computationally powerful	Operates by simple heuristics
Processes accessible to consciousness	Processes beyond the reach of awareness; only products available to consciousness
Essential for coping with novelty, but cannot be sustained for any length of time	Capable of handling routines and recurrences; but often ineffective in the face of change

References

Bartram, D. Dale, H. C. A. (1982). The Eysenck Personality Inventory as a selection test for military pilots. *Journal of Occupational Psychology, 55*, 287-296.

Broadbent, D. E., Cooper, P. F., Fitzgerald, P. & Parkes, K. R. (1982). The Cognitive Failures Questionnaire (CFQ) and its correlates. *British Journal of Clinical Psychology, 21*, 1-16.

Broadbent, D. E., Broadbent, M. H. .P. & Jones, J. L. (1986). Performance correlates of self-reported cognitive failure and of obsessionality. *British Journal of Clinical Psychology, 25*, 285-299.

Broadbent, D. E. (1989). Personal communication.

Empson, J.A.C. (1987a) Error auditing in air traffic control. In J.A. Wise and A. Debons. (Eds.) *Information Systems: Failure Analysis*. Berlin: Springer-Verlag.

Empson, J. A. C. (1987b). Error auditing, workload and ergonomic assessment. *Bulletin of the British Psychological Society, 40*, A29-30 (abstract).

Fisher, S. & Hood, B. (1987). The stress of the transition to university: A longitudinal study of psychological disturbance, absent-mindedness and vulnerability to homesickness. *British Journal of Psychology, 78*, 425-441.

Langan-Fox, C. P. (1983). A preliminary investigation into the use of motor-program error audits in air traffic control tasks. Unpublished M.Sc. thesis, Hull University.

Langan-Fox, C. P. & J. A. C. Empson. (1985) 'Actions not as planned' in military air-traffic control. *Ergonomics, 28*, 1509-1521.

Miller, G. A., Galanter, E. and Pribram, K. H. (1960). *Plans and the Structure of Behavior*. New York: Holt, Rinehart & Winston.

Norman, D. A. (1981). Categorization of action slips. *Psychological Review, 88*, 1-15.

Norman, D. A. & Shallice, T. (1986). Attention to action: Willed and automatic control of behavior. In R.J. Davidson, G.E. Schwarts, and D. Shapiro (Eds.). *Consciousness and self-regulation. Advances in research and theory*. Vol. 4. pp. 1-18. New York: Plenum Press.

Rasmussen, J. (1980). The human as a system component. In H. Smith and T. Green (Eds.) *Human Interaction with Computers*. London: Academic Press.

Rasmussen, J. (1982). Human errors: a taxonomy for describing human malfunction in industrial installations. *Journal of Occupational Accidents, 4*, 311-335.

Rasmussen, J. (1986). *On Information Processing and Human-Machine Interaction: An Approach to Cognitive Engineering*. Amsterdam: Elsevier.

Reason, J. T. 1979). Actions not as planned: the price of automation. In G. Underwood and R. Stevens (Eds.) *Aspects of Consciousness* . Vol. 1. London: Academic Press.

Reason, J. T. (1988). Framework models of human performance and error: a consumer guide. In L.P. Goldstein, H.B. Andersen, and S.E. Olsen (Eds.) *Tasks, Errors and Mental Models*. London: Taylor and Francis.

Shiffrin, R. M. and Schneider, W. (1977). Controlled and automatic human information processing. II. Perceptual learning, automatic attending, and general theory. *Psychological Review, 84*, 155-171.

Stager, Paul. (1991). Error models for operating irregularities: Implications for automation. In: J. A. Wise, V. D. Hopkin, & M. L. Smith (Eds.) *Automation and Systems Issues in Air Traffic Control*. Berlin: Springer-Verlag.

Human Factors in ATC Communication

Ioannis Fokas

Aviation/Aerospace Consultant, Athens, Greece

Introduction

FAA's air traffic control mission is to promote the safe, orderly, and expeditious flow of civilian and military aircraft. Using information processed by computers and displayed on video screens at their workstations, air traffic controllers maintain the required separation between aircraft, provide safety advisories to pilots, and ensure efficient use of airspace (GAO, 1989).

An automated ATC should be able to handle effectively old-fashioned procedures, improved ones, and future concepts as well. Since these features are tied together with the human presence, it is very important that system design consider human weaknesses.

"Human factors" are the interaction of human beings with other humans or objects, which affect the output of work. The following paper is concerned with human factors in air traffic communication.

Issues in Air Traffic Communication

In reports covering 1978-1980, Billings and Cheaney found that nearly 75% of the incidents involved some type of information-transfer problem within the cockpit, among ground controllers, or between air and ground stations. Other reports in 1985-1986 showed 25% of incidents resulted from failures in air-ground information transfer. The most common communication problems in air-ground communications were attributable to pilot misunderstandings of ATC clearances or failure to remember the message (Lee, 1989).

The most critical component of human-machine systems is receiving and understanding voice communications. There are many causes of missed communications, including:

- Inattention
- Poor hearing
- Inarticulate speech
- Competing messages
- Loud ambient noise in the cockpit
- Extraneous noise
- Competing messages co-existing with improper use of electronic equipment (Gas-away, 1986)

NATO ASI Series, Vol. F 73
Automation and Systems Issues in Air Traffic Control
Edited by John A. Wise et al.

Since voice communications are vital to aircraft operations, each crewmember should receive structured training to ensure optimum success. This can be achieved by instruction in radio communications (Gasaway, 1986).

Any common method for enhancing communication between aircraft and ATC is insufficient until both communicators have a written statement as an additional verification, allowing the operators to communicate comprehensively.

By now, computer voice-recognition is available. Thus, both communicators (pilot and controller) can have the appropriate devices, and what they speak can also be printed out. Similarly, vital instructions to any aircraft (such ATC clearance) can be printed, in order to avoid any misunderstanding. If a print-out cannot be achieved, then at least a monitor in both cockpit and ATC station can display the message.

In many cases, a contributing factor to human error is false hypothesis or mistaken assumption. Usually, ATC messages are sent in a form of verbal shorthand; this is problematic in countries where English is not the native language. Additionally, sometimes part of the message is lost as a result of poor quality transmission or reception. In both cases, pilots and/or controllers tend to fill in the missing parts of the communication, based on past experience (Hawkins, 1987).

Fatigue or heavy workload is another contributing factor to human error. Pilots or controllers may use speech which does not express their thoughts exactly. This type of communication resulted a catastrophic accident on Avianca Flight 52 (1990). Because pilots and controllers did not communicate effectively, the fuel quantity was not perceived as an "emergency" situation (Cushman, 1990).

In spite of the inevitability of error, written communication could minimize aviation accidents, since both controllers and pilots could revise the dialogue. Golaszewski, in his research effort (1989), uncovered six generic types of communication errors:

- Incorrect pilot read back of ATC clearance (uncorrected by controller)
- Incorrect controller read-back of pilot communication (uncorrected by pilot)
- Pilot read-back of the clearance intended for another aircraft
- ATC instructions given to wrong aircraft
- Pilot non-compliance with ATC instructions
- Incorrect controller read-back of information from another controller (uncorrected by the other controller)

The most prevalent kind of read-back error identified was a situation in which the pilot read back incorrect information to the controller, and the incorrect read-back was not caught by the controller (Golaszewski, 1989).

Establishing a written communication procedure, and defining for that purpose a specific non-verbal communication channel, could alleviate many ATC problems.

Some advantages are as follows:

- Effective communications without read-back errors
- Minimized stress and workload for both parties
- Time saved from non message repetition (specifically with ATC clearance)
- More aircraft handled simultaneously, since ATC could accept more calls

Disadvantages lie in the facts that pilots may not begin to read and comprehend messages until they have been completely written (voice messages could be comprehended while the controller was still speaking), and that pilots may not read a message immediately if another task is in progress when the message is received (Hinton, 1988).

Air Canada is experimenting with a datalink system. The carrier expects to become the first to gain approval to accept live transoceanic clearances in data form. Since 1985, it has been conducting trials with Transport Canada in which Boeing 767s have received clearances via datalink and never received a corrupted message (Flight international, 1990).

In order for an aircraft to use a data-link system, it must be equipped with a Mode S compatible transponder supporting standard length uplink, downlink, and extended length message;, an airborne data link processor supporting the ATN router function, applications processing functions; and display and input devices (Weed, 1989).

Conclusion

Inevitably, both pilots and controllers will make mistakes along the way. That is one of the inherent weaknesses that go along with membership in the human race (Brenlove, 1987). The identification and the consideration of human factors in the modernized and automated NAS plan will minimize human errors. Written communication between ATC and aircraft can further minimize human error, and therefore should be considered to enhance aviation safety.

References

Brenlove, M. S. (1987). *The air traffic system: a common sense guide.* Iowa State University.

Cushman, J. H. (1990). How a full fuel load ran dry on fatal flight from Columbia to Kennedy. *New York Times.* February 5. pp. A12.

Flight International. (1990). Datalink. *Flight International,* March 21-27, .pp. 26

GAO. (1989, June). Air traffic control: Voice communications system continues to encounter difficulties. GAO/IMTEC-89-39.

Gasaway, D. (1986). Say again! I repeat, say again! *24th Annual Symposium Proceedings.*

Golaszewski, R. (1989). An analysis of pilot-controller read-back errors. *Journal of ATC.* pp. 54-56.

Hawkins, F. H. (1987). *Human factors in flight.* Brookfield, MA: Gower Technical Press.

Hinton, A. D., Lohr, W. G. (1988). Simulator Investigation of digital data-link ATC communications in a single-pilot operation. NASA technical paper 2837.

Lee, A. T. (1989). Data link communications in the national airspace system. *Proceedings of the 33th Annual Meeting of the Human Factors Society.* pp. 57-59.

Weed, A. D. (1989). Emerging air traffic control (ATC) datalink services. *34th Annual Air Traffic Control Association Conference Proceedings.* pp.360-361.

Results of a Perceptual Study on Human Error in Computer Systems Based on Bailey's Research

Esther E. Horne, Jo Lee Pierce, Douglas A. White

School of Library and Information Sciences, University of North Texas, Denton, Texas, USA

Introduction

Position

Human error in computer systems is often involved in system malfunctions. Human error can be reduced by understanding seven causal factors and their implications for system performance. These factors, as described by Bailey, are:

- Environment
- Human computer interface
- Organizational accuracy requirements
- Personal factors
- System design
- Training
- Written Instructions

Of these factors, there are three over which the system designer has no control. These are personal factors, organizational accuracy requirements, and environmental factors. Since these can only be affected in a positive manner through managerial policy and subsequent enforcement, it is critical to the long-term success of the system that management pay close attention to these factors.

Purpose

Initially the purpose for conducting this perceptual survey was to provide a meaningful learning experience during a doctoral seminar. Bailey's study of the role of human error in computer systems prompted the overall question--had the categories and relative percentages of error remained static since Bailey developed his classification scheme? Where, if any, were the changes?

NATO ASI Series, Vol. F 73
Automation and Systems Issues in Air Traffic Control
Edited by John A. Wise et al.

Seven Factors in Human Error

Figure 1 identifies and provides the relative importance of each factor. The identification of the factors is the result of collecting and categorizing errors made by users over a one year period.

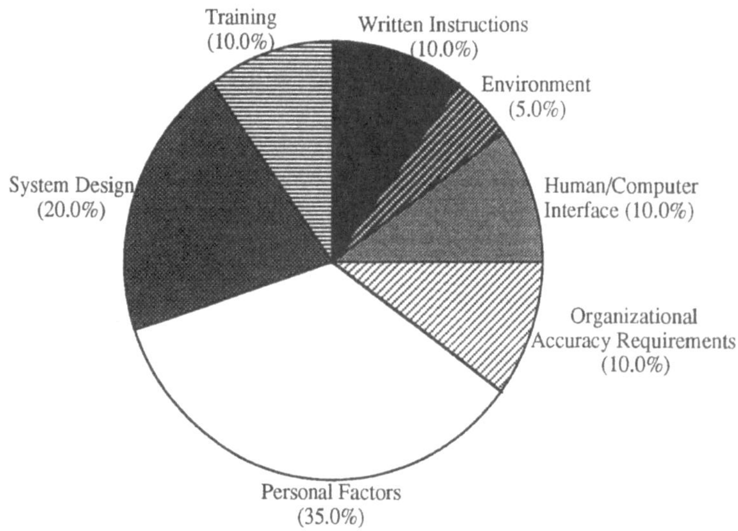

Figure 1. Relative importance of the causal factors, as reported by Bailey.

A brief description of the factors will put them into perspective.

1. *Environment.* These are factors over which the designer has little or no control: for instance, dim lights, room temperatures, noise, disruptions, annoyances, social isolation, overcrowding, and relationships between workers and managers.

2. *Human Computer Interface.* Areas of concern are CRT screens, keyboards, good command languages, masks on forms, and natural language codes.

3. *Organizational Accuracy Requirements.* These are not under the control of the system designer. People set their own personal accuracy criteria. Organizational requirements on data accuracy are informally communicated to operators who attempt to perceive what management feels are acceptable levels.

4. *Personal Factors.* Workers are not under the control of system designers. Actual workers are not the ideal worker; they have individual differences, making it possible to identify high error makers. Users have built-in habits, physiological needs, and psychological

needs. Other personal factors that affect performance are sleep loss, drugs, illness, body rhythms, fatigue, motivation, etc.

5. *System Design.* These refer to any overall system design condition that may ultimately reduce the reliability of the human component and increase the probability of human error. For instance, good function allocation takes into consideration the capabilities and limitations of people and computers. Some problem areas are excessive demands, forcing people to perform faster than they are accustomed, insufficient time, lack of feedback, inappropriate error messages, and inappropriate and inefficient assignment of work tasks.

   According to Bailey, an indication of poor system design is a high error rate. One obvious solution is to design systems that minimize the potential for error among operators. Another solution would be a requirement for high accuracy levels with continuous follow-ups or reviews of system performance.

   More specifically, three major sources of error are:
   1. Errors inherited by the system (those that already exist in the input data)
   2. Errors made by people in the system
   3. Errors made by the computer (software and hardware malfunctions)

6. *Training.* Inadequate training can be costly. Psychology offers a set of principles for developing training materials. Operators need centralized computer-based training, as well as testing for predetermined performance levels and refresher courses.

7. *Written Instructions.* Performance aids are much needed and should store information, such as online help, for ready use. If operators do not have good written instructions, they will ask, guess, or develop their own. Good instructions should be concise ,yet not too brief, easily understood, and indexed for easy retrieval.

Hypothesis

1. Findings from our perceptual study will support the findings of Bailey; that is, the seven factors and their relative importance will be maintained.
2. Managers and computer operators will not differ in their perceptions as to the seven factors and their relative importance.
3. Corollary hypothesis: The years of experience between computer operators and their managers will not differ.

Methodology

Subjects

Fourteen computer operators and their respective managers participated in the study, making a total of twenty-eight participants. For the most part, the subjects came from the students' places of employment. Since the backgrounds and employment patterns of the students were so varied, a wide range of settings was possible for random sampling.

Instrument

The instrument was a questionnaire listing the seven factors and accompanying scope notes (see Appendix A for the complete questionnaire). To eliminate possible bias we listed the factors in alphabetical order and did not inform the respondents of Bailey's findings. The respondents could use an "other" category to describe an error which in their opinion did not fall within Bailey's list.

Instructions

In completing the questionnaire:

- The computer operators were to assign percentage values to each category in accordance with their perceptions of the relative importance of each factor as a contributor to error in their work.
- The managers were to assign percentage values to each category in accordance with their perception of the relative importance of each factor as a contributor to error in the computer work of those whom they supervised.
- In all cases the percentage values were to equal 100%.
- The subjects were to provide background information concerning their personal experience in computer work and the nature of their organization.

Categorization of Data

We sorted the responses into groups of managers and operators, maintaining the pairing of operators with their respective managers. We checked each questionnaire to determine whether the totals for the percentages equaled 100%; the results of two questionnaires did not equal 100%, so we normalized this data. We then entered the data into a spreadsheet, listing the results of each manager, each operator, and Bailey under the appropriate factor.

Statistical Methods and Results

We used a software package called SPSS/PC to generate t-tests (Table 1), paired t-tests (Table 2), and produce one-way analysis of variances (Table 3) on the data which compared the responses of the computer operators with their respective managers for each of the seven factors.

The one-way analysis of variance provided a confidence interval which we used to correlate the results of the operators and managers with Bailey's findings (Table 4).

Table 1. Two Tail T-test: Separate Variance Estimate

Separate Variance Estimate

	F value	Probability	T value	Probability
Environment	2.05	0.208	0.24	0.816
Human Computer Interface	1.28	0.665	0.96	0.345
Organization Accuracy Requirements	9.09	0.000	1.11	0.288
Personal Factors	1.03	0.962	1.15	0.260
System Design	1.61	0.402	1.13	0.268
Training	1.10	0.862	0.33	0.746
Written Instructions	2.50	0.112	0.72	0.477

Table 2. Paired T-test on Causal Factors

Difference Correlation T	Means	Coefficient	Probability	T value	Probability
Environment	0.5807	-0.248	0.393	0.21	0.835
Human Computer Interface	3.2807	0.771	0.001	1.98	0.069
Organ. Accuracy Requirements	4.0171	0.025	0.931	1.12	0.283
Personal Factors	9.4986	0.555	0.039	1.72	0.108
System Design	3.1364	0.053	0.857	1.16	0.265
Training	1.3221	-01.58	0.590	0.30	0.768
Written Instructions	2.8386	0.519	0.057	0.99	0.339

Table 3. One-way Analysis of Variance

.95 Confidence

	Interval	Ratio	Probability
Environment	4.60 - 9.57	0.0554	0.8157
Human Computer Interface	7.37 - 14.36	0.9237	0.3454
Organizational Accuracy Requirements	3.93 - 11.38	1.2350	0.2767
Personal Factors	28.45 - 45.49	1.3230	0.2605
System Design	6.15 - 11.86	1.2851	0.2673
Training	12.28 - 20.51	0.1051	0.7483
Written Instructions	8.03 - 16.01	0.5232	0.4759

Table 4. Comparison of Means and Combined Confidence Intervals for Managers and Operators with Bailey

	Managers	Operators Means	Confidence Intervals Means	Bailey
Environment	6.79	7.37	4.60 - 9.57	5.00
Human Computer Interface	9.23	12.51	7.37 - 14.36	10.00
Organization Accuracy Requirements	5.65	9.67	3.93 - 11.38	10.00
Personal Factors	41.72	32.22	28.45 - 45.49	35.00
System Design	7.44	10.57	6.15 - 11.86	20.00
Training	15.73	17.06	12.28 - 20.51	10.00
Written Instructions	13.44	10.60	8.03 - 16.01	10.00

Table 5. Years of Experience: T-test and Paired T-test

	Sample	Means	Difference of means	T value	Probability
T-test					
Managers	14	7.7143	not reported	1.94	.064
Operators	14	5.6071			
Paired t-test					
Managers	14	7.7143	2.1071	1.94	.075
Operators	14	5.6071			

Table 6. Organizations Participating in Study

Type	Number	Type	Number
Corporate	3	Governmental	2
Educational	2	Manufacturing	3
Financial	2	Non-Profit	2

We also computed the average number of years of experience for both the computer operators and managers, and applied the t-test and paired t-test to the data (Table 5).

Wide sources of work settings were available because of the varied backgrounds of the students. Table 6 depicts the variety and number involved in the study.

Discussion of Findings

The findings of the perceptual study support the findings of Bailey, in that no new category emerged. However, the relative importance of two of the factors, systems design and training, differed from Bailey. Since the data from Bailey's study were not available, the pie chart percentages (Figure 1) were compared to the confidence intervals of the managers and operators. Based on Table 7, the following conclusions are evident: (1) the relative importance of system design appears to have declined, and (2) the relative importance of training appears to have increased.

Table 7. Comparisons of Bailey with Confidence Intervals for Managers and Operators

	Bailey	Managers	Operators	Combined
System Design	20	3.74 - 11.13	5.88 - 15.27	6.15 - 11.86
Training	10	9.66 - 21.80	10.67 - 23.44	12.28 - 20.51

Figure 2 (below) depicts the sample means for managers and operators, as compared to Bailey's findings. It is evident that personal factors are still the largest cause of human error.

Based on the results of the t-test, the paired t-test, and the one-way analysis of variance, the null hypothesis of no difference in perception between managers and operators failed to be rejected.

In the years of experience, the results of the t-test and paired t-test showed no meaningful difference between operators and managers. Their perceptions would not be affected by experience. Nevertheless, a variable seems to be intervening, and that variable may be one of position. The difference we expected did not surface directly for the group as a whole but seems to be surfacing because of the pairing. A sufficient number of pairs seemed not to be in agreement.

A regrouping of the test results reveals a deviation in the probability values of the paired t-test from the probabilities for t-test and one-way analysis of variance for some factors, specifically human/computer interface (.069), personal factors (.108), and written instructions (.339) (Table 8). This effect may be due to differing perceptions within pairs, brought on by position.

In addition, other interesting effects can be seen. Examination of the correlation coefficients (Table 2) revealed negative values for two factors, training (correlation coefficient: -0.158, level of significance: 0.590) and environment (correlation coefficient: -

0.248, level of significance: 0.393). Human/computer interface factor shows a correlation coefficient of 0.711 but .001 probability. These findings also seem to indicate that some differences in perception may be occurring between certain of the pairs.

Further treatment using a larger sample might reveal whether this assumption is statistically meaningful.

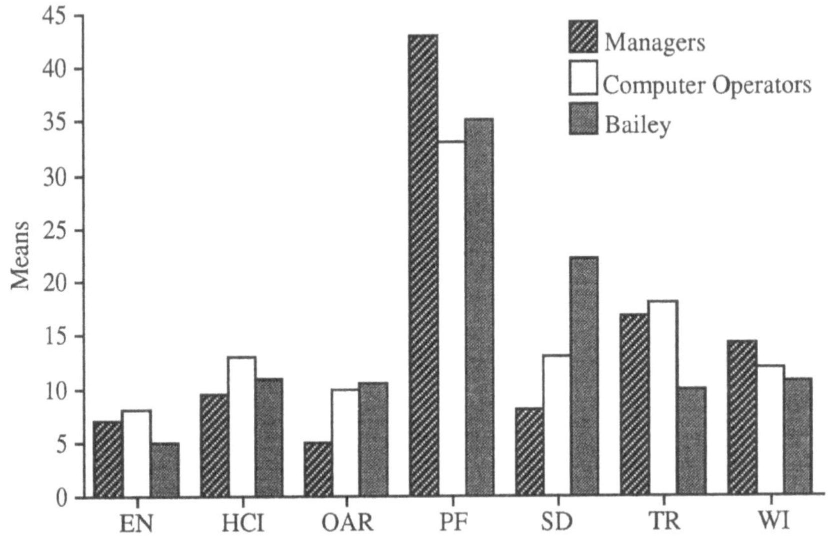

Figure 2. Comparison of the relative importance of the causal factors for managers, computer operators, and Bailey.

Implications

This perceptual study, while constrained by time (one semester), has been able to demonstrate that human error and the seven factors have altered little. Retrievable errors are annoying and time consuming, but irretrievable errors are often disastrous. System designers should be able to reduce human error where they have control.

More research would be helpful in the following areas over which the decision makers and systems designers have control:

Decision makers:

1. Organization Accuracy Requirements
2. Personal Factors
3. Environment

System designers:

1. Human Computer Interface
2. Systems Design Factors
3. Training
4. Written Instruction

Table 8. Comparison of Scores and Their Probabilities

Statistic (One way analysis)	Score	Probability
Environment		
T-test	0.24	0.816
Paired t-test	0.21	0.835
Human Computer Interface		
T-test	0.96	0.345
Paired t-test	1.98	0.069*
Organizational Accuracy Requirements		
T-test	1.11	0.283
Paired t-test	1.12	0.283
Personal Factors		
T-test	1.15	0.260
Paired t-test	1.72	0.108 *
System Design		
T-test	1.13	0.268
Paired t-test	1.16	0.265
Training		
T-test	0.33	0.746
Paired t-test	0.30	0.768
Written Instructions		
T-test	0.72	0.477
Paired t-test	0.99	0.339 *

* Deviations

Summary

What began as a learning experience ended as a research project which we hope fills a need to acquaint or re-acquaint system designers and their respective decision makers with the importance of Bailey's seven causal factors in human error. Hardware and software are becoming more sophisticated, but what of the human in the system? Should system design computerize as many human functions as possible? This has been the trend. But are the functions of the system designer and decision maker to be computerized as well? Since personal factors are seen as extremely important to the successful performance of a system, how much do we understand, or need to understand, of the perceptual and cognitive functions which are as yet to be automated? Can the human be entirely eliminated from the system? If not, the problem stands.

Acknowledgments

The authors would like to recognize the efforts of the members of the research team.Without their efforts, this paper would not be possible. They were: William Robert Miller (Project Leader), Cecilia Armas, Kyungmi Chun, Jimmy-Jack Jackai, Michael Marsh Lauch, Susan J. Page, Iljong Park, Jo Lee Pierce, Rachelle Howarton Smith, Zelina Urias-Barker, and Douglas A. White.

References

Bailey, R. W. (1983). *Human error in computer systems*. Englewood Cliffs, NJ: Prentice-Hall.

Appendix A

HUMAN/COMPUTER ERROR SURVEY

Name of organization:________________________________

Nature of organization: (please circle one or indicate other)
Corporate
Educational
Financial
Governmental
Manufacturing
Non-Profit
Other: ________________________________

Years of computer experience:________

Average number of hours per week you spend at the computer: ______

Of the errors that are encountered in computer work, what percentage, in your opinion, are attributable to the following factors? If you believe some error is due to factors not encompassed by the following categories, please describe as "Other" on the bottom line. Please try to make sure that your figures total 100%.

% Environment: (lighting, noise, temperature, furniture, work space, social environment)

% Human/Computer Interface: (input/output devices, keyboard, CRT, prompts, responses, codes, command language)

% Organizational Accuracy Requirements: (acceptable/unacceptable error limits, reporting, control)

% Personal Factors: (skill requirements, behavior, physiological and psychological needs, motivation, fatigue)

% System Design: (integration of components, feedback, function allocation between human/computer, processing time)

% Training: (formal, on-the-job, computer-assisted, specialized)

% Written Instructions: ('help' screens, accuracy, up-to-date, clarity, documentation, specifications)

% Other:

Appendix B

Journals Citing Bailey (1983)

An online search in Scisearch and Social Scisearch revealed Bailey has been cited for his work, Human Error in Computer Systems, in the following professional journals:

- *Communications of the ACM*
- *Computers & Industrial Engineering*
- *Data Base*
- *Information & Management*
- *Information Age*
- *International Journal of Man-Machine Studies*
- *Management Science*
- *Omega-International Journal of Management Science*

Other Works Written by Bailey

Bailey, R. W. (1972, May). A classification scheme for human error in computer-based business information systems. Bell Laboratories Report.
Bailey, R. W. (1973, February). The reliability of human performance. Bell Laboratories Report.
Bailey, R. W., Demers, S. T., & Lebowitz, A. I. (1973). Human reliability in computer-based business information systems. *IEEE Transactions on Reliability* 22(3). 140-148.
Bailey, R. W. (1974, December). Handprinted characters and human error. Bell Laboratories Report.
Bailey, R. W., Blank, R. G., & Walker, J. T. (1975, July). A study of human error in the operation of the TIRKS F/ECP-1 data system. Bell Laboratories Report.
Bailey, R. W. (1978). The effect of stimulus-duration and error-stage on the ability to self-detect errors. (Doctoral dissertation, Rice University. Dissertation Abstracts International, 39(3-B) 1510.
Bailey, R. W., & Desaulniers, D. H. (1978, March). Directory error analysis. Bell Laboratories Report.
Bailey, R. W. (1979). Effect of stimulus-duration on the self-detection of errors in a modified typing task. *Ergonomics, 22*, 585.
Bailey, R. W. (1982). *Human performance engineering: A guide for system designers*. Englewood Cliffs, NJ: Prentice-Hall.
Bailey, R. W. (1983, May 2). Human performance engineering: Looking through the user's eyes. *Computerworld*, pp. In Depth 1-6.
Bailey, R. W. (1983). Human factors technology transfer. In Proceedings - IEEE Computer Society's 7th International Computer Software & Applications Conference, *COMPSAC 83*. New York: IEEE.
Bailey, R. W. (1989). *Human performance engineering; Using human factors/ergonomics to achieve computer system usability (2nd ed.)*. Englewood Cliffs, NJ: Prentice Hall.

Organizational and Cultural Aspects of Air Traffic Control Automation

Automation, Information and Consciousness in Air Traffic Control

Ron Westrum

Interdisciplinary Technology Department., Eastern Michigan University, Ypsilanti, Michigan, USA

Introduction

Successful air traffic control requires two things: the management of information and the management of consciousness. Both are important, but they are different. Automation can improve one or it can improve both. Human consciousness and human information-processing ability are limited, whereas, in practical terms, the span of attention of automation and its ability to process information is unlimited. Hence automation can greatly improve information flow to human awareness, but this can also overload it. On the other hand, if human consciousness is not part of the loop — i.e., in a completely automated system — it is quite possible for the machinery to process accurately a much larger volume of information. That is, as long as the program for the computer is correct, an important (and not necessarily correct) assumption.

There are two parts to this paper. In the first I will discuss operational issues associated with human information flow, and how automation might affect this flow. In the second part I want to consider some of the larger organizational issues associated with using automation and with organizational change.

Information Flow

Consider the destruction of Avianca flight 52, whose fuel tanks ran dry over Long Island, New York, on January 25, 1990 (Cushman, 1990). The destruction of this aircraft can be laid almost entirely to "communication breakdown." The pilots did not communicate their degree of danger effectively to the ground; the controllers did not communicate well with each other. The destruction of Avianca 52 would seem to be merely a matter of information flow: people did not transmit the knowledge they had. But the reasons for this failure have to do with the nature of consciousness, and its management by the air traffic control system.

Think about air traffic control as a series of interlocking bubbles, each representing the sphere of consciousness of an air traffic controller. Within each sphere, airplanes are identified by a number of parameters: their type, altitude, amount of fuel, and so forth. Between the spheres, communication is tricky. We know from other studies that every time an object passes from one sphere of consciousness to another, there is likely to be a loss of

NATO ASI Series, Vol. F 73
Automation and Systems Issues in Air Traffic Control
Edited by John A. Wise et al.

information, whether the object is a hospital patient or an airplane (Metz, 1981, pp. 154-159). Consider the task of handing off an airplane to another controller: what does the controller acquiring the airplane need to know? In the case of Avianca 52, hand-off was not well handled; key information was lost as the co-pilot talked to the ground and regional controllers handed the plane off to local controllers.

Human consciousness is amazingly finite. When it is overloaded, when mental workload becomes too high, it is unable to manage well. This is what happened with Avianca 52. The plane would no doubt have successfully landed if:

1) bad weather had not closed off runways at the destination, Kennedy International Airport;

2) and therefore the controllers had not had to manage so many planes in the air;

3) the ground controllers had been more able to probe the seriousness of the situation.

One immediately thinks: What could automation have done to help? For instance, could the plane's fuel reserves have been entered into a computer at some point, which would have been able to compute how long the plane could continue to circle? Could an expert system, through extrapolation, have noted that the plane needed to land soon, since a computer would not suffer from information overload as the humans did? Or would the computer simply have provided additional information which would have distracted the already-overloaded humans even further?

We have passed quickly over these several issues. Now we must examine them in more detail:

- What is consciousness?

- How do consciousnesses interpenetrate?

- What is information overload?

- How can automation help with information overload?

What Is Consciousness?

Avoiding metaphysics, I will simply say that consciousness is made up of the field of those objects of which we are aware, to which we can pay attention. How sharp this attention is, is partly a function of our physiological and psychological state, and partly a function of how much we have to pay attention to. It might be useful to think of consciousness as something like a blackboard—the implication is that only so much can be put on the blackboard before something needs to be erased. We choose to pay attention only to certain objects, while others do not get this attention.

In a study of aircraft carrier landings, investigators from the Institute for Governmental Studies at Berkeley noted that their naval informants talked about "the bubble." This was the sphere of awareness a commander had about what was going on with the whole fleet or the whole air wing at a given time (Laporte, 1990; Laporte, Rochlin, and Roberts, 1987). Interestingly enough, one could only "have the bubble" for so long, and then it needed to be taken over by somebody else. Command needed to be handed-off when the commander started to "lose the bubble." One wonders if there is a similar problem with air traffic controllers.

With occupational experience, often very small cues are necessary to allow one to recognize what is going on in a situation. A key function of occupational jargon is to organize information into appropriate chunks, and to make key distinctions which allow more graceful handling. Thus air traffic control at its best involves a "flow state" indicating that the controller is in control of the information necessary to manage the objects in question. Good air traffic controllers, like chess masters, probably use gestalts to a greater extent than do the less experienced and able controllers.

Automation can contribute to this grasp of the situation, or it can interfere with it. The key point is that the controller must be able to "fly the situation" just as the pilot is able to fly the plane. Thus automation must provide information (whether on past, present, or future states of the system) which synchronizes well with the controller's flow (Csikszentmihalyi and Csikszentmihalyi, 1988). If it does not synchronize, it may well place the controller outside the "optimal experience" zone and become a significant stressor.

But the controller must be "in the loop." If the controller has too little to do, he or she may be flying "alongside the loop." This is dangerous. It may promote a feeling of security when in fact the system needs intervention. Recently a study showed that with advanced fetal monitoring, doctors lost more babies: the high tech equipment gave them a false sense of control. Doctors who used stethoscopes, on the other hand, tended to intervene more readily and thus saved more children's lives[11]. One of the important skills of controllers is knowing when and how to intervene.

Consciousness also has a very interesting aspect in relation to action: one tends to become conscious only of what one can do something about. Thus two corollaries:

1) when the capacity to act is limited, consciousness will be restricted also;

2) increasing the ability to act increases the willingness to recognize problems and opportunities.

One wonders if 1) had some bearing on the Avianca 52 co-pilot's failure to inform the ground that there was an "emergency," by using that word in the transmission. In an "emergency" condition, the plane would have been cleared to land immediately. Could the co-pilot have failed to transmit the word because he thought it would make no difference? Did the crowding of the skies around Kennedy lead to a sense of futility and thus a fatal unwillingness to impose?

This observation has an important relation to automation. If automation can provide more options and opportunities for action, controllers may be willing to stretch their consciousness further. But the system is also likely to stretch itself further, until the limit of

11. This story appeared about March 4, 1990, in a large U.S.A. national newspaper. I regret the lack of a specific reference.

the controller's ability to control is again reached. If automation makes the controller's task easier, will not additional tasks be loaded on?

How Do Consciousnesses Interpenetrate?

We can say that two minds interpenetrate to the extent that they are both aware of the same things in the same way. The obvious image is overlapping circles. There are two ways that this can happen, both important to air traffic control: common knowledge and communication.

Common knowledge can come about through general occupational experience. Anyone who has been an air traffic controller for many years will automatically know or assume certain things. It can also come about through specific personal knowledge of the other person's habits of thought. When we work with a specific individual, we come to know their strengths and weaknesses, and can often intuit what the other person is thinking (or not communicating), even when they are silent. Teams tend to operate on this principle, often operating like a well-oiled mechanism because of constant work together. One of the striking points brought out at the conference was the value of controllers' being able to confer off-the-air, so that joint responsibilities could be handled with a better awareness of counterparts located in a remote control center.

Good communication depends critically on a sense of what the other person needs to know, and this is often a difficult thing to appreciate. Willingness to communicate is often a function of knowledge of the other person, and the more organizational, status, language, and ethnic barriers intervene, the less adequate communication is likely to be.

The "hand-off" issue is likely to involve a serious communication problem with any technology. For instance, in a hospital, when a patient is passed from one set of caretakers to another, there is often a severe loss of key information. This problem is even more severe when the patient crosses the hospital boundary itself, as when the patient arrives by ambulance (Metz 1981, pp. 154-159). Fortunately, the air traffic control system is usually not this bad, but there are problems.

For instance, in what might be considered one of the first studies in this area, a sociologist at a remote airbase in England in 1941 tried to discover why the base had a high accident rate (Patterson, 1955). Upon investigation, it turned out that this accident rate was caused by pilot attitudes, and by relations between pilots and ground controllers. The sociologist manipulated pilot attitudes through a subtle conversational campaign. He also worked on changing attitudes through cross-training, by placing pilots and controllers in each other's shoes. Pilots sat in the ground operations center and watched what went on, and controllers trained in the Link trainers. As a result of this cross-training, there was much better co-operation and many fewer accidents. Thus, when the sphere of pilot and controller interpenetrate more, their ability to jointly accomplish the mission is very much improved. One of my colleagues who teaches aviation management tells me that such cross-training is much more rare today than it should be.

This also suggests another basic issue, which is the proper spread of consciousness. While we want people involved in a high tech situation to have some common elements (i.e., they should "'communicate," they need to "understand each other"), they must also necessarily have separate elements, because there must be a division of consciousness---otherwise everyone will be managing the same task, and key functions will be neglected. Just as, in "cockpit resource management," we must make sure that every task that needs managing has somebody managing it, so in air traffic control we must be sure that everything that needs watching is being watched by somebody. This is an area where automation might help; it can be an extra watcher. It can also hinder, by overloading controllers: it may provide something else that needs watching.

Information Overload

We can define "information overload" as a situation in which those managing a technological system can no longer react with the speed and sophistication necessary to assure safe operation (Miller, 1978, pp. 121-202). Such overloading can occur in a variety of ways. Most commonly, it occurs when an individual or a group is faced with an unplanned increase in the inputs they must process, or with a decrease in the resources they have to cope with the inputs. When such overloads occur, the system is likely to adjust in several ways (Miller, 1978):

1) errors are more likely to occur;

2) problems that need actionswill be queued up until they can be addressed;

3) some problems will be omitted;

4) some problems will be delegated elsewhere;

5) there may be a synthesis, or "chunking" of the inputs to provide more efficient processing.

6) Along with this increased load, stress on the members of the team is likely to increase. At some point, the system may break down or have its overall effectiveness reduced as a result of the stress.

7) One might add another item to this list: failure to inquire about evidently hazardous conditions.

The first six mechanisms, error, queueing, omission, delegation, abstraction, and breakdown are all easy to interpret in the context of air traffic control, and all threaten the safety of aircraft. Pilots can be given bad directions, late directions, no directions, be sent elsewhere, become a mere symbol, or can experience extreme danger as the ground control on which they depend disintegrates. The value of ATC dialogue is degraded when the system is stressed (Kreamer, 1987).

With Avianca 52, overload of the pilots obviously led to omission of key data the ground needed. One suspects also that cultural barriers might be involved here. Regional controllers omitted pertinent facts when communicating with local controllers. Queueing was also a serious problem: the controllers had a lot on their minds, and so they stacked up planes until they could land them on the limited runway space. We also note that one of the controllers repeatedly asked Avianca 52 about its alternative destination, a sure sign that the controllers did not want to recognize the seriousness of the situation. This again affirms the postulate that a system's consciousness is unlikely to want to know about what it cannot handle.

The seventh mechanism is a restriction of the willingness to inquire. This is often a direct function of how much additional mental workload the controller can handle. Controllers who are constantly under high stress often do not have time to worry about situations that are not immediately obvious. Information overload can be thus be seen to occur not only at the level of the individual controller or pilot, but as a property of the system as a whole. It is also notable that it is the more experienced controller who is more likely to inquire about the invisible. The knowledge gained by long experience needs to be appreciated, and assisted. Does automation assist the ability to inquire? Or does it presume in advance all the knowledge the controller needs to have?

How Can Automation Help?

Perhaps a better question should be: What needs to be done? Clearly the consciousness of the system needs to expand to cope with more airplanes, greater complexity, and increasing demands. The appropriate question is how a computer might help in this expansion of the system's consciousness, rather than simply increasing information flow. Here are some hypothetical answers:

1) The computer could manage the whole ATC system, with humans monitoring the computer. At present, this seems unlikely. So far, no one wants to fly on a system completely managed by computer, and like the "Star Wars" weapons systems, there is legitimate concern that the software would have some catastrophic glitches. When budgets are cut, won't someone make the argument that, since machines can do it more cheaply, why involve error-prone and fractious humans in air traffic control?

2) Automation could assist through better abstraction. It could combine more aircraft variables than presently, thus providing richer displays. This would continue a long-term trend toward presenting the controller with more information (Philips, 1985). However, it might be possible to use automation to identify, and focus the controller's attention on, situations which require immediate action. The AERA-II system discussed by Jack Edwards (1991) is a system of this kind.

3) Controllers could provide more assistance to the cockpit. Computerization potentially can provide not only more assistance to the controller on the ground, but also to the flight deck. Expert systems could identify courses of action, project scenarios, and evaluate options to pilots as well as controllers.

4) Automated systems could also fly the planes. This would provide for much higher system capacity. It would also very sharply increase the ante in terms of an accident, caused by a programming error or otherwise.

5) Computers might assist in the hand-off function. Can computers assist in providing more information transfer when hand-offs take place?

These changes would have quite different implications, and it ought to be clear that automation can mean very different things. So far the aim of automation seems to be to increase information flow and also to abstract variables and to highlight situations that are of particular importance. How far can automation safely go? One wonders, however, how such changes might lead to the system getting further out on an automated limb. As Gene Rochlin has pointed out, automation has sometimes neglected the important issues involved in protecting high-tech systems (Rochlin, 1986). John Levesley (1990), a trainer of air controllers, noted that there already seems to be a "loss of airmanship." By this he meant two things. The first was that many of the key decision-making and information-sharing skills were starting to atrophy as automation increased. The second was a change in attitude, similar to what has been called "the Peach Bottom Syndrome." The reference is to the notoriously-careless nuclear reactor operators at the Peach Bottom reactors near Philadelphia. These operators would often be found sleeping, for instance, with the reactor on at full blast.[12]

One of the key issues here is standardization. Once a standardized system is imposed for hardware---necessary for getting economies of scale---will there not be a tendency toward passivity on the part of the operators? What kind of input will local controllers have regarding automation of their own location? Some previous studies have suggested, for instance, that automation which is locally customized by a high-performance team may perform in a superior manner to that which is standardized between locations[13]. A reasonable prediction might be that the system will breakdown at some point. If so, what kind of team will be most likely to be able to handle it? The obvious answer would seem to be, the one which did not treat its computers as a "black box."

Another key issue to be considered relates to software for highly automated systems. Earl Wiener (1987) and others have identified the potential for software or programming "bugs" to cause serious accidents[14]. As more of the ATC system becomes automated, human error in the programming and design phase may bulk larger. Human input of information into automated systems, unavoidable at some point, can always be faulty. Thanks to the abstract nature of the data to be entered, it should not be surprising if digits are misplaced or omitted. Such small omissions could have large consequences indeed.

Questions about Teamwork

One of the key questions that this kind of inquiry raises is: What constitutes a high

12. Note also some of the early observations of Nehemiah Jordan (1968) about the negative effects of automation on morale in the SAGE radar system in the 1950's.

13. Marcus and Fox (1988) showed that operations that customized their own safety procedures had fewer "incidents." This is also true of manufacturing operations. There is generally greater customization for "high performers."

14. Wiener (1977) notes that "Over-reliance on automatic devices, and underloading of the operator may lead to complacency and errors during those critical times when human judgment and intervention are required." (p. 176).

performance team in air traffic control? Simply as important background information, it would be valuable to know some of the differentiators of high- and low-performance teams in the ATC role. What can we learn from studying high-performance in ATC? What mechanisms are available to share the fruits of such knowledge, if we generate it? We all now recognize the value of "cockpit resource management." Is something similar needed for controllers? Or is there some analogous group dynamics problem among controllers, so far undetected, which needs to be addressed in a similar manner? It was obvious from the comments of the controllers at the conference that these teamwork skills are often carefully taught, monitored, and appreciated in the ATC operational community.

Another issue is the nature of the new controlling technology in fostering or hindering teamwork. Does the technology encourage collaboration between controllers and pilots in moving aircraft through the air? Between controllers? Since the advent of "groupware" allowing groups to use computer technology, it would seem an open question what applications such groupware might have for air traffic control. Patterson's study suggested that a problem existed about the extent to which pilots and controllers could understand each other's role. There again appears to be some need for additional cross-training in the current aviation environment. Whenever major differences in status exist in a technological environment, they are likely to become barriers to information flow and to co-operation. I wonder if controllers have ever been involved in CRM-style exercises with the pilots?

Organizational Problems in ATC

Just as the management of attention is critical in ATC operations, it is also critical in the organizations that support ATC operations. In designing for the 1990's and beyond, organizations will "push the envelope" in terms of the tasks they are called on to manage. To provide organizations equal to the tasks should be a major goal for us. There are many spheres in which organizations will be asked to perform, but I would like to select, again, the question of information flow.

We can distinguish three rough levels of organizational functioning in relation to information flow: pathological, bureaucratic, and generative. To put this in ordinary language, we could call them "sick, normal, and healthy." An extended definition might be as follows:

Pathological. These organizations are not adequate in handling information flow, even under normal conditions. These organizations actively block information flow, usually due to political infighting. For these organizations, information is highly political, therefore a scarce resource, control of information is the subject of intense political battles. Often information is used to promote or punish individuals, rather than to carry out the organization's task.

Bureaucratic. Such organizations tend to operate systematically "by the book." These organizations do a reasonable job as long as they are not under stress. Information flows according to accepted channels, and as long as the problem is the one for which the organization was designed, they will perform adequately. Stress, novel technology, or new problems will, however, quickly remove these organizations from the zone in which they operate successfully. Often they can only be reformed from the top.

Generative. These organizations have the ability to self-organize. They are highly resilient, because they use their intellectual resources more effectively. They typically exhibit a high degree of leadership and creativity, and may respond to new problems in singularly effective ways. They regularly set goals for themselves beyond usual expectations, and fulfill them because they do unexpected things in unexpected ways.

There are two major criteria which strongly separate these levels of organizations: honesty and permeability. Let us discuss honesty first.

Organizational honesty is willingness to confront problems. If an organization discovers a problem or senses an opportunity for innovation, what does it do? I would suggest a spectrum of six responses, which could be expanded or contracted depending on one's needs:

Suppression. This is the familiar pattern of "shooting the messenger." Those who point out problems or opportunities are punished and the information buried.

Encapsulation. Here nothing particularly bad happens to messengers, but they are ignored, isolated, and otherwise made harmless. "That's just Jones's opinion" is a common expression of this tactic.

Public Relations. In this tactic, the observations emerge, but are watered down by explanatory or deprecatory information. The significance of the ideas or observations are thus very much reduced.

Local Fix. The problem is fixed, or the opportunity is responded to, but not with gusto. Wider implications or implementation are denied.

Global Fix. The problem is admitted to be global, and global action is taken. The innovation is admitted to be worth risking, and substantial resources are devoted to it.

Inquiry. Action on the problems leads to consideration of basic system defects, and inquiry is initiated to reform the system or to change it to respond to the novel opportunities presented by new technology.

Organizations often show a consistent response pattern in use of these tactics, and organizational culture tends to institutionalize the pattern through example, myth, and anecdote. To long-term organization members, the pattern, no matter how pathological, seems "the natural thing to do."

Pressures to Ignore Problems

In a previous paper, I have spelled out some of the ecological pressures against good information flow (Westrum, 1988). Organizations like to think about only what they can do something about. Thus, when organizations feel relatively powerless, they are often unwilling to confront problems. Significant pressures against free information flow can come from many of the factors listed below:

1. Threats of regulatory action

2. Economic pressures for high output rate

3. Declining resources or funds

4. Low organizational morale or internal struggles

5. Arbitrary or unjust actions by organizational managers

The last may seem surprising, but consider that organizations in which arbitrary actions are common are likely to "shoot their messengers." This tends to make potential messangers nervous. Indeed, consider this pattern of action on the part of organizations functioning at the three different levels of competence shown in Figure 1.

Thus, the same behavior that is welcomed by generative organizations may be severely punished by those less adequate in functioning.

I would suggest that the many observations in Figure 1 could be summarized by a single criterion: how able is the system to use observations and knowledge within it? Clearly, the more adequate a system is, the more it will tend to behave according to the following rule:

The organization is able to make use of information, observations, or ideas wherever they exist within the system, without regard for the location or status of the person or group having such information, observations, or ideas.

By contrast, in the less adequate organizations, status, privileges, cultural barriers, and organizational politics will all conspire to restrict information flow. Consciousness, in the latter case, will always be partial and restricted. Unpleasant surprises would be a natural expectation for such an environment, and many of them might come from accidents or other major system failures.

It is obvious that consciousness will be wider in organizations that are more generative. Since information is inherently political in pathological organizations, awareness also will be restricted. One application of this observation is that in organizations which function better, more people know what is going on. Thus more people are likely to have sufficient knowledge to avert dangerous situations that they observe, and will feel free to act on their observations. In less capable systems, knowledge of restricted consciousness will result in a greater reluctance to act and to take responsibility. Willingness to inquire, an important feature of the adequate organization, will generally be lower in pathological organizations; and when inquiry takes place, it will often have a judicial rather than an inventive character — to punish rather than to improve.

Pathological	Bureaucratic	Generative
Don't want to know	May not find out	Actively seek information
Messengers are shot	Mesengers are listened to if they arrive	Messengers are trained
Responsibility is shirked	Responsibility is compartmentalized	Responsibility is shared
Bridging is discouraged	Bridging is allowed but neglected	Bridging is rewarded
Failure is punished or covered up	Organization is just and merciful	Organization encourages inquiry and redirection
New ideas are actively crushed	New ideas present problems	New ideas are welcomed

Figure 1. How Organizations Treat Information

How Can Positive Organizational Culture Be Cultivated?

While it is fascinating to examine the comparative sociology of these kinds of organizations, clearly our task is to ask, especially in view of coming ATC automation, how we can get the new systems to move toward more adequate functioning. There are several key features which will allow organizations concerned with ATC to build a positive culture. Here are some of my candidates:

1) Leadership. Leaders who encourage and reward information flow activities will be important in producing positive organizational culture. One of the prime examples of the kind of leadership I have in mind is the following anecdote about Wernher von Braun:

> During the development of the Redstone missile, there was a failure of one of the expensive prototypes, for no apparent reason. An expensive re-design was about to take place, when Von Braun was approached by one of his engineers, who said that he might unintentionally have caused a short-circuit during the pre-launch testing. A check revealed that indeed this was the problem. The expensive re-design was avoided. Von Braun sent the engineer a bottle of champagne (Von Braun, 1956).

But large systems, as scholars such as David Collingridge (Collingridge and James, 1989) and Thomas Hughes (1983) have shown, have their own momentum. Information which gets in the way of system "progress" is often sidetracked or squashed. In the wake of a number of technological accidents, we are only too aware of the mind-guarding, information blockages, and similar failures of systems under intense political or economic pressures.

Change in ATC systems needs to involve the kind of leadership which encourages such information flow. Superficially, the AERA approaches are presented as taking place in such an environment. But one wonders about the underlying reality. How much is the wider control community really involved? What will be the response once problems crop up, as they inevitably will? These questions need more probing than was possible at the conference.

2) A Culture of Competence. High technology organizations with a low accident rate believe that their own people are competent (the authors note that relatively low-ranking persons on aircraft carriers are often the principal transmitters of the organization's knowledge), or to put it more simply, they believe in their people (Laporte, Rochlin, and Roberts, 1987) . A very large amount of effort is spent at lower levels of the organization to bring all members up to speed, whatever their rank. Often the higher echelons may be only vaguely aware of such training efforts, but they are to be encouraged and recognized. Cross-training and broader education are encouraged. Unlike the classic bureaucracy, in which skills and therefore responsibilities are strictly delimited, the effective organization recognizes a need for what has been called "organic management," avoiding sharp boundaries between adjoining areas of responsibility (Burns and Stalker, 1962).

A variety of remarks by controllers and others at the conference suggested some problems with current organizational cultures. Two participants remarked that often lesser folk in the ATC organizations had trouble getting heard; the same message presented by one of the scientific "boffins," however, often got immediate management attention. Another problem was that controllers who talked on the radio seldom got the opportunity to get together off the radio to compare notes. When these rare opportunities occurred, however, they were very valuable. They should evidently be intentionally brought about to avoid some of the problems inherent in the current situation.

3) Encouragement of Self-Organization Efforts. Effective organizations keep track of their operations, and actively try to improve them (Argyris and Schon, 1978; Schon, 1983; Argote and Epple, 1990; Steinbrook, 1988). Only if the operational sections of the organization are actively involved in such inquiry, however, is it likely to be rapidly implemented. Often response to the new studies or data will be a function of such involvement. Operational parts of the organization, then, must have the ability to self-organize to improve operations. Rather than a passive attitude, blaming "them" for inadequacies, operational personnel should be rewarded for taking responsibility and making whatever changes are necessary to improve operations.[15]

15. Note here that the emphasis is on rewarding the free flow of information over all other considerations.

This means that organizational skills should accompany technological change. Organizational training and awareness may be a necessary element in the new sociotechnical organizations that will be required to make automated ATC successful. Indeed, such knowledge and empowerment is required if the benefits of new technology are to be realized and its hazards minimized.

How much such self-organization will be encouraged in a highly automated organizational environment? In the past, as the participants from the ATC community have made clear, ATC has been a heroic endeavor. The limitations of the equipment, the threats posed by the weather, and the often complicated situations have been sorted out by humans often performing at peak capacity. As John Levesley (1990) said of these peak periods and the controllers' impressive ability to manage them, "After a day when you have moved planes successfully across the sky, you go outside and the sun is brighter, the grass is greener, and you feel like you've really done something." How much of this spirit, one wonders, will be taken into the coming highly-automated era? And if it is not continued, what will the consequences be?

Conclusion

The basic thrust of this paper is that effective future control systems are likely to involve not only effective automation but also effective organizational structuring. It is essential that many of the positive features of the present system be carefully studied. They may contain implicit safeguards which an automated system may not be sophisticated enough to incorporate. Incremental approaches to automation, such as that being developed in Germany by Ewe Volckers (1991), are much more likely to bring out such hidden safeguards than the more far-reaching kinds of automation proposed by the AERA studies. Talks with controllers such as John Levesley (1990), Mike Tonner (1990), and John Lane (1990) brought home the very high amount of skill involved in present systems. It is hard to believe that similar skill and its cultivation will not be a key element in the effectiveness of future systems. Future systems should involve a more effective use of consciousness, not a "loss of airmanship."

References

Argote, Linda and Dennis Epple. (1990). "Learning Curves in Manufacturing," *Science, 247* (23 February), pp. 920-923.

Argyris, Chris and Donald Schon. (1978). *Organizational Learning: A Theory of Action Perspective.* Reading, Massachusetts

Burns, Tom and G.W. Stalker. (1962). *The Management of Innovation.* London: Tavistock.

Collingridge, David, and Peter James. (1989). "Technology, Organizations and Incrementalism: High Rise System Building in the U.K.," *Technology Analysis and Strategic Management, 1*, pp. 79-97.

Csikszentmihalyi, Mihaly and Isabella Csikszentmihalyi (Eds). (1988). *Optimal Experience: Psychological Studies of Flow in Consciousness.* New York:Cambridge University Press.
Cushman, John H. Jr. (1990). "How a Full Fuel Load Ran Dry on Fatal Flight from Columbia to Kennedy," *New York Times*, February 5, p. A12.
Edwards, J. (1991). "Intelligent Dialog in Air Traffic Control." in J. A. Wise, V. D. Hopkin, & M. L. Smith (Eds.) *Automation and Systems Issues in Air Traffic Control.* Berlin: Springer-Verlag.
Hughes, Thomas P. (1983). *Networks of Power: Electrification in Western Society 1880-1930.* Baltimore: Johns Hopkins University Press.
Jordan, Nehemiah. (1968). *Essays in Speculative Psychology.* London: Tavistock.
Kreamer, Tom. (1987). "Listen--Read Back--Comply," *Journal of Air Traffic Control,* (January-March) pp. 24-27.
Lane, J. (1990). Personal communication.
Laporte, Todd. (1990). Personal communication.
Laporte, Todd, Gene Rochlin, and Karlene Roberts. (1987). "The Self-Designing High-Reliability Organization: Aircraft Carrier Flight Operations at Sea," Naval War College Review, Autumn, pp. 76-91.
Lautman, L. G. and P.L. Gallimore. (1987). "Control of the Crew-Caused Accident," *FSF Flight Safety Digest*, June, pp. 1-8.
Levesley, J. (1990) Personal comminication
Marcus, Alfred A. and Isaac Fox. (1988). "Lessons Learned About Communicating Safety-Related Concerns to Industry: The Nuclear Regulatory Commission After Three Mile Island," Symposium on Science Communication, Environmental and Health Research, University of Southern California, Los Angeles.
Metz,Donald L. (1981). *Running Hot: Structure and Stress in Ambulance Work.* Cambridge: Abt Books, pp. 154-159.
Miller, James G. (1978). "Information Input Overload," in his *Living Systems.* New York: McGraw-Hill, pp. 121-202.
Patterson, T.T. (1955). *Morale in War and Work: An Experiment in the Management of Men.* London: Max Parrish.
Philips, C. F. (1985). "Evolution of the Air Traffic Control Beacon System," *The Controller,* December , pp. 19-24.
Rochlin, Gene (1986). "High-Reliability Organizations and Technical Change: Some Ethical Dilemmas and Problems," *IEEE Technology and Society Magazine*, September.
Schon, Donald. (1983). *The Reflective Practitioner: How Professionals Think in Action* . New York: Basic Books.
Steinbrook, Robert. (1988). "Heart Surgery Rates Found High in One in Six Hospitals," *Los Angeles Times*, July 24, Part I pp. 3, 32.
Tonner, M. (1990), Personal communication.
Von Braun, Wernher. (1956). "Teamwork: Key to Success in Guided Missiles," *Missiles and Rockets*, October , pp. 38-43.
Völkers, U. (1991). "Application of planing aids for air traffic control: Design principles, solutions, and results." in J. A. Wise, V. D. Hopkin, & M. L. Smith (Eds.) *Automation and Systems Issues in Air Traffic Control.* Berlin: Springer-Verlag.
Westrum, Ron. (1988). "Organizational and Inter-Organizational Thought," paper presented to the World Bank conference on Safety Control and Risk Management, Washington, October. Available from the author.
Wiener, Earl. (1977). "Controlled Flight into Terrain Accidents: System-Induced Errors," *Human Factors, 19*, pp. 171-181.
Wiener, Earl. (1987). "Fallible Humans and Vulnerable Systems: Some Lessons Learned from Aviation," in J. A. Wise and A. Debons, Editors, *Information Systems: Failure Analysis.* New York: Springer Verlag, pp. 163-182.

Issues in Cultural Ergonomics

Michael Kaplan

Office of Basic Research, U. S. Army Research Institute, Alexandria, Virginia, USA

Introduction

It is time for tomorrow's world...

With its promises and challenges already here to be addressed, we may point to at least two of them in the context of the present deliberations. (1) There is before us an asymptote we are approaching: fulfillment of the promise of total computerization and automation of air traffic control systems. (2) At this juncture, however, when we are earlier in our approach curve than we might wish, the systems remain human-machine systems, still requiring, in the view of many of us, attention to human components of system and organizational design and functioning. We are, thus, in the realm of ergonomics or human factors[16], and it is here that we need to take a world view of our problems — looking at countries in every continent. Also before us, then, is the challenge of cultural diversity, but, as it impacts on ours and related types of systems, how shall we deal with it?

Eighteen years ago this very month, a NATO group much like ours met in The Netherlands. The occasion was a symposium organized by Alphonse Chapanis on "National and Cultural Variables in Human Factors Engineering" held under the auspices of what was then the Advisory Group on Human Factors of the North Atlantic Treaty Organization. Chapanis' hope (Chapanis, 1975) was primarily "to show how and to what extent human engineering (or ergonomic) *principles and practices* have been, or *need to be modified to take account of national and cultural differences*" (italics mine). Our own international deliberations, it may be said, are linked to those earlier concerns of Chapanis and his colleagues.

Pondering the Chapanis concerns, and the growing body of investigations, data, influences, and activities which feature cultural variables and cross-national considerations in many subdisciplines of psychology and in the expanding domain of ergonomics/human factors, I have suggested that we view this entity as an evolving specialization and speak of it as "cultural ergonomics"[17] (Kaplan, in press). This is the position from which the following comments proceed.

16. The terms ergonomics and human factors are used interchangeably in this paper.

17. Specialists in ergonomics may note that the focus on the "cultural" has a certain consonance and parallelism with the *ergonomie heuristique* and ergonomic psychology of Leplat (e.g., 1980a, 1980b, 1981, 1982), the cognitive ergonomics of Faverge (see Ombredane and Faverge, 1955; and Karnas, 1970) and of Montmollin (1984), the psychological ergonomics of Singleton (1989), the cognitive engineering of Rasmussen (e.g., Rasmussen, 1986; Rasmussen, Duncan, and Leplat, 1987), and notions such as macroergonomics (e.g., Hendrick, 1987).

NATO ASI Series, Vol. F 73
Automation and Systems Issues in Air Traffic Control
Edited by John A. Wise et al.

They constitute an outline of some relevant topical considerations in human factors and of some issues requiring further elaboration, analysis, and resolution, if we wish to build a systematic framework for a cultural ergonomics capable of treating many specialized problem areas. At the same time, some are offered for their likely bearing on one such area: human aspects of air traffic control systems.

What follows, then, is a sharing of thoughts and suggestions on cultural ergonomics to be considered for their reference to this Institute and their role in the long term.

Cultural Variation

This section highlights a number of considerations relating to the specification and treatment of culture as a variable affecting human performance in organizations and systems.

The Two Cultures

From within behavioral science, we may view the notion of "culture" in at least two ways (quite unlike C. P. Snow's use of "two cultures" [see Snow, 1964] to distinguish between the scientist and scholars in the humanities). First, as in much of the thrust of the present discussion, there is a focus on nations and other identifiable groups, labeled according to some set of specifiable characteristics, much as we might find them in cultural anthropology (e.g., geographical locus). Second, there is the "macroergonomic" or "sociotechnical" treatment of characteristics and "climate" of organizations (e.g., Hendrick, 1987). It is also the case that the latter, organizational cultures, whose influences are of demonstrated significance, may themselves be objects of cultural/cross-national comparison.

Culture as a Variable

Whether we may speak of culture as a "variable" in the mathematician's or experimental scientist's technical sense of the term, needs to be examined.

Focusing on variation in human performance, examples are observed that do appear to be related to cultural difference. Still, questions remain: (1) whether culture is the proximate cause, (2) whether difference effects are physiologically determined or generated by environmental features or life styles, or (3) whether difference effects are stable or change with age or training (cf. illusions, for example, that are modifiable with practice [e.g., Eysenck and Slater, 1958]).

Another aspect of this issue may be found in the thinking of Hofstede (1980), who argued that, although "Cross-cultural studies proliferate in all the social sciences,...they usually lack a theory of the key variable: culture itself."

In his own substantial examination of organizational cultures, involving comparative studies of 40 countries, he attempted to specify the characteristics (cultural features) of work organizations. His approach involved four dimensions along which he believes "dominant value systems" in the work organizations of his 40 countries can be ordered: Power Distance, Uncertainty Avoidance, Individualism, and Masculinity.

Whatever the theoretical/methodological positions central to one's investigations in this arena, the requirements for adequate experimental design in determining cultural influence are difficult and challenging (cf. Cole and Means, 1981).

Cultural Parameters of Human Performance Functions

How do culturally-influenced aspects of behavior enter into the performance of work-related tasks? For descriptive purposes, at least, it may be useful to conceive of functional relationships that could be characterized by the general equation

$$P = f[(v..., t..., k...), (C...)],$$

where P refers to measures of performance-related variables (e.g., properties of performance, psychophysical judgments, preference ratings, verbal reports, psychometric responses, changes in psychophysiological responding); v... to independent variables; t... to temporal variables; and k... to constants. We may think of C... — representing cultural variables (or conditions) — as a parameter of the equation, influencing properties of the relationships.

Figure 1 shows an ordering of data of this sort. Adapted (and simplified) from findings reported by Gleiss and Renhall (1986), it has some relevance for visual equipment in an Air Traffic Control area. From performance research to evaluate a visual display, the data derive from cross-national investigations under the aegis of the Swedish Telecommunications Administration, where subjects in European laboratories, using the same category estimation scale, judged the quality of pictures carried over telephone-transmission systems under deliberately varied picture-coding conditions. Except for condition 1, these involved different kinds of transmission-system impairments. The authors ascribe displacement of the similarly shaped curves to "subjects' different interpretations of the verbal definitions of the scale grades, *due to difference in their background, instruction, etc*" (italics mine).

The auditory realm provides similar national examples with ATC relevance. Human performance tests of speech-transmission equipment have been conducted cross-nationally by CCITT, the Consultative Committee on International Telegraphy and Telephony, a body of the United Nations' International Telecommunications Union. The curves of Figure 2 are extracted and adapted from an evaluative discussion on wideband speech codes in the report of CCITT Study Group XII-R5 (1985), and each shows category judgments — Mean Opinion Scores — by its listeners of the quality of taped speech in its own language with its own speakers, in relation to properties of test reference equipment systems through which the speech was transmitted in each national laboratory. In these and similar studies, Gleiss (1989) feels that different results arise from individual speaker differences as well as language (at the sending end) and from subject and laboratory differences (at the receiving end). He has also pointed to the problem of distinguishing cultural from individual differences.

Culturally-Affected Sensory Functioning

A problem for cultural ergonomics is whether sensory functioning can be affected culturally. If so, does culturally-affected sensory functioning enter into and influence human task performance or human-machine system functioning to the extent that compensatory changes should be introduced in procedures or equipment?

An approach to these questions can be found in fundamental psychological research bearing on ATC interests. It has been argued, for example, that, for auditory functions, "[w]hat a person perceives depends not only on the signal that reaches the ear...but on what the person *expects* to hear..." (italics mine), a result of cognitive, conceptual processes (Working Group on Speech Understanding and Aging, CHABA, 1988). These also influence speech understanding and intelligibility (Miller and Selfridge, 1950; Miller, Heise, and Lichten, 1951; Pisoni, 1978; Marslen-Wilson and Welsh, 1978; Cole and Jakimik, 1980). This implies the influence of cultural and environmental factors in the chain of events affecting auditory perception, since culture affects cognitive and conceptual processes. Culture may play a contributory role in tone-deafness (Tanner and Rivette, 1964), in the ability to analyze complex waveforms (Soderquist, 1970), and possibly in presbycusis[18] (Rosen, et al., 1962), although in the latter study, subjects' age reports may not have been reliable.

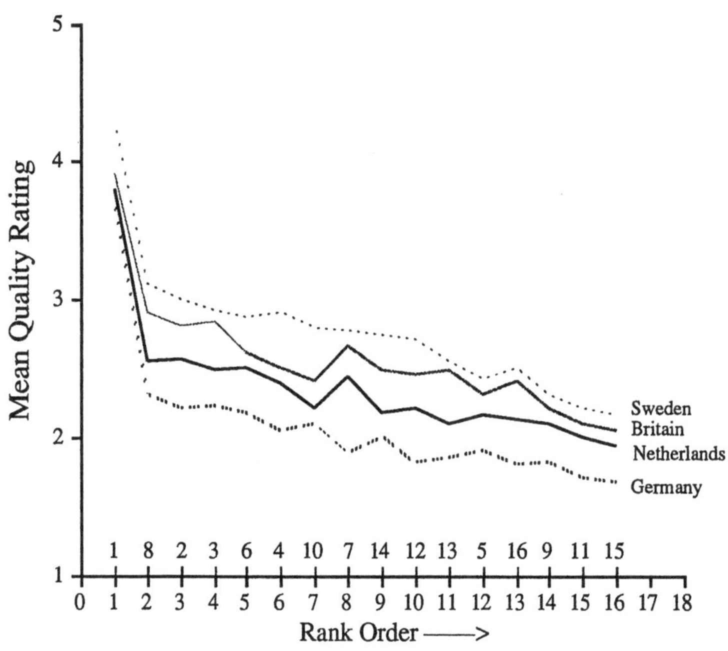

Figure 1. Evaluation of visual display quality as a function of telephone-transmission impairment, with country of judges as a parameter. (Adapted from Gleiss & Renhall, 1986)

Culture as a parameter of visual functions is discussed directly and, in a sense, by implication in a considerable body of literature addressing fundamental perceptual issues and

18. Presbycusis is loss of sensitivity to sound stimuli with increasing age.

their relevant variables. (Independent variables other than culture may themselves be culturally influenced.) An example of a variable that may be regarded as implying potential cultural influence is the role of instructions in shape constancy (Lichte and Borresen, 1967). "Direct" illustrations, as in color discrimination and color naming (e.g., Bornstein, 1985; Rosch, 1975), are numerous.

Relative to the foregoing, Singleton (1990) has argued against any meaningful effect of culture in human-machine systems. He sees human-human interaction as the locus of significant cultural influence.

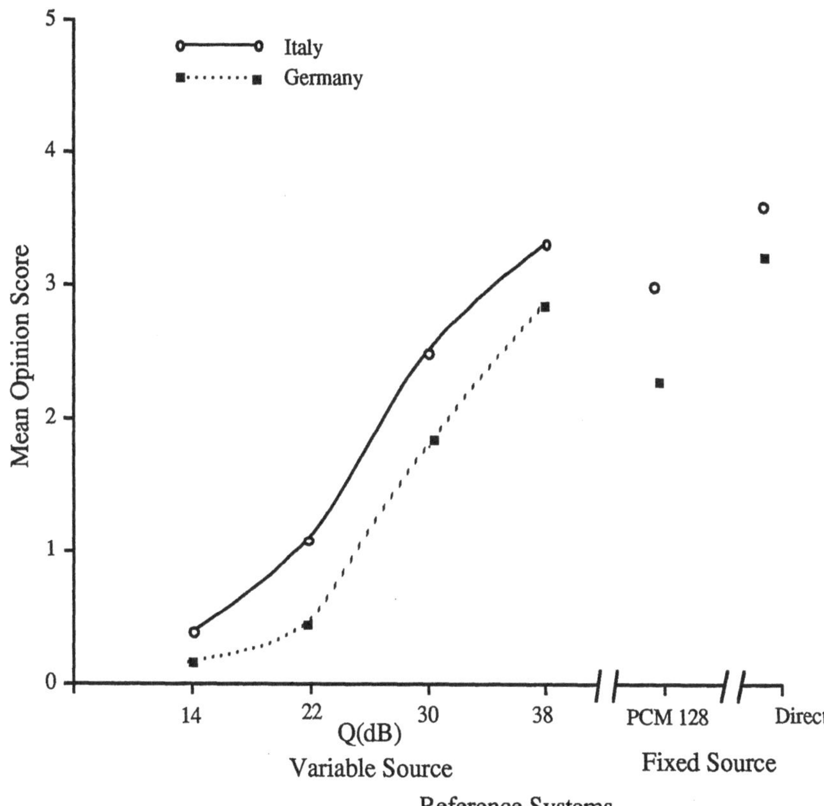

Figure 2. Speech quality judgments in relation to properties of noise reference systems, with country of listeners as a parameter. (Adapted from CCITT Study Group XII-R5, 1965)

Some Human Factors Areas Susceptible to Cultural Influence

A number of human factors areas regarded as susceptible to cultural influence are presented in this section.

Command and Control

Implicit cultural relevance may be seen in certain views of NATO DRG Research Study Group 12. In its report on "Computer-Human Interaction in Command and Control" (1989), the Group calls for NATO basic research support on "how to optimize the performance of the human-computer cognitive system by merging the skills of the person and the computer that constitute the system." It sees as a central issue in exploiting new computer technology "the *needs, skills, and limitations of the human user*" (italics mine). Surely, this should also be a focus for ATC task and system concern.

Other issues of special note for us to investigate appear in the report. Referring to "communications systems for multinational environments" (page 73), it states:

> Of special interest to NATO is the need to ensure that systems which are intended to serve multi-nation forces minimize the probability of difficulties arising from linguistic and other *culture based differences* (italics mine) amongst their users. Problems resulting from mistranslations from one language to another are obvious matters of concern, but much more subtle effects, while less apparent, may be equally important. The issue is particularly significant vis-à-vis systems that can change substantively over relatively short periods of time; anticipation and evaluation of the effects of change must be sensitive to the possibility *that those effects may not be precisely the same for users with different national or cultural backgrounds* (italics mine).

Voice Communication

Cultural influences are significant contributors to issues in voice communication.

To begin with, problems of speech pronunciation and understanding in English received attention in early work, such as the comprehensive investigations of Moser and Dreher (1953, 1955) on phonetic alphabets. These demonstrated — and increased —the superiority of the International Civil Aviation Organization alphabet over that of the US-UK ("Able-Baker-Charlie"), having found sources of "confusability" in factors such as code group word position and unfavorable attitude toward a word.

Speech pronunciation and language understanding, however, have not disappeared as matters of concern, especially when we turn to non-native speakers and listeners. Ruffell-Smith (1975), citing the adoption of English as the common language for aircraft operating purposes in the Western world, pointed to shortcomings in this system and noted that "Marked differences in pronunciation are apparent both when English is the native language of the speaker...or where it is spoken by people who have another native language."

He recommended a solution through data links, but Foushee (1990) has warned of the loss of valuable information when reducing a system's reliance on verbal communication (for example, the urgency of a situation communicated through the controller's tone of voice).

As for cross-national communicators, difficulties between tower and pilot have not been fully erased, and have contributed to incidents and accidents in recent years. Under stress and abnormal circumstances, for example, a non-native English speaker, falling back on his or her native language, could cause confusion.

For Europe, some information on cross-national language difficulties is on file at the British Civil Aviation Authority (Baker, 1989). Similar involvement of U. S. aircraft, reported on a voluntary basis, has been entered in the data bank of the Aviation Safety Reporting System (1990). A printout for end of May, 1990, contained 193 "full-form" (not abbreviated) reports since 1 January 1983, "referencing foreign language or accent problems." Figure 3 (below) provides an analysis of these problems.

	1983	1984	1985	1986	1987	1988	1989
Fgn Language Report Intake	**13**	**25**	**21**	**31**	**24**	**25**	**41**
ASRS Database Intake	**2187**	**4952**	**6226**	**8954**	**13599**	**16320**	**24918**

* Data for 1990 is complete through January 1990

Select Occurrence Breakdown **			
Operational Errors	22	Runway Transgressions	31
Operational Deviations	11	Wrong Rwy Landing	3
Pilot Deviations	101	Wrong Rwy Takeoff	1
Near Midair Collisions	24	Non Adherence to Legal Requirement	131
Conflict/Airborne Less Severe	37	Primary Problem:	
Emergencies	1	Controller	58
		Flight Crew	112

**Categories are NOT mutually exclusive

Fig. 3. Frequency of aircraft problems occasioned by voice communication difficulties involving foreign languages. (Data and analysis provided by NASA Aviation Safety Reporting System, May, 1990.)

On a world scale, new attention to language training and standards can be expected from the International Civil Aviation Organization (Bradbury, 1990), and in Europe,

EUROCONTROL, the European Organization for the Safety of Air Navigation, has recently begun examination of new techniques for improving English speech pronunciation (Baldwin, 1990; EUROCONTROL Institute of Air Navigation Services, 1990). Mandatory bilingual air traffic control in Quebec air space has apparently been effectively implemented through the thorough simulation training and research of Stager (1980), and is said to be satisfactory to this day (Fudakowski, 1990).

With respect to standards, tests of English speech intelligibility designed especially for non-native English listeners (compare the pilot-tower difficulties) have been developed in England by Irvine (1974). Perhaps he has given us the germ of a useful selection approach for relevant personnel, as efforts develop to assure effective voice communication between cross-national speakers and listeners.

Stress

In relation to prediction of stress and ways of assuring adequate performance under stress, two issues have received little systematic attention: (1) How do different cultures vary in the types of events that generate stress reactions in their people? (2) How do they vary in the types of reactions evoked by a given stressor?

The second is not dealt with here, but an example relating to the first has special relevance. It is a comparison of New Zealand and Canadian air traffic controllers by Shouksmith and Burrough (1988). Although they report similar stressors for both groups, there were some differences in degree of stressfulness felt for certain items. Canadians were more stressed by bad weather and — by way of mandatory bilingualism requirements — relations with management, whereas New Zealanders felt more concern over irregular shift work requirements, dependence on only a primary (without a secondary, backup) radar, and the need for too-heavy reliance on memory.

If culture does, indeed, exert specifiable effects on the controller's stress-bound performance, this is important information to obtain and to use for personnel and system management.

The Cultural Background of Decision Making

In the complex world of decision analysis, a seminal finding appeared in 1977 (Phillips and Wright, 1977), showing "cultural differences in viewing uncertainty and assessing probabilities." Having taken root, it led to other research (e.g., Wright et al., 1978; Yates et al., 1989) that has established and extended the finding of cultural variation in probabilistic thinking. Summing up, comparisons along these lines have been made among subjects from England, Hong Kong, Indonesia, Malaysia, the U. S. , China, and Japan.

For example, "Asians were found to adopt a less finely differentiated view of uncertainty both numerically and verbally than did the British sample," although "these differences were found to out-weigh any influence of subculture, religion, occupation, arts/science orientation and sex" (Wright and Phillips, 1980).

Clearly, this type of basic research needs to be pursued further, both for the decision analysis approach to decision making and the more recent class of investigations involving the dynamics of decision making in naturalistic environments (Klein, Orasanu, and Calderwood, in press). It would also appear to have implications for occupations involving both multi-national and decision making components. This may be still another cultural area, then, to be considered in the context of air traffic control.

Organizational Culture and System Integration

Ryan (1990) has developed a unique "Nuclear Organization and Management Concept" (NOMAC), which is being implemented as part of his larger human factors program at the Nuclear Regulatory Commission. A direct link is seen between organizational factors and, via consequent organizational reliability, nuclear safety. In the thinking of his group, heightened importance is given to "what is valued" by the organization; and these factors, along with others of an organizational nature, are seen to influence the operating structures of the organization, its human resource management practices, and the styles of its managers and supervisors (O'Brien and Haber, 1989).

This program has implications for other organizations with responsibility for reliability in high risk settings, e.g., air traffic control. Ryan also sees cross-cultural considerations affecting the intra-organizational relationships delineated in the work of his group.

Some recent work pointing to cultural influences on system integration has also been cited by Booher and Rouse (1990); and of some relevance is much earlier work by Triandis (1967) on interpersonal relationships in international organizations.

Consideration of personal preferences of a cultural nature may also be important in determining the most effective organizational design and operations options, when dealing with a multi-cultural group (cf. the NASA work of Wise and McConville, 1989, on cross-cultural preferences affecting lunar base design).

A Vision of the Future: An International Center for Cultural Ergonomics

How might we deal with the cultural issues noted above — and still others not cited? For a great many of them, the answer lies in increased cross-national cooperation and collaboration to jointly acquire, develop, and utilize knowledge in cultural ergonomics.

Returning to our world view, we may envision gradual establishment of an International Center for Cultural Ergonomics. The Center would be the locus of accumulating classes of projects dealing with the broad spectrum of cultural ergonomics issues (see Kaplan, in press). What might some of these be? In addition to what we might call the "cultural ergonomics of air traffic control," we may see before us:

Cultural/Cross-National Databases

The Center would be a repository or transfer node for continually updated and easily accessible international data on such matters as: foreign language problems affecting air safety; resources on bilingual and multilingual research and instruction; cultural practices affecting behaviors in small, hierarchically structured groups; cultural practices in conflict with organizational or mission requirements; cultural variations in teaching techniques and classroom customs of students.

A useful structuring model/guide might be the Yale Cross-Cultural Survey, known now as the Human Area Resource Files.

Complete Reference Library and Information Clearinghouse

A systematically elaborated framework and fleshed out treatment of topics in the field, together with all relevant "documents of record" and technical material, would make the Center a convenient central source for the entire body of literature in cultural ergonomics. It would clarify what is known and set the stage for needed knowledge acquisition.

It would also be the locus of complete information on applied problems as they develop (e.g., some of the ATC issues considered at this NATO Institute), along with information on approaches to and successes and failures in their resolution.

This effort would make possible rapid information access and exchange, facilitating the activities of various types of workers in the field.

Expert Systems

As appropriate, the Center would be used to facilitate development of expert systems to be used cross-nationally.

The issue to be resolved here is that a given expert system tends to be idiosyncratic, by virtue of factors that include the thinking peculiar to its designer, the information and views of experts interrogated by knowledge engineers, and the questions of the knowledge engineers. (Cognitive bias during knowledge acquisition [see Madni, 1988] might be culturally determined.) Some of those idiosyncrasies could be addressed, it would seem, by way of this approach.

Collaborative Research

Reviewing NATO's 15-year Special Program in Human Factors, Singleton (1984) observed, "The problems of conducting multinational experiments emerged as a methodological and organizational issue." The organization of multinational meetings and of multinational research are themselves problems for cultural ergonomics.

Still, difficulties notwithstanding, it is hard to imagine meaningful research on cross-national issues without collaborative multinational research. The speech research of the CCITT (1985), cited earlier, is an encouraging example, as is the reference from Sweden (Gleis and Renhall, 1986).

An example from a different field, sustained performance in military operations (Allnutt, 1990), may be instructive. The research utilized test materials gleaned from different countries (England, Canada, the U. S.), and were applied in field exercises involving soldiers in their respective countries. It was undertaken by military psychology research establishments belonging to The Technical Cooperation Program, whose terms of reference removed national obstacles from such joint research.

In our vision, the Center, as a resource and a focus for multinational research of mutual interest, would successfully engender such effort in a truly cooperative spirit. So aided, cultural ergonomics would be a productive field, advancing knowledge, resolving problems, and meeting the human factors challenges of cultural diversity.

Acknowledgments

Capt. Paul Chatelier, USN (Ret.), Perceptronics, and Professor Anthony Debons, University of Pittsburgh and Robert Morris College, have provided much generous encouragement of these efforts toward a cultural ergonomics.

I should like to express great appreciation to Mr. Vincent J. Mellone, NASA Aviation Safety Reporting System; Dr. Lawrence Phillips, London School of Economics; Monsieur E. Pierre, Counsellor to the Comité consultatif international télégraphique et téléphonique, Genève; and Dr. Thomas Ryan, Human Factors Regulatory Research Program, U. S. Nuclear Regulatory Commission, for generously and rapidly supplying documents used in the preparation of this paper. Ms Holly L. Spurlock, Howard University, was of great help in preparing the bibliography.

The opinions herein expressed are those of the author and do not reflect official policy of the U. S. Army.

References

Allnutt, M. (1990). Personal communication [Army Personnel Research Establishment, Farnborough, UK].

Aviation Safety Reporting System. (1990). *Foreign language difficulties* (Special Request No. 1815). Mountain View, CA: Battelle Columbus, Division ASRS Office.

Baker, S. (1989). Personal communication [British Civil Aviation Authority, Farnborough, UK].

Baldwin, R. (1990). Personal communication [EUROCONTROL Institute of Air Navigation Services].

Booher, H. R., & Rouse, W. B. (1990). MANPRINT as the competitive edge. In H. R. Booher (Ed), *MANPRINT: An approach to systems integration*. New York: Van Nostrand Reinhold.

Bornstein, M. (1985). Infant into adult: unity to diversity in the development of visual categorization. In J. Mehler & R. Fox (Eds.), *Neonate cognition*. Hillsdale, NJ: Erlbaum.

Bradbury, J. N. (1990). Personal communication [International Civil Aviation Organization].

CCITT Study Group XII-R5. (1985). *Results of category judgment tests on wideband speech codecs* (Report of Study Group XII-R5, Consultative Committee on International Telegraphy and Telephony, Appendix 3, Annex F). Geneva: International Telecommunications Union.

Chapanis, A. (Ed.) (1975). *Ethnic variables in human factors engineering*. Baltimore: The Johns Hopkins University Press.

Cole, M., & Means, B. (1981) *Comparative studies of how people think: An introduction.* Cambridge, MA: Harvard Univers. Press.

Cole, R. A., & Jakimik, J. (1980). A model of speech perception. In R. A. Cole (Ed.), *Perception and production of fluent speech*. Hillsdale, NJ: Erlbaum.

Defence Research Group. (1989). *Computer-human interaction in command and control* (Final Report). Brussels: NATO Document AC/243(Panel 8/RSG.12)D/7.

Eysenck, H. J., & Slater, P. (1958). Effects of practice and rest on fluctuations in the Muller-Lyer illusion. *British Journal of Psychology, 49,* 246-256.

EUROCONTROL Institute of Air Navigation Services. (1990). *Training programme 1991.* Luxembourg: European Organisation for the Safety of Air Navigation.

Foushee, H. C. (1990). Development of a national plan for addressing future aviation human factors needs. Presentation to the ICAO Human Factors Symposium, Leningrad, USSR, April 4, 1990.
Fudakoski, T. (1990). Personal communication [Transport Canada].
Gleiss, N., & Renhall, P. (1986). Measurements of picture quality. *TELE, 40,* 21-25. (Sweden.)
Gleiss, N. (1989). Personal Communication.
Hendrick, H. W. (1987). Macroergonomics: a concept whose time has come. *Human Factors Society Bulletin, 30*(2), 1-3.
Hofstede, G. (1980). *Culture's consequences.* Newbury Park, CA: Sage.
Irvine, D. H. (1974). A new type of speech intelligibility test. *Ergonomics, 17,* 783-788.
Kaplan, M. (In press). Cultural ergonomics: an evolving focus for military human factors. In R. Gal & A. D. Mangelsdorff (Eds.), *A handbook of military psychology.* London: Wiley.
Karnas, G. (1970). Analyse d'un poste de contrôle-décision: les stratégies op´ratoires des contrôleurs de la vente de places d'avion [Analysis of a reservation office: Operating strategies of airline ticket agents]. *Le Travail Humain, 33,* 59-76.
Klein, G., Orasanu, J., & Calderwood, R. (Eds.) (In press). *Decision making in action: Models and methods.* Norwood, NJ: Ablex Publishing Corporation.
Leplat, J. (1980 a). Ergonomic pyschology in French-speaking countries. *French-Language Psychology, 1,* 307-318.
Leplat, J. (1980 b). *La psychologie ergonomique* [Ergonomic psychology]. Paris: Presses Universitaire de France.
Leplat, J. (1982). Personal discussion [Laboratoire de Psychologie du Travail, Paris].
Leplat, J., & Hoc, J-M. (1981). Subsequent verbalization in the study of cognitive processes. *Ergonomics, 24,* 743-755.
Lichte, W. H., & Borresen, C. R. (1967). Influence of instructions on shape constancy. *Journal of Experimental Psychology, 74,* 538-542.
Madni, A. M. (1988). The role of human factors in expert systems design and acceptance. *Human Factors, 30,* 395-414.
Marslen-Wilson, W. D., & Welsh, A. (1978). Processing interactions and lexical access during word recognition in continuous speech. *Cognitive Psychology, 10,* 29-64.
Miller, G. A., Heise, G. A., & Lichten, W. (1951). The intelligibility of speech as a function of the context of the test materials. *Journal of Experimental Psychology, 41,* 329-335.
Miller, G. A., & Selfridge, J. A. (1950). Verbal context and the recall of meaningful material. *American Journal of Psychology, 63,* 176-185.
Montmollin, M. de (1984). *L'intelligence de la tache: Eléments d'ergonomie cognitive* [Task intelligence: Elements of cognitive ergonomics]. Berne: Peter Lang.
Moser, H. M., & Dreher, J. J. (1953). *Comparison of the US-UK and ICAO phonetic alphabets* (Technical Report 519-3, part II). The Ohio State University Research Foundation.
O'Brien, J. N., & Haber, S. J. (1989). Draft of the implementation plan for pilot testing the nuclear organization and management analysis concept. Engineering Technology Division, Brookhaven National Laboratory.
Ombredane, A., & Faverge, J. M. (1955). *L'analyse du travail.* [The analysis of work]. Paris: Presses Universitaires de France.
Phillips, L. D., & Wright, G. N. (1977). Cultural differences in viewing uncertainty and assessing probabilities. In H. Jungermann & G. de Zeeuw (Eds.), *Decision ,making and change in human affairs.* Dordrecht-Holland: D. Reidel.
Pisoni, D. B. (1978). Speech perception. In W. K. Estes (Ed.), *Handbook of learning and cognitive processes.* Vol. 6. Hillsdale, NJ: Erlbaum.
Rasmussen, J. (1986). *Information processing and human-machine interaction: An approach to cognitive engineering.* Amsterdam: Elsevier Science Publishing Co., Inc.
Rasmussen, J., Duncan, K., & Leplat, J. (1987). *New technology and human error.* London: John Wiley.

Rosch, E. (1975). Universals and cultural specifics in human categorization. In R. Brislin, S. Bochner, & W. Lonner (Eds.), *Cross-Cultural perspectives on learning*. New York: Halstead Press.

Rosen, S., Bergman, M., Plester, D., El-Mofty, A. & Satti, M.H. (1962). Presbycusis study of a relatively noise-free population in the Sudan. *Annals of Otology, Rhinology and Laryngology, 71*, 727-743.

Ruffell-Smith, H. P. (1975). Some problems of voice communication for international aviation. In A. Chapanis (Ed.), *Ethnic variables in human factors engineering*. Baltimore: The Johns Hopkins University Press.

Ryan, T. G. (1990). Personal communication.

Shouksmith, G., & Burrough, S. (1988). Job stress factors for New Zealand and Canadian air traffic controllers. *Applied Psychology: An International Review, 37*, 263-270.

Singleton, W. T. (1984). Academic progress in human factors during the N.A.T.O. programme, 1967-1983. *Ergonomics, 27*, 833-845

Singleton, W. T. (1989). *The mind at work: Psychological ergonomics*. Cambridge: Cambridge University Press.

Singleton, W. T. (1990). Personal communication.

Snow, C. P. (1964). *The two cultures and A second look*. Cambridge: Cambridge University Press.

Soderquist, D. R. (1970). Frequency analysis and the critical band. *Psychonomic Science, 21*, 117-119.

Stager, P., Proulx, P., Walsh, B., & Fudakowski, T. (1980). Bilingual air traffic control in Canada. *Human Factors, 22*, 655-670.

Tanner, W. P., Jr., & Rivette, C. L. (1964). Experimental study of tone deafness. *Journal of the Acoustical Society of America, 36*, 1465-1467.

Triandis, H. C. (1967). Interpersonal relations in international organizations. *Organizational Behavior and Human Performance, 2*, 25-55.

Wise, J. A., & McConville, K. (1989). A cross-cultural survey of personal preferences in design and operation of a lunar base. Paper presented at Human Factors Society meetings, October 18, 1989.

Wright, G. N., Phillips, L. D., Whalley, P., Choo, G. T., Ng, K-O., Tan, I., & Wisudha, A. (1978). Cultural differences in probabilistic thinking. *Journal of Cross-Cultural Psychology 9*, 285-299.

Wright, G. N., & Phillips, L. D. (1980). Cultural variation in probabilistic thinking: Alternative ways of dealing with uncertainty. *International Journal of Psychology, 15*, 239-257.

Yates, J. F., Zhu, Y., Ronis, D. L., Wang, D-F., Shinotsuka, H., & Toda, M. (1989). Probability judgment accuracy: China, Japan, and the United States. *Organizational Behavior and Human Decision Processes, 43*, 145-171

Air Traffic Controller Working Conditions and Organization: Suggestions for Analysis and Improvements from a Psychological Point of View

Claudia Harss,[1] Jitzchack Lichtenfeld,[1] Michael Kastner,[1] and Jan Goodrich[2]

[1]IAPAM-EILAT, Munich, Germany; Tel-Aviv; Israel
[2]Embry-Riddle Aeronautical University, Daytona Beach, Florida, USA

Introduction

In the following we want to look at automation and the introduction of new technology in the working sphere of air traffic controllers from a psychological perspective. The following topics will be covered:

- How can air traffic controllers, technology, pilots, and management be seen as a complex system?
- What is the meaning of flight security in this interrelation?
- Which factors determine human action (in our case, the actions of air traffic controllers)?
- What measures can increase flight security?

It seems that the traditional lack of communication between technicians, natural scientists, and social scientists still exists and continues to have a negative effect on the planning and introduction of technical innovations in organizations. A typical example of this is the ultramodern control tower, appearing around the world, and still producing problems not thought about before. At Munich's new airport, the working room in the tower has proved to be much too small. Israel's Ben Gurion airport also has a new tower, which should serve as an example for European countries: its planners thought of everything except the human and organizational problems which arose shortly after its establishment.

Why do these problems occur?

Human Action in Organizations

Before we analyze why problems occur within an organization, we should first consider what an organization is. Mitchell & Larson (1987) give an interesting definition: "Organizations are systems of coordinated behavior." This means that individual behavior is synchronized toward a specific goal — in the case of organized air traffic, "flight security."

NATO ASI Series, Vol. F 73
Automation and Systems Issues in Air Traffic Control
Edited by John A. Wise et al.

Quicker, more effective and secure action is usually attempted by automation, or optimizing of instruments and equipment. The main reasons for installation of new technology are usually complaints and break-downs which indicate that the old equipment does not work effectively any more (see Figure 1).

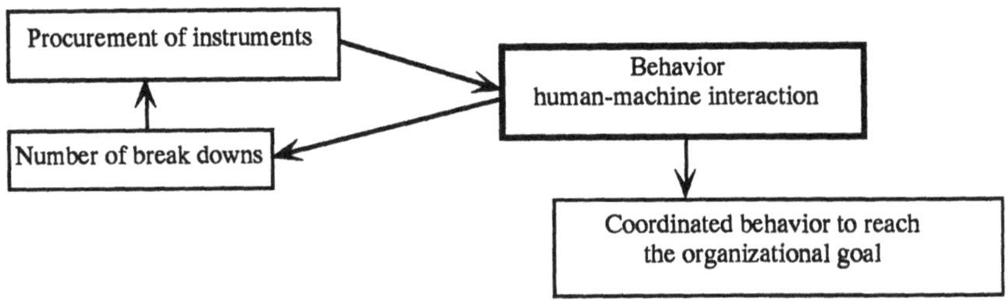

Figure 1. The traditional procedure when inroducing new technology.

The factors determining behavior in the human-machine interaction are, however, much more complex, as demonstrated recently in a study on helicopter pilots (Harss, Kastner & Beerman, in press).

We found that the technical standard (of course) has a large influence on the human-machine interaction. There are, for example, highly -significant correlations between the number of disturbances caused by malfunctioning equipment, and the stress and strain data and performance data during flight. But many other significant correlations and statistical analyses of performance data suggest that other factors within the entire system "organization" influence the actions of the individuals as much as technology. Ignoring those factors cannot lead to more flight security.

For example: when we explored how job satisfaction is related to behavior and stress of helicopter pilots during flight, we found significant correlations between measured job satisfaction and psychological well-being, which very strongly correlated with physical and psychological stress and strain and the efficiency of actions during flight.

So every organization is a system, "a set of interrelated elements" (Mitchell & Larson, 1987), in which changing one element causes change in one or more interrelated other elements. In each organization, a characteristic network of interrelated elements affect human-machine interaction; the main message of Figure 2 is not to show single cause-effect chains, but the complexity of the whole system (see Figure 2).

But in complex systems, we cannot know all that much about their specifics and the implications for future developments. These large information-deficits regarding our complex environment contrast with our blind confidence in the capabilities of technology, which *should* minimize our problems but sometimes does just the opposite (see Kastner, 1989). The results of this type of thinking can be seen all over the world. One example would be the problems with the beautiful, sophisticated new towers mentioned above. They were constructed without regard for the needs of air traffic controllers.

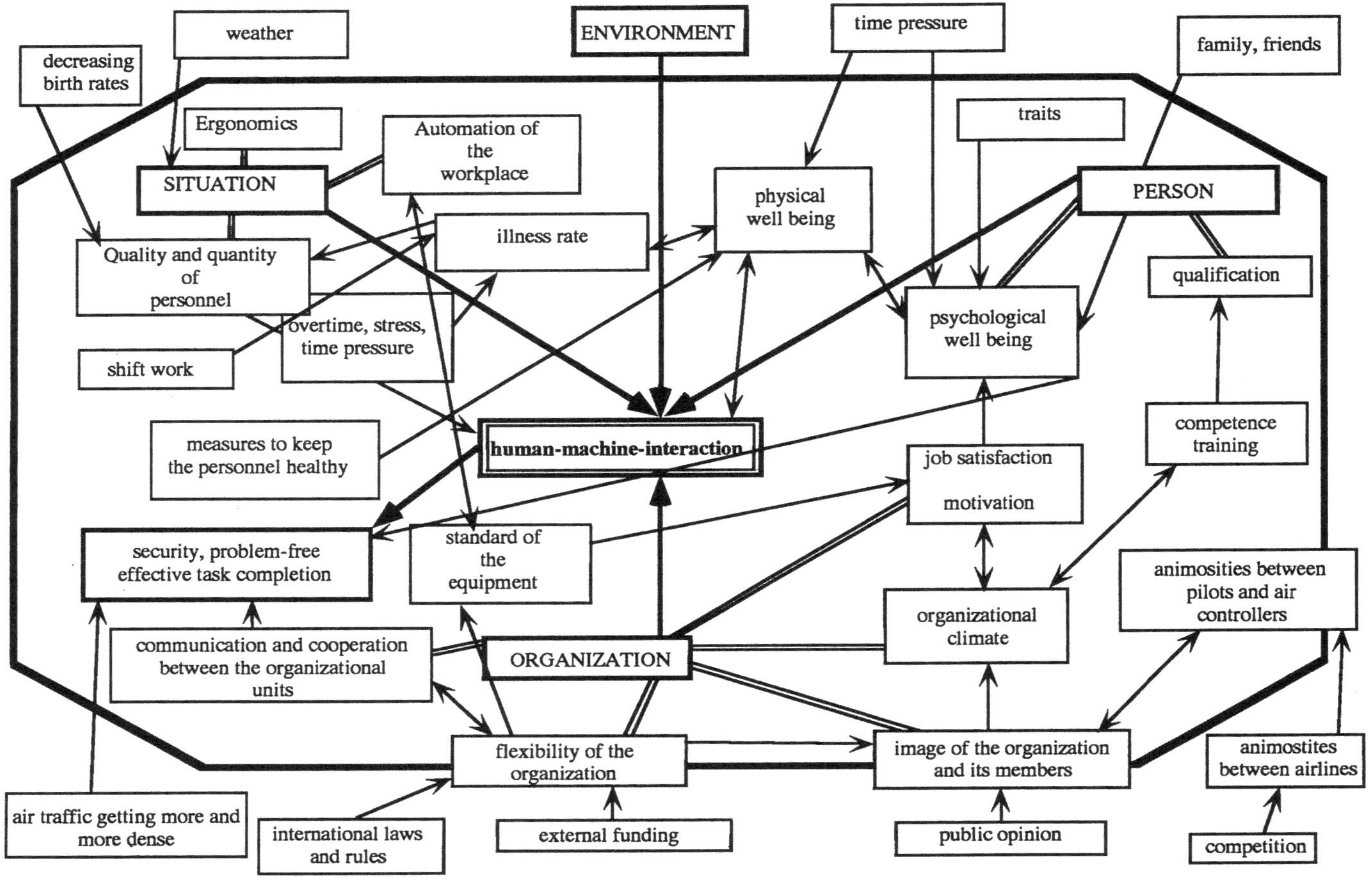

Figure 2. An example of the factors influencing behaviour in an organization.

A Scheme to Analyze the Organization

The question is, how can one make the complex system "Air Traffic Control" more transparent? This can be done by a scheme systematically covering and analyzing all important characteristics of the individuals (person), of the concrete work place (situation), and of the airport (organization).

The Analysis of the Person

Air traffic controllers do highly demanding work. The consequences are strict criteria for their selection and instruction. The air traffic controller should be flexible, independent, intelligent, capable of making quick decisions, able to work in a team, able to cope with criticism, and equipped with quick reactions. As is known from action theory and stress research, these characteristics are not fixed, inborn, unchangeable traits. They are behavioral styles which can be forgotten to a certain degree, if the conditions are not optimal, or can fall apart under illness and stress (see, for example, Alkov, 1983; Alkov, Gaynor & Borowsky, 1985; Broadbent, 1971; Dubos, 1978; Simmel, Cerkovisik & McCarthy, 1987).

It is important to monitor and evaluate the health, the capability to perform, the qualification, and the motivation of the members of an organization.

- How healthy is the individual? Shift workers are more subject to sleep disorders, joint disorders, stomach problems, and psychological problems (such as aggression and tension). How regularly are these people examined, and how often are they informed about topics in the health area (for example, through stress-coping seminars)?

- Which qualifications does the individual have? How many measures of further education are necessary for a new task (for example, working with a new system or taking over a management task) to be accomplished?

- What is the individual's interpersonal style with other members of the organization? Does the person often have conflicts with superiors, co-workers, or pilots? Is it reasonable to send controllers to a regular communication seminar, or to carry out regular discussions about daily problems in the team, in order to avoid recurring, exhausting arguments and fights?

- How motivated and satisfied is the individual? As discussed above, job satisfaction influences psychological and physical well-being, which in turn has a strong influence on behavior. To what extent does the employee have the right to participate in and control the organization of his or her workplace?

- How does the individual perceive information and make decisions? "If you think it's safe, it's safe!" A risky air controller will probably like this sentence better than will an anxious one, who prefers fixed rules and instructions. The decision-making behavior of risky controllers will be quicker and more flexible (but sometimes more dangerous) than that of anxious controllers (who might hinder the flow of air traffic in some critical situations, because they stick to the rules). Both could use regular decision-making training, in order to become clear about the positive and negative effects of their behavior and to learn appropriate alternatives.

The Analysis of the Work Place (Situation)

We define "workplace" as the concrete task, the necessary technical equipment, and the ergonomic conditions. All of these factors have a decisive effect on long-term stress and strain and therefore on performance.

Task and technical equipment: Let us first discuss the difficult demands involved in the controller's task. It is a complex (Simon, 1964) regulatory activity under considerable decision-making pressure (approximately every 30 seconds, a yes/no decision has to be made). This demands a continuously-stable high level of concentration. The medical literature recommends a regeneration period every 30 minutes for activities such as these, because after this time span, the probability of making mistakes increases very strongly (see Fünfsinn, 1984). Having to carry out these tasks for two or three hours without a break (which is unfortunately often the case in practice) cannot be justified either from the medical or from the safety point of view. Recreation periods are also necessary when, in quiet work-phases, there is very little information to process (also in this case, due to monotony, concentration will markedly decrease and the probability of faulty decision-making will increase).

Especially at times when the information barrage is heavy, controllers must use the right measures in time and reduce the complexity of incoming information. The limit of human information processing is no more than five to nine information units at the same time (an information unit is "a perceived phenomenon or complex of phenomena"; for a novice controller, this might mean a single incoming aircraft, for a veteran, six or seven [Linsey & Norman, 1981; Schröder, Driver & Streufert, 1967]). More simultaneous information than that will be either misperceived or processed incorrectly. Air traffic controllers usually use the following strategies in order to reduce incoming information:

- Try to separate relevant from irrelevant information (which aircraft, which tasks, which operations must wait? what has to be done immediately?).
- Delegate some of the tasks to a second person (which presupposes an appropriate number of personnel and a frictionless organization).
- Carry out routine tasks (i.e., handling the apparatus, changing the volume) automatically, in order to keep sufficient mental capacity for creative tasks.

Above all, it is important that the controller be able to take steps fast enough that no danger to flight security results in situations where the demands are very high. Early-warning systems should be adapted to human ability to deal with information overload. This could be accomplished if every technical innovation were developed through empirical studies, and the information-capacity limits of controllers were to be tested in their natural work situation.

Automation of the work place: Automation means that one has to give up a degree of flexibility in order to gain, on the one hand, a fast and efficient task-fulfillment; this would, on the other hand, produce a certain amount of rigidity. In nature, this always functions on low organizational levels: automation, the phenomenon of inherently-programmed activity, occurs throughout nature. A gooney bird, for example, will invariably roll its misplaced egg back into the nest; if the egg is taken away entirely, the bird will still perform the rolling motions: it is inherently programmed to do so.

In human acting and thinking, information or movements which are often repeated become automatic (Fuhrer, 1984). For a beginner, driving a car is a very complex task. The beginner has to coordinate many single tasks on the highest level of information-processing (which takes most capacity). For an experienced driver, the task is accomplished with much less capacity. The actions and tasks take place in a fixed, automatic action-scheme on a very low level of information-processing (Kastner, 1986). The price that the driver paid for flexibility becomes obvious when he buys a new car and constantly looks for the blinker on the wrong side. That is why non-uniform work places and panels cost a lot of wasted capacity, since automation of the routines is not possible ("new car effect").

But automation is a "two-sided sword." It should only be introduced when an unnecessary amount of information-processing capacity is being used, and flexible action which is able to adapt spontaneously to the demands of the environment is not needed. Two examples from the work of air traffic controllers could be:

- The paper strips with the data of current flights could be communicated by computer.
- The entry of the four-digit personal number of the air controller (squaks) could function like an automatic supermarket till.

The air traffic controllers in your organization could probably make this list much longer (if only asked).

But to really improve the situation, we should not ask the controller about only the disadvantages of old equipment and existing procedure. There are actually advantages to transferring information by paper strip: for example, each time a flight is handed over to another person, a short moment of attention is given to the new information. Information transfer by computer would have to produce this effect in another way (for example, by optic or acoustic signals), in order not to lose important information to lack of attention.

Another problem is whether automation decreases the amount and quality of social interactions, which might also have negative consequences.

Old, broken down equipment: The replacement of old and broken-down equipment is very important. Our research indicates a significant correlation between the number of disturbances caused by malfunctioning apparatus and the physical and psychological stress of task completion. In addition to this, performance becomes worse under these conditions. Disturbances caused by poorly-functioning equipment have an effect not only on stress and strain, but also on the quality of the verbal dialogue in teamwork situations (Harss, Kastner & Beerman, 1991).

Getting new equipment does not, however, solve all problems automatically. In the beginning, new equipment actually means more stress for the controller. Automatic actions which have been carried out for years on the old equipment cannot be used any more. Instead, all the actions have to be learned again and done consciously, requiring much more mental capacity and attention. If the expensive new apparatus continues to display old problems (which had already been known for years) and/or new problems (which could have been avoided easily by asking controllers' opinions in time), the controllers who have to work with it will naturally resist.

Before acquiring new equipment, it is very important to carefully observe and question those who will be using it, in order to avoid mistakes and acceptance problems.

Shift work problems: Shiftwork affects not only the long-term health of the controller, but also his/her performance (and therewith, flight security). Numerous medical and psychological studies (see Colquhoun & Rutenfranz, 1980; Hettinger, Kamminski & Schmale, 1980; Klein, Wegmann, Athanassenas, Hohlweck & Kuklinski, 1976) have shown that the human organism can best adapt to shifted day and night activities by holding the same rhythm for at least fourteen days. But long-term shift rotations have negative effects on social life (family, friends, etc.). As Costa (1991) recommends, quick-paced rotation between morning, evening, and night shifts may best prevent circadian-adaptation problems (because the body will not have time to adapt to any one shift). In any case, we dare not forget that shiftwork is, in and of itself, a major stressor.

Ergonomic factors: From the many ergonomic factors that affect the health, stress, and performance of air traffic controllers, we pick out only noise, temperature, and physical narrowness.

Stressors such as heat, cold, noise, physical narrowness, and forced poise lead automatically, by means of an old genetic program, to a stress reaction. They not only add up in their negative results, but also get into a reciprocating interaction with psychological stressors (as produced by the task, interpersonal quarrels, or shiftwork). Under permanent stress and strain, the body produces increased amounts of cathecholamins and glucocorticosteroids, which in the long term lead to many physical and psychological problems, such as gastric ulcer, coronary heart diseases, essential hypertension, insomnia, nervous problems, and concentration problems (Scheppokat, 1984; Silbernagel & Despopoulos, 1979).

In this way, a too-narrow, too-loud, too-warm, or too-cold workplace (which seems to be the case in many towers) affects the health of controllers and has a negative effect on the task to be carried out (Schönpflug & Schulz, 1979). It is amazing how few medical experts and controllers are included in consultations on the design of a new airport. The area of workplace for air controllers in their beautiful new towers is often below the minimum square-meterage prescribed by the European Community for *pig-sties*.

The Analysis of the Organization

Remember that an air traffic control organization is a complex social system consisting of administration, technical department, controllers, management, and (often forgotten) pilots. The task of all these functional units is to guarantee flight security. Here the air traffic controllers take over a function similar to that of the brain: there is a constant input of information to be evaluated, placed into order, and coordinated. The question is: how must the organization be structured in order to fulfill that central coordination and steering function?

Without going into detail about all the interesting approaches of system theory (see, e.g., Capra, 1987; Luhmann, 1985; Maturana & Varela, 1987; Vester, 1987; Förster, 1985) and their application to social systems, we want to mention some ideas which might help solve the problem.

Internal communication and information flow: There are three important aspects of communication and flight security:

1. In order to coordinate a highly complex and dynamic environment, even under time pressure, the internal cooperation and coordination must be frictionless (Westrum, 1991). A worsening of the organizational climate, or arguments between the different organizational units or members, lead to friction, loss of energy and efficiency, stress and strain, and lack of synchronized behavior of individuals with a mutual goal. These disturbances should therefore be taken just as seriously as the problems caused by old and broken-down technical equipment, since they affect flight security just as severely. Training on conflict management should be required for all individuals (at least all managers). Information should flow from bottom up and from top down.

2. The system "air traffic control" consists not only of controllers but also of pilots. Artificially separating these two groups is not reasonable, especially because here we are concerned with two organizational units who have to work together very closely. In Germany, one hears quite often of tensions between air controllers and pilots. We need a precise analysis of the causes of these problems, in order to improve the climate and avoid unnecessary friction in the system.

   Pilots should be better informed about the causes and reasons for a prolonged holding in the air or long waiting periods on the ground, in order not to attribute the blame to the controllers. On the other hand, more information about the compulsions put upon pilots of some commercial airlines could lead to more understanding for the situation of the pilots on the side of the controllers. The right measures to humanize the contact between air controllers and pilots would be common trainings and seminars, social gatherings, or meetings to exchange information.

3. Knowledge from other areas of flight security could lead to improved conversation between pilots and air traffic controllers. We (Harss, Kastner, & Beerman, 1991) found in our studies on helicopter crews, in accordance with other authors (Kanki, Lozito, & Foushee, 1987), that the quality of crew coordination depends less on the quantity and functionality of the information exchanged, than on its clarity and its calm, relaxed delivery. This means replacing wordy, stiff, robot-like exchange of information, in a tense atmosphere, with less but clearer (more precise) information. Small jokes or a personal "hello," "please," or "thank you" may fall — and actually should.

Negative feedback loops: In natural systems, which are the most sophisticated ones (Kastner, 1989), the principle of negative feedback loops dominates positive self-reinforcement tendencies. Both principles are described in figure 3 and figure 4.

In our study, understaffing in the technical department led to the fact that the fewer people who had to do more work suffered from chronic time pressure, overtime, and stress. Under these circumstances, the illness rates further increased as more workers became ill (or at least claimed to be ill), only increasing the pressure on the remaining workers even more (see figure 3).

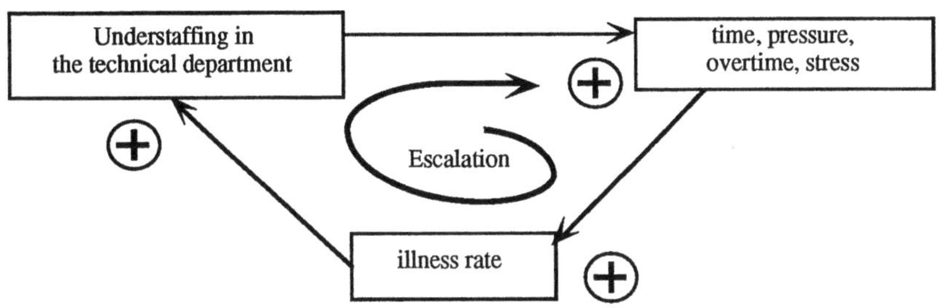

Figure 3: Example of a process with positive reinforcement tendency

In the case of dominant reinforcement tendencies, a negative development will escalate. Vicious circles therefore have to be noticed in time, and appropriate measures introduced (fewer tasks, more people, new technology, training, etc.). The system cannot afford to wait for the breakdown of one of its functional units. An alert supervisor should be able to spot increasing rates of mistakes, decreasing work satisfaction, conflicts between functional units, etc., then diagnose the causes and resolve them. That can be accomplished very easily by obtaining information from all important units in the organization. A good idea here would be a regular confidential short survey, in which all individuals are asked to name weak points of the organization and suggestions for improvement. By doing this, the superior can install negative feedback loops into the system (see figure 4).

Flexibility of the Organization: Complex systems have to adapt flexibly to quickly-changing environmental conditions. Inflexible budgets hinder fast reactions to new demands.

Organizational philosophy: Aside from the weakpoints we described above, less-concrete factors, such as organizational philosophy, can also have an effect. As we know from organizational psychology (for example: Mitchell & Larson, 1987), reward-and-punishment structures in organizations play a large role in reaching internal organizational goals. Flight security which rewards only the non-occurrence of accidents (with, for example, flight security prizes) can mean that no one willingly admits to "almost-accidents" or "close calls." Naturally, what constitutes a "close call" plays an important role here. Presently, this definition leaves a wide gray zone of (nonrecorded) irregularities. Traffic security research shows that "almost-accidents" represent a more reliable security criterion than real *de facto* accidents (Hoyos, 1980). We should therefore try to create an internal, confidential reporting system for irregularities of all kinds and reward honest reports about “almost-accidents.”

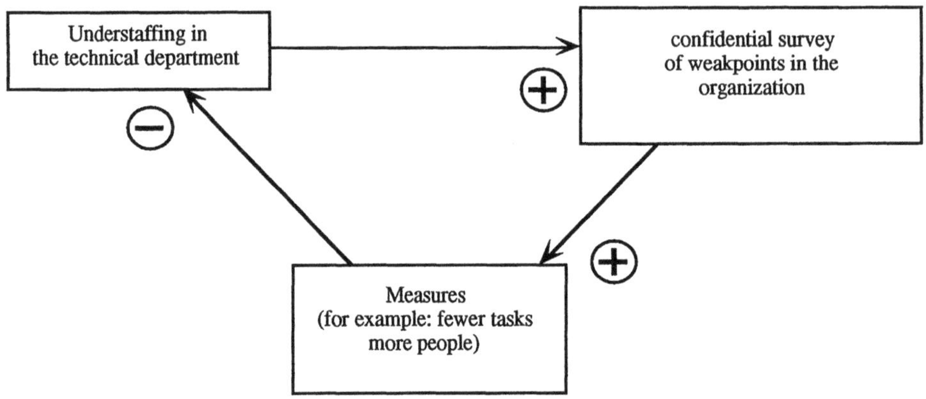

Figure 4: Example for a negative feedback loop in the organization

Image of the organization and its members: While pilots generally have high status in society, the image of the air traffic controller is not as good. The image of the organization and its members affects job satisfaction, motivation, efforts co-workers are willing to make, and last not least, the long-term chances an organization has to recruit qualified personnel. The organization should therefore ask: what status has the co-worker in society? what does the organization do to improve its image?

Qualitative vs. Quantitative Growth

Burgeoning air traffic presents seemingly-uncontrolled quantitative growth, but shall we call this progress? Only if it is acceptable to the environment can a natural system survive. In nature, organisms which show uncontrolled growth destroy their host-environments (as cancer destroys the human body). Because the old Munich airport could not cope with the air traffic flow of the 1990's, a new airport was built on one of the last remaining natural landscapes in the region; the last storks and frogs are gone now. We can no longer afford to accept this suicidal trend. In air traffic, as in all other human endeavors, we must now turn from expansion to improvement of our system. Our salvation lies in better *quality*, not more *quantity*.

Conclusion

From a psychological point of view, it is a mistake to isolate the so-called human-machine interaction from the context in which it is embedded. Human behavior, the key to all of this, is only understandable within the whole system, which includes:

- Person (individuals with specific behavioral styles – in our case, pilots, air traffic controllers, technicans, administrators, managers, etc.)

- Situation (tasks with specific ergonomic and technical equipment, at a specific work-place)

- Organization (management styles, image of the organization, coordination between departments and organizational units, organizational climate, etc.)

All these characteristics influence each other in a network. Therefore it is an illusion to think faults in the system can be localized to only "the person" or "the machine." Replacing the old type of controller with a new type of controller, or old technology with new technology, is not enough! Effectiveness and resources will still be wasted through other weakpoints within the system. We should place more attention on psychological and ergonomic work conditions, climate, conflicts, coordination, and communication, in order to improve flight security and effectiveness.

In the next few years, qualified personnel will become scarce, due to decreasing birth rates. Any organization which does not actively make efforts to improve work conditions will face serious problems finding qualified personnel (especially if the organization does not pay very well).

Organizational psychologists are usually called much too late — or as we say in Germany, "When the child already fell into the well" — when, for example, a new instrument is not accepted, a whole organizational unit is breaking down, coordination problems and conflicts occur, accident rates increase, etc. There are a lot of measures which, if taken in time, can save money and energy in the promotion of flight security.

We must also now take a new look at reducing the complexity and uncontrolled growth of air traffic. The controllers must be the first to say, "This is the limit. You're asking the unreasonable, and any more will destroy our environment and ourselves."

References

Alkov, R. A. (1983). Pilot error and stress management: US Navy experience. IFALPA/Aer Flight Operations Symposium.

Alkov, R. A., Gaynor, J. A. & Borowsky, M. S. (1985). Pilot error as a symptom of inadequate stress coping. *Aviation, Space and Environmental Medicine, 56,* p. 244-247.

Broadbent, D. E. (1971). *Decision and Stress.* New York: McGraw Hill.

Capra, F. (1987). *Das neue Denken.* Bern: Scherz.

Costa, G. (1991). Circadian variations of vigilance and performance and shiftwork. in J. A. Wise, V. D. Hopkin, & M. L. Smith (Eds.) *Automation and Systems Issues in Air Traffic Control.* Berlin: Springer-Verlag.

Colquhoun, W. P. & Rutenfranz, J. (1980). *Studies of shiftwork.* London: Taylor & Francis.

Dubos, R. (1978). Health and creative adaption. *Human Nature, 1,* p. 74-82.

Fünfsinn, G. (1984). *Antwortkatalog Arbeitsmedizin.* Niederselters: JungJohann.

Fuhrer, U. (1984). Mehrfachhandeln in dynamischen Umfeldern. Göttingen: Hogrefe.

Harss, C., Kastner, M., & Beerman, L. (1989). The Impact of Personality and Task Characteristics on Stress and Strain During Helicopter Flight. In *Stress and error in aviation*. London: Grower.
Harss, C., Kastner, M., & Beerman, L. (1991). Interactions Between Person and Task During Helicopter Flights. (paper in press).*First International Journal of Aviation Psychology, 1.*
Hettinger, T., Kaminsky, G. & Schmale, H. (1980). *Ergonomie am Arbeitsplatz*. Ludwigshafen: Kiel.
Hoyos, C. Graf v. (1980). *Psychologische Unfall- und Sicherheitsforschung*. Stuttgart: Kohlhammer.
Kanki, B. G. Lozito, S. & Foushee, H. C. (1987). Communication Indexes of Crew Coordination. In R. S. Jensen (Ed.) *Fourth International Symposium on Aviation Psychology* (p. 406-412). Columbus: Association of Aviation Psychologists.
Kastner, M. (1986). Zur Koppelung von Beanspruchungen im Rahmen der Arbeits-und Organisationspsychologie. In H. Methner (Hrsg.), *Psychologie in Betrieb und Verwaltung* (S. 43-66). Bonn: Deutscher Psychologen Verlag.
Kastner, M. (1989). *Führen als Denken, Erleben und Handeln im System. Sonderbeitrag der Süddeutschen Zeitung. München:* Süddeutscher Verlag.
Klein, K. E., Wegmann, H. M., Athanassenas, G., Hohlweck, H. & Kulinski, P. (1976). Air operations and circadian performance rhythms. *Aviation, Space and Environmental Medicine, 47*, p. 221-230.
Lindsay, P. H. & Norman, D. A. (1981). *Einführung in die Psychologie*. Berlin, Heidelberg, New York: Springer.
Luhmann, N. (1985). *Soziale Systeme*. Frankfurt: Surkamp.
Maturana, H. & Varela, F. (1987). *Der Baum der Erkenntnis*. Bern: Scherz.
Mitchell, T. R. & Larson, J. R. (1987). *People in Organizations*. New York: Mc Craw Hill.
Reason, J. (1989). Organizational Accidents in Complex Technologies. Paper presented at the 18th WEEAP Conference. Brighton: University of Sussex.
Schönpflug, W. & Schulz, P. (1979). *Lärmwirkung bei Tätigkeiten mit komplexer Informationsverarbeitung. Forschungsbericht 79-10501201*, Berlin: Institut für Psychologie der FU.
Schröder, H., Driver, M. J. & Streufert, S. (1967). *Human Information Processing*. New York: Holt Rinehart & Winston.
Silbernagl, S. & Despopoulos, A. (1979). *Atlas der Physiologie*. Stuttgart: Thieme.
Simmel, E. C., Cerhovisik, M. & McCarthy, J. (1987). Source of Stress Affecting Pilot Judgement. In R. S. Jensen (Ed.) *Fourth International Symposium on Aviation Psychology* (p. 190-194). Columbus: Association of Aviation Psychologists.
Simon, H. A. (1964). *The Architecture of Complexity. General Systems Yearbook.* Vester, F. (1987). Unsere Welt - ein vernetztes System. München: DTV.
von Förster, H. (1985). *Sicht & Einsicht*. Braunschweig: Vieweg Verlag.
Westrum, R. (1991). Air traffic control automation: Information and consciousness. in J. A. Wise, V. D. Hopkin, & M. L. Smith (Eds.) *Automation and Systems Issues in Air Traffic Control.* Berlin: Springer-Verlag.

Automation as an Influence on Controller Selection

Selecting for Air Traffic Control: The State of the Art

Hilda Wing

Staffing Policy Division, Federal Aviation Administration, Washington, D.C., USA

Introduction

In developing a selection system for air traffic controllers — or any occupation, for that matter — considerations do not stop with selecting predictors. The paper presented at this conference by Della Rocca, Manning, and Wing (1991) briefly described the present political climate in which selection occurs in the United States, as well as the technical support for procedures the Federal Aviation Administration (FAA) uses to select air traffic controllers. This paper will be more theoretical, or, at least, less empirical and more thoughtful, about the concepts personnel psychologists are currently using in evaluating selection systems.

First, there are predictors, and there are many more to consider than just a few decades ago, and in different ways. There has been an even larger expansion in the ways we view performance, as well. We have also begun to take far more care in developing performance measures. Finally, we are beginning to understand that good performance measures can be useful in more ways than simply providing criteria for statistical validation studies.

Predictors

The use of selection procedures could be traced to the dawn of antiquity, but such tracing will not be done here. Certainly, the formal use of selection instruments by American psychologists was initiated with the Army Alpha in World War I (Zeidner and Drucker, 1988). The U.S. military paid little attention to mental testing and psychometrics between the two World Wars, but the civilian sector used multiple-choice mental tests to a great extent, causing some commentators such as Walter Lippman to question these practices severely (Tuddenham, 1962). A number of academic psychologists, both in the United States and elsewhere, investigated both the contents of mental tests and the statistical procedures used to analyze them, such as Thurstone at the University of Chicago with his seven primary abilities (Tuddenham, 1962).

When World War II began, the U.S. Army built on the work of these psychologists to develop more sophisticated mental ability tests for selection and classification (Zeidner and Drucker, 1988). Since the end of the Second World War, the use of testing in employee selection and classification has waxed (until the implementation of the 1964 Civil Rights Act) and waned (until a slight resurgence during the 1980's). However, psychometric theory has been an international and extraordinarily productive venture, culminating in practical

NATO ASI Series, Vol. F 73
Automation and Systems Issues in Air Traffic Control
Edited by John A. Wise et al.

applications of item response theory during the 1980's (Lord, 1980). Similarly, mainstream psychological theory began to shed some of its behavioristic past with the publication of Neisser's *Cognitive Psychology* in 1967. The acceptance of cognitive psychology signaled the theoretical and scientific acceptance of mental or cognitive ability tests. Indeed, many of the classic experiments in cognitive psychology have used mental test items either as control variables or as experimental materials (e.g., Hunt, 1985; Sternberg, 1985).

Domains

Over the past two decades, a number of psychologists have developed taxonomies of human abilities and skills. The research supporting these taxonomies began during the Second World War and the post-war period in the United States. The purpose of many of these ventures was to have the means of (1) evaluating jobs to determine what knowledges, skills, abilities, and other worker characteristics (KSAOs) are required to perform the job at a specified level of competence; and (2) to evaluate a person for the level or amount of the required KSAOs possessed by that individual. The best known of these is, perhaps, Fleishman's (e.g., Fleishman and Quaintance, 1984); Peterson and Bownas (1982) expanded on Fleishman's to include personality traits.

While there is no one agreed-upon standardized list of measures, American psychologists today appear to agree on three general domains of predictors: cognitive, including mental and perceptual abilities; non-cognitive, including personality traits and interests; and physiological or physical, including medical conditions and physiological indicators. Let us consider each of these in turn.

Cognitive: The Differential Aptitude Tests (DAT; Bennett, Seashore, and Wesman, 1981) can be linked to military and civilian research completed in the late 1930's and early 1940's. Used now primarily for providing vocational guidance to students in secondary schools, the tests cover Verbal Reasoning, Numerical Reasoning, Abstract Reasoning, Perceptual Speed and Accuracy, Mechanical Reasoning, Space Relations, Spelling, and Language Usage. There are, of course, other batteries developed by other test publishers that could equally well provide examples of tests of cognitive skills. These abilities are not independent of each other: a person scoring high on one test is somewhat likely to score high on the others as well. These are the abilities that predict, as well as any one thing can predict, how well a person will do in school or in college or university, or how well a person will do in the job. Spearman, a British psychometrician of the early twentieth century, had argued, on the basis of his early factor-analytic work, that all tests measured but two factors: "g," or the general factor, and a factor specific to the test in question. Indeed, two American psychologists, Jack Hunter and Frank Schmidt, have argued (see Schmidt, 1988) that available statistical evidence supports the use of only the general or common factor, "g," in selection. This is not a generally accepted claim, however.

While many educational programs and occupations may not seem to have specific cognitive ability requirements, there are some that definitely do. For example, technical occupations, such as engineering, electronics repair, and carpentry have specialized requirements for mechanical reasoning and space perception. Air traffic controllers worldwide appear to require a high level of spatial reasoning ability. Successful performers in these occupations may or may not have high levels of skill in verbal reasoning or language usage; individuals earning high scores in verbal reasoning or language usage may or may not do well in technical occupations.

The distinction between cognitive and perceptual skills is not clear. Perhaps cognitive skills are those requiring more internal mental processing, such as reasoning and vocabulary, while perceptual skills are demonstrable with speedier responses. In this light, spatial skills can be cognitive or perceptual; clerical speed and accuracy appear to be more perceptual. The distinction is not likely one of mental complexity, as skills such as attention and listening can be highly sophisticated.

There is potentially quite a difference in the complexity of the instruments used to measure these skills. Most cognitive skills, and some perceptual skills such as clerical speed can be measured more or less reliably without much more than paper and pencil and a stopwatch. With the development of fast, inexpensive computers, and sound production equipment, definition and comprehension of other perceptual skill concepts has exploded. The relevance of many of these to the workplace, particularly to the identification of individual differences for air traffic controllers, has not yet been demonstrated. It is likely that successful controllers are superior in a number of such skills.

Since pilots in World War II were required to use their bodies as well as their brains in flying planes, instruments were developed to evaluate individuals on the requisite psychomotor skills, such as multi-limb coordination. Apparatus for skills required in other occupations, such as eye-hand coordination for assembly, was also developed. While these tools were useful, such equipment could be subject to severe problems of unreliability. While the advent of computers has changed all that, the development of automation suggests that requirements for psychomotor skills, as traditionally understood, may have disappeared. If, however, we redefine terms and include attributes such as speech, keyboarding, and moving a track ball to follow a target, then we can demonstrate the requirements for psychomotor skills in occupations such as air traffic controller.

Non-Cognitive: It is always difficult to define anything by starting off with what it is not, but such is the case with non-cognitive attributes. One positive and useful distinction is that cognitive measures are designed to tap an individual's maximum performance — the best that the person can do or can be, while non-cognitive measures tap the individual's typical traits or states — the average of what that person usually is. Another way of stating this is that the cognitive types of measures refer to what the person can do, while the non-cognitive types refer to what the person does or will do. Mathematically, the relationship between cognitive predictors and performance is expected to be monotonically increasing, if not linear: the more the measured amount of the cognitive predictor, the higher or the better the performance. For non-cognitive measures, however, the mathematical relationships could be u-shaped: there is an optimal level of the non-cognitive trait under consideration, below and above which performance decreases. For example, there could be an optimal amount of motivation. Too little means the job won't get done at all, while too much motivation will lead to so many errors that the job won't get done that way, either.

It is easiest to list the attributes each domain includes, to which we now turn. One important set of non-cognitive measures refers to personality variables, such as ability to handle stress, motivation to succeed, and so forth. The test items refer to feelings, attitudes, activities, and judgments and preferences about the self and others. Most of the instruments available for measuring personality rest on the identification of a set of relatively enduring traits. Knowledge of an individual's standing on such traits can be helpful, both for the individual's own well-being, such as in a clinical or counseling context, as well as for an employer's well-being, in hiring the person "best suited" for the job. The evidence for and usage of personality test data for the latter purpose are more recent than the first. Many professionals have concerns about both the adequacy of the validation as well as the possible invasion of privacy.

In recent years there has been a consolidation of information about what personality tests are measuring, with a consensus in the research that there appear to be five major, complex factors of personality. Digman (1990) identified them as (I) Extroversion/introversion; (II) Friendliness/hostility; (III) Conscientiousness (Will); (IV) Neuroticism/Emotional Stability; and (V) Intellect (Culture). Hough (1988) presented a summary and meta-analysis of the validity of such scales for employment. Based on Digman, it may be that the most important factors for air traffic controller are (III) Conscientiousness (Will) and (IV) Neuroticism/Emotional Stability. Factor III incorporates dependability, task interest, and prudence. Factor IV taps anxiety, adjustment, and emotional stability. These two classes of attributes are evaluated in some way for controller candidates in the United States. The Sixteen Personality Factor Questionnaire (16PF; Cattell, no date), one of the first multi-trait personality tests developed, is used by the FAA to evaluate the ability to deal with stress and anxiety.

Somewhat related to personality traits are interests. Interest inventories have been used more for counseling than for selection. Most seem to be geared to helping an individual think through many of the available options of vocational choice and do not necessarily provide much detail about occupations within a given conceptual area. Inventory questions refer to likes and dislikes for a variety of activities and occupations. For example, an individual might be helped to understand that he or she wants a career in aviation, but not necessarily whether that career is as a pilot or as a flight attendant or as a controller. Holland's Vocational Preference Inventory (Holland, no date) and Self-Directed Search (Holland, no date) are based on a theory of careers falling into six areas that can be placed on the six corners of a hexagon: Realistic, Investigative, Artistic, Social, Enterprising, and Conventional. The RIASEC model is easy for people to comprehend and to use, and evidence suggests that it is helpful to them in sorting out career options. Whether it would be useful for selection is a different question. The largest single category of jobs is Realistic (occupations are not equally distributed about the hexagon's corners).

More specific interest scales could be more useful for selecting air traffic controllers, along the lines of the Strong Interest Inventory (Strong, Hansen and Campbell, no date). Earlier versions of the SII were developed with a strong if not exclusivist empiricist base, although recent versions incorporate much of the theory developed in the past half-century, such as the RIASEC model. However, there is no scale for air traffic controller, nor for pilot. It seems reasonable that some controllers have a history of interest in certain activities, such as air shows and computer games, rather than others. Whether this is so, and what interests and activities these might be, remain to be determined. A cautionary note: determining the interest patterns of current controllers — or of incumbents in any occupation, for that matter — can freeze-in these interests as well as other, concomitant characteristics of the incumbents. For example, controllers in the U.S. are still predominantly white males, while the projected work force (Johnstone and Packer, 1987) is predominantly female and non-white. To the extent that race and gender determine interests in general more than they determine interest in a specific occupation, empirical scale construction can be misleading.

A third category of non-cognitive measures is biographical data, or biodata. It is the case that occasionally it will be difficult to distinguish a given item as coming from a biodata questionnaire, an interest inventory, or a personality test. An important distinction here is the potential for the applicant-supplied information to be verified. The typical instrument for collecting biodata is the employment application form, and the typical datum requested is an item of the applicant's history. Previous employment and educational history are fairly standard.

Recent years have seen more biographical questions pertaining to drug and alcohol abuse. Case law is developing concerning the conditions where such questions are permissible. Since the banning of the use of polygraphs in most private employment selection in 1988, a number of firms have begun to use honesty or integrity tests in employment selection. Such tests include many more or less explicit and specific questions concerning the applicant's admissions of and attitudes towards theft, dishonesty, drug use, and so forth. Much of the information requested could potentially be developed by a full security background investigation, at a much higher cost than the paper and pencil honesty test. While the evidence for the validity of such instruments is only beginning to accumulate, it suggests (O'Bannon, Goldinger, and Appleby, 1989; Sackett, Burris, and Callahan, 1989; Sackett and Harris, 1984) that they could be useful.

Physiological/Physical: Job requirements in the physiological and physical domain can take a variety of forms. One is in terms of strength and stamina, either in general terms, as for an infantryman in the Army, or in specific terms, as for a waitress or surgeon who must stay standing for hours on end. There are pronounced sex differences in physical strength; other physical characteristics may reflect less of genetic or sex-linked characteristics and more of opportunity to develop physical fitness.

A number of occupations have specific strength and stamina requirements, such as firefighter and police officer. Other occupations have a more general requirement of excellent health, such as pilot and air traffic controller, as documented in a specified medical examination. In such occupations, certain medical conditions may be disqualifying. Active air traffic controllers in the United States must begin training before their 31st birthday and have to retire by the age of 56. Since active controllers can retire with larger pensions than other civil servants at the same age and with the same amount of service, it can be seen that the age requirement has political implications as well as medical. Controllers also regularly undergo medical examinations, and at present certain medical conditions can also disqualify them from active service, such as taking AZT for AIDS, or being insulin-dependent diabetics.

Other physiological indicators such, as Heart Rate Variability (HRV) and galvanic skin response (GSR), have shown promise as important indicators of psychological response in enough situations to support their potential for research with air traffic controllers. Since the presence of consistent individual differences in such measures is not well-established, their potential use in selection is unknown. The GSR is a component in the polygraph, however, and the polygraph has been made illegal in much employment selection. The other measures discussed above appear more promising for selection of air traffic controllers than do these physiological indicators.

Methodology

It has become a truism to say that any psychological concept is the method by which it is assessed: "Intelligence is what intelligence tests measure." While mostly false, because the research base in psychology has provided meaning for such concepts above and beyond their measures, the method we choose to obtain information does influence the information we will get. The classic paper by Campbell and Fiske (1959) on the Multi-Trait-Multi-Method approach addressed just this issue. To develop comprehensive knowledge of an attribute or a concept, measure it in more than one way and in more than one situation.

Paper and Pencil: Most psychological measures are obtained with paper and pencil. These include not just cognitive tests and personality surveys, but supervisor ratings and archival records (absenteeism, awards) as well. The concepts measured this way are static; they are recollective rather than creative. Some standardized situation is placed before the respondent; the respondent's reaction to the task does not change the nature of the task itself. We know how to obtain scores with useful psychometric properties from such tasks.

Computers/Simulators: Recent years have seen the development of computerized testing, the increased use of simulators. Some of the approaches have limited themselves to being automatic page turners for the familiar paper and pencil approach. Others, however, have used computers in more dynamic ways. One approach, based on the psychometric development of Item Response Theory (IRT; Lord, 1980; Weiss, 1983) presents tests tailored to individual examinee ability, as indicated by the correctness of responses to more and less difficult items. IRT provides the foundation for assigning scores on the same scale to individuals who did not answer the same test items. Another approach uses the computer to present stimuli and tasks that are impossible with paper and pencil. For example, the U.S. Navy developed a computerized battery that included tasks to evaluate static and dynamic spatial ability (Hunt, Pellegrino, Abate, Alderton, Farr, Frick, and McDonald, 1987). The static spatial tasks were tapping skills that were measured as well by paper and pencil tests, while the dynamic spatial tasks appeared to require the additional ability of dynamic movement processing. This latter skill appears to be important for air traffic control, but would be missed by any assessment procedures limited to obtaining applicant information by paper and pencil.

The question of the possible relevance of computerized tests for selection to a given occupation leads directly to the question of the appropriate use of simulators in selection. Certainly, ATCS video games are available commercially, and their face validity appears high. In the United States, we face the important but solvable problem of using a commercial product for use in a Federal selection program: we can't use commercial products for selection but we could develop our own simulator. More important than whether we can do so is whether we should. Simulators could be tapping both more and less than what we need to measure for accurate prediction of later occupational success. Further, there is the problem of differential experience. We know that such computerized exercises, as well as paper and pencil tests of spatial ability, are fairly subject to practice effects. For training purposes, improvement of scores with practice is the desired goal. For selection, we need to know which score, the first or the second or the third or whatever, is the most valid predictor of some later, post-training criterion. If the first score is the more valid, then providing opportunities for second and subsequent scores is misleading at best. Of course, an even better alternative might be to assess learning rate, and use that as a predictor of subsequent performance. This is not a new approach, but, until recently, research has not provided much support. Ackerman (1987; 1988; 1989) and Ackerman and Kanfer (1989) have suggested ways in which learning ability could be measured and be useful as a predictor of subsequent events. The approach is promising.

Assessment Centers: While computers can provide dynamic tasks reflective of activities and events in the technological or built world, they don't imitate people very well. That is, controllers work with other people, such as other controllers, supervisors, developmental or trainee controllers, and pilots. For selection of controllers based on their skills with other people, it could be best to develop an appropriate task that requires many people, then watch and record. Developing an appropriate task, getting people together, watching, and recording are difficult but not impossible tasks. Whether the measures would be useful, and, if so, for what criteria, also must be determined.

American pioneer work in assessment centers began during the Second World War, to select for, among other occupations, spying behind enemy lines (Thornton and Byham, 1982). An individual was placed in a specific situation and his behavior was carefully observed. An oft-cited if not apocryphal situation was ordering the individual to build a bridge with the assistance of a few troops to be supervised. Of course, one of the troops was a stooge who did his best to thwart every move by the examinee. The score was less of building the bridge and more of keeping one's temper. After the war, American Telephone and Telegraph (AT&T) commenced a massive study of managerial selection and development (Bray, Campbell, and Grant, 1979). In recent years, "assessment center" has come to mean the arrangement of a 1-5 day or so time period where individuals come together to be put through a set of group and individual exercises. The purpose of the assessment center can be either selection or development; it is typically for managerial positions but can be useful for any occupation where interacting with people is critically important. The evaluators in assessment centers tend to be incumbents of the occupations for which the center is directed, who are trained and monitored by experts in the assessment center process.

This type of assessment is expensive: only a handful of individuals can be evaluated at a time and an evaluator can attend to only one or a few candidates at a time. These are similar circumstances to those for assessment with computers and simulators, however. Using computers for selection might require dedicating a small number of them, in a few places, where access and functioning can be tightly controlled. For air traffic controllers, assessment centers could be a way to combine measurement of cognitive and social skills. To the extent that job analyses support the requirements for such abilities for the air traffic controller job, an assessment center approach could be appropriate.

Work Samples: A final class of methodology is work samples, where people are asked to perform, in a quasi-standardized situation, what they do on the job every day. Such measures are highly credible criteria for validation research. For example, in validating the written test portion of the Professional and Administrative Career Examination (PACE) for Federal occupations, Corts and his colleagues developed a number of innovative work samples. For Customs Inspector (Corts, Muldrow, and Outerbridge, 1977), a video tape was prepared that presented four situations requiring a Customs Inspector, such as giving permission for an incoming boat to dock and quizzing incoming international passengers. Each vignette included procedural errors that the examinee was to notice and correct; each vignette had its own special set of forms to complete. For Social Security Claims Examiner (Trattner, Corts, van Rijn, and Outerbridge, 1977), sample claims were developed that the examinee had to process correctly.

During the 1980's the U.S. Department of Defense sponsored the Joint Service Job Performance Measurement Project. The four military services developed work sample measures for approximately 25 enlisted occupations, including the U.S. Air Force Air Traffic Control Operator. Critical oversight of this project was provided by a committee of the National Academy of Sciences. Technical summaries of both individual Service efforts (e.g., Campbell, 1990; Campbell, Ford, Rumsey, Pulakos, Borman, Felker, de Vera, and Reigelhaupt, 1990; Hedge and Lipscomb, 1987; Hedge and Teachout, 1986; Kraiger, 1990) and of the total project (e.g., Green and Wing, 1988; Green, Wing, and Wigdor, 1988) are beginning to appear. Two important components of the impetus for this project and other performance measurement efforts began during the 1970's. First, civil rights organizations and others were less than satisfied with test validation studies that used training grades or supervisor ratings as the criteria to predict. Training performance is not necessarily the same thing as job performance, and supervisor ratings were at best global and at worst biased and uninformed. Second, the U.S. military learned during the late 1970's that the Armed Services Vocational Aptitude Battery (ASVAB), the test battery used to select applicants for enlisted occupations, had been misnormed. Approximately 250,000 young men and women had been allowed to enter the service who actually demonstrated less aptitude than required. Military selection test batteries have traditionally been validated against training criteria. While many of these young men and women did not succeed in the military, a large number of them were successful on the job. One can argue that the size of the available validity coefficients should support the expectation that some percentage of low-scoring applicants would indeed be successful. Employment selection is at best an imperfect process. However, individuals both in and out of the military services recognized that validation against job performance measures, and against samples of actual work, would be far more credible than training measures. This is reflected in contemporary legal (U.S. Equal Employment Opportunity Commission, U.S. Civil Service Commission, U.S. Department of Labor, and U.S. Department of Justice, 1977) and professional (American Educational Research Association, American Psychological Association, and National Council on Measurement in Education, 1985; Society for Industrial and Organizational Psychology, Inc., 1987) guidelines and standards.

The use of work samples as criteria for validation of selection tests is different from the use of work samples for selection. The latter is appropriate and fair when all candidates have had equivalent opportunity to learn to perform the job. It is not appropriate for entry-level jobs where differential candidate experience in situations similar to the target occupation may cast uneven bias on all candidates. However, a fair selection procedure could result when all candidates are provided the same opportunity to learn to perform the job, or some important part of the job. One name for this approach is the training and test situation, and it is the model the Federal Aviation Administration currently uses as a second-stage screen for air traffic controllers. As described in a preceding paper (Della Rocca et al., 1991), during the nine-week FAA screen, candidates are taught the principles of non-radar control and are evaluated by their performance on a series of academic and laboratory exercises. While performance on the screen predicts subsequent attainment of journeyman or FPL status, the screen is much misunderstood. It has been criticized for inadequate provision of training, yet its function is selection, not training. It has been criticized for not using radar training and for being less than adequately job-related, yet its function is to predict subsequent achievement. More importantly, it has been criticized for being expensive, since the cost to the FAA of each candidate is $10,000. Such criticisms of the FAA screen — that it is not training, that it isn't job related, that it is costly — are actually criticisms of the training and test model itself. For a small number of occupations, that require long and intensive training to perform complex activities, however, the training and test model may be an attractive option. The progenitor of the current screen was attractive to the FAA in the mid-1970's, when the U.S. Congress suggested that field training attrition of developmental controllers could be reduced by more intensive initial selection.

Limits of Validity Generalization

The theory of validity generalization (Schmidt, 1988) provides models for demonstrating that selection validity coefficients generalize across different organizational settings for the same occupation, if not for different occupations, and across different forms of the same test, if not different tests. Some versions of this theory (e.g., Hunter and Schmidt, 1982) suggest fewer limitations on the extent of such generalizability than do others (Hartigan and Wigdor, 1989). For entry level selection, there may be only a few meaningful differences among occupational classes, and these differences could be mapped to an equally small number of different classes of cognitive tests. However, there are at least two circumstances that mitigate against accepting validity generalization without major qualification. First, and less important, is that the statistical procedures used to fit validity generalization models are based on multiple regression, an approach well known for ignoring the relatively minor differences among the different cases in favor of identifying the similarities. A few occupations with relatively unique requirements, such as air traffic controller, would escape notice. The statistical methods of validity generalization make it virtually impossible to identify any small relative contribution. Second, and more critical, is that most of the validity coefficients used in validity generalization studies are based on either training measures (somewhat less than half) or supervisor ratings (somewhat more than half). Virtually none have included studies with work sample measures. To the extent that a predictor is general, as is the case with most commercially available cognitive tests, it will predict a general criterion, such as training or a supervisor rating, at a moderate level. This is validity. The more the predictor becomes like the criterion, however, the higher the validity coefficient can become. The issue of validity becomes the issue of reliability. To be sure, whenever the selection question is one of predicting for entry-level applicants what their job performance will be at some subsequent time, validity can never become reliability. However, conceptualizing validity as imperfect reliability can lead to a different and useful perspective on how we develop and choose both predictors and criteria.

Performance Criteria

Any job, or any part of a job, is a mental construction. Virtually all people engage in continuous activity, which a given observer will categorize as his or her purposes permit. There is no one right way to apportion a controller's activities into units of performance, but some ways are more effective than others for specific purposes.

This conference has seen a variety of ways to categorize controller job performance. Benoît (1991) and Völkers (1991) looked for controller actions that could be transformed into computer functions, requiring the controller to perform somewhat differently, and hopefully more economically, to accomplish the same ends. Stager (1991) studied those aspects of controller behavior that could be called systemic errors, and applied the power of cognitive theory and models to explicate them. Our resident controllers, on the other hand, specifically Lane (1991) and Levesley (1991), took issue with this scientific categorization because the Stager taxonomy was not identical to the legal taxonomy of controller errors. Tattersall (1991) looked at performance in terms of the influence of stress. Edwards (1991) viewed controller performance in metaphoric terms from artificial intelligence. Della Rocca's (Della Rocca et al., 1991) interest was in controller performance that should be predicted from selection procedures. And so on.

The participants in this symposium shared an interest in the performance of air traffic controllers, but each looked at this performance differently. Because this is to be expected, it is necessary to be explicit about our categorizations and our purposes. Sharing the different perspectives can enrich all of them.

Selecting People to Perform

When assessing individual differences in performance to serve as criteria for selection procedures, or in any way for any purpose, certain constraints do hold. The first is that the sample of incumbents is a selected one. Perhaps many individuals could perform the job; perhaps only a few individuals could do it. In point of fact, however, only a few individuals are doing the job, and generalizations outside of the incumbent population, while necessary, must be cautiously made. For example, varying experience might be the only attribute linked to varying performance within the group, because the group includes only high ability people. This instance of range restriction is fairly common in criterion-related validity studies. While statistical corrections are both possible and recommended, qualifications are also required.

A second constraint reflects the effectiveness of the training required for performance. Completely effective training would theoretically wipe out any individual differences in performance. This also can reduce validity coefficients. For example, validity studies for U.S. military occupations are typically lower than those for civilian occupations (Hunter, 1983). Perhaps this is can be attributed to military training being generally more effective than civilian training, including public school systems. Third, the type of performance measure to be obtained is also important. In the United States, as mentioned above. a strong preference for work sample measures is present, both among the scientific community as well as among managers, policy makers,and civil rights organizations.

When it comes to arranging for the sample of incumbents and going to them to obtain their performance data, other characteristics become critical. First, the individuals must have comparable training and experience. It wouldn't be appropriate, for example, to assume that the performance of an advanced developmental controller is comparable to that of a five-year veteran. Second, it would be both unsound and unfair to request performance on tasks that people aren't otherwise required to do. In the United States, controller assignment — and performance — varies by type of facility and complexity of operations. Requesting a center controller to demonstrate airplane takeoff and landing would be inappropriate at best.

Third, different types and methods of performance measurement will shed light on different aspects of the controller "performance space," as it were. Job knowledge tests and work samples will speak to what the controller can do. Supervisor ratings and archival measures address what the controller has done and will do, in the eyes of supervisors and others. The selection of which aspects are more important, and how much so, is a decision of high policy import. We can help inform this decision as we keep our terms, procedures, and findings clear.

There are additional, strictly measurement issues that influence our findings. It seems fairly trivial to state that the completeness of the description and the necessity of defining terms for the elicited performance are crucial, or that the stimulus and response must have complete clarity. However, there are enough instances to the contrary in the professional literature to make this statement necessary.

Second, there are a number of statistical or psychometric concerns requiring attention that may have conceptual implications as well. For example, specific performances that are either very common or very rare may or may not be interesting, but they will always have bad psychometric behavior. Different variables are rarely independent of each other, and the qualities of their interrelationships will play more or less havoc with multi-variate analytic procedures that assume only minimal relatedness. The relative weight of different measures will depend very heavily on their relative variances, or comparatively how far apart people find themselves on the scale. Weighting an interview score four times as heavily as a test score is meaningless if nearly everyone passes the interview while there is great dispersion on the test. Finally, as stated above, the method of measure influences the meaning of the measure greatly, although not completely (Campbell, McHenry, and Wise, 1990).

If we have acceptable selection procedures, such as cognitive tests and good biodata measures, and appropriate psychometrics, what values do we get for validity coefficients? As might be expected, it does depend on the criteria we use. For supervisory ratings, a validity of about 0.25 is average. For training measures and job knowledge tests, the validities are higher — about 0.35. Average validity coefficients for work samples or simulations are yet higher, averaging about 0.40 (Hunter, 1983; Hunter and Hunter, 1984). An average value for archival measures is not readily available, as this is a most heterogeneous group.

While it is possible to predict performance better, particularly were we to have better measures, it is unlikely we could approach perfect prediction. First, performance has to change as the work place changes. The FAA will be fielding new air traffic control systems over the next ten to fifteen years that will require changes in controller performance. Some of these changes appear to be profound, and it would be unwise to assume that the best performers on the old equipment will be the best on the new. Second, management priorities shift. At present the air traffic systems of many countries place a very high premium on air safety. With more sophisticated equipment both on the ground and in the air, the amount of traffic to be accommodated could increase. Whether the difference between safety and the next highest value would decrease is a new question. Third, otherwise extraneous characteristics of the incumbents influence job performance in ways we will only begin to understand as different kinds of people become incumbents. Young white men from West Texas are in decreasing numbers as applicants for air traffic controllers, while the numbers of young women and minority men are increasing. Certainly there will be subtle and not-so-subtle changes in the ways managers find effective to motivate their employees, but how the performance might change is even less clear. Fourth, geopolitical events have been changing the map of the world, and of the air space. In 1990 there is an entity called Eurocontrol that provides what American controllers would call flow control over 14 (or 22) European countries. Such an entity would have been inconceivable fifty years ago, yet at this conference we imagined the expansion of Eurocontrol past what used to be the Iron Curtain into countries of Eastern Europe. As I sit revising this paper, the countries of the Middle East are in the throes of the revision of one country(Kuwait)'s border by another country (Iraq), with world-wide repercussions. Finally, there are major life events for an individual that can influence performance but which are impossible to predict. These include marriages, births, illnesses, and deaths of the performer or the performer's significant others. We could list a variety of statistical excuses, but it is less arrogant to concede that life remains mostly a mystery.

Developing Performance Measures

Before one can develop performance measures, it is necessary to obtain information about the performance in a systematic manner, or to perform a job analysis. Obtaining information systematically is a general description of scientific observation; During this century a variety of methods to collect data on jobs has evolved. Gael (1988) edited a two-volume set of such job analysis methods. Much of the diversity reflects the different uses to which job information is put. For example, job-task analysis is exhaustive and detailed, and is necessary for developing training programs. Critical incident methodology, on the other hand, focuses only on those components that differentiate the clearly superior from the clearly inferior. Equipment design, compensation, and job restructuring would focus on still different aspects of the job. For any focus of job analysis, however, certain perspectives are gained while others are necessarily lost. We can understand the simpler abstractions from experience better than the experience itself, but the gain in understanding is purchased with a loss of fidelity. Obviously, the more we know ahead of time about what we choose not to investigate, and what we choose not to measure, the more likely we are to produce effective and efficient job analyses.

Given that a job analysis method has been selected to identify components of interest to our purpose, choices are still necessary. For example, say we have a job-task analysis for the air traffic controller job that has been completed for the personnel administration functions of selection, training, and performance assessment. Most occupations incorporate several hundred tasks, depending on the level of detail in the task description. There are more tasks present than are done by any one incumbent. Which tasks are to be trained when? Which are the tasks for which we should select? Which tasks should be incorporated into performance measures?

To answer such questions, we require information about the relative frequency, importance, and complexity or difficulty of each task. How severe are the consequences of performing a task incorrectly? What are the tasks that differentiate the top performers from the poor or average performers? At this stage, the different personnel functions would diverge. Training might focus on the frequent and important tasks, regardless of their difficulty. Selection might focus on the tasks distinguishing good from poor performers, even though most performers might be able to do most tasks adequately. Performance measurement might attempt both, looking for those tasks that most incumbents perform that are also critical to the job.

There is no one "right" focus, but different perspectives depending upon purpose. Selecting aspects of job performance is neither an arbitrary nor a capricious process, however. First, the process should be both explicit and documented. Second, prior approaches to the same or similar problems can be consulted and, when appropriate, cited. Third, judgments about importance, criticality, and so on, while subjective in nature, are empirical data collected with reapplicable procedures and whose reliability can be estimated. Such judgments, like supervisor ratings, can provide useful and unique information when appropriately used.

When the investigator has determined what particular aspects of occupational performance are of interest, the next issue is determining how these components should be evaluated and under what conditions. The influence of methods and psychometrics on performance measurement has been discussed above. It is vital to consider the conditions of performance assessment because people do not behave in vacuums. An appropriate synonym for conditions is scenarios. An American controller may first learn how to vector an airplane in early training, and there is but one plane in the developmental's airspace, with good weather and no mechanical problems. As the controller progresses, the scenarios will become more complicated, with the addition of more aircraft with different performance characteristics on different paths, possibly with bad weather, and with certain segments of the airspace off limits. The sequence and timing of aircraft could vary as well. None of these complications would be added if it were not believed, with overwhelming evidence, that they influence performance. Potentially important, and not done very well at this stage, are parametric studies of scenarios for air traffic controller performance, to document what environmental conditions make the task more difficult for what types of controllers.

No one, including controllers, does just one task. Most incumbents perform many tasks over periods of time. In the discussion of job analysis methods above, attention was given to selecting performance components for different purposes. Information about frequency and importance was mentioned as providing knowledge on which selection of performance components could be based, and it is now important to suggest that another choice to be made is that of who provides such ratings. Incumbents (controllers) and their supervisors can be considered Subject Matter Experts (SMEs), individuals whose technical knowledge of the job in question is unsurpassed by almost all others. They can provide judgments for which "accuracy" is a relevant characteristic. For "usefulness" and "value," other classes of people are also able to provide judgments, people such as managers, politicians, and members of public interest groups.

The importance of considering these other types of judgments may become more clear as the next step of performance measures is considered, that of combining them. How should we obtain one measure to represent the performance of an individual across several components of performance? How should we combine performance measures across different scenarios? How should we combine performance measures of several people in the same specialty, such as terminal or center controller? How should we combine across specialties? There are no right or wrong answers to these questions, either; it is wrong only not to ask them.

Because an entire job can never be measured in total, some sample of it has to be selected for observation. In developing performance measures for validation of selection procedures, both legal constraints and technical requirements state that the measure must select important or representative parts of the job. "Importance" is a value judgment, and we could have different groups of people tell us what of the occupation's performance components are more important than others. Perhaps these different groups would agree, although that is not a certain outcome. "Representativeness," on the other hand, is capable of a more technical definition through our understanding of statistical sampling theory.

Shavelson (Shavelson and Webb, 1984; Shavelson, Webb, and Rowley, 1989) suggested that thinking about the representativeness of a performance measure could be helpful in developing more credible instruments. For example, a job or occupation consists of many tasks. The first stage is conceiving of a random sample from this population of tasks. However, some of the tasks might not be observable, because setting up the situation in which to measure them would be too time-consuming or too risky. Those tasks for which we can develop valid simulations would not be in this category. Second, we might wish to stratify the sample of tasks, so that, for example, the more important and more frequent tasks had a higher probability of being selected for inclusion in the performance measure. Third, we would obtain our sample of tasks and then determine the most appropriate ways of measuring them. The most important feature of our resulting set of performance measures is that the findings are generalizable: they are generalizable to the total job, to the overall performance of the individual, and to the average performance of groups of individuals evaluated with these measures.

It is very difficult to develop such representative performance measures, but it is possible to analyze our current measures to determine how far from representativeness our measures might actually be. We can also use this "thought experiment" to develop a representative model of performance for the occupation.

Performance Beyond Validation

There are many other uses for performance measures, particularly for those that can serve as representative or quasi-representative samples of the occupation and of the person. In personnel administration, performance measures permit the setting of performance standards, and performance standards are useful in a variety of endeavors.

Start with the use of performance measures as criteria for validation of selection procedures. When representative performance measures are used, supervisors can identify score levels on those measures that they can label as "superior," "average," and "unacceptable." The regression and other psychometric analyses and other techniques that are used to link predictor measures to these performance measures can also be used to link the labeled performance values to values on the predictor scales. Thus, cut scores on the predictor scales could be based on more direct linkage to predicted performance than is currently the practice.

The next consideration of performance measures is in training. Many sets of performance measures might be developed, each conforming to a different stage of training or learning. Again, psychometric and judgmental procedures can be used to establish cut or passing scores. Achievement at or above a given level on each measure in the series would provide convincing evidence that the student is ready to proceed to the next stage (Jaeger, 1989). Such measures could also be used to evaluate and select individuals coming into training at advanced levels. In the United States, for example, the FAA presently has formal arrangements with a few collegiate programs in air traffic control, whose graduates will enter the FAA above the initial entry level. Staged performance measures could be an important component of an effective selection procedure for them. This leads easily to another use for representative performance measures, that of evaluating competence, for licensing and other legal endeavors.

Other specialized performance measures have existed and will continue to do so. Equipment design engineers need to know how their prototypes work. They can ignore many dimensions of the human performer who tries the machines out, but their findings could be more useful were psychometric issues such as reliability and validity paid more attention. Giving the concept of "representativeness" more heed could provide insight into the creativity with which people relate to new equipment of all kinds, creativity that can have both good and unsafe outcomes.

Those users who bear some responsibility for performance — the selection officials, the trainers, the licensers, the engineers — are not the only people who can find performance measures useful. Consider their supervisors, and the managers and the second and higher level decision makers. Their decisions and delegations of authority could be more effectively made when each occupation has an explicit, analytic, rational definition of its performance. They have better understanding of what to expect and not to expect.

Further, explicit definitions lend themselves to explicit theories and models. Costs and benefits can be assigned and compared. For example, consider again the procedures of personnel selection, for air traffic controllers. There are costs involved in hiring and retaining controllers, such as recruiting, testing, time delays in hiring, investigations, training, salaries, and moving expenses. The benefits are air safety and savings in time and fuel for travelers and airlines. Every country that has air travel has learned that it costs money to control it. The question is one of how best to spend money to obtain the safest and perhaps the fastest air travel. One can spend one's resources on controllers, on training, on equipment, or (more likely) some combination of the above. How might one determine the best mix?

Psychologists (e.g., Cascio, 1987; Eaton, Wing, and Mitchell, 1985) and other social scientists have addressed these issues. A promising approach is suggested by some econometric modeling provided by scientists at the Rand Institute (Armor, Fernandez, Bers, and Schwarzbach, 1982) to determine the optimal enlistment standards for Army infantryman. It is first necessary to define some unit of performance. For the infantryman, this was the Quality Man Month (QMM). For each month an infantryman serves after he achieves a passing score on the Skills Qualification Test (SKT), he is credited with one QMM. Despite the QMM's being a psychometric nightmare, as a unit of performance it is the means to learn more interesting things. The unit incorporates both ability level, as indicated by the test score, and longevity, as indicated by the number of months of service. For the Army and many other employers, ability and longevity are both important.

Next, one has to develop an equation that can balance the marginal unit of performance against the marginal cost of selection. For Army infantryman, "quality" recruits had high aptitude test scores, high school diplomas, and other opportunities for employment besides the Armed Forces. They were more likely to produce more QMM, but they cost more to recruit. Less costly to recruit were individuals with high test scores but no high school diploma, who would earn high SQT scores but were more likely to leave the Army before completing their tours of service, as well as individuals with low aptitude scores but high school diplomas, who were less likely to leave but also less likely to pass the SQT. The Army would like to have more of the quality recruits, but each successive quality recruit cost more than the last. The marginal costs were high.

Armor et al. (1982) developed econometric models that determined the optimal cut score or enlistment standard that would provide either the highest level of performance for the existing number of soldiers in the force, or the same amount of performance for a reduced number of soldiers. There is no question that this and subsequent modelings of the selection process, against performance, make a lot of assumptions that have unexpected consequences, that are more or less incorrect, and that are at the least infuriatingly simplistic. It is also true that these analytic models provide the opportunity to ask very specific questions about personnel procedures. For example, there is an international problem of too much air traffic to control with our current controller resources. Finding persons with adequate aptitude to master an arduous training process is becoming increasingly difficult. New equipment systems are on the way but will provide little relief in the immediate future. We can conceive of new equipment that would require fewer controllers, and perhaps even less able controllers, but we are skeptical that such equipment would provide the margins of safety to which we are accustomed.

Econometric modeling will not provide the solution; no one thing will. But the clearer our ideas are about the performance we need to have, and how we can achieve it, the better able we will be in allocating our resources for the future. How much should we spend for selection, and how much for training? How much should we spend for new equipment, and how should it be designed? So while performance measures are very useful if not vital for selection of air traffic controllers, they are obviously important for a great many of the other questions we can ask about air traffic controllers. Since the characteristics of performance measures influence our findings, it behooves us to understand them well.

References

Ackerman, Phillip L. (1987). Individual differences in skill learning: An integration of psychometric and information processing perspectives. *Psychological Bulletin, 102*, 3-27.

Ackerman, Phillip L. (1988). Determinants of individual differences during skill acquisition: Cognitive abilities and information processing. *Journal of Experimental Psychology: General, 117*, 288-318.

Ackerman, Phillip L. (1989). Within-task intercorrelations of skilled performance: Implications for predicting individual differences? (A comment on Henry & Hulin, 1987). *Journal of Applied Psychology, 74*, 360-364.

American Educational Research Association, American Psychological Association, & National Council on Measurement in Education (1985). *Standards for educational and psychological testing*. Washington, DC: American Psychological Association.

Armor, David J., Richard L. Fernandez, Kathy Bers, & Donna Schwarzbach (1982). *Recruit aptitudes and Army job performance*. R-2874-MRAL. Santa Monica, CA: Rand Corporation.

Bennett, G. K., H. G. Seashore, & A. G. Wesman (1981). *Differential aptitude tests (DAT)*. San Antonio, TX: The Psychological Corporation.

Benoît, A. (1991). Close ground/air support in dynamic air traffic control. In: J. A. Wise, V. D. Hopkin, & M. L. Smith (Eds.) *Automation and Systems Issues in Air Traffic Control*. Berlin: Springer-Verlag.

Bray, D. W., Campbell, R. J., & Grant, D. L. (1979). *Formative years in business: A long term AT&T study of managerial lives*. Huntington, NY: Robert E. Krieger.

Campbell, Charlotte H., Patrick Ford, Michael G. Rumsey, Elaine D. Pulakos, Walter C. Borman, Daniel B. Felker, Maria V. de Vera, & Barry Reigelhaupt (1990). Development of multiple job performance measures in a representative sample of jobs. *Personnel Psychology, 43*, 277-300.

Campbell, D. T., & D. W. Fiske (1959). Convergent and discriminant validation by the multi-trait-multi-method matrix. *Psychological Bulletin, 56*, 81-105.

Campbell, John P. (Ed.) (1990) The Army selection and classification project (Project A). *Personnel Psychology, 43*, 231-378.

Campbell, John P., Jeffrey J. McHenry, & Lauress L. Wise. (1990). Modeling job performance in a population of jobs. *Personnel Psychology, 43*, 313-333.

Cascio, Wayne F. (1987). *Costing human resources: The financial impact of behavior in organizations* (Second edition). Boston: Kent.

Cattell, Raymond B. (No date). *The sixteen personality factor questionnaire (16PF)*. Champaign, IL: Institute for Personality and Ability Testing.

Corts, Daniel B., Tressie W. Muldrow, & Alice M. Outerbridge (1977) Research base for the written test portion of the Professional and Administrative career Examination (PACE): Prediction of job performance for customs inspectors. Washington, DC: U.S. Office of Personnel Management, Personnel Research and Development Center.

Della Rocca, Pamela, Carol A. Manning, & Hilda Wing (1991). Selection of air traffic controllers for automated systems: Applications from today's research. In: J. A. Wise, V. D. Hopkin, & M. L. Smith (Eds.) *Automation and Systems Issues in Air Traffic Control.* Berlin: Springer-Verlag.

Detterman, Douglas K. (Ed.) (1985) *Current topics in human intelligence. Volume 1: Research methodology.* Norwood, NJ: Ablex.

Digman, John M. (1990) Personality structure: Emergence of the five-factor model. In Mark R. Rosenzweig & Lyman W. Porter (Eds.) *Annual Review of Psychology, 41*, 417-440.

Dunnette, Marvin D., & Edwin A. Fleishman (1982). *Human performance and productivity: Human capability assessment. Volume 1.* Hillsdale, NJ: Lawrence Erlbaum Associates.

Edwards, J. (1991). Intelligent dialog in air traffic control. In: J. A. Wise, V. D. Hopkin, & M. L. Smith (Eds.) *Automation and Systems Issues in Air Traffic Control.* Berlin: Springer-Verlag.

Eaton, N. K., Hilda Wing, & Karen J. Mitchell (1985). Alternate methods of estimating the dollar value of performance. *Personnel Psychology, 38*, 27-40.

Fleishman, Edwin A., & Quaintance, Marilyn K. (1984). *Taxonomies of human performance: The description of human tasks.* Orlando, FL: Academic Press.

Gael, Sidney (Ed.) (1988) *The job analysis handbook for business, industry, and government.* New York: John Wiley & Sons.

Green, Bert F., Jr., & Hilda Wing (1988) (Eds.) *Analysis of job performance measurement data: Report of a workshop.* Washington, DC: National Academy Press.

Green, Bert F., Jr., Hilda Wing, & Alexandra K. Wigdor (Editors) (1988). *Linking military enlistment standards to job performance: Report of a workshop.* Washington, DC: National Academy Press.

Hartigan, John A., & Alexandra K. Wigdor (1989) (Eds.) *Fairness in employment testing: Validity generalization, minority issues, and the General Aptitude Test Battery.* Washington, DC: National Academy Press.

Hedge, Jerry W., & M. Suzanne Lipscomb (1987) . Walk-through performance testing: An innovative approach to work sample testing. (AFHRL-TP-87-8). Brooks Air Force Base, TX: Air Force Systems Command, Training Systems Division.

Hedge, Jerry W., & Mark S. Teachout (1986). *Job performance measurement: A systematic program of research and development. (AFHRL-TP-86-37).* Brooks Air Force Base, TX: Air Force Systems Command, Training Systems Division.

Holland, John L. (No date). *Self-directed search.* Odessa, FL: Psychological Assessment Resources.

Holland, John L. (No date). *Vocational preference inventory.* Odessa, FL: Psychological Assessment Resources.

Hough, Leaetta (1988, April). Personality assessment for selection and placement decisions. Workshop presented at the Third Annual Conference of the Society for Industrial and Organizational Psychology, Dallas, TX.

Hunt, Earl (1985). The correlates of intelligence. In Douglas K. Detterman (Ed.), *Current topics in human intelligence. Volume 1: Research methodology.* Norwood, NJ: Ablex.

Hunt, Earl, James W. Pellegrino, Ronald Abate, David L. Alderton, Simon Farr, Robert W. Frick, & Thomas P. McDonald (1987). *Computer-controlled testing of visual-spatial ability*. NPRDC-TR 87-31. San Diego, CA: Navy Personnel Research and Development Center.

Hunter, John E. (1983). A causal analysis of cognitive ability, job knowledge, job performance, and supervisor ratings. In Frank Landy, Sheldon Zedeck, & Jeanette Cleveland (Eds,), *Performance measurement and theory*. Hillsdale, NJ: Lawrence Erlbaum Associates.

Hunter, John E., & Ronda F. Hunter (1984) Validity and utility of alternative predictors of job performance. *Psychological Bulletin, 96*, 72-98.

Hunter, John E., & Frank L. Schmidt (1982) Fitting people to jobs: The impact of personnel selection on national productivity. In Marvin D. Dunnette and Edwin A. Fleishman (Eds.), *Human performance and productivity: Human capability assessment. Volume 1*. Hillsdale, NJ: Lawrence Erlbaum Associates.

Jaeger, Richard M. (1989). Certification of student competence. In Robert L. Linn (Editor), *Educational Measurement* (Third Edition). New York, NY: American Council on Education, Macmillan Publishing Company.

Johnstone, William B., & Arnold E. Packer (1987). *Workforce 2000: Work and workers for the twenty-first century. HI-3796-RR*. Indianapolis, IN: Hudson Institute, Inc.

Kanfer, Ruth, & Phillip L. Ackerman (1989). Motivation and cognitive abilities: An integrative/aptitude-treatment interaction approach to skill acquisition. *Journal of Applied Psychology, 74*, 657-690.

Kraiger, Kurt. (1990). *Generalizability of performance measures across four air force specialties. (AFHRL-TP-89-60)*. Brooks Air Force Base, TX: Air Force Systems Command, Training Systems Division.

Lane, John. (1991). The implementation and impact of automatic data processing on UK military ATC operations. In: J. A. Wise, V. D. Hopkin, & M. L. Smith (Eds.) *Automation and Systems Issues in Air Traffic Control*. Berlin: Springer-Verlag.

Landy, Frank, Sheldon Zedeck, & Jeanette Cleveland (Eds.) (1983). *Performance measurement and theory*. Hillsdale, NJ: Lawrence Erlbaum Associates.

Levesley, John. From under the headset: The role of the air traffic controlers professional association in present and future air traffic control systems development. In: J. A. Wise, V. D. Hopkin, & M. L. Smith (Eds.) *Automation and Systems Issues in Air Traffic Control*. Berlin: Springer-Verlag.

Linn, Robert L. (Ed.) (1989). *Educational measurement* (Third edition). New York: American Council on Education, Macmillan.

Lord, Frederic M. (1980). *Applications of item response theory to practical testing problems*. Hillsdale, NJ: Lawrence Erlbaum Associates.

Neisser, Ulric. (1967) *Cognitive psychology*. New York: Appleton-Century-Crofts.

O'Bannon, R. Michael, Linda A. Goldinger, & Gavin S. Appleby (1989). *Honesty and integrity testing: A practical guide*. Atlanta, GA: Applied Information Resources.

Peterson, Norman G., & David A. Bownas. (1982) Skill, task structure, and performance acquisition. In Marvin D. Dunnette and Edwin A. Fleishman (Eds.) *Human performance and productivity: Human capability assessment. Volume 1*. Hillsdale, NJ: Lawrence Erlbaum Associates.

Sackett, Paul, Laura Burris, & C. Callahan (1989). Integrity testing for personnel selection: An update. *Personnel Psychology, 42*, 491-528.

Sackett, Paul , & Harris, M. (1984). Honesty testing for personnel selection. *Personnel Psychology, 37*, 221-245.

Schmidt, Frank L. (1988) Validity generalization and the future of criterion-related validity. In Howard Wainer and Henry I. Braun (Eds.) *Test validity*. Hillsdale, NJ: Lawrence Erlbaum Associates.

Shavelson, Richard J., & Noreen M. Webb (1984). Generalizability theory: 1973-1980. *British Journal of Mathematical and Statistical Psychology, 34*, 133-166.

Shavelson, Richard J., Noreen M. Webb, & Glenn L. Rowley (1989). Generalizability theory. *American Psychologist, 44*, 922-932.

Society for Industrial and Organizational Psychology, Inc. (1987). *Principles for the validation and use of personnel selection procedures*. (Third Edition). College Park, MD: Author.

Sternberg, Robert J. (1985). Componential analysis: A recipe. In Douglas K. Detterman (Ed.) , *Current topics in human intelligence. Volume 1: Research methodology*. Norwood, NJ: Ablex.

Stager, Paul. (1991). Error models for operating irregularities: Implications for automation. In: J. A. Wise, V. D. Hopkin, & M. L. Smith (Eds.) *Automation and Systems Issues in Air Traffic Control*. Berlin: Springer-Verlag.

Strong, E. K., Jr., Jo-Ida C. Hansen, & David P. Campbell (No date). *Strong interest inventory*. Palo Alto, CA: Consulting Psychologists Press.

Tattersall. Andrew J., E. W. Farmer, & A. J. Belyavin. (1991). Stress and workload management in air traffic control. In: J. A. Wise, V. D. Hopkin, & M. L. Smith (Eds.) *Automation and Systems Issues in Air Traffic Control*. Berlin: Springer-Verlag.

Thornton, George C. III, & William C. Byham (1982). *Assessment centers and managerial performance*. New York: Academic Press.

Trattner, Marvin H., Corts, Daniel B., Paul P. van Rijn, & Alice M. Outerbridge (1977). *Research base for the written test portion of the Professional and Administrative Career Examination (PACE): Prediction of job performance for claims authorizers in the social insurance claims examining occupation*. Washington, DC: U.S. Office of Personnel Management, Personnel research and Development Center.

Tuddenham, Read D. (1962) The nature and measurement of intelligence. In Leo Postman (Ed.), *Psychology in the making: Histories of selected research problems*. New York: Alfred A. Knopf.

U.S. Equal Employment Opportunity Commission, U.S. Civil Service Commission, U.S. Department of Labor, & U.S. Department of Justice (1978). Uniform guidelines on employee selection procedures. *Federal Register, 43*(166), 38290-38315.

Weiss, David J. (Ed.) (1983) *New horizons in testing: Latent trait test theory and computerized adaptive testing*. New York: Academic Press.

Zeidner, Joseph, & Arthur J. Drucker. (1988) *Behavioral science in the Army: A corporate history of the Army Research Institute*. Alexandria, VA: U.S. Army Research Institute for the Behavioral and Social Sciences.

Selection of Air Traffic Controllers for Automated Systems: Applications from Today's Research

Pamela Della Rocco[1], Carol A. Manning[1], and Hilda Wing[2]

[1] Civil Aeromedical Institute, Federal Aviation Administration, Oklahoma City, Oklahoma, USA
[2] Staffing Policy Division, Federal Aviation Administration, Washington, DC USA

Over the next two decades, the Federal Aviation Administration's (FAA) National Airspace System (NAS) Plan for new automated air traffic control systems will radically change the role of the Air Traffic Control Specialist (ATCS)[19]. While introduction of automation is not new to the occupation, the NAS Plan calls for automating more critical job tasks than could have been previously supported by technology. The challenge of identifying the characteristics of the controller to operate these future automated systems is currently facing the FAA.

A number of questions must be raised to respond to this challenge. Having served for some time as a guide to meeting such challenges in air traffic related human factors research, Hopkin has been particularly adept at looking into the future to identify important issues, the answers to which are best identified through research prior to implementation of new technology. Hopkin (1988) has recently articulated many of these issues.

One issue concerns an optimum balance between selection and training. That is, can anyone be trained to be an ATCS, or is selection the key to identifying individuals with the best aptitude and ability for air traffic control (ATC)? The FAA has historically chosen a combination of selection and training. However, if the job changes substantially, with increasing automation the question must be revisited.

The FAA's experience with current selection systems may provide some insight for designing better predictor and criterion measures for the future. This paper will address areas related to the changing job and applications from current selection research.

Introduction

The first step in designing a selection system for current or future jobs is an analysis of the tasks required to perform the job. In approaching such analysis, a would-be designer of a selection system unfamiliar with air traffic control faces a challenge. Hopkin (1988) has captured first impressions of the job in the following comments:

To a naive observer, the workspace of controllers seems complex and well nigh

19. The term ATCS and controller are used interchangably throughout this paper. ATCS is the formal job title, whereas, controller is the more informal title.

NATO ASI Series, Vol. F 73
Automation and Systems Issues in Air Traffic Control
Edited by John A. Wise et al.

meaningless. It is not intuitively and immediately obvious what any of the data mean, what the tasks are, how they are done, or what would constitute successful performance."

Thus, the first step toward a successful selection system involves analysis of tasks not immediately observable and perhaps not easily measured.

By definition, the controller is tasked with promoting the safe, orderly, and expeditious flow of air traffic. This is accomplished through accurate, effective application of rules and procedures for separation in a real-time, dynamic environment. Thus, the current ATCS's job consists of a complex set of tasks that demand a high degree of skill and active application of certain cognitive abilities, such as spatial perception, information processing, reasoning, and decision making. As an increasing number of ATC tasks become automated, the role of the controller as an active participant may gradually evolve to a more passive role of system monitor and manager.

The current work force has been rigorously selected through a two-stage selection system. The first stage consists of a paper-and-pencil aptitude test battery. The second stage is a performance-based measure consisting of a condensed training-testing ATC task sample in a nonradar environment. The result is the selection of individuals with many of the cognitive abilities identified above. As the job itself changes, the question arises as to how long a selection system based upon the current selection criteria will be effective and what factors will be important in designing the replacement.

To address these questions, the job of the ATCS in its current form will be reviewed and contrasted to projected changes in job tasks from implementation of the Initial Sector Suites (ISSS) through the Area Control Computer Complex (ACCC), and particularly, the Automated En Route Air Traffic Control (AERA). The current selection system will then be examined to determine the extent to which current criteria might predict success in future ATC jobs as automation causes the tasks to evolve.

Automation and Air Traffic Control Services

Over the 50 year history of the ATC system's evolution, a mixture of equipment of varying technological generations and types was used. Computer automation techniques were first applied to flight data processing systems, utilizing UNIVAC and IBM computers, during the early 1960s. The systems were installed in six northeastern Air Route Traffic Control Centers (ARTCC). Through this effort, the FAA demonstrated that computers could be used to improve safety and increase productivity of the controller. By 1968, an integrated plan to automate ATC functions was approved. This plan was NAS Stage A. By 1981, the NAS Plan for Facilities, Equipment and Associated Development was issued to define an orderly, rational evolution of the system as a whole This plan is updated annually.

The NAS Plan established three principal objectives for the en-route and terminal systems. The first was consolidation of more than 200 facilities existing today into less than 30 by the year 2000. The second objective was to install common modular computers, software, and controller work stations (i.e. sector suites). The en-route and terminal radar approach control facilities will be consolidated into area control facilities (ACF) with essentially identical equipment, including the HOST computer and Initial Sector Suite System

(ISSS). The third objective was the improvement of safety, fuel efficiency, and productivity through higher levels of automation. The NAS Plan includes improvements such as Mode S data link, ASR-9 and advanced Doppler weather radar, navigation by Global Positioning System (GPS) satellites, and advanced stages of automation processing, such as Automated En-route Air Traffic Control (AERA). Other advances in the air traffic system which will impact the controllers' functions include point-to-point navigation and direct routing from point of departure to destination.

The NAS Plan provides for a staged delivery of the various equipment and software developments. The HOST computers have been installed, and facilities await delivery of the first Sector Suites beginning in the mid-1990s.

NAS Plan and the Air Traffic Control Specialist

Two automation advancements in the NAS Plan will particularly affect the environment and job tasks of the en-route controller. These are the Sector Suites and the implementation of AERA.

The Sector Suites provide an entirely new controller workstation. Briefly described, each Sector Suite is a collection of one to four common consoles which contain data entry and display equipment. These will be the first improvements to have a large impact on the en-route controller's environment. The Sector Suite consoles will provide information to the controller on multiple display screens. A common console will consist of a main and an auxillary display, voice switching, and control panels. Display screens will have multi-color capability. Data entry devices will include a keyboard and trackball. Information presented will include a situation display of the air traffic and weather, alpha-numeric flight and weather data, as well as other pertinent information, such as notices to pilots. The most substantial change to the controller's environment presented by the sector suites is the replacement of paper flight progress strips with an electronic display of flight data (Gutmann, 1988).

The second NAS Plan project which will substantially impact the en-route controller's job tasks is AERA. AERA will be designed to support the ATCS through predicting and resolving problems along an aircraft's route of flight, and in planning and maintaining traffic flow and aircraft separation. This project involves development of a software system which will be implemented as an enhancement to the earlier ATC hardware and software system.

The NAS Plan proposes three stages of AERA implementation, AERA 1, AERA 2, and AERA 3.

Under AERA 1, the computer will detect certain common problems, such as conflicts between aircraft, violations of protected airspace and nonadherence to ATC-imposed traffic flow restrictions.

AERA 2 will provide enhanced assistance to controllers through proposed resolution of detected problems. Under AERA 2, several resolutions for a problem will be generated. The highest ranked resolution will be presented to the controller currently controlling the aircraft. The controller will assess that resolution, examine others if necessary, and determine which resolution to choose, depending on his or her plans for use of the airspace. This activity will be complicated by the increased use of direct routings instead of currently established jetways. For aircraft equipped with data link, clearances will be delivered by the computer, once approved by the controller. Clearances for aircraft without data link will be delivered by the controller. Certain coordination functions between controllers will also be automated. The controller remains responsible for all control actions under AERA 2.

However, under AERA 3, automated decision-making and control will be introduced, with capabilities which include a degree of autonomy for the computer to detect and resolve ATC problems without human intervention.

The NAS Plan establishes a clear path toward increased automation of air traffic control. If the current role of the controller is to ensure the separation of aircraft, as well as the safe, expeditious flow of traffic, then by the time AERA 3 is implemented, much of that role will shift to the computer. To design a selection system for the evolving role of the controller or to assess aspects of the current selection system that might have utility for the future, it is important to understand the job tasks of the controller and the changes that will be made in those tasks with increased automation. The following section examines the current functions of an ATCS and reviews studies analyzing the impact of ISSS and AERA on those functions.

ATCS Job Task Analyses

A job task analysis (JTA) provides a detailed itemization of the tasks required by a specific job. Such a detailed analysis can help us systematically map changes in job functions with each phase of the NAS Plan. From there we can begin to assess the knowledges, skills, and abilities required of the human operator. In addition, the JTA can help us understand what we are currently accomplishing with current selection systems.

In preparation for the transition to the Advanced Automation System (AAS) under the NAS Plan, Computer Technology Associates, Inc., (CTA) was commissioned to conduct a job task analysis in today's ATC environment, as well as for the planned automation capabilities. The result of this effort was a multi-volume documentation of Operations Concepts for ATC personnel in operational environments at different stages of AAS evolution. The focus was placed on the interaction between the controller and the automated system during operational tasks (Ammerman, Becker, Jones, Tobey, and Phillips, 1987).

Data were collected through field observations, through interviews conducted in a large cross-section of ATC facilities, and through regular involvement of a controller validation team. Thus, the JTA was completed and validated with substantial input from subject matter experts (i.e., ATCSs).

The CTA methodology derived a task list utilizing a stimulus-process-response model in which a controller performs actions in response to air traffic events. Therefore, the job analysis began with identifying and defining the air traffic events to which a controller responds. After identifying the response, the controller's action was then decomposed to the task level--the central analysis result. Within this framework, the work performance actions were classified into three increasingly detailed levels: activities, sub-activities, and tasks. “Activities” were the highest level of description of operational job functions. The operational duties of a position were typically described with five to nine activities. “Sub-activities” were a more detailed statement of the work performed, and a “task” was the most detailed unit of description. The task level specification is a meaningful unit of work performance.

Within the task level specification, four task types were identified, as follows: 1) entry tasks (i.e., entry of data into the system), 2) receipt tasks (e.g., receipt of information from the displays), 3) analytical tasks, and 4) verbal coordination tasks (coordination with other individuals or the system).

Task characterization analyses were also performed which described tasks on four separate dimensions. Two of these dimensions are of particular interest in establishing a selection system. These task characterizations included 1) the cognitive/sensory attributes required for a task, and 2) identification of task performance requirements.

For purposes of this paper, only the tasks required of the en-route controller (ARTCC) will be reviewed (Ammerman, Bergen, Davies, Hostetler, Inman, and Jones, 1987). Six activities were identified for ARTCC/Host en-route controller in the current ATC environment, as follows:

1) perform situation monitoring,
2) resolve aircraft conflicts,
3) manage air traffic sequences,
4) route or plan flights,
5) assess weather impact, and
6) manage sector/position resources.

Within the six activities, 46 sub-activities and a total of 348 distinct tasks were identified. Clearance delivery was identified as a set of tasks that was relevant to many sub-activities, and was thus defined as a "macro" (performed as a function of a number of tasks).

The cognitive/sensory attributes required for performance of tasks of extreme or high criticality were identified. Fourteen attributes were found to be relevant to performance of computer and radar workstation jobs associated with current ATC. These were as follows: coding, movement detection, spatial scanning, filtering, image/pattern recognition, decoding, visualization, short-term memory, long-term memory, deductive reasoning, inductive reasoning, mathematical/probabilistic reasoning, prioritizing, and verbal filtering (definitions for these attributes are included in Appendix A).

Further analysis of each of 161 critical tasks revealed that analytical attributes such as visualization (42 tasks), short-term memory (30 tasks), deductive reasoning (41 tasks), and mathematical/probabilistic reasoning (32 tasks) were key requirements cognitive attributes. Additionally, message filtering (40 tasks) and decoding (57 tasks) were important.

Impact of ISSS and AERA on Job Tasks

In 1988, CTA completed a report of the impact of ISSS on ATC procedures and training (Ammerman and Jones, 1988). Expanding on a prior report, the analysis provided an early identification of current ATC procedures projected to be impacted by ISSS. The analysis for the report was conducted by comparison of the Operations Concepts for the current job described above and the Operations Concepts prepared for ISSS utilizing the same job analysis methodologies (Alexander, Alley, Ammerman, Hostetler, and Jones, 1988). The study estimated that 48 of 373 (13%) of the job tasks of the enroute controller would be substantially changed from the current ATC system, while 237 (87%) would be essentially unchanged.

A comparison of displays of information reveals a distinct parallelism between current sector workstation physical displays and written records as compared to the Sector Suite workstation. In general, the ISSS will provide the controller with greater flexibility and capability in data handling. In addition, there will be increased data processing and flight data will be available more rapidly. Entry of data into the system will be increased over today's requirements.

CTA observes that the ISSS introduces substantial changes to the en-route controller's operations. The manner in which controllers will manipulate the data will be significantly changed. The most substantial change with ISSS is the conversion of paper flight progress strips to electronic flight data entry and electronic display and manipulation of the data. Today's flight progress strips are manually added to the bay, strip-marked, sorted, flagged, and removed. Under ISSS, this will be accomplished electronically. CTA notes that display management will be much more critical with ISSS than with today's system. More information will be available under ISSS, and this will need to be displayed for most efficient use. They also express a concern that electronic presentation of data may impact the silent coordination between the radar and "D-side" controllers.

In summary, the 48 radically and significantly different tasks under ISSS pertain to the controllers' interaction with information displays and the input of control data and messages. CTA's analysis of changes in the cognitive/sensory attributes determined that "coding" will be substantially increased because of the expanded availability of messages. CTA concluded that all other cognitive and sensory attributes would be comparably required for the more critical controller tasks in both the current ATC system and under ISSS. In terms of performance requirements, accuracy of data entry and receipt from the system would be increased the most in terms of frequency required for critical tasks.

Researchers at MITRE, in conjunction with teams of subject matter experts, prepared several papers assessing the changes in the en-route controllers' operational functions after implementation of AERA. AERA 1 introduces primarily problem detection capability and some decision aids for the en-route controller. However, AERA 2 expands these capabilities in several important areas. These areas include resolutions to problems detected by the automated problem detection software, display of proposed resolutions to the appropriate controller, and finally provision of additional aids to assist a controller in decision-making for separation of air traffic (such as Trial Planning) and coordination.

Carlson and Rhodes (1990) described the changes in the controllers' work activities in AERA 2. In today's environment, controllers rely on their own mental abilities to detect and resolve situations occurring in their airspace. ATCSs must recognize potential loss of separation between aircraft, based upon available data in flight strips, the plan view display's presentation of current and recent past aircraft positions, and their knowledge about the operational ATC environment. Upon recognition of a potential situation, the ATCS must formulate and evaluate resolutions for the situation. The human being has finite limits on his or her capacity to process information. Limits include the maximum number of aircraft an invididual can safely and efficiently control at any one time, as well as the look-ahead time to project potential conflicts. Introduction of direct routing of aircraft and reduced use of established airways may, in fact, overload a controller's information processing capabilities.

AERA 2 should expand the capacity of the ATC system by automating detection of problems up to 20 minutes prior to predicted loss of separation and for an aircraft's entire route of flight. The Automated Problem Resolution (APR) function will produce a number of resolutions, based upon a number of the aircraft's projected trajectory, geometry of the conflict, airspace characteristics, and other aircraft or system requirements. This capability is projected to improve detection and resolution of problems because they will be detected earlier and more optimal resolutions will be considered.

The impact of AERA 2 on the controller's job has been described as shifting the emphasis away from tactical decision-making to more strategic management of the ATC environment. Carlson and Rhodes (1990) suggest that many routine tasks will be automated, and more complex tasks will become easier because of the availability of computer proposed resolutions that a controller in today's environment must generate. They suggest that controllers will rely less on their own mental abilities to monitor flight data and identify situations requiring control actions. The focus of the controller's attention will shift from the flight data and situation display to the problem alert and resolution displays. Finally, Carlson and Rhodes (1990) suggest that because problems will be resolved earlier than they are in today's system, the controller's time orientation for detecting and resolving ATC problems will change.

Another perspective on the time orientation of the controller under AERA was provided by McKinley and Jago (1984). In an early evaluation of the en-route controller's skills requirements, they suggested that the automated control capability of AERA would only be possible for long-term (greater than 20 minutes in the future) and some medium-term (0-20 minutes) problem prediction and resolution. For the short-term (0-5 minutes) and some medium-term problem detection, reliance on the information from the situation display and electronic flight progress data would continue to be required.

Because AERA 2 and AERA 3 are still under development, it is difficult to assess the cognitive skills and abilities required for the job tasks. However, it is clear that the implementation of AERA 2, and particularly AERA 3, will significantly impact the role of the ATCS in controlling traffic, and that research on the impact must continue.

The previous sections reviewed information on job tasks required of an en-route ATCS. CTA's methodology also provided information about the cognitive/sensory attributes required to accomplish the tasks in the current ATC environment as well as performance requirements. The following sections review the current selection system to evaluate the extent to which principles applied and learned in development of this system will apply to the future air traffic controller's job.

Background on Selection of Personnel in the United States

This discussion of selection of controllers should be considered within the context of selection, in general, in the United States. That context is highly regulated, considerably politicized, and normally very public. Prior to the Civil Rights Act of 1964, selection was a backwater of personnel psychology, including only a few individuals of talent and sensitivity among its ranks. Civil rights activists perceived, mostly accurately perhaps, that the selection procedures followed by many companies bore little relationship to the jobs at hand while they excluded unacceptably high percentages of minority applicants. Much of the early case law on personnel selection was built from such cases. It could be argued that many of the resulting judicial opinions overgeneralized and, by inference, developed procedural requirements that few if any of the best selection systems could meet. Since the 1960's, selection procedures have become the subject of scrutiny by many different groups. This has meant that the procedures must both be extensively documented and technically acceptable. Because of this, personnel psychology left its position in the backwaters and has become one of the more exciting areas of psychology.

The Uniform Guidelines on Employee Selection Procedures (UGESP), published in 1978, apply to public as well as private employers. They include two major requirements: the determination of adverse impact and the demonstration of job-relatedness. If an employer's selection system shows adverse impact against one or more selected groups, then the job-relatedness of this system must be demonstrated.

The debates surrounding the proposed Civil Rights Act of 1990 have focused on defining these terms: "adverse impact" and "job-relatedness." For now, however, "adverse impact" can be understood as the amount or degree of differential effect that a selection procedure has against a protected group, that puts the employer on the defensive and initiates an investigation of the selection procedure. This is primarily a legal or administrative matter, although psychologists are concerned about representativeness of samples, statistical significance, and so forth.

Job-relatedness is the concern of psychologists. Over the past twenty-five years it has been understood as validity, with criterion-related validity providing the most persuasive evidence. Showing a significant statistical relationship between the scores on a predictor test and the scores on some criterion, typically training grades or supervisory ratings, has been convincing in many cases. Content validity can be acceptable in certain circumstances, such as using a typing test to select for clerical occupations. The FAA's Screen for ATCS, described below, is based on both a content validity strategy—the Screen is a miniature training and test situation of how a controller learns the job—and a criterion-related strategy —Screen grades predict field training performance. A third strategy is construct validity where, say, the construct of spatial ability is posited as necessary to the ATCS occupation. Construct validation strategies can be risky legal defenses unless their empirical grounding is well-articulated.

A major contribution to the understanding of validation in personnel settings is the concept of validity generalization, initially proposed in 1977 by Schmidt and Hunter. These authors, and others following, collected many criterion-related studies and used the study results as data in meta-analyses. Validity coefficients vary widely from one study to the next, a phenomenon many scientists had interpreted as supporting "situational specificity," that the use of one predictor test cannot be generalized from the situation where the criterion-related validity study was performed. Hunter and Schmidt showed that this variability in correlation coefficients might not be due to variability in situations, or employment settings, but to the varying numbers of subjects in the studies, the differing reliabilities of criterion measures, and other artifacts. That is, criterion-related validities generalize.

There have been two major outcomes of the validity generalization movement — and it does have some of the characteristics of a movement.

First, it has restored both the faith in and reliance on paper-and-pencil tests of general mental or cognitive abilities as entry-level predictors of success in most jobs. While some argue that a test of general mental ability is useful for any occupation, studies show that the historical distinctions among test types (e.g., verbal, quantitative, spatial) reflect small but important differences in predicting job success. Further distinctions among test types and the mental processes required to perform well on them, such as those being explored by cognitive psychologists, are of much less interest because these distinctions are not reflected in predictiveness. Generic paper-and-pencil tests are still a major component of personnel selection. However, those responsible for the selection of air traffic controllers may have more interest in the cognitive approach because of test practice effects especially prevalent in spatial ability tests. Fairness issues as well as validity are paramount for this approach.

The other impact of validity generalization has been to focus more on the criteria that tests have been used to predict. Most of the validity generalization research is based on only two classes of criteria, training success and supervisory ratings. By the time the validity generalization research began in the late 1970's, both civil rights activists and some personnel psychologists were questioning these two criterion categories. While training is important, so went the argument, it is not the job. Training may require more cognitive processing, which paper-and-pencil tests predict fairly well, than may the job. As for supervisory ratings, these are typically generic, probably hodge-podge, and possibly biased. Why, argued these critics, can we not validate our selection procedures against closer surrogates to actual job performance?

The use of job performance measures in criterion-related validity research is fraught with conceptual and logistical problems of high order. On the conceptual side, should we deal with "typical" performance, that which the employee does on a regular basis and that might be measured by detailed supervisor ratings? Or should we try to measure "maximum" performance, what the employee can do, by developing specialized situations in which the desired performance can be evaluated? What types of performance should we measure? Should we measure performance with a test of knowledge, or a demonstration, or a simulation? How useful are archival measures, such as absenteeism? Logistically, most employees are far more scattered in time and place than are applicants for a job. Further, since job incumbents tend to specialize, how can a single, albeit complex, performance measure be fair to all?

The research on performance measurement is not as advanced as that on tests and predictors, but during the last decade there has been progress. The major effort has been a joint-military-service effort sponsored by the U.S. Department of Defense. Preliminary answers to the questions raised above are (1) job performance can be measured; (2) job performance measures can be used in criterion-related validity research and can be predicted by multiple abilities test batteries; (3) the most appropriate performance measures will differ from job to job, e.g. knowledge tests would be appropriate for claims examiners but not for mechanics; (4) all available performance measures, including archival measures, can be used to describe a "performance space" to provide a more complete picture of the occupation and its requirements.

An emphasis on performance assessment, or on criterion-referencing, has been growing steadily during the last two decades in many areas of psychological research. The most prominent example is that of educational assessment. It no longer is sufficient to have a comparison to "who," the traditional basis of norms-referencing. It must be a comparison to "what," or criterion-referencing, explaining why that "what" is important. Competency requirements for high school graduation are an example.

Performance measures may also be used in ways not previously considered for training grades and ratings, and it is this potential that is most exciting. For example, let us say a work sample test were developed for a subgroup of full performance level (FPL) air traffic controllers. Say that this measure could be judged as a representative sample of controller performance by an appropriate sample of controllers and their supervisors, and that the test scores could be scaled into ranges such as outstanding, exceptional, fully satisfactory, marginal, and unacceptable. If this performance measure has a valid predictor that we would like to use in our selection procedures, we can link the score levels on the work sample to score levels on the predictor in both a more rational and more empirical manner than is the current practice.

Thus, the state of the art in selection research is: (1) strictly regulated, very public, heavily documented, and of mostly high quality; (2) using paper and pencil tests of general abilities has much technical justification; and (3) performance measures are becoming much more necessary in validation of selection procedures. It is likely that computers will play an increasing role in selection research and in selection, as they have in other areas of human endeavor. However, the question for the technology will be the technical adequacy of the selection research program in which it is embedded. Research in selection of air traffic controllers of the future will take place in this environment and must meet the existing standards and regulations for employee selection.

Selection of Air Traffic Controllers

History of the development of ATCS selection procedures

As noted in the introduction, in the debate of selection versus training, the FAA chose to utilize a selection system prior to investing in training individuals as ATCSs. The reason for the decision was a high attrition rate from the occupation when no formal selection system was utilized prior to entry into training. The attrition was generally due to training failure and occurred, on the average, two to three years into a person's career in air traffic (Cobb, Mathews, and Nelson, 1972). Such late attrition resulted in considerable costs to both the FAA and the ATCS, who found little opportunity to apply the skills in other occupations. Toward the goal of reducing attrition as well as identifying the individual characteristics required for success in the occupation, the FAA developed an active selection research program.

Research on ATCS selection dates back to the 1950s and a contract by the Civil Aeronautics Administration (CAA--precursor to the FAA) for the development of a paper-and-pencil aptitude test battery that could be used to select ATCS trainees. The results of the study indicated that some aptitude tests reflecting content areas identified from a job analysis could potentially contribute to the selection process. This study provided the format for an Air Traffic Problems (ATP) test (Brokaw, 1984).

In a joint effort with the CAA, the United States Air Force Personnel Laboratory administered thirty-seven aptitude tests to ATCS trainees in 1956 (Brokaw, 1984). The findings indicated that a composite aptitude test score — created by adding scores on tests of arithmetic reasoning, symbolic reasoning, code translation, and the ATP test — effectively predicted instructors' ratings of training performance and supervisors' ratings of job performance approximately one year after training.

Research continued in 1960 with the establishment of the FAA Civil Aeromedical Research Institute (presently the Civil Aeromedical Institute — CAMI; Collins, Boone, and VanDeventer, 1981). CAMI research on Civil Service Commission (CSC) tests of similar content to the tests originally studied by Brokaw led the CSC for the first time to establish selection tests for screening ATCS trainees. Test requirements for selection were implemented in 1962. The initial selection test battery consisted of the following: Arithmetic Reasoning, Spatial Relations, Following Oral Directions, Abstract Reasoning, and Air Traffic Problems. During the period between 1962 and 1972, in addition to continuing validation research, aviation psychologists studied a number of factors relevant to ATCS selection and screening: attrition from the profession, age, prior experience, education, sex, and military ATC training.

Implementation of the CSC test battery alone failed to decrease the attrition rate sufficiently. Prior to 1976, the attrition rate from field training was 38%. A number of studies completed in the early 1970s, as well as Congressional hearings of the Committee on Government Operations, resulted in several recommendations. As a result of the reports, the FAA developed a standardized, centralized, validated program designed to identify, as early as possible, and remove from training those candidates demonstrating insufficient aptitude to become ATCSs. The program was designed to decrease the costs of attrition and improve the selection of ATCSs. The FAA's aviation psychologists collaborated to provide a scientific framework required for the design, implementation, and evaluation of the program.

The result was the implementation of a second stage hurdle designed to follow the CSC test battery. This second stage hurdle consisted of two option-specific (en-route and terminal) pass/fail programs utilizing courses similar to those used in the nonradar separation portions of the existing Academy ATC training curriculum. Grades in these courses had been found to be predictive of future field attrition (Cobb, 1974). The programs were implemented in 1976.

During the latter half of the 1970s, a new initial selection test was developed. Research focused on studies relevant to the Uniform Guidelines on Employee Selection Procedures, developing field training performance ratings, development of a longitudinal database for continuing validation research, and refining optimal combinations of old and new aptitude screening measures described below (Collins, et al., 1981).

The current ATCS selection system

The current ATCS selection system consists of two hurdles or stages of assessment. The following section presents these procedures in more detail, as well as statistics on each method's ability to predict other measures of ATCS performance.

The OPM Selection Battery

The first stage of the selection process is the Office of Personnel Management Air Traffic Control selection test battery. This stage is a paper-and-pencil format aptitude test battery consisting of three tests: the Multiplex Controller Aptitude Test (MCAT), Abstract Reasoning (included from the prior selection battery), and the Occupational Knowledge Test (OKT). It provides the usual benefits of paper-and-pencil tests, allowing initial assessment of large numbers of potential candidates at relatively low cost.

The MCAT was developed to establish a paper-and-pencil test with higher predictive validity than tests included in the CSC selection battery. The design of the MCAT was intended to approximate simulated air traffic activities. Simulated airspaces are presented with several aircraft traversing the space. An altitude, speed, and route chart is also presented. Candidates are asked to identify situations that may result in conflicts between aircraft, based upon a limited set of separation criteria. Other items require computing time-distance functions, interpreting information, and recognizing spatial relations. The test is in a timed, multiple choice format. The Abstract Reasoning measures two principles of logical development exhibited by sequences of figures and letters (Collins, et al., 1981). The

Occupational Knowledge Test (OKT) was developed to provide a more objective and reliable measure of a candidate's ATCS job knowledge than was provided by ratings based upon job history. Items on the OKT cover seven knowledge areas related to air traffic control.

This battery, as is true of many others, including spatial ability tests, is highly subject to practice effects. Generally, a second, higher score has lower validity than the first score. To compensate for practice effects, applicants earning 70 or above presently cannot retake the exam for 18 months.

The FAA Academy Screen

The second stage is the FAA Academy Screen, a nine-week assessment program equivalent to a miniaturized training-testing environment, in which candidates are taught rules for aircraft separation and are assessed in laboratory simulation problems on their ability to learn and apply the rules. This stage is very resource-intensive, but provides selection of candidates likely to succeed to full performance level from among those who have passed the initial OPM battery.

The goal of the second stage pass/fail program, implemented in 1976, was to reduce field attrition through early identification of candidates with little chance of succeeding in field training. Because the Screen is a miniaturized training-testing assessment procedure, the purpose of the program is to teach a candidate, with no knowledge of air traffic control, enough about the job to assess whether he or she has the potential to succeed to the full performance level. This is accomplished by training the candidate in aspects of the job and assessing a task sample (i.e., separating aircraft in laboratory nonradar simulation problems). The complexities of the laboratory problems are structured to escalate quickly, generally beyond most candidates' ability to master the complexity levels being tested. This allows for a certain level of discrimination of the better candidates from the poorer candidates.

The program includes four weeks of academic, classroom training on nonradar ATC rules and principles. Content includes rules of separation (vertical, longitudinal, and lateral), cooperative agreements, protection of special use airspace, phraseology, and strip marking. Some practice in application of the rules is provided in the classroom. For the final five weeks, training focuses on application of the rules and principles in a nonradar laboratory. Each student performs nonradar controller job tasks in one or two 30 minute scripted scenarios each day. During each scenario, an individual instructor is assigned to each student. Following the scenario, the instructor reviews the student's performance on the problem, providing feedback on errors, correct application of rules, and other required skills.

Thirteen performance measures are made on each candidate during the nine-week period. Five of these assessments are multiple choice tests on the academic portion of the Screen. These assess the candidates' ability to learn and retain the basic knowledges required in the job. A sixth measure is a map test, assessing the students' ablility to learn a map of the relatively simple synthetic airspace. These measures account for 20% of the final grade. Six of the performance measures are formal evaluations of the students' performance during a 30 minute standardized scenario. The score on a laboratory simulation problem consists of a technical assessment (number of errors) and an instructor assessment (subjective evaluation of student performance). Only five of the six laboratory problems are included in calculation of the final score. These five scores, however, comprise 60% of the final grade. Finally, a timed multiple choice test, the controller skills test (CST), is a paper-and-pencil assessment of the students' ability to apply ATC rules and procedures. The CST comprises 20% of the final composite grade. A final grade of 70 is required to pass the Screen. Failure results in removal from the ATCS job.

How effective has the two-stage selection process (OPM battery and Academy Screen) been in meeting the goals identified above, especially since the October 1985 modifications to the program? The following sections discuss the utility of the program for decreasing field attrition rates and the correlations between OPM and Screen measures and field training status.

The Academy screening programs and field attrition rates

As noted the Academy screening programs were implemented in 1976 to provide early identification of candidates with little chance of success in the occupation and to reduce field attrition rates. Prior to 1976, the attrition rate in field training was 38%. After implementation of the screening programs, attrition rates at the Academy averaged 30%, and field training attrition was decreased to 8% (Manning, et al., 1988). Following the 1981 strike and subsequent increase in the population of Academy entrants with no prior ATC experience, the attrition rate at the Academy increased to 40%, while the field attrition rate increased to 11%. Thus, the Academy screening programs were effective in decreasing attrition from field training.

To determine the extent to which OPM and Screen measures predict subsequent measures of performance in field training, Manning, et al., (1989) compared correlations between the OPM ATCS selection test scores and scores from the FAA screening programs with measures of field training performance for competitive Academy students who entered between September 1981 and September 1985. Of the OPM measures, the MCAT was found to be a better predictor of field training status in the en-route option than other OPM tests. Academy laboratory performance scores were found to be more predictive of field training performance than the academic test scores. Finally, for the en-route option, Academy performance measures, particularly laboratory scores, were better predictors of supervisor/OJT instructor ratings and training status than were OPM scores.

To assess the validity of the Academy Screen, analyses were conducted to examine the distributions of OPM and Academy scores for first-time competitive students who entered the Academy between October 1985 and September 1986. (Appendices B-E show statistics related to the distributions of OPM test scores and Academy performance measures. Means and standard deviations of OPM test scores for all applicants who took the OPM test between April and October 1985 are shown in Appendix B. Means and standard deviations of OPM test scores for those who entered the Academy from October 1985 and September 1986 are shown in Appendix C. Means and standard deviations for Academy performance measures are shown in Appendix D. Appendix E shows correlations between OPM test scores and Academy performance measures.)

In October 1985, a change was made to the program which, among other things, resulted in reweighting the performance measures. Because the average time for an en-route ATCS to attain FPL status is just under 3 years, we are just now able to evaluate the impact on predictive validity of this change. The analyses reported here address only the en-route option.

Tables 1 and 2 show correlations of OPM and Academy performance measures with a variable measuring status in field training, both unadjusted and adjusted for restriction in the range of predictor scores, respectively. Field training status is an ordinal variable having low values for ATCSs who succeeded in training and remained in an er-route facility. ATCSs who left en-route at various stages of training had higher values on this variable. Negative correlations occur when ATCSs with higher scores on OPM and Academy tests succeeded and remained in er-route field training. Thus, the greater the negative correlation the better the selection measure was at predicing success in field training. The tables compare the correlations reported in the Manning, Kegg, and Bryant (1989) study with those computed for the FY-1986 sample of Academy Screen graduates assigned to the en-route option. It can be seen that the correlations are similar. Examination of Table 2 shows that the correlation between the OPM rating and field training status, corrected for restriction in the range of OPM rating, is higher than was found for the 1981-1985 sample, although the correlations of individual OPM component scores with the criterion was similar.

Table 1
Correlations between OPM and Screen Performance Measures
and Measures of Field Training Performance
For 1981-1985 En-route and 1986 Screen Entrants
(Uncorrected for restriction in range)

	1981-1985 En-route (n=2992)	1986 Screen (n=402)
OPM TEST MEASURES		
Multiplex Controller Aptitude Test	-0.12*	-0.09
Abstract Reasoning	3.00	-0.03
Occupational Knowledge Test	0.00	-0.05
Transmuted Composite	-0.10*	-0.08
OPM Rating	-0.05*	-0.09
ACADEMY SCREEN MEASURES		
Block Test Average	-0.05*	-0.06
Controller Course Test	-0.04	-0.07
Ave. 5 of 6 Laboratory Problems	-0.24*	-0.21*
Ave. Instructor Assessment	-0.25*	-0.22*
Ave. Technical Assessment	-0.22*	-0.21*
Controller Skills Test	-0.08	-0.16
Comprehensive Score	-0.24*	-0.24*

*Correlations were significantly differently from 0 at p<.01.

Table 2
Correlations between OPM and Screen Performance Measures
and Measures of Field Training Performance
For 1981-1985 En Route and 1986 Screen Graduates
(Adjusted for restriction in the range of predictor scores)

	1981-1985 En Route (n=2992)	1986 Screen (n=402)
OPM TEST MEASURES		
Multiplex Controller Aptitude Test	-0.28	-0.24
Abstract Reasoning	0.05	-0.04
Occupational Knowledge Test	0.00	-0.04
Transmuted Composite	-0.25	-0.24
OPM Rating	-0.13	-0.30
ACADEMY SCREEN MEASURES		
Block Test Average	-0.09	-0.10
Controller Course Test	-0.05	-0.09
Ave. 5 of 6 Laboratory Problems	-0.42	-0.36
Ave. Instructor Assessment	-0.46	-0.37
Ave. Technical Assessment	-0.33	-0.30
Controller Skill Test	-0.14	-0.26
Comprehensive Score	-0.46	-0.44

Also, the correlations of the Laboratory Average (average score on best 5 of 6 laboratory problems) and Laboratory Instructor Assessment (average score on instructor subjective rating in laboratory problems) with en-route field training status decreased somewhat, while the correlation between the CST (Controller Skills Test) and en-route field training status increased somewhat.

Overall, the similarity in correlation coefficients for the two samples suggests that the validity of the program in predicting status in field training has remained stable in spite of the changes made to the ATC Screen.

Use of Nonradar ATC Tasks to a Predict Radar-based Job

The Screen is a program based upon nonradar air traffic separation rules. It was initially designed to select candidates for the en-route environment. At the time it was developed, nonradar procedures were used more frequently. Today, radar separation procedures are utilized almost exclusively in the majority of the centers. Thus, the question arises about the extent to which a program based on nonradar procedures may effectively predict success in a radar environment.

Comparison of the tasks trained in the Screen to the tasks required of an en-route controller suggests that many of the behaviors measured in the current Screen are similar to those required in the radar environment. Of the six activities of an en-route controller identified in the (CTA) job task analysis (1. perform situation monitoring, 2. resolve aircraft conflicts, 3. manage air traffic sequences, 4. route or plan flights, 5. assess weather impact, and 6. manage sector/position resources), the Screen assesses tasks associated with five of the six activities. Only weather is not tested in the Screen program.

The first sub-activity under "perform situation monitoring" is checking and evaluating separation. With the exception of the tasks that require checking the plan view display (PVD) or utilizing radar separation procedures, most of the tasks in the composition graph apply to the nonradar training taught and evaluated in the Screen. The sequence of tasks proceeds as follows: i. Review flight progress for present and/or future aircraft separation; ii. Review PVD (flight progress strips in nonradar) for potential violation of airspace separation standards; iii. Review location of flight progress strip in bay (if unsatisfactory, resequence manually); iv. Mentally assess potential conflict (project mentally an aircraft's future position/altitude/path); evaluate mental flight plan projection for appropriateness; v. Range/bearing data assist assessment of possible conflict--request range/bearing/time (by contacting the pilot in nonradar). Of eighteen tasks described for the sub-activity, twelve tasks or nonradar correlates are assessed in the Screen.

This is an example of the extent to which the tasks taught and assessed in the Screen can be related to the tasks required of an en-route controller, even though radar equipment has added an automated dimension and new specific air traffic separation procedures. It is important to understand that the principles of air traffic control are taught and the candidate's ability to see and separate air traffic is assessed.

The knowledge that performance on nonradar tasks has some utility in predicting future success in the more highly automated radar-based occupation may have application for the development of selection procedures for future automated systems. Perhaps selection procedures designed for the current occupation will predict performance in the evolving occupation.

Because of the ATCS strike in 1981, development of a radar-based selection program was delayed. Recently, however, interest and enthusiasm for addressing questions of new and improved selection procedures for ATCSs have been revived.

Computerized Selection Systems

Since the development of the FAA's selection procedures, the use of computers in assessment has become more feasible and may have some utility for future selection of ATCSs. Much interest and effort during the last two decades has been devoted to creating computer-based assessment and selection systems. A number of computer-based assessment systems are commercially available. Some are directed towards clerical occupations, while others are marketed for both vocational guidance and selection using computer-adaptive multiple cognitive abilities batteries. Further selection applications have been few.

However, several problems complicate the application of computerized testing to selection of employees.

First, typical test construction problems of standardization also apply to computerized tests, as do the computer-specific problems of calibration and user-friendliness.

Second, when an organization has computers, it is not hard to justify the use of some of them for testing. When the selection function requires its own set of computers, it is hard for the organization to justify the use of all of them in testing. Thus, large-scale computerized selection programs, such as might be expected in the military services, have been difficult to implement.

Third, the selection research base for abilities measurable only by computer, such as dynamic spatial ability and processing speed, remains comparatively slim.

The three military research laboratories in the United States have each developed computerized assessment systems, which measure abilities different from those assessed by other paper-and-pencil selection tests. As this research base develops, computers are likely to play a larger role in selection.

Future Selection Systems

This paper has presented an overview of the NAS Plan for automation of the air traffic controllers' job over the next two decades. To the extent that changes in job tasks have been projected so far, we have attempted to map them for ISSS and AERA. At this time, it appears that the job of ISSS controller will require many of the same skills and cognitive attributes required today. Implementation of AERA, however, should begin to substantially change the role of the controller from a tactical mode of operation to a more strategic mode.

It is clear that much of the controller's activity when operating AERA 2 will involve examining data, testing resolutions, and communicating with the computer, aircraft, or other controllers. While a controller will still need to be able to visualize interrelationships between aircraft, the need to make rapid decisions using the information will be reduced. The description of controller activities provided by MITRE suggests that it might be appropriate to select controllers with different types of skills, emphasizing strategic instead of tactical planning. This would suggest that a different battery might be appropriate for selecting ATCSs.

The problem with the description of the future controller exclusively as one who executes the computer's suggestions is that different skills may be required if control actions are required in a short-term time frame (perhaps a shorter time frame than the computer can resolve) or during emergencies. For example, if events (e.g., VFR popups) occur which are not identified by the computer, then the controller may be required to respond quickly to avoid loss of separation between aircraft. These actions are complicated by the proposed "direct routing" of aircraft from one destination to another, without using designated airways. If emergencies requiring controllers to engage in tactical control activities occur only rarely, and if tactical ATC is be complicated by direct routing, then perhaps the future controller will require skills at separating traffic which exceed those required today. As long as the controller retains responsibility for the movement of all aircraft in his or her assigned airspace (even though the computer ordinarily provides resolutions to potential conflicts), then he or she will have to possess and be able to maintain the KSAs to "manually" separate those aircraft, in case of emergency. It can be seen that in this case the selection of ATCSs must be based on the requirement to perform under the worst-case scenario. If, over time, the AERA software is accepted as accurate, and is found to be able to handle all emergency situations, then the responsibility for the aircraft may be removed from the ATCS and assigned to the computer, and the controllers' KSA requirements may be changed, resulting in modifications to selection procedures.

As plans for AERA 3 become more fully developed, the tradeoff of separation responsibilities between the ATCS and the computer will become more pronounced. Preliminary projections are that the AERA 3 software will undertake all control activities, perhaps leaving the ATCS to issue clearances to aircraft which do not have data link capabilities, or perform traffic management activities. The controller may also be responsible for ensuring that the computer is operating properly. In all cases, even through AERA 3, the computer is dependent on timely, accurate input of information from human operators and is, thus, vulnerable to the reliability and accuracy of the human. Again, if the controller has responsibility for taking over the control of air traffic when (if) the computer fails, then the selection procedures must identify those who have the ability to control traffic, even if they only do so on rare occasions. If, on the other hand, the full responsibility for aircraft separation lies with the computer (perhaps another level of software), then selection procedures can be targeted to the KSAs required to perform the duties most frequently performed by the ATCS using AERA 3.

Development of any selection procedure depends on an accurate description of the tasks to be performed and the knowledges, skills, and abilities (KSAs) required to perform those tasks. The problem with applying this approach to the development of selection procedures for the "automated controller" is obtaining a sufficiently detailed job/task analysis prior to implementation of the new system, since no controllers will have used the system extensively enough to qualify as subject matter experts. A fallback position would be to work with controller members of the Air Traffic AERA Concepts Team (ATACT), who are as knowledgeable as any controllers about the way the new system will work. These "near SMEs" could speculate about the requirements to perform the tasks. Preliminary work can be done early in the development cycle to identify or develop tests to measure the KSAs, as they are currently identified. Review cycles will have to be built into the process to allow periodic review of the KSAs with SMEs, as more controllers are brought into the loop to test the AERA software and procedures, and as the software and procedures become increasingly well-defined. Using such an iterative process, a tentative set of selection procedures would be available upon implementation of the new system.

When contemplating the development of selection procedures for employees who will be ATCSs under the AERA levels of automation, it is clear that work must begin now in order for the procedures to be ready when AERA 2 software is installed. At the same time, the division of duties between the software and the controller have not yet been finalized, so the final choice of a selection procedure is premature.

Because of the criticality of the ATCS occupation, it will not be possible to pass everyone who applies for the job for a period of time until the validity of the new selection battery can be established. When converting to the AERA system, it may be necessary to use the old selection procedures for a while, until the validity of the new procedures can be evaluated. The correlations between measures of performance in the nonradar Screen and success in radar-based field training suggests that using a selection procedure validated, for example, for the radar job to select controllers for the new ATC job under ISSS may be an acceptable interim solution while validation of the new procedures occurs.

Furthermore, several years may be required to establish the validity of new selection procedures. Most of that time is needed for those selected to complete field training so that their performance on the job can be measured, and the relationships between predictors and criteria can be assessed. Measures of performance in training can be used as interim criteria, but, as previously mentioned, are less desirable than measures of job performance.

Discussion

The research from the past 30 years has resulted in some findings that can contribute to the development of selection procedures for automated systems. First, paper-and-pencil aptitude tests alone have been found insufficient to predict performance on the complex procedural tasks involved in ATC in its current environment. Second, the addition of a second-stage miniaturized training-testing program improved the utility and validity of the ATCS selection process and reduced field attrition. This selection program, based upon nonradar separation, has been useful for predicting success in the radar environment.

As with selection research in other occupations, the research on air traffic control specialists suffers from lack of availability of criterion measures of ATC job performance. Because the Screen measures task samples via the laboratory problems, the experience gained in developing measures of training performance for both the Screen and the Academy basic radar training courses could contribute to future efforts to develop criterion job performance measures for the automated ATC environment.

The current selection system has been found to have utility in decreasing the field attrition rate; however, it may not be the best selection system for radar ATCSs or for the automated occupation as it evolves, and it is certainly very costly in terms of monetary and human resources. Therefore, the FAA is beginning to explore alternatives to the current methods. The goals are to have a shorter, less costly selection process with similar predictive validity to the present system in place within the next five years.

Clearly, the development of selection procedures for automated systems will be a difficult process, complicated by a lack of information about the job, varying expectations about eventual job task requirements, extensive time frames for validation, and a need to develop appropriate criterion measures of job performance. The time to start addressing these issues is now, so that development of selection procedures for the changing occupation can parallel the evolution of the job.

References

Alexander, J. R., Alley, V. L., Ammerman, H. L., Hostetler, C. M., and Jones, G. W. (1988.) FAA Air Traffic Control Operations Concepts Volume III: ISSS En Route Controllers. DOT/FAA/AP-87-01. Washington D.C.: Federal Aviation Administration,

Ammerman, H. L., Becker, E. S., Jones, G. W., Tobey, W. K., and Phillips, M. K. (1987.) FAA Air Traffic Control Operations Concepts Volume I: ATC Background and Analysis Methodology. DOT/FAA/AP-87-01. Washington D.C.: Federal Aviation Administration

Ammerman, H. L., Bergen, L. J., Davies, D., K., Hostetler, C. M., Inman, E. E., and Jones, G. W. (1987). FAA Air Traffic Control Operations Concepts Volume VI: ARTCC/HOST En Route Controllers. DOT/FAA/AP/87-01. Washington, D.C.: Federal Aviation Administration.

Ammerman, H. L., and Jones., G. W. (1988) ISSS impact on ATC procedures and training. Contract DTFA-01-85-Y-01034, CDRL C108. Colorado Springs, CO: CTA Incorporated.
Brokaw, L. D. (1984) Early research on controller selection, 1941-1963. In Sells, S. B., Dailey, J. T., and Pickrel, E. W. (Eds.), Selection of air traffic controllers. FAA Office of Aviation Medicine Report No. FAA-AM-84-2.
Carlson, L. S., and Rhodes, L. R. (1990) Changes in Controllers' work activities in AERA 2 and implications for controller training. MITRE technical report No. MP-89W00021, January.
Cobb, B. B., Mathers, J. J., and Nelson, P. L. (1972) Attrition-Retention rates of air traffic control trainees recruited during 1960-1963 and 1968-1970. FAA Office of Aviation Medicine Report No. AM-72-33.
Collins, W. E., Boone, J. O., and VanDeventer, A. D. (Eds.). (1981). The selection of air traffic control specialists: History and review of contributions by the Civil Aeromedical Institute, 1960-1980. *Aviation, Space, and Environmental Medicine, 52*, 217-240.
Gutmann, J. (1988) Advanced Automation System, Human computer interface: An overview. In Proceedings of Symposium on Air Traffic Control training for tomorrow's technology, Oklahoma City, OK, December 6-7.
Hopkin, D. V. (1988) Training implications of technological advances in Air Traffic Control. In Proceedings of Symposium on Air Traffic Control training for tomorrow's technology, Oklahoma City, OK, December 6-7.
Manning, C. A., Della Rocco, P. S., and Bryant, K. D. (1989) Prediction of success in FAA air traffic control field training as a function of selection and screening test performance. FAA Office of Aviation Medicine Report No. FAA-AM-89-6, May.
Manning, C. A., Kegg, P. S., and Collins, W. E. (1988) Selection and screening programs for air traffic control. In Jensen, R. S. (Ed.) Aviation Psychology. Aldershot, England: Gower Technical.
McKinley, J., and Jago, R. (1984). AERA En Route Controller Skills Requirements. Unpublished manuscript, September.
Schmidt, F. L. and Hunter, J. E. (1977) Development of a general solution to the problem of validity generalization. *Journal of Applied Psychology, 62*, 529-540.
Uniform Guidelines on Employee Selection Procedures, Federal Register, (1978), 43, 38290-83815.
U. S. Department of Transportation, Federal Aviation Administration. (1989) National Airspace System Plan: Facilities, Equipment, Associated Development, and other Capital Needs.

APPENDIX A

Cognitive/Sensory Attribute Definitions from Computer Technology Associates' Job Task Analysis (Ammerman, et al., 1987, p 3-33).

CODING	Transformation or translation of information for entry into the system; Converting textual information to graphics or symbols.
Example:	Entering a PIREP (pilot report); Composing a flight plan amendment.
DECODING	Transformation or translation of information received.
Example:	Recognizing a symbol for a handoff; Reading a Flight Data Entry.
DEDUCTIVE REASONING	Ability to reach a conclusion that follows logically from the known facts or data; Selection from among alternative answers or methods.
Example:	Concluding that two aircraft are on intersecting paths.
FILTERING	Selection of inputs on which to focus attention in the presence of distracting stimuli or high workload; Selective attention; Overload accommodation.
Example:	Identifying communication transmissions for attention during a period of heavy radio traffic.
IMAGE/ PATTERN RECOGNITION	Perception of spatial patterns and relations among static or dynamic visual inputs. May involve orienting oneself to the picture or configuration.
Example:	Forming a picture of the traffic situation by reviewing Flight Data Entries on the Flight Data Display.
INDUCTIVE REASONING	Generation of an explanation for a set of specific data or instances, giving structure and meaning to the information; Generalization of working hypotheses from specific events; Discerning basic differences and relationships among symbols, figures and figure patterns; Generating a new solution to a problem; Ability to make a knowledgeable assumption using incomplete data.
Example:	Checking the adequacy of a proposed aircraft maneuver.
LONG-TERM MEMORY	Mental storage of knowledge over a period of time and selective recall of items relevant to a situation.
Example:	Remembering aircraft characteristics; Remembering procedural instructions or letters of agreement relevant to an uncommon situation, such as an airshow or large flight formation.

MATH & PROBABILISTIC REASONING	Translation of uncertainty into probability; Assigning a subjective probability regarding the likelihood of an event occurring; Ability to use probabilities to identify optimal courses of action.
Example:	Assessing the risk of an aircraft maneuver.
MOVEMENT DETECTION	Recognition of the physical movement of a visual object; Estimation of its direction or speed.
Example:	Observing aircraft on the Situation Display responding to a clearance or advisory.
PRIORITIZING	Ordering of events in sequence; Establishing priorities.
Example:	Deferring a request for flight plan changes in the presence of more urgent activity.
SHORT-TERM MEMORY	Mental storage and selective recall of of relevant information over a brief period of time.
Example:	Briefly retaining and entering an aircraft call sign.
SPATIAL SCANNING	Rapid identification or detection of objects or events displayed in a wide or complicated visual field.
Example:	Observing the Situation Display for new aircraft; Searching for data in a table.
VERBAL FILTERING	Same as FILTERING, but limited to voice communications
VISUALIZATION	Observation of spatial patterns and subsequent mental transformations into other spatial patterns.
Example:	Determining the effect of a proposed aircraft maneuver on other aircraft; Comparing intended time-position profiles for intersection in position/altitude/time.

APPENDIX B

Applicant OPM ATCS battery performance

Mean Scores on OPM ATCS battery component tests for
OPM open period from April-October 1985
N=8,826

Measure	Mean	Std. Dev.
Multiplex Controller Aptitude Test (MCAT)	73.0	16.6
Abstract Reasoning (ABSR)	31.5	9.3
Occupational Knowledge Test(OKT)	29.2	11.6
Transmuted Composite (TMC)	75.5	12.5
OPM Rating (RAT)	76.4	13.3

APPENDIX C

Academy Entrant OPM ATCS battery performance

Mean Scores on OPM ATCS battery component tests
for October 1985 through September 1986 Academy Entrants
N=402

Measure	Mean	Std. Dev.
Multiplex Controller Aptitude Test	94.3	5.8
Abstract Reasoning	40.6	5.5
Occupational Knowledge Test	36.5	14.6
Transmuted Composite	91.8	4.3
OPM Rating	93.7	4.0

APPENDIX D

Academy Performance Measures

Mean Scores for Academy Performance Measures
October 1985 through September 1986 Competitive Entrants
N=402

Measure	Mean	Std. Dev.
Block Test Ave.	95.7	4.4
Controller CourseTest	93.4	5.4
Ave. 5 of 6 Laboratory Problems	74.0	7.1
Ave. Instructor Assessment	84.6	5.5
Ave. Technical Assessment	54.1	10.8
Controller SkillsTest	82.0	7.5
Comprehensive Score	79.7	5.0

APPENDIX E

Correlations between Academy and OPM battery component scores
FY 1986 competitive entrants assigned to en-route option
N=402

	OPM measures				
Academy measures	MCAT	ABSR	OKT	TMC	RAT
Block Test Ave.	0.02	0.03	0.14*	0.02	0.11
Controller Course Test	0.01	0.02	0.25*	0.02	0.19*
Ave. 5 of 6 Lab. Problems	0.22*	0.04	0.00	0.20*	0.15*
Ave. Instructor Assessment	0.26*	0.05	0.03	0.25*	0.19*
Ave. Technical Assessment	0.23*	0.05	0.01	0.22*	0.17*
Controller Skills Test	0.16*	0.12*	0.03	0.19*	0.18*
Comprehensive Score	0.24*	0.08	0.05	0.23	0.21*

*Correlations were significantly different from 0 at $p<.01$.

Automation Issues for the Selection of Controllers

Gill Nyfield

Saville & Holdsworth Ltd, Surrey, UK

Introduction

This paper provides some ideas on issues related to the selection of air traffic controllers in an automated environment. It is mostly based on Saville & Holdsworth's (SHL) work with the Civil Aviation Authority (CAA) in the UK, where we are retained to provide advice and research into the selection of controllers.

The paper is divided into three sections:

- The need for the development of systematic and objective selection systems to ensure that the most suitable people are selected
- The consequences of further automation in ATC on selection
- The use of automation in the selection procedure to provide better assessment in the future

Systematic Selection Procedures

The present system used for the selection of controllers by the CAA was based upon an extensive analysis of the role, carried out in 1984. This identified a number of key issues that should be assessed or considered in the selection of controllers, summarized in Figure 1.

On the basis of this information, a selection procedure was designed with the purpose of providing relevant yet objective information on which to make the decision. The protocol included:

- Minimum screening by application form on the basis of basic academic attainment and age
- First-stage selection on the basis of psychometric tests and screening interview
- Final selection based on test results, plus a panel interview.

The relevance of the tests was initially evaluated by a study conducted on established controllers whose test results were correlated with obtained measures of job performance. These showed that better controllers scored more highly on ability tests measuring:

NATO ASI Series, Vol. F 73
Automation and Systems Issues in Air Traffic Control
Edited by John A. Wise et al.

- Spatial reasoning
- Clerical checking
- Diagrammatic/logical reasoning

The more highly-rated controllers also showed some significant personality differences (as measured by a work-oriented personality questionnaire). Important scales included the personality dimensions of being democratic, optimistic, and willing to accept constraints and not act independently.

The tests were then administered at time of selection, but used very cautiously until a sufficiently large number of controllers had been taken on and exposed to the college-based training program. The program took one year to complete, though it has been revised recently.

The predictive nature of the tests was evaluated, and again the results were very positive, showing a significant reduction in failure rate for the higher scorers on the tests. The most useful ability test results and personality questionnaire data were combined to form a prediction equation which is currently used in the first stage of selection. Candidates who score low on this are recommended for rejection.

The utility of this process is shown in Table 1, which summarizes the progress through training of 150 cadets.

Core Skills

Ability to absorb information simultaneously from multiple sources
Ability to absorb new information while making decisions
Ability to project forward on the basis of current information
Ability to adjust constantly the whole picture

Temperamental Factors

Readiness to work within a system
Preference for working to set standards
Cooperativeness
Convergent thinking
Decisiveness and confidence
Conscientiousness
Structured thinking
Self control

Contextual Factors

Speed of decisions
Sporadic time pressure
Sudden high level demands on the individual
Distractions
Fluctuations between routine and non-routine
Checking/up-dating information
Short cycle repetitive work

Figure 1. Model of ATC Core Skills

Table 1. Relationship between Test Results and Progress Through Training

Test Results	Failed/Resigned	Successful	Total	N
Low Scorers	45%	55%	100%	76
High Scorers	7%	93%	100%	74
Total	26%	74%	100%	150

Multiple R = .53 (significant at 0.01 level)

This table shows that high scorers have a 93% chance of being successful, versus only a 55% chance for low scorers. Since the full implementation of the tests within the selection procedure, the success rate during training has risen from about 50% to nearly 80%.

The procedure continues to be evaluated and the progress of successful trainees at their work units monitored. Further improvements to the selection procedure, using computerized tests, are currently being evaluated.

These results show the benefits of a systematic approach to selection. The techniques used are not novel. They are based on well-researched psychometric principles:

- Thorough, objective job analysis to define attributes required for effective performance
- Use of selection methods emphasizing *reliable* assessment of *relevant* attributes
- Where possible, research, follow up and monitoring of results to evaluate the effects and to continue to improve the selection process.

This methodologyhas been shown to work in other countries with controller selection, and indeed is used by major organizations worldwide in the selection of professional staff. As research continues to indicate, unless selection follows these principles, there is a strong chance of failure or an inability to replicate its effectiveness in a different location.

The Selection of Controllers for a More Automated ATC System

As work on automated ATC systems continues, decisions will continue to be made on how they will be implemented, with the subsequent effects on the way the controller carries out his/her role. The questions for those involved in the selection and training of controllers include:

- Will increased automation require different skills/abilities than those required today?
- Will it require a different emphasis or balance of skills?
- Will research conducted on present controllers, who will be moving into increased automation, be applicable to those who will experience only the more automated environment?

I do not propose to answer these questions, but instead to suggest that practical solutions to future selection will come from

- Better systems for analyzing jobs and activities, to identify which underlying attributes are required to perform them
- Better understanding of how present selection systems work, so that these can be evolved and developed in the correct direction as the role changes
- More detailed and objective assessment of job performance, to allow for more detailed evaluation of the impact of different selection methods on the different aspects of the role

Intrinsic to these solutions will be the use of objective selection methods and systematic evaluation to ensure that future ATC is efficiently staffed by effective controllers.

However, in conjunction with this, there is a need to ensure an adequate supply of individuals who wish to carry out the role and who will be effective. Certainly in the UK and elsewhere in the world, the demographic trends are such that, in the 1990s, there will be increased competition by all employers for a decreasing number of young people.

If there is a shortage of candidates, then employers need to ensure that

- They are not placing unnecessary restrictions in terms of academic qualifications, age limits, etc., on who can apply
- That the selection procedures are fair and relevant to ensure that suitable candidates are not rejected for unjustified reasons
- That the employers will have to market themselves and the job more assertively to compete successfully in attracting the right applicants

This last issue is related to the more complex concept of job design, which in turn is related to job satisfaction and related issues.

Automation as a Route to Improved Selection

Besides having a major impact on, and benefit to air traffic controller work, automation potentially should have a major impact on the tools that will be available for the selection of controllers, and indeed on the management of the whole selection/training system.

Most nations still rely on a large proportion of pencil-and-paper tests, but some, including the UK, have introduced or are experimenting with more automated systems. There is a great attraction for using more automated methods, particularly as the job itself becomes more automated. But some caution is appropriate. The low costs attached to pencil-and-paper systems, and the higher development costs of automated ones, indicate that clearly defined benefits must justify the automated route.

Automation does allow greater choice of information and potentially much better use of whatever is gathered; and without doubt, it will continue to benefit those involved in selection. A few of the potential areas for gain are explored below.

Pre-Selection

Selection today is becoming more of a two-way process. Not only is there emphasis on the employer's finding out whether the candidate will become a good controller, but candidates are also concerned to find out whether the role, and indeed the employer, is right for them. Automation potentially provides a more cost-effective route for the candidate to find out more about the role and indeed, in an informal way, whether they might be suited for the job. An interactive system which the individual interrogates at will, can make this procedure readily available in as many sites as required, with minimal support. This could have the effect of opening up the potential applicant pool to those who would not initially have considered it, while providing the employer with candidates who have, at least to some extent, "screened themselves."

Interview Improvements

The interview is the most common selection method. However, a mass of evidence shows that in untrained or inexperienced hands, it is a very poor predictor of success. Automation can, by the use of expert system technology, possibly reduce the emphasis on the quality of the interviewer in getting the correct result. This could provide interviewers with a structure for the interview of a particular candidate, suitable questions to ask, and, depending on the response, potential follow-up questions or guidance on how to evaluate the responses.

Given the enormous amount of time spent on interviewing, often involving highly paid professional staff, the cost-benefit of this work may easily be shown to be advantageous.

Test Administration

Most nations still emphasize pencil-and-paper testing. It is certainly cost-effective, given the benefits in improved selection, and particularly where tests are administered to groups.

Most tests could be adapted to be administered by computer, providing greater standardization and ease of scoring. However, the cost benefits of this in 1990, at least for ability tests in large selection campaigns, can be questionable. Where scoring is more complex (such as with personality questionnaires), large numbers of tests are used, or small numbers of candidates are processed at any one time or place, then it is likely to be more beneficial. It will provide users with accurate results very quickly, so that immediate use can be made of the information. However, the benefits are likely to be greater if the automation provides better — more accurate — information in a potentially shorter time. Adaptive (or tailored) testing, where the test is adapted to the candidate is being evaluated by a number of large employers in the US and UK. The results should be monitored to indicate whether this route is likely to be of use.

Another potential advantage of automation could be the assessment of skills not readily assessed through pencil-and-paper.

Work by SHL, in conjunction with the CAA, showed that the important skills of multi-tasking and dynamic spatial ability were not clearly assessed by the present selection procedure. Computerized tests utilizing standard cognitive-processing paradigms showed no significant improvement in prediction over the pencil-and-paper tests, as discussed in the first section. Therefore, work has now commenced on the development and trialling of automated procedures designed to specifically measure these attributes. The aim is to provide additional relevant information to improve selection decisions, with particular reference to the use of radar.

Simulations

Perhaps the greatest attraction of automation is the possibility of providing simulations for both selection and training. Automated simulation is now well-established in training, but the present costs of equipment indicate that, for a while yet, it will not be available to any great extent for selection purposes.

These factors will change over time, and cheaper, simpler systems will also be available. But perhaps the complexity of the simulation is its greatest enemy. So much information can be obtained that it will be difficult to ascertain which salient features will be predictive of job performance and will withstand the test of time. It may be better to design discrete scenarios to assess specific skills, similar to those described above in automated test administration.

Simulations could potentially provide very useful information on trainability. Trainability testing has shown potential in a number of occupations where semi-complex repetitious tasks are performed, and the results are quite positive. However, the disadvantage is that they tend to be very expensive to administer — in terms of both requiring relatively expensive equipment and being time consuming. Here, we may have to weigh the balance of costs of selectivity versus training.

System Management

One of the main advantages of automation in selection and training must be the better use of available information on candidates and successful controllers. Better database management should allow, among other improvements:

- More efficient and effective use of the selection information
- Use of expert system technology to utilize skills of expert recruiters while releasing them for other activities
- Better feedback of selection information to candidates and to trainers
- Improved monitoring of training and job performance.

Summary

In summary, automation will bring considerable change to ATC selection. With systematic methodology and available technology, the future should bring improved selection tools to help in the recruitment of competent controllers.

Automation in ATC: How Does It Affect the Selection of Controllers?

Hinnerk Eißfeldt

German Aerospace Research Establishment DLR, Hamburg, Germany

Introduction

As air traffic control work is closely related to its technical environment, it has been influenced strongly by the technological developments of the last decades. Nowadays the controller benefits from digital radar-data, flight plan management, and other assistance by computers. High resolution monitors also improve working conditions in air traffic control. If we remember the usual situation of a controller in the early 60's — sitting in a dark room looking at the flickering display of an analog radar, and writing flight strips by hand — we can appreciate the changes and improvements in today's working conditions. This comparison might also be helpful if we consider the changes expected in the next decades: even though the development will continue in small steps towards automation, the discrepancy between today's status and that in about another 30 years will be enormous.

Automation in Air Traffic Control

It is not my purpose to spend many thoughts on the level of automation to be reached by a particular time. The only basic assumption in this context is that, according to expert opinion, 100% automation will never be reached. Human beings will always be the central decision-makers in ATC systems, assisted by evernmore powerful tools but never replaced by them. The task of selecting the right men and women to do this very demanding job will therefore always be crucial.

In this context, we must consider whether the changes in ATC work might be accompanied by changes in the required aptitudes of controllers, and whether modifications in the test systems applied today should be undertaken. In other words, the task is "... to adapt selection to new requirements" (Hopkin, 1982, p. 100).

Selection of Air Traffic Controllers

Test systems used today for the selection of ATC-applicants work on the same principle everywhere in the world: they try to identify applicants with good performance in aptitudes

NATO ASI Series, Vol. F 73
Automation and Systems Issues in Air Traffic Control
Edited by John A. Wise et al.

considered to be basic requirements for the ATC profession. However, what is considered to be a basic requirement and how it is measured differs from country to country, and every test system is modified over the years. Goeters (1987) reported new psycho-diagnostic developments in aviation psychology, with special consideration of the benefit of computers for test application.

The task of the psychological selection of air traffic controllers is first to predict the success of an applicant in ATC training, and second to guarantee that the checked-out controller will be able to cope with the high job demands until retirement age. In addition to operational skills, aspects of interpersonal behavior and emotional stability have to be taken into account. Therefore, the psychological selection of ATC staff in West Germany provides a multi-modal evaluation covering biographical data as well as performance tests and personality questionnaires (Eißfeldt, 1989).

The following list shows traits assessed in the DLR Test System for the selection of air traffic controllers :

Performance
- Basic Knowledge:
  - English
  - Technical Comprehension
  - Mathematico-Logical Thinking
- Operational Aptitudes:
  - Memory (Auditory/Visual)
  - Perception and Attention (Auditory/Visual)
  - Spatial Orientation (Auditory/Visual)
  - Multiple Task Capacity

Personality
- Achievement-Oriented Traits:
  - Motivation
  - Rigidity
  - Mobility
  - itality
- Interpersonal Behavior:
  - Extraversion
  - Dominance
  - Aggressiveness
  - Empathy

- Stress Resistance
  - Emotional Stability

Future Changes in Required Basic Knowledge

The degree of basic knowledge required to become an air traffic controller depends on the content of the training. In many cases the curriculum will remain the same for the foreseeable years. But far as technical comprehension is concerned, we have to expect an increasing influence of computer-oriented knowledge in future training of air traffic controllers. Therefore, we can expect a shift from physics (i.e., mechanics, electronics) towards computer sciences. It might be of value, for the assessment of job motivation, to collect some information concerning the attitude of future applicants towards working with computers.

Future Changes in Required Operational Aptitudes

We can also expect a further increase of information presented to the controller at work. The increase in auditory information due to high traffic density (and thereby crowded frequencies) might be balanced by the development of systems able to exchange standard information automatically, so that radio information can be reduced.

The increase of visual information is manifested by the increasing number of screens at a controller's working position, with several "pages" of information presented on the same monitor. Such systems are planned to help controllers to do their work. However, those who develop such systems should bear in mind that there is a maximum in a human being's capacity to recognize, memorize, and process visual information. Beyond a certain point, additional information will result not in a further increase, but in a significant decrease of efficiency in the controller's work. Although a successful selection system channels only those applicants with high capacities for information processing into ATC training, "information overload" must be strictly prevented.

Another aspect of future air traffic control systems that might influence the selection of controllers is linked to the problem of "information underload." Future systems might be able to operate low density traffic more or less automatically by "auto-control" function. The controller's task in that case will be more to monitor the system than to operate it. As shown in several experimental studies (Zetzmann, 1970), monitoring always includes problems of vigilance, or maintaining high attention in reaction to infrequent target signals over long periods of time.

Although the DLR Test System already selects only applicants with good vigilance capacity, the problem of unrealized signals will occur to some degree, as Maschke and Finell reported (1988). To require vigilance capacity of ATC applicants will become more important in the future.

Accuracy of information processing is very important for air traffic control. Unrecognized errors might cause an accident. The present selection procedure dismisses those applicants who make many errors on the test items. Perhaps in the future those same applicants could be recommended for training, since their errors will be checked and corrected by computers. Similar ideas might arise concerning deficits in operational aptitudes (for instance, memory capacity). Computer programs to check the accuracy of the controller's input and correct it if necessary might become real in the future. This development encourages lowering the requirements in selection.

However, certain objections become clear: The first and main objection is the vulnerability of computer systems. Even an optimum of redundancy cannot fully avoid system failures, and for that reason controllers must be able to work at their own aptitudes at any time. For the same reason, regular simulator-training must be provided to keep the controller's skills on a high level, so that he is prepared to take over full control immediately.

Another important objection is based upon the usual way in which a technical system reacts to recognized errors of the human operator: it provides visual and/or acoustical warnings in different intensity, related to the kind of action that is required. Although this seems an effective strategy, we should remember that warning indicators in some cases are more likely to result in confusion than in correct action. In many cases, warning signs are ignored, simply because people are used to them. Therefore the correct function of the ATC system should be based not on supportive error-avoiding software but on the capacity of the controller to work correctly. Consequently, the requirements concerning operational aptitudes should remain high in the future.

Future Changes in Required Personality Traits

A recent improvement of test systems in several countries has been the evaluation of applicants' personalities as well as their aptitudes. At the DLR, personality is evaluated in several traits: achievement motivation, interpersonal behavior, and emotional stability. Only those applicants will be selected, who are able to apply their good operational aptitudes without being disturbed by deficiencies in their personality. This selection strategy was formed by analyzing the reasons for training failures of trainees who seemed to have high aptitudes.

In future, the evaluation of personality will become even more important for various reasons. The first is that air traffic control is teamwork, and the higher the density of air traffic, the higher the workload for the whole team (and, sometimes, the impairment of interpersonal relationships). The second reason is the tendency to segregate the working positions of controllers. In order to present more information to a single controller, the trend is to build up several control-islands, instead of long lines of controllers sitting side by side. Reducing the number of controllers who are actually working together makes it more important to ensure that their work is not influenced by interpersonal difficulties.

The evaluation of personality traits during selection is a reasonable way to avoid such difficulties, by selecting applicants who tend to be sociable and tolerant even when they encounter heavy workloads.

Conclusion

Many trends in the future development of air traffic control systems (the increase of visual information, the importance of vigilance capacity, the reduction in the number of team members, etc.) seem to parallel the development in modern aircraft cockpits. It might therefore be a promising solution to some of the problems mentioned above to establish dual working positions in ATC, as it is done in most aircraft. Like "pilot flying/non-flying," there could be "controller operating/non-operating," with interchanging roles and a similar "team resource management": the controller non-operating assists the controller operating, and each is monitoring the other. To reach sufficient redundancy on the technical side as well as on the side of the human operator, every display is presented twice.

The present state of each selection procedure suggests how much it may be affected by automation in ATC. We have good reasons to assume increasing emphasis on some traits (the ability to recognize, memorize, and process visual information; vigilance capacity) in future selection of controllers. Traits of interpersonal behavior will have increasing importance as teams get smaller, and will be crucial to the success of dual working positions. If those traits mentioned above are not currently represented in the test system, they should be introduced soon — to ensure selection of safe candidates for the next decades.

References

Eißfelt, H.& Maschke, P. (1990). Bewährungskontrolle eines psychologischen Auswahlverfahrens für den Flugverkehrskontrolldienst anhand von Kriterian der Berufsausbildung. Hamburg: DLR-FB-316-90.

Eißfeldt, H. (1991). The DLR Selection of Air Traffic Control Applicants - Predicitive validity of the System. *Proceedings of the 18th Conference of the Western European Association for Aviation Psychology.*

Goeters, K.M. (1987). New Psychodiagnostic Developments in Aviation Psychology. *Proceedings of the 17th Conference of the Western European Association for Aviation Psychology*. Wien, Austria, 141-145.

Hopkin, V.D. (1982). Human Factors in Air Traffic Control. Advisory Group for Aerospace Research and Development, AGARDograph No. 275. Farnborough, Hampshire, UK.

Maschke, P. and Finell, G. (1988). Differentiellpsychologische Analyse eines computergesteuerten audiovisuellen Vigilanztests. DFVLR-Forschungsbericht 88-23. Koln, Germany.

Zetzmann, H.J. (1970). 'Workload' and 'Performance Limiting Factors' of Air Traffic Control Operators. Paper read on the Specialist Meeting of AGARD Aerospace Medical Panel, 13-15 May 1970. Oslo, Norway.

Consequences of Automation for Training and Resource Management

Training Requirements for Automated ATC

Rod Baldwin

Eurocontrol Institute of Air Navigation Services, Luxembourg

Introduction

Training is not an objective in itself. It is required in order to supply a sufficient number of persons, of sufficient quality, to perform specific jobs.

This is particularly pertinent in the ATC field where most controller training is undertaken by national authorities — many now required to be self-sufficient — who require good success rates and low loss of personnel from operational positions.

These are exciting days because the nature of ATC is changing, new procedures are being demanded, new technologies are being offered, and personnel are asking "what is the nature of my job in the future?"

During the Institute many matters concerning automation in ATC were raised and discussions ranged over likely problems, possible implementation, and how staff will be affected. I consider that the trainer will have to address these problems now, and therefore this paper is written with the whole Institute in mind. I have made it a very discussible paper, so that the training problems raised can be discussed when other papers touch on the human aspect and the need to prepare the individual for his future job.

As mentioned previously, the trainer is there to provide qualified personnel who will operate the new automated systems — but we do also expect the system designers to understand the aims and objectives of the trainer.

Present Position and Possible Alternatives

Entry Criteria

The age at which an Air Traffic Controller Licence can be held will be set by the civil aviation authority of each country, but following ICAO guidelines and a training course of three to four years duration, it is usual to set 18 years as the minimum age for entry to an ATC training program.

Some national administrations also set a maximum age, in order to ensure that there will be enough operational years in which to gain the benefits invested in a long, paid training program. Such a limitation, whilst advisable for some professions, should be regularly revised as the personal characteristics being sought for may take time to mature; they could be more clearly identified when candidates will have a longer work track record. In this case higher success rates could be achieved in the training phase, with consequent savings of lost investment.

NATO ASI Series, Vol. F 73
Automation and Systems Issues in Air Traffic Control
Edited by John A. Wise et al.

Whilst comparisons are often wrongly made in comparing requirements for an ATC licence with those for a private pilot's licence, it is interesting that the minimum age for the latter can be quite young — a recent example being a French licence obtained by a fifteen-year-old. In this case the person's attitude to safety and responsible behavior was obviously acceptable.

The qualifications required for entry to a training program can vary widely from country to country. Many authorities accept an average pass in the school examinations that occur at about eighteen years of age, with subjects such as mathematics, science, and geography being regarded as essential in order to cope with the training program.

The level of qualification would seem to be sufficient, or more than sufficient, since there seem to be few failures due to a trainee's inability to master the knowledge content of a course. However, the extent to which the knowledge is mastered may be just sufficient and leaves no room for higher tasks. Here an administration has to accept that a training program, besides its short term aim, also has a long term purpose of preparing future supervisors and managers. Consequently some administrations aim at university graduate entry, thus solving their future management resourcing problems, but only if they have enough higher positions to fulfill the ambitions of all the "high flyers"!

In some cases authorities cannot meet the aspirations of so many well qualified staff, so they must consider whether they should be aiming at a lower level of qualifications and allow career development to be based on performance in the job and perceived potential for higher level work. This could be a sensible solution in an authority that accepts mobility of staff based purely on performance assessment, but difficulties will arise in more rigid and hierarchically based organizations that strongly grade staff by age and qualifications.

Since the problem of career development arises because of the high proportion of controllers to other staff in the air traffic services, a more acceptable solution might be to recruit at scientific graduate level, for all technical and operational staff, and all must practice as controllers for a certain minimum of years. At this time they could then opt to remain operational or move to other scientific or technical disciplines. Whether it would be possible to find sufficient persons with the required breadth of personal characteristics is questionable, but the suggestion should be kept in mind when considering required controller characteristics for automated ATC.

In general, the entry qualifications will serve to indicate how well the trainee can learn the new knowledge that he needs to acquire and the extent to which he can understand the knowledge for application to practical problems.

The skills needed for a trainee controller are obviously dependent on those required when licensed. However, difficulty arises in trying to estimate the extent to which those skills can be leant, how long the learning process should be, and the level of skill performance that is required to commence a course. Trainability tests have been designed for many other occupations, and should be possible for controllers.

There are a number of examples where indicative tests are being applied (for instance, the SNAKE test at the Eurocontrol Institute) but validation will take several years.

One of the problems for selectors is that all the tests and interviews are time consuming. Checking age and qualifications is relatively straightforward, but skill assessment must be done carefully with reasonable time given. We should ask why schools do not give a number of basic tests in motor skills and thus provide each school learner with an estimation of his or her capabilities for practical activities. The advantage of this would be that candidates would apply their own filtering system when considering careers, and also place their enthusiasms in realistic directions.

The need to concentrate on trainability is especially important for a candidate who may have left school with certain skills but not exercised them for a number of years. In Europe this particularly applies in the use of languages, where skillful use of at least three languages is probably essential for many quality jobs. In the ATC field use of English is needed, but many candidates could be regarded as weak, simply because they have not had the opportunity to practice English. In some areas, whereas it may be possible to listen to radio or see films in English, it might not be so easy to practice speaking. And of course, there could well be no need for English before a person became aware of ATC as a profession which he or she might like.

Another skill often demanded is the ability to visualize in three dimensions. This is certainly necessary and causes difficulty for some trainees, but usually the problem has to emerge during the training phase, as a test seems to be rarely applied to candidates.

The attitudes and characteristics being looked for are based on those exercised in performing the job of a controller. For the qualities demanded, the candidate is asked to meet certain requirements — such as ability to take responsibility, work under pressure etc. — when he or she may never have had the opportunity to show these talents, or if so, not often enough to develop them. This will particularly apply if the age is dropped, but it may be that the parents should be interviewed (as they will have observed their offspring over many years, and their children's responses to even minor events could be indicators of future behavior in an ATC environment). Of course there are difficulties with such a suggestion. But it does high-light the fact that personal characteristics must be appraised when entering a profession.

The method most generally used is to call for interviews on the basis of paper qualifications and then conduct a series of assessments for skills and personal qualities. Some authorities have a very rigorous procedure for these stages, but the process can be costly due to the length of time involved, the number of candidates, and the number of professional staff used on the assessments. This might not be a major problem where the ATS service can utilize the facilities and personnel of a larger group — say, the full civil or military aviation authority, or even an airline. This might additionally be useful in avoiding duplicate applications from candidates who have a general interest in aviation and would willingly work in any of its branches. However, putting the selection process to a larger group can result in less commitment to the ATS need, and we certainly often hear the cry that this general group "do not understand our problem"! Of course this might be correct if the selected staff can move around within the larger group, who must then select against broader, and maybe more general, criteria. If group movement is not involved, then there is no reason why the professional assessment staff should not be successful in selecting against recognized, agreed criteria.

The use of peer assessors (i.e., active or past controllers) can be most successfully applied, but it is a problem to get the right persons with a suitable balance of education, training, experience, and further training in subjects appropriate to the selection and recruitment process. Too often there is lack of agreement on the required attributes to make a "good controller" — after all, very few authorities actually rate controllers as to their level of performance — so who is identifying the good ones and then noting the common characteristics? Again, a panel of peers can select in a consistent way, but they do need to be carefully trained and briefed on the agreed characteristics being looked for. Here, ATC is no different from many other disciplines, where it is often considered that those who do a job understand the skills and abilities that they bring to it. Such misconceptions are obvious to those running management courses with the aim of improving performance by getting persons to understand their characteristics and how to use them in the roles that they play.

Finally, for this section, how consistent are organizations in using their assessments? For instance, if a candidate is rated as only just suitable in maybe one characteristic, is that information passed on to the course instructors so that individual corrective tuition can be offered? One argument often made is that such information should not be passed on, in order to avoid a self-fulfilling failure. This implies that the instructors would then start with a negative approach to the student, rather than accepting the challenge of enhancing the student's chances of success. Of course this process does occur, but the instructor takes time to get to know the students, and so, due to the nature of ATC training, time is lost for giving the weaker student additional help. This is particularly serious, since early problems in a course can be carried on to later stages, perhaps cumulatively, causing failure at the end (and after much expense incurred) due to overload.

If consistency is to apply, it would be interesting to know how many authorities use the same set of selection criteria, or at least include them, when making annual (or biennial) assessments of staff. Or is that done by a different department?

Present Courses

Since air traffic control is a relatively new discipline, the sequence of courses has moved from ad-hoc in the early days to more formal stages such as general ATS, non-radar (aerodrome and approach), and then radar control.

For each authority the duration of each course is fairly common, following ICAO guidelines, but the duration of in-between practical stages can vary widely, depending on the authority concerned and the type of work to be performed. A common procedure is for staff to qualify as aerodrome/approach controllers and then proceed to practice that job for as long as desired. After a while, some would continue training for other aspects, such as approach radar control or area control (procedural or radar).

Each course developed the necessary knowledge, skills and attitudes in different proportions, but the sequence of mainly knowledge, then skills and then attitudes served quite well whilst trainees underwent extensive on-the-job (OJT) practical training and many years' career practice for each type of controlling.

Some authorities developed the Cadet scheme, which provided trainees with courses in all the disciplines, spread over three years or more.

Obviously, the structuring of courses would be varied, since trainees on later courses would not have the many years of practical experience acquired by those following continuation courses. However, the concept of a well-rounded, experienced controller with all the skills — or an appreciation of them — continued and made the job an exciting career with regular new developments that could be handled by the knowledge, skills, and attributes that had been acquired.

More recently, pressure has built up with respect to shortage of controllers and the levels of skills being demanded now and for the future. Consequently attention has focused on the content and duration of courses, with respect to both their relevance to the job performed by the controller and training success rates.

Some authorities have made exhaustive studies and drastically restructured the training program, whilst others are in the process of doing this. Many new questions are raised — for instance, the extent to which an area radar controller needs to be proficient in procedural control. And how much knowledge of aerodrome and approach control is needed? As usual it is difficult to change the old ways of doing things, but is is a fact that we are training a new generation of young people, and present practices can have quite a different context from those of, say, twenty years ago. With the increasing use of automation, we will see the context change again.

Example

Let us consider the stage of a typical aerodrome and approach course, consisting of theory and simulator practice in general ATS, aerodrome control, approach control, final examinations, automated data processing, and then on-the-job training.

The general ATS module, of eight weeks' duration, usually consists of much theory and little action, despite the fact that the trainees have been chosen to be "doers." One wonders how many begin to question whether they have chosen the wrong career, and if they do so does that questioning stay with them as they move onto more exciting stages. If so, it could explain some of the inexplicable drop-outs that occur. This could be particularly pertinent, since confidence feeds on itself, causing high confidence trainees to have better success rates.

Study of the module's content shows that many simple practical exercises, such as developing ranging and bearing skills using low cost personal computers, can be commenced very early in the course. The time required can be achieved by delaying theoretical work until later in the course. For instance, much of aerodynamics and meteorology is not required when dealing with simple practical skills, and of course when these are mastered, the trainee is ready for more complicated exercises that are caused by the real world of winds and visibility, aircraft performance characteristics, etc.

In fact, it is necessary to review how much of the theory must be actually learnt and how much must be understood and appreciated. With modern displays available much of the old detailed knowledge can be called up when required, so long as the significance of use is appreciated. In this sense, air traffic control is like flying, in that a lot of aviation knowledge must be appreciated so that aspects can be taken into consideration when necessary and essential. This aspect has certainly increased in flying modern high-tech aircraft, and will increase for ATC as automation increases.

The skill of using this knowledge in this way needs to be developed at an early stage, in order to set a pattern that can cope with a complex system. Computer based learning techniques will prove invaluable as the trainee can then explore his or her capabilities and skills, and decide on task levels to be undertaken and strategies for dealing with the tasks and work scheduling. In this way, the instructors will be free to act in a consultative role, helping to build up attitudes for success, and monitor progress.

Changes

The trainers have the dual responsibility of keeping up-to-date with ATC developments and also regularly improving training methods and techniques. The two aspects naturally overlap, but each must also regularly look to other disciplines to ensure that usable ideas and techniques are not overlooked.

The ATC trainers concern is also how the future ATC system will be used, how the controller will control it and how to ensure that the controller actually believes he is in control.

Alternatively, a bottom up approach can be based on what the new system can do for the user (controller) in achieving the tasks that he has been set.

With any new systems, there are always foreseen impacts, especially where a new system is replacing an older one, in order to perform the same task. Training for these situations is usually a matter of changing course content in order to address the different ways in which the same tasks may be achieved.

The more difficult problem is to prepare trainees for unforeseen impacts and how to deal with them.

Thus, new training courses will have to consider how to train controllers for working with automated systems in which errors of human or machine must be perceived, understood and dealt with. This, in turn raises the questions of whether controllers will require a more advanced understanding of automation technology, to what extent, and how it can be achieved for (often) non technically educated persons. It was suggested earlier that possibly all ATS staff should be recruited at degree level with technical qualifications and then all act as controllers for a number of years. Perhaps this should be considered again, but in terms of helping the controller control the system, rather than the solving the earlier problem of finding sufficient recruits.

If the selection criteria are changed, and training courses altered, in order to produce controllers who will be operating in the new automated systems, and probably new complex organizations, then we must consider the organizational and management structures that are appropriate to the requirements of these staff. If higher grade staff are employed, then management will have to allow for this, particularly by developing team skills and participative approaches.

Traditionally management skills in the air traffic services have followed traditional hierarchical methods — probably copied from the military organizations which were the source of most recruits in the early days. However, the continuation of European peace leaves recruits with less or even no military experience, and they now expect a civil management structure that compares with other organizations. Consequently, there is a need to study other high-tech organizations and how they manage highly qualified, highly motivated staff.

For the trainer, it may be that *ab-initio* courses should include more on management theory, or that new controllers understand the organization, how their colleagues are working in the system, and how they themselves are expected to play their part.

Perhaps we are worrying too much. Should we continue as at present and allow undiscovered problems to emerge with time and operational experience? Then when there are enough problems, a refresher course can be designed for those controllers who feel they wish to be brought up-to-date with the new system. Assuming it is not too late for them to learn new skills!

The last sentence is a reminder that basic skills are learnt in the early days, they must be learnt properly for a well designed system, and refresher courses should not be used to compensate for poorly designed systems or ill-thought out training programs in the early phases. We owe it to new staff to give them the best training that experts can devise, and then supply refresher courses to meet natural mid-career changes of equipment and procedures.

Training Techniques

Institutional Courses

The training techniques used in any course are important in motivating the trainee and creating a feeling that lessons have been well learn, and skills nicely acquired, leading to the pleasure of applying the new knowledge and skills to practical situations.

At one time, there were courses whose students were told that the program was so tough that only 50 % would pass. This certainly produced a high level of motivation to succeed, but one wonders if the 50 % who did pass subsequently acquired an arrogance about their abilities in life — especially if the training program was divorced from contact with another discipline!

With the forthcoming shortfall in availability of young persons, it is essential to select carefully and then expect high success rates — especially in view of the costs involved for each failed trainee.

Future programs must assure candidates, that following stringent assessment, they have been considered suitable to be a controller, provided they work at the training program. Also, they will be prepared for increasing automation, but that change should be regarded as an opportunity for further development — especially if refresher courses are offered.

I mentioned earlier that ATC is a relatively young profession, and because of this, the experience and skills of the older controller are still often sufficient to give him the edge on the newer controller. However, the increase of further automation will probably see this decrease, as it has done in so many other professions subject to technological change; consequently, ATS management must be prepared for the inter-personal problems that will arise. Ideally, if the older controllers are well prepared, they should not fear the future, since they will be helping to make it happen the way they would like it to. Of course, it will depend on the degree to which team efforts are successful in enabling older and younger staff to work on common problems with new technology.

Unfortunately, these efforts may be spoilt by the old problem of insufficient positions for senior staff, even though automation will create many jobs for those with operational experience and appreciation of the automated system.

Designing future training programs could become easier if an earlier retirement age were adopted, but can we afford to lose so many experienced staff, when only 2% of population growth is predicted for Europe by the year 2000? Alternatively, perhaps the jobs should be graded according to operational skills required, and staff move to less demanding positions as they grow older. Of course, who wants to move from the best (i.e., most demanding) operational posts? And will the salary reduce also? Perhaps younger controllers won't mind older ones' receiving more salary for less demanding work — because their time will come.

Should it matter? This problem has been tackled by many other industries and the ATS field should now give it equally serious attention.

Perhaps we need a new philosophy of how to deal with older staff in the high-tech industries. We need to take account of the changes in knowledge, skills, and strategies — but, above all, utilize aspects that have developed and improved with experience.

Although training courses in an institute or school form the basis of a training program, it must be kept in mind that there are drawbacks to instructional training. For instance, the school might be some distance from an operational center, and staff and students thus not have the close contact that could make the job more realistic. On the other hand it must be remembered that operational staff are at their job essentially to operate and not to spend time helping trainees. I will discuss this point further, but first it should be remembered that during the on-the-job (OJT) phase, it is still possible to give knowledge training to trainees through remote learning techniques. This is one way in which trainees could be put to OJT at an early stage, although there is a drawback of producing controllers who are capable of solving specific problems rather than having mastered the groundwork of general principles and practice.

On-The-Job Training (OJT)

The training techniques used in OJT are as wide and varied as the number of staff involved. In classroom instruction, there is a certain amount of control involved (i.e., there is a lesson plan which the instructor intends to follow and the trainee accepts), but in OJT the situation is very fluid, relying on the operational situations that arise and, of course, on the coach/trainee relationship.

Controllers have selected their job because they wish to control, and then found that some coaching is necessary in order to bring on the next generation. This is reasonable in any job where we all have a responsibility to ensure that there is a steady influx of new staff to replace those leaving or so that the establishment can cope with expansion. A major problem arises when there are a lot of trainees, due to late recruitment policies or rapid expansion programs. In these cases, controllers can find that they are doing little controlling but mainly acting as coaches. This situation must surely worsen if automation leads to reduced retirement age or increased air traffic.

The actual OJT techniques will change substantially with the advent of automation and the use of expert systems or artificial intelligence. If controllers become more of monitors, then trainees can actively join in the monitoring and there will be time for discussion about the situation. Of course, it might transpire that trainees are more in tune with the solutions offered by the computer, but this can be an advantage if controllers see themselves as managing a team, and prepared to report favorably on one of the team — so long as no one is shooting themself in the foot by putting themself out of a job.

At present, team-building is an important part of coaching courses, but in future, it will play a larger part in operational control and will include automation and the system as part of the team to be controlled. Certainly the controller will need to be trained to a deeper understanding and philosophy of the total system:, what it can do, how it does it and the role that each of us — the team — plays in the system.

Continuation Training

Continuation training is always a difficult problem for management, who can be reluctant to spare operational staff away from the job. After holidays, sick leave, and occasional attendances at conferences, no further allowance is usually made for staffing ratios.

The tradition is that staff keep up-to-date through guilds, professional organizations, books, magazines, or self-teaching courses. Sometimes management may provide briefings by experts on new systems, methods, ideas, or procedures — but these are usually ad-hoc and few and far between. At present, guild meetings seem to be fairly popular, but if the speakers are practicing controllers, how often is their message one of what should happen (in their interest), rather than what is going to happen (in the interest of various other parties)?

Moving ATS into a high-tech organization will have to change the whole nature of keeping staff up-to-date and progressive. It must become more organized, regular, and pertinent to the needs of each individual. Management must accept the responsibility for these activities, in order to ensure that the continuation training is relevant and appropriate to the progress of the organization.

It is interesting that several computer firms — definitely high-tech — now expect, even insist, that all professional staff undergo at least two weeks upgrading training per year. They see this as essential if the company is to keep in the forefront of the market-place. There is no reason why the air traffic service authorities should not see themselves as being in the forefront of the market-place in providing the best ATS for the traveling public.

Training Tools

Having reviewed some of the problems that automation in ATC will cause for the trainer it is appropriate to ask what new tools are available for making the training programs more effective. By "effective" I mean "higher success rates, more motivated and happier trainees." I include the latter because, although learning is hard, it should be enjoyable, and the time spent by a trainee (may be three years) should be looked back on with pleasure. I think this can be achieved with interested trainees, concerned instructors, and good selection procedures which offer high success rates.

Psychological Tools

We have discussed the psychological tools to be used in the selection and recruitment stage, but I believe there is still much to be done in counseling the trainees and instructors towards positive altitudes, and also to prevent the building up of mental barriers, and blocks and lack of self-motivation. Preferably, the managers should be taking a stronger role in this direction, but as many are more concerned with the organizational or operational task, we must employ specialists to do this work.

It is well known that trainees with problems in the earlier stages of a course tend to carry them forward. These unresolved problems keep reappearing. We must design future courses in such a way that there is time and staff availability to resolve these problems, at each stage, so that the trainee moves on with confidence.

These requirements are just as important at present, but more automation will increase the need to generate the feeling of being part of the system. Machines will still be machines, but more subtle, and the controller will need to be more mentally subtle to control them.

Technical Tools

If the air traffic services are to become high-tech organizations, then trainers should not only use the tools but also brief the trainees on the significance of the tools that they will be using. Let us look at some of the tools available.

Personal Computers: Gradually moving into everybody's life-style, PCs continue to reduce in price whilst offering more computing power. They have created an everyday vocabulary such that many people have bought one just to maintain their level of technological literacy. They are probably the basic building block for any understanding of an automated system. Their use in ATC simulation is becoming extensive and offers a wide variety of techniques.

Graphical Work Station: Mid-range computing power can be used for a lot of skill training devices. The Eurocontrol Institute has developed a basic ATC skills trainer on a VAX® system, EUROSIM, which is so neat and compact that it is envisaged that future courses may have one EUROSIM workstation to two trainee controllers. By extension, it can be used for pre-OJT, thus reducing the (often proclaimed) gap between institutional courses and actual on-the-job training.

Computer Networks: Networking has developed to such an extent that ATC simulators can now offer training facilities for either groups working on the same exercise or each trainee undergoing an individual exercise on his/her own display. The flexibility produced means that future programs should allow courses to be tailored to individual requirements, allowing for speed of learning, skill development, and other features.

Hardware Supports: The above three levels of computing power can be used to develop systems providing computer based training, interactive video, three-dimensional displays, speech processing, and artificial intelligence.

Each of these systems will certainly be used to help the trainee in future automated systems, and this potential certainly provides food for discussion.

Learning Objectives: The present training programs will need to move from instructional centered techniques to individual learning capability within the framework of group development. Controllers will have to learn about personnel development within the team, and this will equally apply to trainees who must learn their skills and knowledge at a pace suitable for themselves but still within the scope of the team's development.

Here the skills of the instructors will be paramount in producing structured learning techniques which can ensure that all trainees receive the necessary attention to drive for the training goal.

Performance is the end result, and consequently it will be necessary to ensure that training objectives are appropriately matched with operational performance.

New Functions: As mentioned previously, the implementation of new systems or procedures — in this case automation — will always produce both expected and unexpected consequences. This section is to remind us that some new training requirements will or may surface, and the trainers must produce suitable programs. This is reasonable and indeed welcome, as it keeps trainers on their toes, continually questioning what is required in the operational environment? However, time and effort are needed to address these "unexpected" matters, and it is available — but only if the trainees are not involved in still trying to sort out the expected problems. Again:training should not be used to solve poor system design in the early stages of a program.

Conclusion

ATC trainers, with their many ideas and their command of institutional techniques, appreciate the new concepts of automation. They are ready to address the problem of creating training courses that will both suffice for the present and prepare controllers for the automated systems of the future.

Despite the usual problems of financial and staff resources needed to meet the challenge, the biggest problem may be to create an appreciation for and the part that training can contribute to ensuring that future ATC systems are staffed with high quality, well managed personnel.

Adaptive Training to Accommodate Automation in the Air Traffic Control System

Marvin L. Smith

Aeronautical Science Department, Embry-Riddle Aeronautical University, Daytona Beach, Florida, USA

Introduction

The Federal Aviation Administration (FAA) is responsible for the operation of the National Airspace System (NAS). The NAS is the world's largest and most complex network of airports, airways, and air traffic control (ATC) facilities. The NAS is designed and operated to accomplish three goals:

- Safety of flight
- Expeditious movement of aircraft
- Efficient operation

The sharp increase in aviation service demand since the Airline Deregulation Act of 1978 has created an unprecedented demand on the aging ATC system and caused turbulence throughout the supporting components of the airspace system. In 1981, the FAA responded to this demand with the National Airspace System Plan (NAS Plan).

As a result of the NAS Plan, the FAA has entered a period of significant technological transition that is changing the way it will perform its primary responsibilities and manage its resources. One of the most challenging demands that parallels this technological transition is the impact the new operational environment will have on the training of air traffic control specialists. The billions of dollars being invested in the NAS upgrade will have limited payoff in terms of improving the safety and capacity of this national resource, unless highly-skilled air traffic controllers are available to operate the complex system evolving from this massive commitment.

Systems Planning

The NAS Plan was initially published in 1981 and periodically updated with subsequent editions of the "brown book," as the plan was called. In the most recent (1988) update, the following statement was published:

NATO ASI Series, Vol. F 73
Automation and Systems Issues in Air Traffic Control
Edited by John A. Wise et al.

> The NAS Plan was developed to expand and modernize the technology that supports the people who achieve the FAA mission. The Plan has now matured to the point that it is time to introduce the human resource management component. (Federal Aviation Administration, 1988: VIII-2)

Many would argue that the better time "to introduce the human resource management component" is during the planning phase, *not 8 years later*, as the statement above would imply. The speed and efficiency of the transition into the highly-automated NAS environment will be impacted by the availability and *proficiency* of trained personnel prior to placing the new systems into operation. In industry, where the competitive edge is essential for survival, the concept of "concurrent engineering" has grown in use to obtain the cost-effective convergence of related products and services. In the U.S. Air Force, a planning concept known as "concurrency" is used to insure trained personnel, simulators, logistics, and maintenance systems are available at the same time a new weapon system becomes operational.

The point to be made here is that the best time to design appropriate training to meet the technological capabilities of the NAS has already passed. As a result, ATC training is now cast in the role of playing "catch-up" in a culture that has a tradition for reliance on equipment as a solution to operational problems.

The widening gap between the resources (best minds and money) that have been invested in hardware/software development and the investment in human resource efforts (e.g., ATC training) can probably be attributed to the "hardware-oriented bias" that Sells and Pickrel (1984) identify in the following:

> The failure to develop systematic training to accommodate the operation of the new automated equipment is most likely a reflection of "a hardware-oriented bias" in the agency, characterized by preoccupation with technical and engineering aspects and neglect of human aspects, particularly human engineering (the interface of the human operator with the system of which he or she is a part) and personnel management. (p. 585)

The acceleration of change, such as the automation of the air traffic control system, will have a resultant impact on traditional human structures, particularly training. The old standards of competence are no longer adequate because technology has changed the nature of the job. The challenge facing the FAA in designing a training program that will meet the needs of the future ATC system is characterized by Toffler (1981) in his book *The Third Wave:* "the character of the work itself is different, and it takes a completely different kind of worker" (p. 35).

ATC Training Challenges

Three major studies conducted in 1988 identified training deficiencies in the current FAA training system. These problems need to be resolved in order to meet the demands of a rapidly changing ATC system. The FAA Technical Training Review Group (TTRG) studied the overall FAA training program, Northern NEF performed an exhaustive study of the management of air traffic training, and the HumRRO study was directed at air traffic trainees and FAA training procedures. These studies identified the need for comprehensive changes in the total FAA training system.

FAA Administrator James B. Busey, in recent testimony before the House Committee on Science, Space, and Technology, Subcommittee on Transportation, Aviation, and Materials, Concerning Advanced Training Technologies, explained the FAA's intent to keep pace with the advances of automation and the dynamic changes of the deregulated air transportation system. Busey (1990) summarized the findings of the investigative groups that had been assigned the task of finding out what needed to be done to improve ATC training:

> We found that our training efforts were too disjointed, with varying agendas and variable results. Among the problems identified were organizational fragmentation, the unavailability of training when needed, lack of long-range training requirements, recruitment practices and staffing allocations inadequate to meet system training demands, training for air traffic controllers overly focused on screening out failures rather than on improving the skills of trainees to meet FAA's standards, inadequate teaching skills by our instructors, an overreliance on On-the-Job Training, outdated and inadequate training delivery systems, and the unavailability of modern training technology.

Economic Impact

In terms of dollars spent, the NAS Plan is one of the largest single civil procurement programs ever, costing more than $16 billion by the year 2000 (McLure, 1987:15). The substantial capital investment required for the NAS Plan is clearly needed in light of the demands from the aviation community. However, for this investment to pay off, human resources must also be accommodated with the same degree of commitment.

There is a growing public perception that the present ATC training system is neither efficient nor cost-effective. The most direct measures of the effectiveness of any educational system are the outcomes of the instruction provided. These would include valid assessments of student performance of learning objectives, course completion rates, and post-training performance of the graduates. The mere size of the NAS investment will almost certainly create persistent pressures from the aviation public, as well as political groups, to produce high payoffs in terms of improved training efficiency as well as system capacity and improved safety.

The High Cost of Controller Training

The annual cost to the Federal Government for staffing actions that screen and manage initial training of air traffic control candidates is $40-60 million. The attrition rate (failures and withdrawals) for ATC candidates at the Academy has been a consistent 40% or higher for many years. This attrition rate remains constant even though 44% of the over 6,000 active controllers surveyed by the Government Accounting Office (GAO) perceive the quality of new developmentals significantly lower than before the 1981 Professional Air Traffic

Controller Organization (PATCO) strike (GAO, 1987). This extremely high attrition rate equates to an investment by the Federal Government of $15-20 million annually. This represents a significant loss of tax dollars that could be used far more effectively if either the screening or initial training were more efficient.

In addition to these costs, the FAA loses another 15% of those Academy candidates who *do* graduate and go to a terminal assignment. Another 35% of the Academy graduates are eliminated from assignments in the en-route facilities.

A by-product of this extremely high failure rate is reflected in the large percentage of high-potential, low-risk job seekers who are dissuaded from pursuing a career in ATC because of the unparalleled opportunity to become instantly unemployed at any time during the three to five year probationary period it takes to reach the sanctuary of full performance level (FPL). The threat of "sudden death" elimination in the non-radar-graded laboratory problems at the Academy is viewed as a particularly-aversive factor for career-minded military controllers and potential university applicants.

Controller Shortages

In addition to the heavy financial losses incurred because of this high attrition rate, the consistent loss of human resources delays the recovery time for the shortfalls in staffing that have persisted since the PATCO strike in 1981. The GAO, in testimony before the House aviation subcommittee, stated that the FAA will be short 1,900 experienced controllers this year if the agency follows through with its current hiring plans. Kenneth Meade, GAO director of transportation issues, reported that "FAA has not been able to achieve overall hiring goals for controllers....Even if FAA receives adequate funding for hiring a sufficient supply of qualified recruits, it will take several years to fully develop the performance skills of this work force" (*Business Aviation*, 1990:91).

The projected retirements of eligible full performance level (FPL) controllers in the near future will put further demands on the FAA's recruiting and training capabilities. The additional pressures on older controllers to upgrade their skills to function in the new automated system is expected to accelerate the normal retirement rates in the early 1990s.

The GAO, in a 1989 report addressed to the Ranking Minority Member on Investigations and Oversight Committee on Public Works and Transportation, summarized the relationship between traffic volume and controller staffing deficiencies:

> The volume of air traffic has increased every year since the strike, but controllers still number about 1,800 fewer than before the strike: 14,437 as of September 1988, as compared with 16,244 prior to the strike. Further, the number of FPLs--the core of experienced controllers--in fiscal year 1988 was about 3,300 short of the prestrike number. Flight operations in fiscal year 1988 exceeded 80 million, compared with 60 million in fiscal year 1982. (p. 7)

The GAO has also reported that, in FY 1987, air traffic controller attrition rates were 9% higher than trainee hiring in the terminal option, and 4% higher in the en-route center option. Hiring more controllers to meet these shortages is not the simple solution it might appear to be. The structure of the FAA recruiting, selection, hiring, screening, and certification process is slow, inefficient, and expensive. It has also become very difficult to find applicants of the quality the FAA has traditionally attracted, primarily from the ranks of former military air traffic controllers who brought extensive prior experience to the agency.

Staffing problems relating to recruiting and training are further exacerbated by the lingering effects of the PATCO strike in 1981. The PATCO strike wiped out 75-80% of the controller workforce, and FAA staffing, as seen in the preceding GAO report, has never regained prestrike staffing levels. Replacing the 12,000 striking controllers in careers that take from three to five years to reach FPL is an overwhelming task for an obsolete training system designed in the 1950's to meet the needs of a manual (non-radar) operating system and the technology in use at that time.

Demographic Changes

The demand for more technically-competent air traffic controllers comes at a particularly austere time in respect to available workforce quality. Finding the quantity and quality of controller trainees needed to reach acceptable staffing levels has become increasingly difficult because of intense competition from industry, the military, and the educational sector for the type of people needed for this highly-specialized work.

Some of the social and demographic realities that must be accommodated in future training systems include the following:

- By the year 2000, 85% of new entrants to the work force will be women, racial minorities, and immigrants.

- Within the next five to ten years, 75% of our current workforce will need total retraining or technical skills to meet "high-tech" changes. (*Network Orange*, July 1990)

A significant drop in the number of high-quality recruits between the ages of 18 and 30 is occurring now and will continue until the end of the century. By the year 2000, there will have been a 20% decrease in both males and females available for work. Consequently, business, industry, and the military have begun developing programs that will respond to the training needs of whatever human resources they can recruit. The successful organizations are also developing innovative ways to improve their human-resource position. Incentives, competitive pay scales, health benefits, educational funding, and in-house training programs are being used to improve recruiting, reduce employee turnover, and improve productivity.

Bob Whittington, FAA Associate Administrator for Administration and Human Resource Management, made the following observations at a Council on Aviation Accreditation meeting May 9, 1990:

> The FAA's concern is the decline in the work force which impacts a number of women and minorities which, combined with an all time-high-median age of employees, dictates a renewed emphasis on both recruitment as education and training. (Whittington, 1990)

Technology Transfer

While the technological explosion has significantly changed our everyday lives and working environments, it has not changed the way we provide learning. Technology has been infused in the operational world of ATC but is only slowly finding its way into ATC training. The success of the NAS upgrade and the increasing improvements in hardware and software capabilities is offset by the widening gap in human resources needed to operate the new NAS technologies as they become available. The technology of instruction is, unfortunately, being underemphasized and underfunded.

The general system approach used in the planning, design, and implementation of the NAS technology has not been applied to the ATC training structure. FAA training, in contrast to the FAA operational environment, has not been affected by the $16 billion investment in technology. Today's ATC classrooms bear a striking resemblance to their ancestors of 1947, when air traffic control specialist (ATCS) training began in what is now the Mike Monroney Aeronautic Center. Instructors still stand in front of the blackboard and delivery highly-structured lesson plans to classes that are paced by group-lock-step design.

The transition from the 50-year-old manual ATC system into the automated ATC system of the 21st century will involve redefining the role of the human controller in the automated environment. There is growing support for a redefining of the Academy mission as well. Instead of being an expensive "screening" facility, it would become a cost-effective *training* facility if it were redesigned to meet contemporary educational criteria.

The FAA Plan for Training

In response to the findings and recommendations generated by the reports and hearings cited previously in this paper, the FAA has published a comprehensive management plan to develop the "revitalized" training capability needed in the evolutionary NAS operational environment. This management plan, entitled *Flight Plan for Training,* has become the handbook for the design, development, implementation, and evaluation of the future air traffic control training program.

The Introduction and Overview to *Flight Plan for Training* provide a summary of the dynamic changes in the external environment, some of which have been mentioned earlier in this report, that are driving the need to restructure the ATC training system:

- A near-doubling of airline enplanements in the past 10 years.

- Significant increases in the number and diversity of aviation operations since deregulation.

- The requirement to hire and train more than 12,000 new air traffic controllers since the strike in 1981.

- System-wide implementation of completely-new ATC equipment as the NAS Plan is implemented

The FAA is responding to the need to achieve a higher degree of parity between hardware technology and human skills-development and has acknowledged the limitations of the current training design:

> Unfortunately, the current training system, designed for a calmer and more stable era, is simply unable to cope with profound technological change on the one hand and overwhelming operational demands on the other. (FAA, 1988:2)

Specific actions that the FAA plan addresses in identifying deficiencies in training design and development include the following areas of interest to instructional designers:

- Training design and development have been overly-focused on job-centered training.
- Training, especially for air traffic control, has been overly-focused on screening out failures rather than on improving the skills of trainees to meet FAA's standards.
- Training curricula are, in many cases, inappropriately sequenced, poorly received, and inefficient.

FAA management has also addressed the need to change the training system through restructuring and integrating modern training technology. This approach is both appropriate and critical to the continued success of the FAA organization. As Robert Branson, one of the leading authorities on instructional systems design (ISD 1978), states:

> Systems can fail because they are obsolete, because they are overloaded, because they have fundamental design flaws, and because they have inadequate management. (p.18)

Changes of the magnitude of the NAS upgrade do not come easily, and the FAA acknowledges it is facing the most serious challenges in its entire history. How well the system adapts to the changes in the environment, and the degree of effectiveness obtained by managers, will determine the ultimate effectiveness of the NAS Plan.

The implications indicated in subsequent statements found in the *Flight Plan for Training* are encouraging for those who favor the systematic design of educational programs. The statement in this document that deserves particular attention is the following:

> The current training systems are outmoded and inadequate to meet the training demand of the future. Modern training technology (automated design systems, interactive training, simulation, etc.) is not available at a level required to meet current and future training needs. (Federal Aviation Administration, 1989:5)

One-Size-Fits-All Training

Deficiencies in the ATC training system have been widely documented by recent research agencies and in congressional hearings. While it is both easy and perhaps fashionable to criticize the efficiency of the FAA training system because of the obvious (high costs, high attrition, and high profile), it should be pointed out that the basic problems are caused by policies, not people. There are many capable, hard-working, totally-committed people involved in ATC training. They are working under federal mandates and policies, however, that restrict flexibility and enforce standardization. This results in a rigid, "one-size-fits-all" training structure that may insure equal treatment for everyone in the system, *whether they need it or not*, but it does not permit the use of individualized, adaptive training delivery systems.

Changing the Focus

There are three fundamental components in any learning situation. These are the

- Student (them)
- Faculty (us)
- Training system (it)

A review of the literature on ATC reveals a rich source of research on students as a variable in the learning system, but practically no data on the faculty and only limited data on the training system prior to 1988. In view of the changing conditions previously discussed, the remainder of this paper will argue for consideration of reorienting the way these three fundamental components interact.

The old mandate has been to pick individuals who fit the delivery system. If the institution is a given, then all recruiting and selection is focused on obtaining students who "fit" the design of the institution. Recent research in the cognitive, social, and instructional sciences has developed a new understanding of how this framework can be used more effectively, if the curriculum is designed from the learner's perspective rather than a content-centered approach.

Integration of Instructional Technology

While there are no quick and easy "fixes" for complex problems of the nature discussed in this paper, there *are* a number of innovations from the educational sciences that have proven highly-successful in a variety of settings comparable to the FAA training mission. Abedor and Sachs (1978) explain the potential benefits of applying instructional technology to the types of problems cited in this paper:

> The emergence of educational technology as a field of specialization provided both personnel and tools to address the complex problems involved in improving learning and teaching. Second, many institutions, as a result of changing enrollment patterns, new clientele, shrinking resources, and burgeoning knowledge in the disciplines, realized the improvement of teaching was not a simple task. (p. 49)

Adapting innovative educational systems through the integration of technology is not a simple task, nor is it easy to "sell" to conservative, noncompetitive institutions. Continued high-level FAA management support will be critical to the successful transition from the existing system of training to a new, more effective concept. Any system that has been in use for 40 years has become institutionalized. Any "tinkering" or "patching up" of the old structure to meet new demands will surely fail. Resistance to change is particularly strong in bureaucratic organizations that have operated satisfactorily over the long haul and survived several crisis situations, such as the PATCO strike.

There is a persistent threat that, in spite of the mandates published in the *Flight Plan for Training* that call for "modern training technology, interactive training, simulations," etc., the amount of actual change that will be made to the curriculum will be minimal. Substantive change would require a lengthy, and expensive, systematic instructional design of a training vehicle that was originally put together on the intuitive skill of former controllers turned into instructors. Since it was never *designed*, it labors under the inefficiencies of its piecemeal heritage. Front-end analysis to determine the end-user knowledge, skills, and aptitudes might seriously change the content and delivery systems that have been solidified over years of use.

One possible scenario that could result from the normal (for organizations) resistance to change, would be a lengthy period of study of alternative actions, followed by some "tinkering around" with the curriculum and, finally, the purchase of some expensive hardware that has high-level *physical* fidelity (i.e., it *looks like* the operational equipment it is simulating), and then back to "business-as-usual." Hopefully, this scenario will not happen, since any system that has reached its upper-limit capacity will not respond to "add-on" technology with any appreciable boost in productivity.

Training Theory and Innovation

In response to the FAA's stated intent "To get on the leading edge of training theory and innovation -and to stay there..." (FAA, 1989:12), the following discussion will attempt to show how the principles of learning theory and instructional technology, if integrated into the ATC training system design, would provide an excellent training vehicle that could improve both the quality and quantity of controller training.

The innovations that are being used in industry and the military to obtain higher returns on training investments are a product of several trends that have evolved over time, perhaps borne out of necessity to train whatever human resource was available. The approach that characterizes many of the recent advances in educational technology is a reflection of three different tracks that have influenced the field since the late 1960's: (1) use of a variety of resources for training; (2) emphasis on individualized and personalized learning; and (3) use of the systems approach.

A significant contribution of modern cognitive psychology to the field of education is the theory of cognition which explain the learning process in terms of external experiences and internal conditions. This theory allows the instructional designer to view the learner as an *information-processor* and to aim at problem-solving and decision-making as learning objectives. Knowledge-only learning objectives, based on the older theories of behaviorism, do not adequately support the actual job requirements of today's air traffic controller.

Instructional technology tends to mean different things to different people. The definition chosen for this study was first introduced by the Definition and Terminology

Committee of the Association for Educational Communications and Technology (AECT) in 1972, and has become the preferred concept for many contemporary writers and instructional designers.

> Instructional technology: a *systematic way of designing, carrying out, and evaluating the total process of learning and teaching* in terms of specific objectives, based on research in human learning and communication, and employing a combination of human and nonhuman resources to bring about more effective instruction. (AECT, 1970:21)

In 1964, Lumsdaine made the point that instructional technology could be thought of in two ways. He indicated that, in one sense, instructional technology referred to "the use of equipment for presenting instructional materials" (p. 372); in a second sense, he stated, instructional technology could be thought of as the application of scientific principles, particularly theories of learning, in order to improve instruction.

Learner-Centered Instruction

From the perspective of modern instructional theory, the purpose of all instruction is learning. Therefore, the central focus for rational development of instructional practices must place the *human learner* as the central focus of the design. The development of sound instructional procedures must take into account learner characteristics such as innate capacities, experiential maturity, and current knowledge states (Gagné, 1987).

One of the major changes that has come about through recent learning research in the educational sciences is a learner-centered view, or what is frequently called the "cognitive-shift." Instructional designers need good data on which to base instructional delivery, but the gathering of data should begin with the knowledge states of the learner and build from there (Office of Technological Assessment, 1987:19).

Individualized Instruction

A major concept associated with the field of instructional technology is the concept of individualized instruction. The AECT Committee has emphasized the importance of the individual learner in the following definition:

> The educational technology approach has been directed toward expanding the range of resources used for learning, *emphasizing the individual learner and his unique needs* [italics mine], and using a systematic approach to the development of learning resources. (AECT, 1972:36).

Adapting training to accommodate individual differences is not a new idea. Tobias found the first record of instruction being adapted to individual differences among students in Plato's description of Socrates' attempt to instruct the slave boy Meno in the mysteries of rectangles (Tobias, 1981:111). Knowledge of individual differences among learners is used by instructional designers to produce interactive delivery systems that meet the learner at the level s/he is at, rather than making the learner fit the level of delivery. The issues to be introduced for discussion and elaboration include the challenge for integrating modern cognitive learning theory and instructional technology into course designs and delivery systems that adapt to the individual differences and needs of learners.

Dailey and Pickeral (Sells, 1984) recognized the importance of individual differences in their research work with the Civil Aeromedical Institute and the Office of Aviation Medicine (CAMI), but it appears that, for the most part, individual differences have been considered a nuisance by the FAA, especially if sex or race might be implied from the differences. Instead of adapting the learning system to these individual differences, the FAA has usually eliminated the individual from the school.

Cronbach, one of the recognized leaders in adapting instructional treatments to individual differences, offers five strategies (see Table 1) for adapting to the large number of differences found in learners to get an idea of the variety of individual differences that can have an impact on the learning process, please refer to Table 2. This is only a partial list of the hundreds of variables that have been identified by the various branches of psychology.

Conclusions

Changing demographics in the work force; increasing shortages of high-potential, low-risk candidates; and intense competition for talented new hires are examples of the external changes that must be accommodated by new, highly-adaptive training delivery systems.

The traditional selection practice that implies "The institution is a given; try to pick persons who fit it" will contribute to the growing shortages of controllers. A new training mandate should be "Design enough treatments to allow everyone to succeed using their own preferred modalities of learning."

Instructional technology should be integrated into the entire spectrum of ATC training, from *ab initio* to continuing, and be adapted to individualized delivery systems that extend, but do not replace, the human instructor.

The institutional culture should be supportive and facilitate the development of positive attitudes toward the profession and its values. Training practices should be modeled on cognitive learning theory rather than the Darwinian theory of "survival of the fittest". If we revise the formula and reverse it so that, instead of the creature being adapted to the environment, the environment had to be adapted to the creature, we could use Darwin's theory in a positive construct.

The term "individualized instruction" (along with "adaptive," "personalized," and "self-paced") should be a common denominator in all curriculum design and content development. Individualizing instruction to accommodate the learner's individual differences expands the window of opportunity for more applicants, and significantly improves the cost-effectiveness of the training investment.

Students whose talents, motivation, and previous experience warrant acceleration through the training program should be allowed to progress faster and at a level well above what is available when we "dumb-down" the delivery to accommodate the "average" student in a class.

Table 1. Adapting Instruction to Individual Differences

Strategies:

1. Adapt the curriculum to match the needs of the individual so that different individuals learn different things.
2. Eliminate those students unable to profit from the experience.
3. Assume that certain curriculum content must be studied until mastery is reached.
4. Use remedial loops to attend to learner deficiencies before returning them to the main track.
5. Teach different students by different methods with the intent of reaching a common goal.

(Adapted from Cronbach, 1967)

There is an urgent need for applied research to evaluate the effectiveness of individualized, adaptive instructional designs. In addition to the need for improvements in training efficiency and cost-effectiveness, there is the matter of providing access to the air traffic control profession for the growing number of women and minorities who, traditionally, have not fared well in the "one-size-fits-all" culture.

Instructional technology is far too broad a topic to cover adequately in a short paper. In an attempt to extend the argument for innovation in ATC training through the application of current educational concepts, a list of recommended "shifts" from traditional to contemporary practices is provided in Table 3.

Table 2. Individual-difference variables that impact learning

- Aptitude
  - Intelligence
    - Fluid
    - Crystallized
    - Spatial
  - Achievement
    - Grade point average
    - Standardized (norm-referenced) achievement tests (e.g., SAT, GRE, LSAT)
    - Criterion-referenced tests

- Prior knowledge
  - Pretests and diagnostic tests
  - Word association tests
  - Cognitive mapping techniques

- Cognitive styles
  - Field-independence/dependence
  - Reflectivity/impulsivity (cognitive tempo)
  - Breadth of categorizing
  - Scanning/focusing
  - Leveling/sharpening of memories
  - Visual/haptic perceptual style
  - Tolerance for unrealistic experiences
  - Cognitive complexity
  - Serialistic/holistic style
  - Cognitve style preferences
    - Sensory preferences
    - Temporal preferences
    - Reasoning preferences

- Personality Variables (styles) — Motivation
  - Intrinsic — Achievement motive
    - Achievement via independence
  - Extrinsic
    - Achievement via dependence (affiliation need)
  - Locus of control
  - Anxiety
    - State/trait
    - Content related
    - Computer anxiety
  - Introversion/extroversion
  - Neuroticism
  - Risk-taking

Table 3. Adapting Instruction Design to Improve Learner Performance

FROM AN EMPHASIS ON	TO AN EMPHASIS ON
• "One-size-Fits-All"	• Adaptive/Individualized
• Closed systems	• Open systems
• Traditional design	• Instructional Systems Design
• Program-controlled	• Learner-controlled
• Time-based	• Mastery-based
• Lock-step	• Self-paced
• Normative grading	• Criterion grading
• Teaching	• Learning
• Traditional design	• Innovative design
• Lecture	• CBI/CBT/CAI/CMI, etc.
• Passive answer-absorbing	• Active answer-seeking
• Memorization	• Learning to learn
• Antisocial culture	• Cooperative culture
• Text-based media	• Hypertext
• Memorization	• Learning strategies
• Emphasis on text delivery	• Multi-média delivery
• High attrition	• Remediation
• Static aptitude testing	• Dynamic selection
• Certified instructors	• Certified/Credentialed
• High stress	• Supportive
• No prior knowledge	• Prior knowledge
• Process	• Outcome
• Behavioristic theory	• Cognitive theory
• Stimulus/Response	• Information Processing
• Old ATC-skills training	• New ATC-skills training
• Fragmented structure	• Continuous progress
• Labor-intensive staffing	• Peer instructors
• Screening	• Learning
• Classroom teaching	• Independent learning
• Teacher as source of knowledge	• Teacher as coach
• Standardization	• Accommodation
• Competitive	• Cooperative
• "Bells and Cells" control	• Flexible scheduling
• Knowledge as objective	• Skills/Problem-solving as objective
• Cognitive focus	• Cognitive, Affective, Psychomotor integration

References

Abedor, Allen J. & Steven G. Sachs (1978). "The Relationship Between Faculty Development, Organizational Development, and Instructional Development: Readiness for Instructional Innovation in Higher Education". In Bass, Ronald K. (Ed.) *Instructional Development: The State of the Art.* McAlester, OK: Best Book Publishers.
Air Traffic Control Association. (1988). "Washington Newsbriefs", *ATCA Bulletin.* Arlington, VA: September.
Branson, Robert K. (1975), Interservice Procedures for Instructional Systems Development: Executive Summary and Model, Tallahassee, FL. ERIC 122 022.
Carrier, Carol A. and David H. Jonassen. (1988), "Adapting Courseware to .Accommodate Individual Differences" in Jonassen, David H. *Instructional Designs For Microcomputer Courseware,* Hillsdale, New Jersey: Lawrence Erlbaum.
Department of Transportation. (1988). *Federal Aviation Administration .Operational Appropriation,* OPS-185 Report.
Federal Aviation Administration. (1989). *Flight Plan for Training.* Washington, .D.C.: Federal Aviation Administration.
Federal Aviation Administration. (1988). *National Airspace System Plan.* Washington D.C.: U.S. Government Printing Office.
Gagne', Robert M. (Ed.) (1987). *Instructional Technology: Foundations.* Hillsdale, NJ: Lawrence Erlbaum.
Human Resources Research Organization International, Inc. (HumRRO). (1988). "ATC Training Analysis Study: Design of the Nest-Generation ATC .Training System." Washington, D.C.: Office of Personnel .Management.
McLure, Herbert. (1987). "How Safe is the System", *Journal of ATC,* October- December, pp 15-18.
Northern NEF, Inc., (1988). "Flight Plan For The Future: Training Controllers for the National Air Traffic System". Colorado Springs: Office of Personnel Management.
McArtor, T. Allan. (1989). "Letter of Introduction." *Flight Plan for Training.* Washington, D.C.: Government Printing Office.
Tobias, Sigmund. (1987). "Learner Characteristics" in R. M. Gagné (Ed.) *Instructional Technology: Foundations.* Hillsdale, NJ: Lawrence Erlbaum.
U.S. Air Force. (1970). *Instructional System Development (ISD).* Air Force Manual 50-2.

Resource Management Training for Air Traffic Controllers

Daniel A. Herschler

Team Performance Laboratory, University of Central Florida, Orlando, Florida, USA

Introduction

The concerns about controller effectiveness in today's highly automated environment suggest that new training approaches may be warranted. Just as airline pilots have had to develop new skills for dealing with cockpit automation, so now must controllers also expand their repertoire to include the skills which comprise effective resource management. Resource management training refers generally to technical training in the efficient use of system resources in the performance of a job. While air traffic controllers are not as explicitly team-oriented as flight crews, there are job parallels which suggest that performance could be improved through resource management training for skills such as leadership, planning, communication, assertiveness, situational awareness, and decision making that are often a part of flight crew resource management training. Today's high-density and complex operations require efficient coordination among air traffic facilities and effective use of system automation. Satisfactory system performance of our newly automated air traffic systems will depend on controllers' resource management skills.

Air traffic control (ATC) systems are evolving rapidly, and the technology required to fully automate air traffic control may be here soon. However, there is a continuing belief that human controllers will have a central role in future ATC systems because their problem-solving skills will be needed (Hopkin, 1989). If future ATC systems are to include human controllers, training techniques must be developed to ensure that controllers work effectively as managers of the automated systems. Training must also contribute to the development of the team skills necessary for efficient ATC task performance. Fortunately, the aircrew coordination research literature of the last decade has identified resource management training approaches which might be successfully applied to training air traffic controllers to perform effectively using the new automated systems.

Managing Automated ATC

While automated systems perform functions that are impossible or difficult for humans to perform, or augment or assist the human operator, these systems also impose additional workload (Wickens, 1984). Operators of automated systems must be skilled in anticipating system malfunctions, in systems-level problem solving, and in responding to anomalies quickly and effectively. The task of monitoring the performance of automated systems has

NATO ASI Series, Vol. F 73
Automation and Systems Issues in Air Traffic Control
Edited by John A. Wise et al.

often proved challenging for human operators in such settings as the aircraft flight deck and the nuclear power plant control room, and it is reasonable to expect that air traffic controllers will find their systems equally difficult to manage. Resource management training for such skills as planning, situational awareness, and decision making has been included in many successful aircrew coordination training programs (Lewis, Oser, Lassiter, & Stees, 1989), and might be adapted for training air traffic controllers.

Team Coordination

According to Hopkin (1989, p. 104), "air traffic control is a team activity." Interactions among controllers at different facilities can be especially critical when the facilities concerned are in different countries. Automation may make effective controller interactions even more critical than they are at present, since task performance may be constrained by higher workload, fewer assistant and supervisory personnel, and less task flexibility. Controllers may find it difficult to aid each other and to adjust their strategies to compensate for emergent events when automated systems are in place (Hopkin, 1989). Training for effective communication among controllers (and between controllers and pilots) could militate against the misunderstandings, complacency, and loss of technical skills which can accompany advanced automation systems (see Morgan, Herschler, Wiener, & Salas, in press).

Resource Management Training Research

The importance of resource management for success in a complex task environment was first identified by H. P. Ruffell Smith (1979). In his study, performance differences among several crews who flew simulated commercial flight scenarios were attributed to the ability to manage resources effectively. Ruffell Smith determined that vigilance, planning, and decision making were the management skills which were related to flight success. Ginnett's (1987) research on aircrews conducting actual line flights indicated that effective leadership skills are critical to performance; he also found that more and less effective performers can be reliably identified from the behaviors they exhibit. For example, captains who encouraged crew interactions and defined crew task requirements performed better. In addition, Ginnett cited the work of Harper, Kidera, and Cullen (1971, cited in Ginnett) who found that the assertiveness of flight crew members determined whether a simulated flight would be safely completed. Ginnett also reviewed several flight mishaps in which inadequate crew performance compromised flight safety. He noted that advanced automation and safety systems aboard aircraft are frequently inadequate to compensate for a lack of effective teamwork.

In their studies of the effects of crew composition and workload on crew performance, University of Illinois researchers examined crew interactions and found that crew function was hindered by non-task communication and that the ensuing impairment of task coordination reduced team effectiveness at high workload levels (Straus & Cooper, 1989; Wickens, Marsh, Raby, Straus, Cooper, Hulin, & Switzer, 1989). Non-task communication, ratings of crew interactions, and task performance were also related in a recent study of team performance on a low-fidelity helicopter simulation (Lassiter, Vaughn, Smaltz, Morgan, &

Salas, 1990). It seems that effective communication in advanced technology settings is essential for successful task performance. Indeed, team performance researchers have often found that coordination is a necessary part of teamwork (Glickman, Zimmer, Montero, Guerette, Campbell, Morgan, & Salas, 1987).

Applying Resource Management to Controller Training

Determining ATC "Team Requirements"

Can recommendations for controller team skills training be derived from the resource management research literature? While there is evidence that team skills are essential for success in advanced technology settings analogous to air traffic control, it is unclear which of the team skills identified will be most critical for successful controller performance. And, since controller resource management and team interaction are neither assessed nor trained at present, it will be difficult to determine how to adapt new controller tasks to facilitate team performance. Despite this difficulty, research should first be conducted to determine which team behaviors are most effective. Subsequent studies could be conducted to compare training approaches which are designed to help controllers perform more effectively.

Developing ATC "Team Guidance"

Theoretically, controller team performance requirements and training guidelines could evolve from workload studies, as they did for commercial aircrew (e.g., Ruffell Smith, 1979). Published ATC procedures and rules could also be developed to formalize team roles for specific traffic situations, much as regulations such as the "sterile cockpit" rule (U.S. Federal Air Regulation 121.542) permit only task-related information and the presence of essential crew at higher levels of workload. Such regulations help to direct resource management training to the range of system operations which are dependent on team performance effectiveness, i.e., the "team performance zone" (see Figure 1).

In addition, research could (a) identify the degrees to which teamwork is involved in ATC task; (b) suggest how training could be used to enhance performance on specific tasks; (c) determine the workload costs incurred by team interactions, including the adverse effect of team inefficiency (which Steiner (1972) labeled "process loss"); and (d) determine the relative efficacy of alternative team training methods for air traffic controllers.

Conceptual models of the relations between workload and team performance could provide support in the identification of team training needs, and help to guide the development of training to meet situational requirements for teamwork. Such models could also provide a framework for research to identify the relative contributions of the various ATC team performance dimensions. Data generated from team research could quantify teamwork requirements and potential training payoffs, which would in turn provide a basis for full-scale investment in controller resource management training. An example of the type of model suggested for ATC team research is presented in Figure 2. The figure depicts a conjectural conceptual model, in which teamwork requirements are defined in terms of the Effort Required for Success less individual Effort Expended. For the sake of argument, if the line depicting Effort Required for Success were a true representation of traffic density (workload) on a typical hectic weekday morning, there would only be a few minutes in which a single controller would even be able to achieve success, were she to operate at peak available effort.

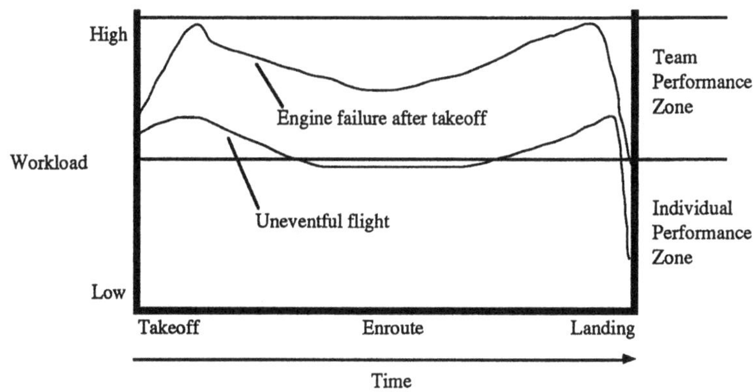

Figure 1. Team performance zone.

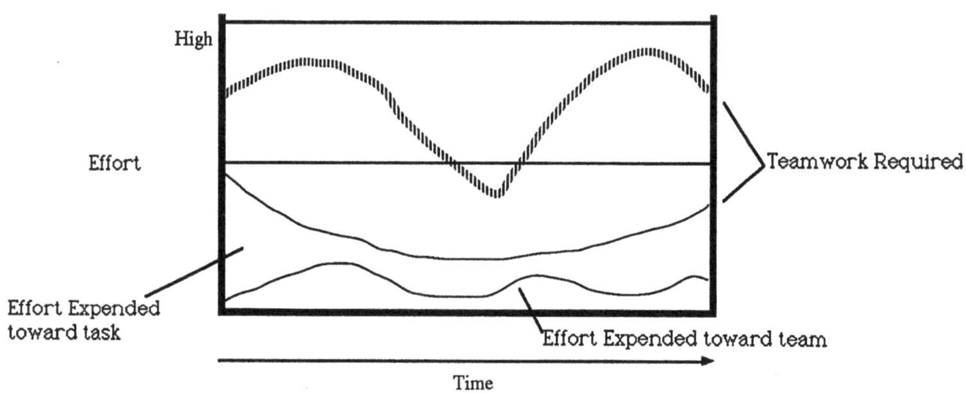

Figure 2. Conjectural conceptual model

However, teamwork would still be required to achieve success at this minimum workload level because the controller could not be expected to expend peak effort. Indeed, team performance might be even more important at a traffic lull, because control responsibility transfer (which includes briefing the relieving controller) usually occurs at such lulls.

Current Research

Human factors scientists at the Center for Applied Human Factors in Aviation (CHAFA), a University of Central Florida and Embry-Riddle Aeronautical University cooperative research center have begun to identify the nature and importance of individual and team variables affecting ATC performance and to develop a suitable model of teamwork in air traffic control. The subjects are highly motivated students who have demonstrated success in air traffic control training by attaining a 93% performance criterion in a simulation of the Daytona Beach Terminal Radar Control (TRACON) facility. The study consists of an orientation and two air traffic control scenarios using TRACON (Wesson, 1988), a microcomputer-based air traffic control simulation. Demographic and aviation experience data are requested from the subjects, and their task aptitude is assessed using a variant of the Federal Aviation Administration's Multiplex Controller Aptitude Test.

The first scenario is performed individually in order to establish a level of individual skill, and the second scenario is performed by both controllers jointly, with task responsibilities divided between the two subjects. For both scenarios, objective performance measures such as type and number of errors and a composite measure (total points) are obtained from the simulation software. Team performance is rated on five team skill dimensions deemed relevant to air traffic control (cf. Franz, Prince, Cannon-Bowers, & Salas, 1990). Answers to the following research questions are sought from this initial study:

- *Can individual or team performance on TRACON be predicted from demographic, task aptitude, experience, and training recency factors?*

This question addresses variables which bear on the relatively intractable problem of controller selection (cf. Manning, Kegg, & Collins, 1989), as well as the issues of skill retention and the need for recurrent training

- *Are team skill ratings related to team performance on an air traffic control simulation?*

Aircrew interaction and ratings on several team performance dimensions have been related to task performance (Ginnett, 1987; Glickman et al., 1987; Lassiter et al., 1990; Ruffell-Smith, 1979), but are the team dimensions and team rating procedures used in aircrew studies appropriate for air traffic controllers?

- *Are there systematic relations among the measured task aptitudes and performances of individuals and their performances as a team?*

It will be useful to know if there are minimum individual performance levels and particular control strategies (Sperandio, 1978) which are required to obtain benefits of team coordination.

Conclusions

The results of the present study will help to guide future decisions concerning resource management training for air traffic controllers. It is likely that teamwork will be necessary for successful performance of future ATC systems, unless controller tasks are completely redesigned to preclude team interaction (Hopkin, 1989). In any case, air traffic control presently is a team task. It is time to determine which team interaction and resource management skills might be appropriate for air traffic controllers, and to train controllers accordingly.

Acknowledgements

Preparation and presentation of this paper were supported by grants from the University of Central Florida, the National Science Foundation, and NATO Scientific Affairs Division. The author gratefully acknowledges the helpful comments and suggestions of Professors Ben B. Morgan, Jr., Director of the Team Performance Laboratory, and Richard D. Gilson, Director of the Center for Applied Human Factors in Aviation.

References

Franz, T., Prince, C., Cannon-Bowers, J., & Salas, E. (1990). *The identification of aircrew coordination skills*. Unpublished manuscript available from Naval Training Systems Center, Orlando, FL.

Ginnett, R. C. (1987). *First encounters of the close kind: The formation process of airline flight crews* (Yale University Doctoral Dissertation). Colorado Springs, CO: United States Air Force Academy, Department of Behavioral Sciences and Leadership.

Glickman, A., Zimmer, S., Montero, R., Guerette, P., Campbell, W., Morgan, B., & Salas, E. (1987, November). *The evolution of teamwork skills: An empirical assessment with implications for training (*Report No. TR87-016). Orlando, FL: Naval Training Systems Center.

Harper, C. R., Kidera, G. J., & Cullen, J. F. (1971). Study of simulated airline pilot incapacitation: Phase II. Subtle or partial loss of function. *Aerospace Medicine, 42,* 946-948.

Hopkin, V. D. (1989). Implications of automation on air traffic control. In R. S. Jensen (Ed.), *Aviation psychology.* Brookfield, VT: Gower.

Lassiter, D., Vaughn, J., Smaltz, V., Morgan, B., & Salas, E. (1990, March). *Evaluation of the effects of training interventions on team communication, attitudes, and performance* (Technical Report). Orlando, FL: Naval Training Systems Center.

Lewis, M., Oser, R., Lassiter, D., & Stees, Y. (1989). *Aircrew coordination training methodologies: A review with training implications.* Unpublished manuscript available from Naval Training Systems Center, Orlando FL.

Manning, C., Kegg, P., & Collins, W. (1989). Selection and screening programs for air traffic control. In R. S. Jensen (Ed.), *Aviation Psychology.* Brookfield, VT: Gower.

Morgan, B., Herscvhler, D., Wiener, E. L., & Salas, E. (in press). Implications of automation technology for aircrew coordination and performance. In W. Rouse (Ed.) *Advances in man-machine systems. research, Vol. 6.*

Ruffell Smith, H. P. (1979). A simulator study of the interaction of pilot workload with errors, vigilance, and decisions (NASA Report No. TM-78482). Moffett Field, CA: NASA Ames Research Center (NTIS No. N79-14769).

Sperandio, J. (1978). The regulation of working methods as a function of work-load among air traffic controllers. *Ergonomics, 21*, 195-202.

Steiner, I. D. (1972). *Group process and productivity*. New York: Academic Press.

Straus, S. G., & Cooper, R. S. (1989). Crew structure, automation and communication: Interaction of social and technological factors on complex systems performance. *Proceedings of the Human Factors Society 33rd Annual Meeting*, 783-787.

Wesson International. (1988). *TRACON Air Traffic Control Simulator* (commercial computer game for the IBM-PC/XT). Austin, TX: Wesson.

Wickens, C. D. (1984). *Engineering psychology and human performance*. Columbus, OH: Charles E. Merrill.

Wickens, C. D., Marsh, R., Raby, M., Straus, S., Cooper, R., Hulin, C., & Switzer, F. (1989). Aircrew performance as a function of automation and crew composition: A simulator study. *Proceedings of the Human Factors Society 33rd Annual Meeting*, 792-796.

Non-Technical Training for Technocrats: Resource Management at Canadian Airlines

Nicole Svátek

Flight Operations Resource Management, Canadian Airlines International, Mississauga, Ontario, Canada

Introduction

Ask ten people to define cockpit resource management (CRM) and you will get ten different answers. There seems to be no nice, succinct way to say "the enhancement of flight safety through the optimal use of all resources and the synergistic efforts of the flight crew as a team." Undaunted, I will now embark on the eleventh attempt to define CRM.

Cockpit resource management owes its origins in part to the work of Hugh Ruffell-Smith (1979), and John Lauber, in association with NASA-Ames during the late 1970's. At that time it was clearly demonstrated through full mission simulator studies that problems with crew coordination result in reduced performance and poor decision-making. Of course, there had long been a widely held belief that the human element was the least definable and most unpredictable component of an aircraft operation. But accident analysis and investigation reports did little to further our understanding of what was going wrong. The term "Pilot Error" was employed as a blanket explanation of anything that was not obviously mechanical. It did not describe *how* or *why* pilots erred. There was little doubt that the aircraft were less and less at fault; their safety, and functionality were steadily improving, to the extent that one could reluctantly submit that an aircraft was far less likely to be the source of an accident than was the operator. Today, this statement is unhesitatingly accurate. The 1989 IATA accident statistics show a full 65.5% of last year's losses were attributable to human error.

Even with our increasing knowledge, industry awareness, and training, there continue to be hull losses and lives lost due to reasons of totally human origin: aircraft attempting take-off without flaps, runway incursions, excursions, engine flame-outs from fuel starvation, and, tragically, controlled flights into terrain. C.F.I.T.'s alone claimed 575 lives last year and most of these collisions with the ground were due to non-adherence to rules and procedures, or lack of cockpit discipline.

The study of the human factor in flying is multi-layered and far-reaching. It goes beyond the ergonomics of cockpit design and computer friendliness; and focuses instead on the *attitude* as well as the aptitude of the operator behind the controls. It is an attempt to gather a collective intelligence on the non-technical province of flying; to understand why year after year, human performance is responsible for more fatalities than any other causal factor.

Cockpit resource management begins where technical training leaves off. At a time when aircrews can reasonably expect to transition from three, to two-man cockpits, to cross oceans with only two engines and to fly all-electronic aircraft, it is essential to focus on the changes these new technologies demand of pilots and their ability to adapt. It is not unreasonable to

NATO ASI Series, Vol. F 73
Automation and Systems Issues in Air Traffic Control
Edited by John A. Wise et al.

suggest that we are at a critically vulnerable moment in the evolution of flight. And that our cockpits are occupied by two entirely different generations - who were arguably first drawn to flying for entirely different reasons. There is often a chasm in understanding between the old piston-pilot and the video whiz-kid. One of the first mandates of CRM is to take a proactive role in preparing our pilots for the personal and group dynamics of flying and for the vagaries of the human condition; for it is in recognizing the weak links that we can hope to resolve them.

For example we now know that what was once considered the classic pilot profile, as so aptly described in *The Right Stuff* (Wolfe, 1984), is in fact the "Wrong Stuff" for the commercial cockpit. We are also aware that many of our flight decks are to this day populated by holdovers from this post-war legacy. It would seem obvious that the Captains with an over-developed sense of individualism, accustomed to barking commands and totally lacking in sensitivity, are not going to be a team players, nor will they be willing to give much credit to the input of their crew. In moments of high stress or emergencies they are likely to hunker down and take over. What was perhaps both desirable and necessary during the war has today become absolutely deadly. These Captains fly solo, to the limits of their ability, in an aircraft designed for two, or three, people. They are doomed to function within the parameters of their knowledge level and capacity for stress. The person sitting beside them is rendered useless and merely "along for the ride."

The problem presented here is subtle. What do you do with those Captains and how do you encourage their co-pilots to be more assertive? What about the "Young Buck" who starts out this way? And how do we design a programme that speaks to pilots and persuades them that good behavioural skills *add* to flight safety?

There have been many answers to these questions over the past eleven years. And many more questions.

It is now an accepted fact that resource management is both needed and here to stay. But that is about the only thing on which we are agreed. Airlines and agencies all have different theories on the method and type of training, and the assessment process. Everyone can describe the desired result but the subject is still so new that there is no mapped route to achieve that end. Our own approach is pragmatic and designed according to the pilot's needs. Rather than management insisting "This is what they will get" we tried to listen to what the pilots wanted.

For all of us, the acceptance of a new idea is often conditioned by the manner in which it is conveyed. In *The Medium is the Massage*, Marshall McLuhan (1967) submitted the position that we are, by virtue of television, advertising, P.R., and media, conditioned to react to the patina of what we see or hear. (He goes on to state that the medium has in fact taken on more significance than the message). Now I wouldn't want to translate that into a teaching philosophy but, nor am I willing to ignore such well documented evidence on the way we consume and retain information today. In our own case, we were highly mindful that in order to change a behaviour, you have to first deal with the attitude. Attitudes are deep-seated, closely held beliefs that determine a person's disposition to respond in a given way. Routine training is neither sufficient nor likely to have any lasting impact. Indeed, an excellent analogy is that you can't convert Conservatives to Liberalism by forcing them to go to a Liberal convention. You might however, one day persuade them through the use of consistent, fair debate, an elegant argument, and persistence without manipulation. Most importantly, you must never let them feel that up until now, they have been *wrong*.

We have also had to recognize that we are dealing with the social psychology of a "group." If you accept that groups of people form their own social characteristics and, given long enough, eventually come to be defined by them, then you will accept that the pilot culture is very distinct and particular. Schoolchildren could describe how they visualize a pilot — brave, independent, cool, calm, and collected — and they wouldn't be far wrong. By and large, that's how pilots see themselves. For over sixty years the group culture has been personified by these qualities, possibly self-perpetuated by the whole pilot mythology.

This thesis becomes relevant when considering training methodologies. It is not an exaggeration to say that most pilots actively dislike anything that remotely smacks of psychology. Granted, this is also true of many people who work in highly mechanical or technical fields. There is no shortage of hypotheses as to why, however, it is interesting to note that an unusually large percentage of airmen (and aviation mechanics) are left-handed, and therefore probably right-brain dominant. They are good at conceptualizing, problem-solving, and intuitive skills — but prefer fact rather than fantasy. Their entire training, and employment history has, until recently, supported if not nourished these traits.

Technical training and simulator sessions are a recurring, six monthly regime in every pilot's life. Over the course of a career they spend hundreds of hours practising and honing reflex reactions, ready to be called on in times of need. In the classroom, they learn the equipment and systems, in the simulator they recreate emergency situations and test their skill. The point is that: technical training produces technical proficiency. If you spend an entire career learning to react to a fire warning panel, you might be slow to respond to a fire alarm coming from where you don't expect it. There have been actual instances where the flight crew did not act on the cabin crew's verbal alert of a fire because there was no corresponding fire warning bell in the cockpit.

Commercial flying is a highly regimented world. Training is specific, dealing with concrete, specific issues that usually have a logical conclusion. The flightdeck is an environment necessarily governed by standard operating procedures, checklists, cross-checks and mutual monitoring. It is a world of cause and effect. The laws of physics are unforgiving and there is no room for erratic thought processes or "original" thinking. You fly the aircraft within an operational envelope, in the most economical manner practically possible. And that is that.

Then one day these scientists come up clutching sheafs of statistics in their hands and tell you that pilots are to blame for all the crashes and that there's a safer way to fly!" That, sadly, was exactly how many perceived CRM.

In developing our own programme, we felt it essential to "consider our customer". Who was this person we wished to speak to? What was their world, their culture, their vernacular? And how could we design a non-invasive training method that would persuade rather than admonish. Given the content of CRM and the amorphous nature of whole topic of the human factor in flying, it was almost certain that if it were mishandled, CRM training could actually backfire and undermine flight safety.

The medium really was as important as the message.

Programme Design

Initial Training

There is a great deal of debate about the ideal course length for initial training. Programmes vary widely, from two to four or five days in duration. Some are presented in the traditional lecture format whilst others include elaborate role playing scenarios. Our own initial training is one day only. In the belief that the strength of the message lies in constant reinforcement and frequent repetition, it was felt that initial training should function more as an introduction. The course was thus designed to serve as an overview, with the emphasis on learning the language of CRM (i.e., Assertiveness, Inquiry, Advocacy, Situational Awareness etc.); becoming familiar with the historical and statistical basis for CRM and seeing video footage which more graphically describes the need for CRM. One day is easily digestible and presents the topic in relatively benign terms. During this session, there is no attempt to either identify nor modify individual errant behaviours. The purpose is not to orchestrate an evangelical experience with the characteristic short-term effects but to present the topic, provoke discussion, and initiate a thought process.

Newsletter

Meanwhile, in order to keep CRM alive, lively, and visible, we put it in writing. The *Right Track* newsletter was born and is published every two weeks. The contents of the articles vary from extremely serious to, on occasion, the slightly irreverent. In a calculated attempt to address a large audience with multiple learning styles and cultural differences, the tone of the writing changes from issue to issue. The articles are kept short, presented in newspaper-type columns, and are easily read.

The advantages of the newsletter are numerous. Most obvious is that due to the frequency of its publication, the material is pertinent, and deals with current concerns, problems or even gossip. The newsletter is widely read and receives a lot of feedback. It serves as a source of conversation amongst the crews and has at times, helped people experiencing difficulties with their colleagues to initiate otherwise "hazardous" conversations. It has, in fact, served as a catalyst. Topics covered have included; "Performance versus Technique," "Conflict in the Cockpit," "Checklist discipline," "Teamwork and Technology," "Aggression," and, "Management is not a Dirty Word" plus many more. To our surprise and pleasure we have found that these newsletters also travel and are frequently passed around in cockpits other than our own. Eighteen months after its formal beginnings, the *Right Track* newsletter has a mailing list and has even been translated into foreign languages.

From the CRM perspective, it is an extremely important means of keeping the issue current and out in front. From the pedagogical point of view, it provides another medium to convey the message and appeal to different learning preferences.

CRM Recurrent

Once a year there is Ground School, as unavoidable as death and taxes, and for some, probably as much fun. In order to lessen the tedium induced by the technical material, CRM, and Flight Safety are placed in the middle of the curriculum. At this juncture it might be germane to add that we perceive CRM to be a composite and inseparable part of Flight Safety and it is presented as such during recurrent training.

This course lasts approximately two hours and comprises an in-depth examination of a carefully selected and relatively recent major accident. Photo-slides, overheads, CVR transcripts, and Jeppeson's transparencies are used to heighten the graphics. The Flight Safety Manager begins by reviewing the accident findings from a safety perspective. The same material is then revisited from the CRM aspect. The class is punctuated by discussion which is of course, aided and encouraged by the instructor. The flight crew are asked to contribute personal, similar experiences; they describe their responses to these situations and critique their own past performance. From a purely subjective point of view, this activity would appear to generate a great deal of "cross-fertilization" as the crews do not hesitate to comment about themselves or each other. The lesson then leads into a typically volatile discussion on whether technical training by itself conditions technical responses and is followed, more loosely, by a look at the impact and outcome of varied human reactions under abnormal conditions.

The objective of this kind of delivery is to present CRM in realistic terms and a manner which allows crews to see the practical application of CRM at work. Rather than alienate them to the concept by overworked psychological debate and lengthy blackboard explanations on the same theme, there is relatively little, if any theory, and total focus on skills.

We are taking master technicians with highly honed skills and asking them to embrace a different and more abstract response under stress. It is our belief that in order to persuade them, flight crews must actually see this work and by themselves, reach the conclusion that the Resource Management approach has merit, and a place in their cockpits.

CRM Cockpit Flights

The combination of initial training, ground school recurrent training, and a communications medium seemed to offer a well-rounded, diverse approach that could reach a substantial number of people. But the question remained: How do you speak to the individual? The answer was simple: You put CRM into the cockpit! The benefits were many and the complications far fewer than one would suppose.

Taking CRM into the cockpit effectively circumvents the lack of enthusiasm factor associated with the classroom. Instruction can be tailored to suit the individual and the one-to-one contact allows better opportunity to overcome trainee prejudices or resistance. The trainer has the advantage of experiencing first hand the cockpit dynamic, atmosphere, and "real time" situations. Further, the trainer benefits by always being current on the issues that are uppermost amongst the flight crew population (and can later address them in the newsletter). Finally, the credibility of CRM is enhanced if only by virtue of its presence.

Without doubt the idea was quite radical and before actually starting, some obvious concerns were raised that this would be construed as a check-ride or at least a management ghost-ride. In order to allay these very real anxieties we first solicited the agreement and assistance of the chief pilots who in turn asked for the cooperation of the captains. It might be added that a lot of very polite and careful letters went back and forth. However, once a clear understanding of the purpose of the exercise was established, there was a great deal of support offered by all the flight crew.

A strict etiquette was set in place and has been scrupulously observed ever since. A written request to ride is always sent days prior to a flight, allowing the Captain the opportunity to say No. The instructor does not fly during check-rides and all personal information gleaned during the flight remains absolutely and utterly confidential. Total confidentiality is the linchpin to the continued success of this venture. The trainer must make neither reference to, nor comparisons with, other flights, other cockpits, other people. It is interesting to note that over the course of almost two years we have heard but two or three objections.

Debriefing

The effectiveness of the cockpit flights was quickly apparent and the results rewarding. However, it gave rise to a major and unexpected question: Do you debrief?

If the answer is yes, it begs a litany of further questions: Where? When? And what do you provide for remedial?

It is true that pilots are acutely aware of and indeed nervous about any perceived threat to their license and right to fly. It would be an abuse of privilege, and of horrific consequences if they felt that "failure to pass CRM" might have negative results. Permission into the cockpit must not have the connotation of an examination, or the potential to FAIL. Neither statement was one we felt we could afford to make. Therefore, upon reflection, the issue appeared to resolve itself. There would be no formal debriefing. If, however, an individual or a crew asked for input or comment, it was available. In those instances where some corrective measures were obviously required, additional flights were scheduled.

Future Plans

It is our belief that the most important person in the implementation and continued management of CRM is the Check Pilot. Check pilots are not only instructors but also serve as role models. They have enormous and lasting impact on the attitude of the line pilot and should thus be offered additional training to learn about interpersonal skills, teaching techniques and dealing with people in stressful situations. Post-simulator ride debriefing sessions often require a degree of tact, consideration and insight that is neither taught nor necessarily talked about. Check pilots are most often appointed due to the distinction of their airmanship and it is hoped that their ability to cope with people and ensure operational standards will occur by osmosis. Of course, we know this doesn't happen. There is a real need complement their technical proficiency with behavioural skills. Once trained, the check airmen would then seem the natural means of ensuring that CRM is supported and practiced on a continuous basis. CRM concepts would become standard operating procedure. We hope to begin check pilot training in the fall.

The Human Factor In Flying

There is no question that the fast developing technological sophistication of today's aircraft has dictated changes of equivalent magnitude for the pilot. The fundamental cognitive requirements and sensory qualities necessary for operating an aircraft have remained the same, but it could be claimed that their function has expanded and come to include completely different applications. Coding, decoding, filtering, image recognition, deductive, and inductive reasoning, visualization, etc. are now predominantly focused on the computer and all its trappings. While technology has opened new frontiers in aviation, the relationship between human and the machines is rife with question marks, gray areas, and lack of certitude. What kind of individuals must we select to best cope with this new environment? Will their technological proficiencies come at the cost of the more emotive, creative skills? Will flight be computer assisted or computer driven?

For humans, we know that performance is affected by non-technical issues such as, committment, motivation, temperament, and the ability to communicate. In fact, airmanship is a combination of attitude as well as aptitude. To date, technical training and simulator experience have provided well for the development of rote technical skill. Comparatively little, however, has been done to address the human strengths and weaknesses that determine our respective levels of excellence.

What is a bad attitude? What is a lack of command aptitude? How do they make airplanes crash? And why does the human element play such a pivotal role in flight safety?

Perhaps such questions can best be answered by asking another; what makes the human a creative, original thinker?

In the face of today's technology and looking forward towards a future that confounds the imagination, we must increasingly turn our attention to the relationship between mankind and machinery. Do we "select people who are similar, extract those who are not, nurture specific traits and try to get people who are interchangeable— because technology demands it?" Or do we instead build error tolerant systems and decide that that is the price of the sensory/perceptive skills and sometimes exquisite ingenuity of the human operator.

The human factor in flying is about how we recognize and apply the best of our multi-dimensional human qualities, how we combine them with technological advances and hopefully attain a mutually supportive environment; one where we can all thrive.

Conclusion

In the beginning our objectives were as follows:

- *Enhance safety*
- Enable flight crew to better recognize and use the variety of resources available to them
- Teach the conceptual and interpretative skills required to effectively manage flight deck personnel
- Undo the habits of the "Macho Pilot" syndrome
- Recognize the early warning signals indicating complacency, reduced situational awareness, error chains, displaced perception and stress
- Introduce management concepts into the cockpit
- Complement cognitive ability and high task orientation with an understanding of the human factor in flight
- *Enhance safety*

Our approach was to use varied media to reach and touch as many people as possible. The method was somewhat different, but the results have been gratifying. It has been our experience that once cockpit resource management is established and the initial suspicion and hesitation overcome, the majority of flight crew are extremely keen and receptive. This supports our first theory; that the method we employed to introduce the subject would be of paramount importance to its future acceptance.

The operation of an aircraft has reached staggering levels of complexity and requires from the operator vastly different job skills than just a few short years ago. Unfortunately, we have not changed our teaching methodologies to adapt to the new environment. We are teaching space-age wizardry using Victorian techniques, with chronically little respect being paid to the process that occurs when human and machine are mutually dependent. There are new words in our language, unrecognizable to our forefathers. We speak in terms of "automation saturation," "computer complacency," "fly-by-wire," and "glass cockpit" — nice, smart-sounding terms we have coined to describe an environment dominated by machinery, where human operators often feel they are being led by, rather than in control of, their equipment.

Accident statistics have pointed a stern finger at how we are going about the business of preparing our aviators to fly in this new world. Between 65% and 80% of airline fatalities over the past 31 years were due to human error, and so far, we are not getting any better.

It is our job to find the means to either teach or inspire an understanding of the human factor. Whether the trainees are maintenance, ATC or pilots, attention should be focused on the human-machine interface and how best to achieve synergy between the two. It is incumbent on us to find the method and the vernacular that reaches out and speaks to all. We can no longer afford to employ the old attitude that said: "You come into class, you behave yourself, and you learn!" It is a fact that people simply do not learn that way, and right now, we have something of monumental importance to teach them.

Most accidents are likely to be situations that the crews have NOT trained for. They are also practically guaranteed to be the result of compound error: not one person slipping-up but everyone contributing either by commission or omission. If this is the case, then we have to train our people to *think*. We cannot have a technically-trained response to every situation, but, with strong teamwork, effective leadership, and good communication, we have gone a long way to improving the odds.

As John Kenneth Galbraith once said, "If all else fails, immortality can always be assured by spectacular error."

References

Hopkin , V. D. (1991). The impact of automation on air traffic control systems. In: J. A. Wise, V. D. Hopkin, & M. L. Smith (Eds.) *Automation and Systems Issues in Air Traffic Control*. Berlin: Springer-Verlag.

McLuhan, M. (1967). *The Medium is the Massage*. New York: Simon and Schuster.

Ruffell-Smith, H. (1979). A simulator study of the interaction of pilot workload and strass, vigilance and decisions. NASA Ames, CA.

Wolfe, T. (1984). *The Right Stuff.* New York: Bantam.

Critical Issues for Decision Makers in Providing Operator and Maintainer Training for Advanced Air Traffic Control Systems

Richard B. Chobot and Mary C. Chobot

Mary C. Chobot and Associates, Annandale, Virginia, USA

Introduction

The introduction of advanced air traffic control systems presents a number of challenges to senior decision makers. The objective of this paper is to present an overview of six critical issues that decision makers must consider relative to training. These issues are:

- Method of acquisition
- Use of specifications and standards
- Need for integrated work team of system engineers and training professionals
- Training delivery
- Courseware validation and measurement of proficiency
- Configuration management

Training can be procured either through contract or through the efforts of in-house staff. Each method has its advantages and disadvantages. A partnership of external developers and knowledgeable in-house staff is recommended.

Guidelines are needed to insure that: materials are developed which meet an agency's requirements; the utility of specifications and standards is explored; a detailed statement of work is used; and carefully selected, tailored supplemental specifications are recommended.

The need for early involvement of training professionals in system design is next discussed. Consequences of sequential development of the system, followed by the training courseware, are presented. Involvement of training professionals from the inception of a project is strongly recommended.

While media selection falls within the purview of the instructional technologist, decision makers have several broad options which they can consider. Each of these options is briefly discussed. Choices must be made based on the specifics of the situation.

The need to establish criteria for courseware acceptability is next considered. The development of validation standards covering technical accuracy of course material and the performance of students in the field trials is suggested. The need for performance-based testing to determine trainee mastery of course content is also recommended.

Finally, the need for system configuration management and the efficient and timely translation of engineering changes into modified training courseware is stressed. Again, however, the mechanism is left up to the reader.

NATO ASI Series, Vol. F 73
Automation and Systems Issues in Air Traffic Control
Edited by John A. Wise et al.

While the context for this paper is advanced air traffic control systems, the concepts presented are equally applicable to any training acquisition action.

Overview

Quoting from the FAA Flight Plan for Training (1989), Smith (1991) notes that "poor and inappropriate training" has caused serious problems in the fielding, operation, and maintenance of complex systems. This problem can only be addressed when decision makers begin to recognize the significance of sound training for successful systems implementation, and translate this recognition into appropriate procurement, fiscal, and personnel actions at the inception of systems design, rather than waiting until immediately prior to system evaluation and preliminary fielding to focus on training issues.

This paper briefly addresses some critical issues which senior managers must consider, to lay the foundation upon which professionals tasked with design, development, and delivery of system training can build.

We begin by placing training in the broader context of systems acquisition and development. The paper then moves on to brief considerations of decision options that directly impact training efforts. These include the need to:

- Decide whether the development of training should be part of the system acquisition contract, be provided for under separate contract, be an in-house activity, or represent some combination of the above.
- Decide what specifications and standards will govern the training acquisition activity.
- Provide for some degree of integration of the systems engineering process and the instructional systems design process.
- Decide, in general terms, how initial training will be delivered and how proficiency will be maintained.
- Establish the standards and level of validation that will be required for acceptance of training materials, and procedures for confirming trainee mastery of course content.
- Decide how system design/engineering changes, both during and after fielding, will be accommodated relative to training materials.

While the context for this paper is advanced air traffic control (ATC) systems, the ideas presented are applicable to any training acquisition action, either as a stand-alone event or as part of a total systems buy.

Elements in Systems Acquisition

Figure 1 is taken from a 1986 U.S. Navy publication that critically examined approaches to the acquisition of complex systems, in this case, weapon systems. The figure shows the major elements of the [weapon] system acquisition process. For purposes of this paper, we are concerned with training materials and equipment that, in this process, fall under the general heading of Logistics (Figure 2).

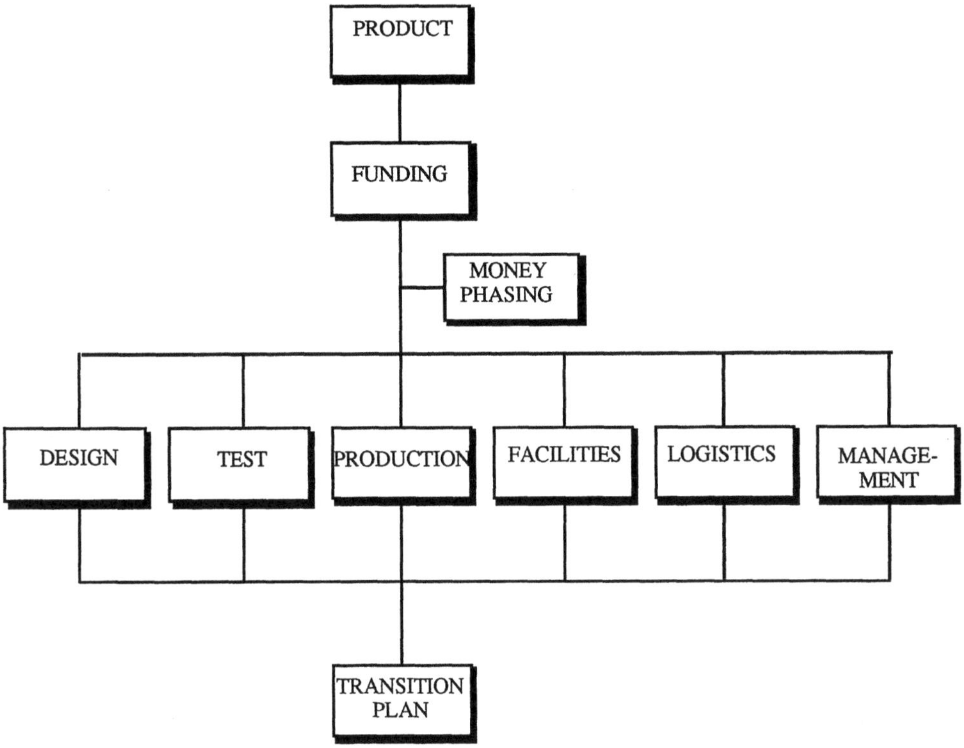

Figure 1. System acquisition process.

Training Acquisition Relative to Systems Acquisition

Three basic options exist with regard to training system acquisition relative to total system acquisition:

- As a component of the systems acquisition contract
- As a separate acquisition
- As an in-house staff function

Consolidated Acquisition

The major advantage of consolidated acquisition, also referred to as the total systems concept, is that the agency only has to go through the issuance of one solicitation. The preparation, review, and issuing of solicitations, particularly for large/complex systems, can be an expensive and time consuming process. Having to develop and issue only one solicitation can save both time and money.

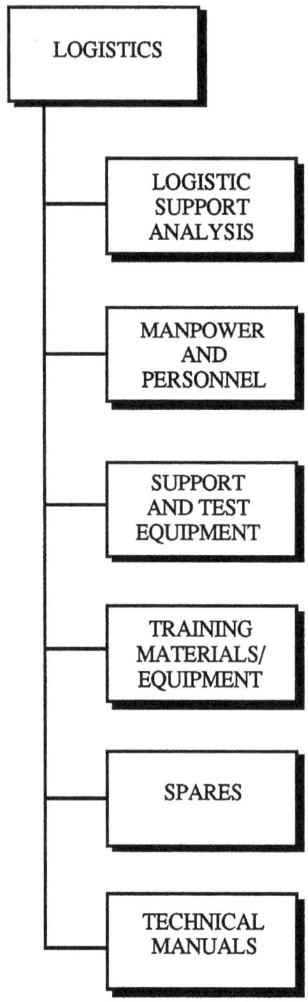

Figure 2. Logistics components.

Further, an integrated team of contractors, including training developers, increases the probability (but does not guarantee) that coordination and exchange of system information necessary for sound training design will occur.

The downside is that training, particularly in the context of large scale systems development activities, often has low program visibility. The problem is exacerbated when a long term project is incrementally funded. This often precludes the involvement of training professionals during the early stages of system design, when most of the decisions/constraints affecting training development and delivery occur.

Separate Training Acquisition

A separate procurement vehicle for training is often justified when development/revision of materials for an existing system is required. For a new system, a separate acquisition is often warranted where a special training technology or capability is required.

One major advantage of a separate training procurement is that it allows more fiscal and technical control of the training development process by the agency. Fiscal control is enhanced because a separate procurement requires more precise costing than when training is simply a line item within a much larger contract. Technical control is facilitated because technical oversight for the effort is usually vested in the training office of the procuring agency. In large scale efforts, training oversight may not always be delegated to a training professional, or, for that matter, to the cognizant training office.

The major disadvantage in this arrangement is the potential limitation of access by the training contractor to system specific information necessary to guide the development of the training materials, particularly when the system developer is working to meet contract deadlines.

Type of Contract: In either contracting scenario, a further decision involves whether a fixed price or cost plus fee contract will be awarded. A contracting professional usually makes this decision. The decision is based largely on the amount of risk inherent in the venture. The type of contract is important to training decision makers only because, if a contractor feels a fixed price vehicle has been improperly imposed, they will usually inflate their cost estimate to cover the perceived degree of risk.

Evaluation of Offers: In establishing the evaluation criteria whereby an outside contractor will be selected, the decision maker must be sensitive to both technical and cost considerations. Within the U.S. Federal procurement system, there have been increased competitive pressures that have driven down contractor bids. While government agencies should benefit from these competitive pressures to the maximum degree possible, there has been an increasing incidence of non-performance in cases where unreasonably low bids have been accepted. Good practice suggests that bids be evaluated against an independently derived estimate of the cost to develop the required materials; and that, besides cost, significant weight be given to the demonstrated technical experience of potential contractors and their records of past performance.

In-House Development

If classroom training using supplemental visual media (e.g., vu graphs, slides, etc.) is required, the need probably can be met using instructor staff. However, ATC training has increasingly utilized computer-based technologies (e.g., computer-based training (CBT), simulation, etc.) in the delivery of operator and maintainer instruction. The design of interactive training media requires personnel with specialized skills.

The decision whether to hire, train, and maintain an in-house courseware development capability must be based on life cycle costs for training, and not simply initial development costs. The life cycle of an item is its anticipated useful life span. Life cycle cost is the total of the direct, indirect, recurring, non-recurring, and other related costs incurred, or estimated to be incurred, in the design, research and development, investment, operation, maintenance, and support of a product over its life cycle.

When just the development and initial fielding of training materials are considered, the use of contractors to produce such materials is usually defensible from a cost/benefit perspective. However, when life cycle costs are computed, the decision is less clear. For example, supportability and maintainability of the training materials are significant elements of life cycle costs. If these costs are considered, especially for a dynamic system where training materials must be updated frequently, serious consideration must be given to either in-house courseware development/support capability, or fixed price maintenance contracts to control costs.

Another important factor to consider in planning for the acquisition of ATC training materials is the need to avoid total dependence on a single external source for training maintenance and updating support.

Training Acquisition as a Partnership

Ideally, the procurement of training materials, either as a component of the total system buy, or as a separate procurement, should be a partnership. The responsibility of the agency is to prepare a realistic description of the requirement and to monitor compliance with that requirement actively. An agency that simply leaves a contractor to its own devices, or that assigns unqualified staff to provide technical oversight for a major training development project, is taking a chance on the outcome.

The contractor has an obligation to develop materials using established practices and procedures, and to produce training materials that meet the customer's specified need. Further, the developer has a responsibility to document fully the training materials (e.g., both manuals and source code for CBT) and to insure that the client can use and maintain the training materials, once delivered, without further contractor intervention.

Total Quality Management

Finally, whatever acquisition modality is selected, provisions need to be made for some form of quality assurance.

Traditionally, quality assurance/quality control has revolved around the inspection of products and/or the evaluation of services during and after the production process. The discovery of defects results in reactive efforts to correct the situation suspected of having caused the problem.

Within the past decade, the concept of Total Quality Management (TQM) has gained increasing acceptance in the U.S. private and public sectors. Based on the independent work of several individuals, W. Edwards Deming and Joseph Juran being two of the most prominent, TQM is an initiative of top management designed to bring about an improved level of customer satisfaction through prevention of problems before they occur. There are varied approaches to TQM. However, they share the common characteristics of active top management support, ongoing employee involvement, (e.g., in quality circles), active solicitation of customer feedback, and continuous measurement (e.g., statistical process control).

TQM is implemented differently in each organization. Decision makers should consider the inclusion of some form of TQM in any courseware development activity.

What Specifications and Standards Should Govern Air Traffic Control Training?

Guidance in the training development process can be provided in the form of work statements and specifications and standards.

Work Statements

The development of training materials should not occur in a vacuum. Guidance should be given to the developers in a clearly written statement of work (SOW), which specifies exactly what is to be produced. A statement of work should be prepared even if the work is to be performed internally.

A work statement:

- Is a clear, written definition of need
- Describes what products or services are desired
- Describes how it will be determined that the products/services provided meet the specified need
- Is individually tailored to the requirement.

Normally, the SOW is prepared by a group of technical specialists, and, if training is to be acquired from an external source, is incorporated into a larger procurement document (e.g., Request for Proposal). Obviously, a greater amount of detail will be provided in a SOW related to a training acquisition than will be available when training is included with the total system acquisition.

Specifications and Standards

A critical issue for decision makers is how much latitude to permit training system designers/developers. Besides the SOW, reference can be made in the acquisition document to documented models of the instructional systems design (ISD) process, such as the U.S. Federal Aviation Administration Order No. 3000.6C: Training.

Within U.S. military training acquisitions, further guidance is given to training contractors in Data Item Descriptions (DIDs). These documents prescribe the format and content of products to be produced as the result of a procurement. For example, there is a DID (DI-ILSS-80092A) that contains the format and content specifications for a Simulator Software Handbook. This DID, in turn is tied back to Military Standard (MIL-STD) 1379D, which represents U.S. Department of Defense guidelines for contractor developed training materials (comparable to FAA Order 3000.6C, cited above).

There is a potential risk, however, in the imposition of too many guidelines on the training development team. The use of valid, proven standards and specifications increases the likelihood that an acceptable product will be produced. On the other hand, invoking such guidelines can serve as a barrier to creativity, innovation, and experimentation with newer training technologies and approaches.

Also, the use of product (deliverable) descriptions such as DIDs can actually increase costs in that:

- At least in our experience, superfluous DIDs are sometimes called out in a contract; and
- Product descriptions, such as DIDs, are often too general, and usually need to be customized for specific applications.

Minimally, we recommend that every training development effort be defined through a well written statement of work, which references an established ISD model. Including the model will insure that major stages in the ISD process will be considered. Beyond this, an agency should be judicious in providing supplemental guidelines such as DIDs, unless they can be shown to relate materially to the quality or utility of the training material.

There are instances where the use of international standards must be considered. For example, if a training project envisions the use of Compact Disc-Read Only Memory (CD-ROM), the international standard for mastering the medium (ISSO 9660) should be referenced at the inception of the project, so that costs related to adherence to the standard can be projected.

Other Related Issues

There is an increasing concern within the U.S. about the need to have the capability to exchange text and graphic data. This includes training materials. Within the U.S. defense community, increased emphasis is being placed on Computer-Aided Acquisition and Logistics Support (CALS). CALS is a U.S. DOD and industry program to facilitate the use and integration of digital technical information in system acquisition, design, manufacture, and support. Two current standards for the digital interchange of technical information are:

- MIL-STD-1840A: Automated Interchange of Technical Information.

- MIL-M-28001: Mark-up and Generic Style Specifications for Electronic Printed Output and Exchange of Text.

There also has been a growing concern, again fostered by the U.S. defense community, about the lack of portability of computer based courseware. The problem has been caused by lack of standardization in hardware, operating systems, and development/delivery software. Several initiatives have been undertaken over the last few years that address possible solutions to this problem. They include the work of the Interactive Video Industries Association (IVIA) (1989) and the Portable Courseware (PORTCO) Architecture Project (Thomason et al., 1990). The above issues have potential relevance for decision makers if the exchange of text and computer-based ATC training materials among NATO members is envisioned.

To What Degree Should the Systems Engineering and the Instructional Systems Design Processes be Integrated?

The early involvement of training professionals (e.g., instructional designers, human factors specialists, etc.) during the systems analysis, design, and prototype phases can provide valuable support for software and hardware engineers.

Early Involvement of Training Specialists

In many projects, training is an afterthought. Often, the funding pattern of a multi-year project precludes the involvement of an instructional design specialist during the crucial period of system design. This is a mistake!

Ultimately, it is the tasks that trainees must perform on the job, and their related knowledge and abilities, which are the principal inputs to the training development process. Often, especially with new systems, operator and maintainer tasks evolve during the design phase of the effort. Given the usual delay in the involvement of a training specialist, the linkage to data from early personnel and training analyses are lost.

Maher and Ingram (1989) suggest that systems engineering and instructional systems design follow similar processes. One major difference, however, is perspective. The systems engineer tends to focus on the internal elements of the system; the training specialist is concerned more with the users of the system (in this case, operators and maintenance personnel) the tasks that they will be required to perform, the environment in which the system will be used, and the constraints which this environment imposes on the user.

It is strongly recommended that management make provisions for the involvement of training specialists from the inception of an advanced ATC systems acquisition and throughout the systems design phase.

Usability Considerations

Decision makers should allow, in any acquisition planning, for usability analysis of the system prior to initial fielding. Usually, systems are evaluated on specific performance criteria (e.g., mean time between failure, minimum standards for uninterrupted operation, etc.). However, similar provisions usually are not made to assess the usability of the system from the operation and maintenance perspectives. For example: Are knobology, switchology, and symbology conducive to operator decision making? Do written or embedded information aids provided with the system help the operators in performing their duties? To what degree do use of colors, symbols, and screen message design facilitate system use?

Concurrency

Where systems development and training are viewed sequentially, training developers are often required to prepare training materials to support system tests and initial system fielding with relatively short lead time. This may be feasible for classroom or print-based instruction, where development ratios may run 20 to 50 hours for every contact hour of instruction. However, for computer-based training, which also may involve the production of analog (e.g., videodisc) or optical digital (e.g., CD-ROM) media, development may require 200 or more hours for every contact hour or unit of instruction. Simulation requires an even higher level of human capital investment. If training professionals are not involved as part of the system design team, it is unlikely that they will have the lead time or knowledge of the system necessary to produce the required materials in the time available. This can cause a delay in system implementation or lead to operational problems.

A special case of the concurrency problem involves the question of whether some training, job aids/decision aids, or context sensitive helps should be embedded in the system software. If this requirement is envisioned, is should be addressed at the inception of the system acquisition process, as opposed to attempting to retro-fit the system with this capability after hardware and software specifications have been completed.

How Should Initial and Recurrent System Operator and Maintainer Training Be Delivered?

Training Options

A major consideration for senior managers is the resource implications of training decisions. Time is money. The goal is to provide initial and recurrent training that allows the maximum number of trainees to achieve and maintain required levels of proficiency in the minimum amount of time, and at the lowest possible cost (e.g., labor costs, travel, and subsistence at the training site). This equation varies for each nation, given some of the demographic variables identified by Baldwin (1991).

Training delivery options can be broadly categorized as follows:

- Resident versus job site training
- Classroom/lockstep instruction versus self-paced instruction
- Hands-on training versus simulation
- Dedicated training systems versus embedded training
- Single training stations versus networked training systems

A separate paper could be written just explicating the elements of each of the above addressed options. It is important that these issues be addressed, and basic choices made at the inception of the system acquisition process.

Training Decision Factors

For any potential instructional medium specified for ATC training (e.g., computer-based training), there should be some basis for the decision. Major factors upon which such an *a priori* decision might be based include:

- Instructional Effectiveness: Does the instructional medium produce an acceptable level of learning?
- Time on Task: Does the medium help students spend more time on learning and on practice?
- Retention: Do students using the medium demonstrate an acceptable level of knowledge and skill retention?
- Cost Effectiveness: Does the medium represent a cost effective alternative to current practice?

The research on each of these decision factors is extensive, particularly the research that compares more traditional instructional approaches, such as classroom training, with computer-based training. Each agency must address these factors in a manner appropriate to its current and projected personnel and resource base. It also must consider training alternatives in light of their projected life cycle costs, including the cost of supporting and maintaining the courseware after initial fielding, as opposed to looking simply at development costs.

How Will Course Material Be Validated and Trainee Mastery Assessed?

Two important elements to be considered by senior decision makers are the criteria that the agency will use to determine that the training materials are acceptable and ready for fielding, and the standards and mechanism by which trainee mastery of course content will be assessed.

Validation Standards

Required levels of technical accuracy for course content and standards of trainee performance should be established at the beginning of any training development project and used for determining the initial acceptability of training material. These levels should be based, largely, on the criticality of the tasks that the materials are designed to train, and the risks associated with trainee error in performing these tasks in the work place.

For example, the SOW governing an ATC training development effort might specify that the course materials will not be accepted until they are totally free of technical errors and a group of representative students have demonstrated mastery of 90% of the critical instructional objectives in field trials of the material.

Decision makers must keep in mind, however, that setting validation standards unreasonably high may drive up the cost of the material and consume a significant amount of additional time in their development.

Minimally, course materials should be reviewed by one or more subject matter experts for technical accuracy. Draft materials should be pilot-tested with a class of representative trainees prior to their release. Also, the training development team should be required to maintain an audit trail of all major decisions, so that, if problems are encountered with the materials during the pilot test, probable causes for such problems might be identified.

Trainee Assessment

It has been our experience that a client organization can spend hundreds of thousands of dollars for a training course and resist the development and validation of meaningful performance-based tests to assess trainee mastery of course content. Traditionally, testing has been predominated by norm-referenced or criterion-referenced, paper-based, multiple-choice testing. For subject matter like ATC system operation and maintenance, such tests usually assess knowledge. From the results of such tests, it is necessary to infer the ability of the trainee to perform the requisite operator and maintainer tasks.

Performance tests (see Priestly, 1982) are designed to measure performance in an actual or simulated setting. Ideally, such tests should consist of a representative sample of all the job elements, or at least the set of job elements that are deemed to be most critical from the perspective of safety, operational effectiveness, and/or economics. Computer technology makes performance testing in a simulated work setting highly feasible.

The foundation for such tests is training front-end analysis, which identifies the tasks which trainees are required to perform on the job once training is completed, and the knowledge and skills required to accomplish these tasks.

It is recommended that, in any ATC training development effort, specific requirements for front-end analysis and the development of valid performance tests be included.

How Are System Design/Engineering/Software Changes Integrated into Training Materials in a Timely Manner?

Systems, particularly complex systems such as advanced air traffic control, evolve. Configuration management is a discipline applying technical and administrative direction and surveillance to identify the functional and physical characteristics of a configuration item, control changes to those characteristics, and record and report change processing and implementation status (U.S. Naval Aviation Systems Command, 1980).

As a rule, trainers are not on the distribution list for engineering change proposals (ECPs) and other documentation related to configuration management. Thus, system changes usually are not systematically incorporated into the training courseware. If enough of these changes accumulate, the validity, utility, and credibility of the training materials can be seriously impaired.

As we stated above, an advantage of an integrated system design team, including training professionals, is that the potential for active configuration management and the continual upgrading of training courseware is enhanced. However, after the training materials are accepted by the agency and the contractor's responsibilities have ended, attention must continue to be paid to this issue.

Decision makers responsible for ATC system training acquisition should require contractors to provide configuration management and courseware maintenance during the life of any contract. Provisions for carry-over of such procedures to in-house training courseware support also must be made at the inception of the project.

Summary

This paper has highlighted some issues which senior decision makers must consider relative to the acquisition and/or in-house development of training materials in support of advanced air traffic control systems. These issues have included: acquisition strategy; the need for specifications and standards; the need for integration of system engineering and instructional system design activities; selection of an approach to training delivery and proficiency maintenance; standards for validation and acceptance of training materials; the recommendation that performance-based assessment be used to assess trainee mastery of course content; and the suggestion that configuration management be proactively addressed, and a mechanism created for translating engineering changes into updated training materials.

Decision makers do not get directly involved in the development of training materials. However, the choices they make at the inception of a project have as much, and probably more of an impact on the potential for a successful outcome as the work of training professionals.

References

Baldwin, R. (1991). "Training Requirements for Automated ATC." In: J. A. Wise, V. D. Hopkin, & M. L. Smith (Eds.) *Automation and Systems Issues in Air Traffic Control.* Berlin: Springer-Verlag.

Federal Aviation Administration. (1989). *Flight Plan for Training.* Washington, DC: Federal Aviation Administration.

Federal Aviation Administration. (1987). Training. Order No. 3000.6C. Washington, DC: Federal Aviation Administration.

Interactive Video Industries Association. (1990). *Recommended Practices for Interactive Video Portability.* Washington, DC: The Interactive Video Industry Association.

Maher, J.H. and Ingram, A.L. (1989). "Software Engineering and ISD: Similarities, Complementarities, and Lessons to Share." Software Engineering Institute, Carnegie Mellon University, Pittsburgh, PA. Paper presented at the Annual Meeting of the Association for Educational Communications and Technology, Dallas, TX, January 31-February 5, 1989.

Priestly, M. (1982). *Performance Assessment in Education and Training: Alternative Techniques.* Englewood Cliffs, N.J.: Educational Technology Publications.

Smith, M. (1991). "Adapting Air Traffic Control Training to Accommodate Changing Conditions". In: J. A. Wise, V. D. Hopkin, & M. L. Smith (Eds.) *Automation and Systems Issues in Air Traffic Control.* Berlin: Springer-Verlag.

Thomason, B., Van de Wetering, B. & Booth, R. (1990). "A Portable Courseware Architecture (Revision 1.0)." Unpublished paper distributed by Systems Engineering Associates, 2204 Garnet Avenue, Suite 203, San Diego, CA 92109.

United States Dept. of Defense. (1990). Military Training Programs. MIL-STD-1379D.

United States Dept. of Defense. (1988). Automated Interchange of Technical Information. MIL-STD-1840A.

United States Dept. of Defense. (1988). Markup Requirements and Generic Style Specification for Electronic Printed Output and Exchange of Text. MIL-M-28001.

United States Navy. Naval Air Systems Command. (1980). Configuration Management Manual. Washington, DC: Department of the Navy (NAVAIRINST 4130.1A).

United States Navy. Reliability, Maintainability, and Quality Assurance Directorate. (1986). Best Practices: How to Avoid Surprises in the World's Most Complicated Technical Process. Washington, DC: Department of the Navy (NAVSO P-6071).

Speculations on Automation and Air Traffic Control

A Modest Proposal for Future Systems

John A. Wise

Center for Aviation/Aerospace Research, Embry-Riddle Aeronautical University, Daytona Beach, FL, USA

Introduction

Mitroff (e.g., Mitroff, 1971; Mitroff & Beltz, 1972) and others have argued quite persuasively that when trying to predict an uncertain future[20] one of the most effective techniques is the dialectic argument. In the dialectic approach, proponents of the two most extreme views on an issue argue the strenghts of their position and the weaknesses of the other. As a result, the strengths and weaknesses of each will be identified, and a third point-of-view, hopefully including the strengths of each and the weaknesses of neither, will be created by the observer.

In my earlier paper (Wise, 1991), I suggested that it would be of value to formulate sets of assumptions that deviate significantly from those currently utilized in air traffic system design and operation (i.e., antitheses), and then to investigate the types of system that might evolve from their application.

Toward that end, several "blue sky" papers that attempt to look at the future in this way were developed by the participants (e.g., Hopkin, Levesley, David). Because I issued the challenge, I felt a minor moral obligation to develop at least one alternative worldview.

This paper is a first attempt to describe an alternative type of air traffic system that might evolve if a different set of assumptions were employed. The assumptions for this paper basically grow out of the modern notions in computer and management science. In computer science, the concept is often referred to as "distributed processing." This involves assigning various processing tasks to independent processors, which perform a task and feed the results to the another processor for further analysis, rather than trying to perform all the analysis on a single large processor and using the outlying processors only as dumb terminals. In management science, people like Tom Peters have been suggesting that industrial systems would be much more effective and efficient if more decision making and problem solving tasks were assigned to the people who actually do the work, rather than being closely held by a centralized management team.

My air traffic correlate to the above concepts assigns a considerable amount of the air traffic decision making currently performed by controller to the pilot. The underlying assumption is that systems are more effective when responsibilities for low-level control tasks (e.g., maintaining separation) rest as close as possible to the place where the actual action takes place (i.e., in the cockpit).

20. In reality, no part of the future is certain. No one can predict absolutely what is going to happen two seconds from now, let alone five years in the future.

NATO ASI Series, Vol. F 73
Automation and Systems Issues in Air Traffic Control
Edited by John A. Wise et al.

En-route Separation and Formation Flight

The first new assumption addresses with the best way to maintain a safe distance between aircraft. Aggressive fighter pilots and military demonstration teams, like the U.S. Navy's Blue Angels, fly formation with wings overlapped. In air traffic parlance, this might be considered as negative separation (i.e., negative two meters' separation). U.S. Air Force student pilots are required to demonstrate the ability to fly with one meter separation between wingtips. In both cases, this spacing is maintained not only during straight and level and shallow banks, but also in what the FAA clearly defines as acrobatic flight.

However, when we ask air traffic systems to do essentiallly the same task — i.e., maintain minimum safe separation — the distance grows from one or two *meters* to eight to ten *kilometers*. The increase in minimum separation standards is primarily a function of the controller's sensor characteristics (e.g., radar resolution, update times), the fact the controller is dealing with a larger number of aircraft, and the fact that the controller is farther out in the control loop than the pilot is. The pilots flying formation have much better sensors in their eyeballs than do controllers in their radar, in terms of both resolution and update times. The eyeball has it all over the radar, in this case. Formation pilots have only a few things to monitor: maintaining the appropriate separation and looking out for other aircraft. The controller must do the same things for many sets of aircraft. It is a well-established fact in human control theory that, the farther humans get out in the control loops, the worse they perform. Therefore, controllers are at a significant disadvantage.

Various forms of "desired separation standards" are utilized by military pilots, depending on the mission. For example, on long missions, the standard might be eight to fifteen meters' wingtip separation. During inflight refueling, it would be one to two meters in trail. In air combat maneuvering, it might be ten to twenty meters in a cone behind the lead aircraft. During certain tactical maneuvers, it might be one or two kilometers in trail. The point is that maintaining even extremely close separation is fairly easy, even for relatively low time (i.e., less than 150 hours) military student pilots.

Given the above to be a reasonable assumption, the proposed future system will assign all separation duties to the pilot. This will allow much smaller separation standards — perhaps one or two kilometers — using current operational procedures. In order to accomplish this, the pilots will need their own display of air traffic data of at least their immediate surround. Each pilot would be assigned a protected block of airspace in which to maneuver as required to maintain the required separation from other aircraft. For example, USAir 116 might be assigned to fly a one kilometer trail on TWA 213 until Wink vortac and then to transition to a two kilometer trail on United 415. The controller would still be responsible for sequencing and overall organization, but the nitty-gritty task of maintaining adequate separation would be left to the pilot.

Let Pilots See Where They Are Going

An assumption held by most designers of large aircraft is that air traffic control will provide complete separation. Therefore, they feel secure in minimizing the size of windscreen that the crew uses to search for other aircraft. Not long ago (before the widespread use of microprocessors and electronic displays and controls in the airline cockpit), almost every square inch of cockpit real estate was covered with toggle switches and electro-mechanical displays. Cockpit windscreens were small just to provide more space for switches and displays. However, the modern cathode ray tube based cockpit has removed the need for dedicated controls and displays and freed up large amounts of cockpit space. However, none of the major airframers used that extra space to make the windscreen any bigger.

In reality, no matter how good the air traffic control system, the final defense against a midair collision is the pilots' eyeballs (for example, analysis of the San Diego midair involved a study to determine how long the light aircraft was hidden from the crew by windscreen posts). Therefore, I propose that all future airliners be required to have bubble canopies (i.e., completely transparent coverings like those found on modern fighters such as the F-16). This one single change would significantly increase safety of the entire national airspace system, by minimizing the chance of crews not being able to see aircraft missed by the air traffic system. It could also improve efficiency by making it easier for airliners to maintain their own separation when conditions permit.

A Different Approach to Departures

A perceived limitation in the national airspace plans is the difficulty of developing new airports capable of handling passenger demand for more departures to more places, all departing at about the same time. Not only are the costs for new airports high, but environmental and other political issues (e.g., Not in my backyard, you don't!) must be overcome. Current air traffic assumptions relevant to departures involve the required use of sequential separation with controllers-in-the-loop. However, if we extend the above en-route assumptions to the departure process, then formation take-offs become a potential way to double or triple airport departure capacity.

This tremendous increase in capacity involves only the cost of doubling or tripling the width of the runway to allow two or three ship formation take-offs. There would be no need for surveys of hazards, analysis of winds, most environmental impact studies, etc., normally required for a new airport

Formation takeoffs would not be a very big problem, as most airline pilots are former military pilots who are very familar with the procedure. The remainder could quickly learn (it is not really all that hard, and it is a lot of fun). The biggest "technical problem" would be matching aircraft with similar performance characteristics to depart together. But even having to wait one or two slots to be matched up with a similar aircraft would seem to be much more desirable than waiting in a long queue for single departures.

The other potential problem would be passenger comfort. Having been brainwashed for years that another aircraft within a mile or two is inherently dangerous, and that formation flying is something done only at air shows, some passengers will naturally have some apprehension. One way to ease passenger apprehension would be to eliminate windows for the passengers (a solution probably close to the heart of all structural engineers). A higher-tech solution would be to present images of a virtual world, showing a beautiful clear day, on each window during the takeoff phase. A cheaper solution would be to dream up a safety reason for having the window shades closed during takeoff and departure.

A Departure from Current Approach Methods

If formation takeoffs can reduce departure queues, why not employ formation landings to reduce holding patterns during arrivals? The U. S. Navy's Blue Angels landed six aircraft in diamond formation for years. While such a large formation landing may be out of the question for airlines, two or three ship formation landings would be not only safe but highly efficient. They could be performed either to widened runways discussed above or to minimums, with the wing aircraft breaking off to circle to land at a parallel runway too close for simultaneous parallel approaches.

Formation landings would also be a means of building high internal incentives for pilot quality. I always used to remind my student pilots in the Air Force that everyone on the ground could see how well they flew during formation landings. If they could pick one point in the mission to excel, the formation landing should be it.

Hopkin (1991) has pointed out one significant potential problem with this increased efficiency. He observes that there would be a tremendous jam at the baggage claim, the rental car offices, the taxi stands, etc. But those are relatively simple problems, compared to building new airports.

Flow Control Should Go

A basic assumption of modern air traffic control is that ground based "flow control" (i.e., the government saying who can depart and when) is an effective solution to the air traffic jams. This seems rather questionable, if we assume that the problems are really a result of an inefficient and outdated air traffic system concept. Current flow control concepts are basically a small version of the current centralized absolute control system (i.e., more of the same). Flow control systems simply collect the relevant data (e.g., weather, traffic, construction) and then the flow control staff make *all* the decisions regarding departures, arrivals, etc.

The best analogy to the current flow control system would be the *requirement* for you to call your local fast-food franchise's flow control office to get a slot time to come down and buy a hamburger. You can imagine how long that fast-food franchise would be in business. Another analogy of the current flow control concept is the failed economies of Eastern Europe, where the government "economic flow control" attempted to make all economic decisions from a centralized bureaucracy. Given their spectacular failures, why would any organization build an air traffic flow control concept parelleling that model?

I would like to suggest an alternative concept to flow control, called "flow enhancement." In the flow enhancement system, the relevant government aviation agency would continue to collect all the necessary data, but instead of the agency making the decisions, they would simply process it and pass it along to all interested users. The agency could provide *advice*, but it would be up to the users to make their own decisions.

For example, airlines might hire a company like AT&T®, to develop a system that would allow them to re-route their flights (the way AT&T routes their phone calls) as a function of efficiency (e.g., least fuel burned), total quality (getting passengers to destination on time. regardless of cost), or some other corporate or consumer goal. General aviation flights might schedule their operations into feeder airports via different routes, or they may decide to wait till after lunch.

Flow enhancement is built on the concept that the person doing the actual job knows best how to solve problems related to it (the concept that appears to be the basis of most successful modern corporations). There is no reason to expect a result similar to the "commons" syndrome[21], because each airspace user would have the relevant data on current system performance. If you know airport XYZ is saturated, you will not send your aircraft there to waste its expensive fuel; rather, you will develop a solution that meets your customer's needs.

Conclusions

The ideas in this paper are not presented as an optimal set. They are intended only as alternative positions to the current assumptions held by designers and operators of air traffic systems. However, if our aviation system had started off differently, or if our governments and corporations were organized differently than they are, the air traffic system described above could be operating safely today. (And if that were the case, this paper might have described an alternative that would be very much like our current air traffic systems.)

My purpose was to propose a different model that might have one valuable idea buried in it. I only ask you to consider the above arguments and ideas in this light and to create your own assumptions for a new and improved air traffic system. I also want to stress that the above ideas do not represent the opinions or official positions of the sponsors of the Institute, my employer, my family, or my friends. They are mine alone. And at times, *I* wonder about them.

Acknowledgements

I want to thank all participants at the Institute (especially my antagonists) for providing the kind of discussions that help even a muddled mind recognize its weaknesses.

References

Mitroff, I. I. (1971). A communication model of dialectical inquiring systems — A strategy for strategic planning. *Management Science, 17*, B-634 – B-648.

Mitroff, I. I. & F. Beltz. (1972). Dialectical decision theory: A meta-theory of decision making. *Management Science, 19.*

Wise, J. A. (1991). Assumptions and automation in air traffic control. In: J. A. Wise, V. D. Hopkin, & M. L. Smith (Eds.) *Automation and Systems Issues in Air Traffic Control.* Berlin: Springer-Verlag.

21. The over-grazing of common land by farmers because each was trying to maximize profit without concern for the capability of the commons to support another animal.

The Blue Sky Challenge: A Personal View

John Levesley

Bransgore, Hampshire, UK

Introduction

The boldest challenge of the NATO Advanced Study Institute (ASI) on "Automation and Systems Issues in Air Traffic Control Systems" was to look at so-called "blue sky" solutions to future ATC systems, unprejudiced by precedent, convention, or existing concepts of ATC system evolution. Given that most ATC services are already committed to programs up to about 2005, the ATC prophet's task is to go forward to 2040, fifty years from now. From the view point of 50 years ahead, and with a system in mind, the visionary can look back over one shoulder and identify what developments will be needed to take ATC from 2005 to 2040, if indeed there is ATC in 2040. I do not pretend to be enough of a systems expert, or an engineer in either hardware or software, to make a detailed projection; but allow me as a simple lay controller to throw my hat into the ring, as part of the debate.

How Good Will Progress Really Be?

To anticipate progress, the trick is to examine the past. Read 1930's science fiction, the wonderment at unavailable wrist radios, digital clocks, television, radio, death rays, supersonic speed, atomic power. I wonder where the wonder went?

The sad fact is that a scientific success is for many only today's fad and tomorrow's kitsch — remember the fiber optic table lamp? The true achievement remains, but the recognition of that achievement by public and mass media is short lived. These advances become a series of nine day wonders which are either soon forgotten by the person in the street, or perhaps worse, taken for granted. With that loss of wonderment goes the inspiration to develop the end product, or even worse, this new discovery is diverted to a trivial but commercially-lucrative product. Anyone remember the 1988ish shortage of speech synthesis chips for industrial and research programs, because the toy manufacturers had got there first for the Christmas market?

A technology current affairs program called "'Tomorrow's World" has run for over 20 years on the BBC. Any optimist for achievements by technological advances would be sorely tested by the section of the program entitled "Whatever Happened To ...." a catalogue of missed opportunities, unkept promises, disappointments, and lack of inspirational exploitation which condemned good idea after good idea to the dustbin.

NATO ASI Series, Vol. F 73
Automation and Systems Issues in Air Traffic Control
Edited by John A. Wise et al.

Whatever we forecast today for aviation, will anyone who matters actually take it up? Lesson One of forecasting is that cost and commercial interest have always been major factors in system implementation, and accountants are not paid to be techno-freaks (rather the reverse).

What Have We Achieved So Far?

Before we go fifty years into the future, let's first look at what we had already achieved fifty years ago.

1. VHF radio telephony and automatic relay stations
2. Centimetric wavelength primary radar
3. Secondary radar (IFF)
4. Monopulse radar for naval fire control
5. Cathode ray tubes
6. Gyroscopes
7. Automatic pilots capable of maintaining height, heading and speed
8. Data communication in the form of radio teletype
9. Alan Turing's work on the mathematical theories of machine intelligence (AI)
10. Development of the design concept for Colossus 1 and 2, the analog computers used for code breaking in the ULTRA program
11. Long-range and accurate radio-based area navigation aids in aircraft over Europe
12. Telephone communication
13. Development of the transistor (the first integrated circuit)
14. The jet engine
15. Rocketry and inertial guidance
16. Stellar navigation's development of the basic theory of navigation utilized by global positioning navigation satellite systems
17. Multi-engined metal-skinned aircraft

So What Has Been Our Achievement Since?

With the exception of orbiting satellites for surveillance and navigation, we have added nothing new to the tools potentially available to ATC since the 1939-45 war. We have produced new applications for systems designed for other functions, improved their speed of operation and reliability, miniaturized them, made them more sophisticated, introduced ergonomic design improvements, and improved system redundancy. Today's electronic tools are also undoubtedly cheaper in real terms than those of 1940. These tools have improved considerably, as has our utilization of them; but there is nothing in use in ATC today that could not be easily explained to, and understood by, people from 1940. We have come a long way, but it's hardly revolutionary — just progress. If that is the best we can do in fifty years, can we really expect so much during the next fifty?

What Price the Hero?

The public loves mythical heroes and villains in public service, transport, and the armed forces. That has been a convention from the days of oral tradition. Whether Captain Cook or Casey Jones, the transport industry has had its heroes. Comic book heroes for the kids, romantic lean-jawed handsome pilots and cruise ship officers for the women, gritty John Wayne "Flying Leatherneck" role models for the men. Remember the succession of disaster movies in the 60s and 70s, with their improbable heroes? All these characters are traditions in our society.

Such characters have a real role in our society, as people to be identified with, who will make the sacrifices and assume the responsibility for other lesser mortals. They will take risks, incur displeasure, gain honor — all in the public service. Somehow computers find it difficult to compete with these role models, albeit that the role models are sometimes as synthetic as the computer. The public perception is of the computer as an evil *it*, the establishment (who are always to blame) as *them*, and of heroes and heroines as those who fight and win again *it* and *them*. Any attempt to replace popular heroes with automation will have to fight the romantic prejudices of people (inevitably bolstered by the most sensational sections of the media) as well as the more logical debates. It takes a bold person to totally disregard public prejudices and hysteria, however illogical, especially as it is the public who buy airline tickets and vote.

Is There a Brighter Side?

Probably yes, and I hope the rest of this paper will be less negative than the preceding paragraphs. This is a bad year to make forecasts about aviation. The public attitude of governments to aviation remains a commitment to growth, extra capacity, more runways, etc. When asked to investigate the hole in the ozone layer, our response is still to fly two gas guzzling jets back and forth in the hole for several weeks. The environmental facts of life have not yet begun to bite in aviation. Dramatic political changes taking place will, God willing, lead to a massive reduction of tension and confrontation, with major reductions in the medium term in military needs and activity. With these variables, it is difficult to make forecasts, but let's make our best guess towards 2040.

1. Air travel will grow but environmental issues will slow that growth, as more people become intolerant of the associated noise and pollution. Aviation leisure activities will grow, as will holiday travel. The communication revolution will *finally* start to reduce the need for business travel.

2. Military operational and training demands will reduce, giving more airspace and extra airfields into civilian use.

3. In Europe and Japan initially, but later in North America and Asia, high speed railways will take a significant proportion of the former airline passenger market. In June 1990, a catamaran ferry crossed the North Atlantic from New York to the

Scillies in 40 hours. The ferry is capable of carrying 300 passengers and 70 cars. A sign of things to come?

4. No smart alec is going to perfect a teleportation system before 2040, or controls for the weather.

5. As usual, the world will *just* avoid cataclysmic environmental or nuclear disaster, so someone will be around to fly in 2040.

Who Will Want to Fly?

Flight today is enjoyed by a variety of types of people. There is the commercial air transport passenger and air freight industry, and the business executive aircraft fleets. The military operational and training organization, civilian flying schools; leisure pilots in anything from the Confederate Air Force to a Cessna spamcan; gliders, microlights, hang-gliders, and parachutists. They all want to fly anywhere and everywhere, at any time, shortest routing, no delays and no formality, at minimum cost, without hitting the ground except on landing, and without hitting another aircraft. Fine.

Navigation, Control, and Separation

Birds have an inherent skill, so they know where they are going, most of the time. Human pilots don't have an inherent skill to find their way. Instead, they use an acquired skill called navigation, so that they too can find their destination, most of the time. Thankfully, one benefit of the computer is its ability to take the same information as a pilot, and to tell the pilot where it is, where its been, where it's going, and how long it will take to get there. Whether the information comes from ground beacons, area navigation services, or satellites is irrelevant, as the computer has a friendly graphics presentation which shows the pilot where he is. Such systems are cheap, getting cheaper, and reliable. In 2040, we can be certain that navigation to very fine tolerances, using satellite navigation systems and laser gyroscopes, will allow exceptional accuracy in position fixing in four dimensions.

The flight control of aircraft like the A320 gives us a hint of aircraft control in 2040. Automated, non abusable flight control systems will, I hope, still allow the pilot tactical freedom, but otherwise be protective of the aircraft. Air ground data link will permit identification of all other aircraft in the proximity, and as all aircraft intentions will be known, conflict resolution will be achieved by sensitive adjustments to the aircraft's flight profile.

Where aircraft need to have total tactical freedom, for military or leisure purposes, there will be large free airspace reservations for their use, particularly in the lower and middle airspace. At higher level, aircraft will fly direct routings, with cruise climb and cruise descent profiles, for maximum fuel economy and minimum flight time. Global scheduling will replace antiquated airline scheduling, airspace management, and flow control techniques. The slot time will be the scheduled departure time.

Assuming that aircraft will not be surrounded with force fields to prevent collisions, some form of separation between aircraft will be called for. A separation will be applied based on speed of reaction, time available for resolution, appreciation of the situation, weather conditions, limitations of control actions, and any communication delays. It will probably be a composite separation: some vertical, some lateral, some speed variation. It should never need to be achieved by violent action for tactical purposes, but by gentle adjustment as part of a strategic scheme designed to optimized airspace capacity.

Ripples on the Pond

The scenario above is not science fiction. It is an achievable goal based on today's systems evolved to the nth degree. The real questions are, how flexible and safe is it? Unless we control the weather in 2040 (pretty unlikely), snow, wind, rain, fog, and thunderstorms will continue to cause delay. Diversions will take place, emergencies will occur, system failures or degradations will happen — Murphy's Law strikes again. Sometimes it's pretty difficult to come up with combinations of rules to cope with all these eventualities, when the solutions are normally derived from a combination of applied intelligence, common sense, urgency, and inspiration. The ASI has addressed these problems *ad nauseam*, and I don't intend to reiterate the statements made.

What I actually predict will happen is that the lessons of the last fifty years will be repeated in the next fifty. Airlines will still prefer to spend $500 on aircraft for every $1 spent on ATC. Will the cost of potential super-systems actually prohibit their introduction, as they prove totally cost-ineffective? If I survive to the age of 93 and I fly somewhere in 2040, I suspect that there will still be a human problem solver on the ground in control of my flight, who will rejoice in the title of "the controller." And I don't think that controllers will be there because they are irreplaceable, or because the public wants someone there. I think that, with the right tools to help, the controller will still be there as the most cost effective, flexible system solution to the problem of safely guiding pilots and passengers to their destination. And *that* is what air traffic control is really all about.

Automation and Systems Issues in Air Traffic Control: Blue Sky Concepts

V. David Hopkin

RAF Institute of Aviation Medicine, Farnborough, Hampshire, UK

Introduction

This paper speculates about the evolution of air traffic control over the next fifty years or so, with particular reference to human factors aspects. While current air traffic control practices and known developments within that timescale are not ignored, these speculations do not merely extrapolate beyond them but consider some new factors that could appear in the meantime and influence the future evolution of air traffic control. It is assumed that aviation will remain a very large industry, and that fuel resources will not be a main limiting constraint.

Capacity Considerations

Several changes will have been made to circumvent capacity limitations, including the following:

- More short take-off and landing (STOL) traffic
- Helicopters or STOL aircraft used more to convey passengers between local and hub airports, especially international airports
- Multiple final approach and departure angles and directions to minimize wake turbulence constraints on traffic handling capacities
- More remoting and dispersal of functions such as baggage handling, check in, and security away from airports to alleviate congestion
- Large airports sited at environmentally acceptable locations

Regarding the last point, air traffic control procedures will implement policies more influenced by environmental factors, particularly in relation to fuel conservation, noise abatement and reduced disturbance of population centers. Associated ground transportation systems will be much more speedy and automated. Perhaps many nations will have implemented, rather than aspired to, an efficient co-ordinated transport system linking air, sea, and surface travel. Distancing airports from large metropolises will alleviate many restrictions on night flying.

The continuing trend towards greater leisure will increase demands for air travel and for leisure pursuits involving aviation. Minority aviation interests — hang gliding, pleasure

NATO ASI Series, Vol. F 73
Automation and Systems Issues in Air Traffic Control
Edited by John A. Wise et al.

flying, microlight aircraft, and their numerous successors fifty years hence — will be more popular and have more powerful political lobbies demanding airspace. Despite recent publicity about dislike of flying, for many it is a pleasurable experience which more people will wish to enjoy, particularly by flying aircraft themselves. The need to retain some divisions of airspace to separate different and potentially incompatible categories of air users seems likely to persist.

Accuracy of Data

Few technical constraints similar to those experienced now will remain, though new types of constraint will arise. Highly accurate, frequently updated and reliable data will be available routinely throughout the air traffic control system, mostly derived from satellites but integrated with other data sources, including radar, whenever this is advantageous. This improvement will extend to other kinds of data, such as meteorological data, and to computations and predictions based on those data.

Legal Considerations

The legal status of air traffic control and of those who work within it will have been changed and clarified, though some apparent anomalies will be difficult to resolve. In particular, will people or machines be blamed for any safety hazard? It would be unrealistic to hold pilots or controllers responsible for technological failings which they may have no means of influencing or even detecting. (It is believed that there will still be pilots and controllers.) Much equipment, hardware and software will be produced by machines and not people. One of the most difficult problems to resolve in the next decades will be to reconcile legal responsibilities and concepts of blameworthiness with developments of computer technology and of artificial intelligence.

Human Factors

The principles enshrined in the classic paper on human factors in air traffic control by Fitts and his colleagues forty years ago are still applicable today, because they are about abiding human characteristics. These will not change significantly in the next fifty years, though expectations and the means to satisfy human needs and aspirations at work will change. People will still need to know the consequences of their actions so that they can learn, will prefer jobs with opportunities to develop and exercise skills, will generally prefer to work in contact with other people and not in isolation, will seek self esteem and the esteem of their colleagues, and will look for evidence that their job is worthwhile and valuable and not lacking in human dignity.

Most stress at work will be preventable, and there will be better understanding of boredom and its effects on safety. It would be expected that jobs will not be boring, for humanitarian and safety reasons, and in order to minimize staff turnover and recruitment rates, with their associated high costs. As well as knowledge of the causes of boredom, there will be the technological means to prevent boredom, particularly through adaptation.

Human and Machine Flexibility

Most air traffic control functions will be capable in principle of fulfillment in several ways: by the human alone; entirely automatically; or by human and machine jointly, with a variety of feasible allocations of workload and responsibilities between them. The various essential requirements —ranging from safety, cost-effectiveness, and orderliness on the one hand, to human needs and aspirations on the other — can probably all best be served by a flexible system in which task demands are not a limiting factor, and human and machine collaborate to minimize their own and each others' errors or failures. The forms of computer assistance provided will ensure, if they are used as planned, that human workload is not a constraint on system capacity.

Advantage will be taken of individual strengths by using machines that adapt to them. Air traffic control will become almost wholly strategic. Many sources of potential hazards may be better understood and more predictable, with problems prevented rather than having to be solved (though the unpredictable can never be wholly eliminated). Ultimate safeguards against what cannot be predicted must probably still rely, in some instances, on human as well as machine flexibility, although how far the latter will have evolved is an imponderable. Guidelines should have evolved and been proved to optimize human-machine relationships, when the human and the machine can both be adaptable, innovative, flexible, and intelligent. Such human attributes will become machine attributes also.

International Uniformity

Air traffic control is becoming more international, and this trend may be expected to continue. Tasks, jobs, procedures, and equipment may be more standardized internationally, and controllers may have closer ties with their colleagues in other countries. International considerations will become more important in relation to management-controller relationships; the role of trade unions; and general conditions of employment, including hours of work, rostering, work-rest cycles, and retirement ages. Divisions of airspace may reflect air traffic control needs rather than national boundaries.

Job Uniformity

A probable consequence of more uniform sources of information, and particularly of satellite-derived information, will be to reduce or eliminate many of the current differences among jobs in air traffic control. Distinctions between enroute and terminal area control will largely have gone, and the integration of traffic around airports with enroute and terminal area traffic will be more seamless. The advent of plan views derived from satellite information will make oceanic and enroute traffic more similar, and the distinction between radar and non-radar control may go, because satellite-derived information has replaced both.

Within the next few years, there may be more specialization in air traffic control, with different selection and training procedures for various jobs which are not interchangeable. In the longer term, this trend will be reversed because of greater commonality of information sources and tasks at every air traffic control workspace and internationally.

Professional Mobility

It seems less probable, in future, that air traffic control will normally be a job for life. Large human-machine systems will become more similar to each other in the roles for human beings within them, and will allow increased mobility of employment between air traffic control and other large systems. Demographic trends towards a smaller workforce, together with more jobs requiring the specialized skills associated with large human-machine systems, will combine to pose problems in recruiting and training the right caliber of people for air traffic control jobs. Those who win the competition to recruit and keep the most suitable people will offer good pay and conditions, and provide work which is satisfying and enjoyable, and which meets human needs and aspirations. Efficient and safe system performance will rely largely on a wide range of technological options, and there will also be more emphasis on providing the right job conditions.

The almost exclusive emphasis on the controller will have been redressed, with much technical, human factors, and other expertise devoted to the problems of planning and designing air traffic control systems, supervision (which will itself have changed radically), engineering (including software engineering), maintenance (which will become mainly automated), quality control, and numerous other roles in relation to air traffic control, including administration and career development.

Staffing

Probably air traffic control will become a more continuous process. Numerous expedients have been suggested above to handle more traffic at airports, but if the demand goes on expanding — and if the trends continue towards more aircraft, steeper rates of climb and descent, airports located away from centers of population, and more feeder services — an inevitable consequence seems to be much more flying at night. It is therefore envisaged that, although the need for flexible staffing in air traffic control to take account of gross changes in demand will remain, to some extent, it will not be as important as it is now, because the range of demand will diminish. Strategic planning of traffic flows should curtail peak demands on the individual controller, and more night flying will increase minimum demands, so that future workspaces may not have to cope with such gross changes in demand as current ones.

The flexibility of computer assistance will permit more consistent staffing levels throughout changes in traffic demands.

Public and Professional Image

Air traffic control will have become better known. Even more people will fly and will be aware of it. The need to attract people to work in air traffic control implies that greater efforts will have been made to present a positive image of it to the general public. By then, a positive image should match the reality. Although for different reasons, future jobs for controllers will probably be as satisfying as current jobs are, in terms of the work rather than necessarily of the conditions under which it is done.

Probably in fifty years' time there will still be arguments about how air traffic control should be funded, and complaints that controllers are having to use old equipment and that their needs are not understood. Such management problems seem endemic in large human-machine systems, including air traffic control. The satisfactions of working in them seem difficult to comprehend by outsiders, especially if the penalties of not funding them adequately are partly disguised because professional dedication helps to remove some of the worst potential consequences of underfunding. Whatever the prevailing political system, complexion, or climate, there never seems to be enough money to satisfy professionals who want the best equipment to achieve the very best standards.

What If... We Integrated Both Training and Monitoring Functions into Air Traffic Control Workstations?

Marvin L. Smith

Embry-Riddle Aeronautical University, Daytona Beach, Florida, USA

Introduction

One of the unique characteristics of an ASI is the interaction and confrontation that takes place when a group of recognized experts from diverse professions are gathered for the purpose of addressing complex societal issues. This strategy was especially effective with the particular group of participants that came together at the Hotel Villa Del Mare for the 14 days of this 1990 ASI. The iterative dynamics of the group's formal and informal contacts facilitated a comprehensive understanding of automation and the air traffic control system that would not have been possible under any other conditions. Each of the participants made valuable contributions to the existing knowledge of the issues addressed and was rewarded, in turn, with a broader perspective of the problems and potential solutions through the integration of different views. It was this integration of unique theoretical orientations with professional experience that many of us considered the most important product of the conference. Gaining new perspectives on air traffic control problems through the cross-fertilization on many disciplines — *that* was the real value of the ASI!

This paper is submitted in response to John Wise's challenge to all participants for a "Blue Sky" concept paper. It combines the knowledge-base created by the presenters to propose a totally new strategy for providing training and quality assurance for air traffic controllers in the 21st century.

Somehow, the expression "Blue Sky" implies a greater-than-usual allowance for speculation and less demand for extensive empirical data to support a proposed course of action. An addition caveat that should be made clear is that I do not claim to be an expert in the area of artificial intelligence (AI), even though that subject is essential to the concept presented. Although I have been the project coordinator for an AI research project in which a rule-based air traffic control expert system was created, I have not been involved in the actual programming of this technology. What I will address is the potential of AI — as demonstrated in many other domains — to significantly impact some of the widely-publicized problems facing the ATC system now and in the future.

The purpose of this paper is to suggest that both the developmental training and monitoring of trained air traffic controllers could be accommodated by developing and integrating existing technology with existing operational equipment. While this is admittedly a radical departure from the traditional conventions for training and quality assurance, there is growing evidence that the technology has been demonstrated in other applications and the potential is there.

NATO ASI Series, Vol. F 73
Automation and Systems Issues in Air Traffic Control
Edited by John A. Wise et al.

These key questions are posed for the purpose of providing a frame of reference:

- What are the critical needs, problems, and opportunities in the air traffic control system?
- What are some possible alternative ways to respond to them?

Critical Needs

Never in the history of the Federal Aviation Administration (FAA) has there been a greater need for a more efficient, cost-effective training system. Recruiting, screening, and training of air traffic controllers are staffing problems that have been increasingly difficult for the FAA since the PATCO strike in 1981. The decline in the number of high-potential ATC candidates available has made the shortage of trained controllers a long-term problem that can not be solved with a "business-as-usual" approach to training.

The need for a more effective training system was acknowledged by Robert Whittington, Executive Director of Policy, Plans and Programs, in his introductory address to the FAA Symposium on Air Traffic Control Training, December 6 and 7, 1988:

> The FAA wants nothing less than to revolutionize the way it trains its professionals, especially its controllers. We can no longer afford the nearly three years it takes through time-consuming on-the-job training, to produce a full performance level controller. (Whittington, 1988, p 1)

Problems

The sharp increase in air traffic since the Airline Deregulation Act marked the beginning of what is projected to be a continued growth in aviation service demand. The U.S. General Accounting Office (GAO) summarized the effects of the pressures being put on the FAA controller resource:

> The volume of air traffic has increased every year since the strike, but controllers still number about 1,800 fewer than before the strike: 14,437 as of September 1988, as compared with 16,244 prior to the strike. Further, the number of FPLs — the core of experienced controllers — in fiscal year 1988 was about 3,300 short of the pre-strike number. Flight operations in fiscal year 1988 exceeded 80 million, compared with 60 million in fiscal year 1982. (GAO, 1989, p 7)

The relationship between these staffing problems and the success of the national airspace system (NAS) modernization is clearly stated by Whittington in the following:

> The billions of dollars that we are investing in state-of-the-art control and navigation equipment is meaningless without the trained personnel we need to operate and maintain it. Air traffic control is more than just a trade, a craft that can be instilled almost totally by on-the-job training. It's a high-tech profession that requires both

training and extensive education. And we have to begin the transition to developing the air traffic controller of the future right now. (Whittington, 1988, p 2)

With those challenges in mind, let us turn to an examination of the opportunities we could explore in finding ways to bring the controller resources into a less-disadvantaged posture.

Opportunities

Technology has changed virtually every aspect of our lives. Computers, for example, have become ubiquitous — invading our homes, cars, and everyday gadgets. The integration of technology into the way we in the USA learn, however, has spread much more slowly. With the exception of the military, where instructional technology has been in use for over 20 years, and recently within the business community, the use of even low-end computer-based training (CBT) within the federal government has had only limited acceptance.

The increasing use of artificial intelligence (AI) in ATC applications, such as Edwards (1991) discussed in his ASI paper, is an example of what is perceived by some as a high-potential opportunity for making significant gains in controller productivity.

Borrowing the familiar line from the Hewlett-Packard® television commercial, "What if..." we created knowledge based systems (KBS) appropriate to both the training of developmental controllers *and* the monitoring of operational air traffic control? Developmental controllers have satisfactorily completed the "knowing about" phase of training and are now in the "doing it" phase of training. Three major studies conducted in 1988 to analyze ATC problems found the present method of OJT both costly and ineffective. Highly-paid full performance level (FPL) controllers provide the one-on-one training, using very expensive single-operator DYSIM stations on an "as available" basis. This is a very labor-intensive and inefficient method, because students are taught by controllers who have little, if any, instructional qualifications, and there is frequently little standardization between instructors or ATC facilities. Resources for training (e.g., instructors and equipment) are shared with operational requirements, this can create training backlogs and keep a controller from being upgraded at the earliest date possible based on ability.

The creation of *expert systems* that monitor and interpret the actions taken by controllers to identify conflictions (e.g., separation errors) is an application that is currently being researched in both the university environment (i.e., Embry-Riddle Aeronautical University) and by large private research institutions such as the Mitre Corporation®. Expert systems, designed to capture and store the operational knowledge and control technique of an expert, have been used successfully in the medical field, among others, for diagnosis, analysis, and other types of troubleshooting situations. Expert systems have a significant advantage over most other computer systems, and even some instructors, in that they have the ability to *explain how they solved a problem.* While educators do not agree on much, they all reach a consensus on the importance of feedback in a good learning system.

Research on AI's reliability compared to a system's human monitor provides support for the contention that it would be both feasible and effective to integrate an AI expert system into operational ATC workstations and use the system to both train and monitor controllers:

> Expert systems are able to make decisions as well as experts in various fields, because their structures reflect the manner in which human experts arrange and make inferences from their knowledge. (Michaelson, Mickie, and Boulanger, 1985)

Other samples of AI development include some of the explicit models that support intelligent dialogue between the controller and the ATC system that Edwards (1991) reports. He notes examples of models that include the task, individual controllers, and the system itself (see Edwards and Mason, 1989; Kass and Finin,1988; and Mason and Edwards,1988, as referenced in Edwards' paper).

Conclusion

There is growing concern within the United States' aviation community that the massive investments in the national airspace system modernization will not deliver the improvements in safety and capacity that were promised in the NAS Plan proposal. Unless there is some immediate and success-oriented plan to train and safeguard the operation of the rapidly-changing ATC system, the potential for disasters and continued system delays is high and will remain that way. The "blue sky" proposal contained in this paper is one approach to making radical changes in the status quo that has a high probability for leveraging significant improvements in controller productivity.

References

Brown, J. S., R. R. Burton, and A. G. Bell. (1975). "SOPHIE: A step towards creating a reactive learning environment." *International Journal of Man-Machine Studies, 12*, pp.259-282.

Edwards, J. L. (1991). "Intelligent dialogue in air traffic control systems." In: Wise, J. A., V. D. Hopkin, & M. L. Smith (Eds.). *Automation and Systems Issues in Air Traffic Control.* Berlin: Springer-Verlag.

Hopkin, V. D. (1988). "Training implications of technological advances in air traffic control" a keynote address for Symposium on Air Traffic Control Training for Tomorrow's Technology. Oklahoma City, Oklahoma.

Michaelson, R. A., D. Michie, and A. Boulanger. (1985). "The technology of expert systems". *Byte. 10*, pp. 303-312.

Whittington, R. (1988). "Introduction: Training initiative overview." *Symposium on Air Traffic Control Training for Tomorrow's Technology.* FAA. December 6 & 7.

Some Final Thoughts

Closing Remarks

V. David Hopkin

RAF Institute of Aviation Medicine, Farnborough, Hampshire, UK

Introduction

When I started to compile material for this summary, it rapidly became obvious that the task of summarizing the ASI exemplified a point made during it: although a task can be defined and the actions required to accomplish it can be specified, it does not follow that a satisfactory task performance can be achieved or is actually feasible. The task of considering every paper, presentation, plenary discussion, group discussion, and informal meeting — in order to record every significant point, to capture the gist and enthusiasm of the discussions, and to synthesize everything into a structured and balanced summary — seemed impossible in such a short time. Good ideas were prolific, but a bald listing of them could not convey their value. Even the simplest summary of all of them would be a long document. It might also be dull, which the ASI was not. In any case, most of the ideas will appear in the published Proceedings. Therefore these remarks attempt to review the ASI from a broader perspective.

Aims and Objectives

The products of this ASI can be related to the NATO objectives of Advanced Study Institutes, and, within that framework, to the particular objectives of this one on "Automation and Systems Issues in Air Traffic Control." A NATO ASI seeks to consider theories, constructs, concepts and ideas as the foundations for applying knowledge, for interpreting findings, for gaining insights, and for identifying gaps in knowledge. Before the meeting, the Directors were keen to ensure that the theoretical coverage would be broad enough and that its applicability to automation in air traffic control would be clear enough. Any misgivings proved to be unfounded; in the event, the theoretical coverage was broad, its relevance was clear, and its insights were valuable and often innovative, albeit controversial.

An ASI should fulfill a tutorial role, and introduce those present to new ideas and new approaches. Everyone not only heard new ideas but was receptive to them, to the extent that some discussions developed into "Think Tanks" in exploring their implications. As a result, we believe that everyone present has acquired new knowledge which will lead to an appreciation of new problems and a new perspective on familiar ones.

An ASI is expected to provide a setting where professional contacts can be made and strengthened. Not only has it been highly productive in this regard, but the range of professional expertise recognized as relevant to the main theme has been expanded. We all

NATO ASI Series, Vol. F 73
Automation and Systems Issues in Air Traffic Control
Edited by John A. Wise et al.

have a better appreciation of how our problems appear from the perspective of other professions, some of which are now acknowledged as relevant for the first time.

The opportunity to hear lucid expositions of theories, points of view, and approaches not encountered before has forced us not only to expand our frames of reference but also to reexamine some of our basic assumptions. Most of the NATO nations, many different disciplines and numerous points of view — from highly theoretical to mainly applied — have all been represented by those attending this meeting. The relevance of all the contributions has been apparent, and they have melded gratifyingly well.

The venue and its facilities have provided an environment which has fostered the fruitful and lively interchange of ideas throughout all the official sessions of the ASI and far beyond its nominal times, with avid discussions at all hours. The eagerness of everyone to participate took the form of a willingness to listen, to learn, to think, and to question. A key to its success was professionalism. High standards were set in the presentation of papers and in the questioning. No speaker was ever lost for an answer to a question, a tribute not only to the speakers but to the perspicacity and lucidity of the questioners. The speakers were well prepared, knew their subjects thoroughly, and could withstand rigorous questioning, to the extent that it was taken for granted that questioning should be probing and persistent.

This professionalism has resulted in a greater awareness of, and respect for, other disciplines, but also in higher expectations about what they can contribute and achieve. There has been recognition that questioners who are probing are not being obtuse, and questioners who challenge assumptions are not being obstructive, but that such questioning comes from seeing problems differently because of different constructs, approaches, and theoretical perspectives. For each of us, assumptions which may have been taken for granted have had to be defended, and we have been forced to reappraise and clarify them in order to answer challenging questions about them. The extent of the common ground between those with very disparate points of view has been consistently rewarding and insightful.

The Directors are grateful for, and impressed by, all the participants' work, ideas, determination, and positive attitudes throughout this ASI. We have also been delighted by, and benefited from, the happy atmosphere that was created at the Institute. This was part of the achievement, but what we can learn from it? We would want the newly recruited trainee controller to enter into a happy work atmosphere and to work with high enthusiasm and motivation comparable to what has been achieved at this ASI. Can we identify the key factors which promoted such friendliness and total absence of animosity throughout this meeting? The Directors tried to plan a meeting that would be productive and enjoyable, very hard work and fun too, but they did not and could not control the atmosphere of the meeting, which depended on all who attended. How might the key factors that determined the atmosphere of this meeting be identified and introduced into air traffic control?

New Ideas

As Keynote Speaker I was in a privileged position, because it is much easier for the first speaker than for subsequent speakers to put forward ideas which seem to be new. The longer a meeting lasts, and the more speakers and discussions there are, the more difficult it becomes to be original. I was therefore impressed and pleased that all our speakers, including Rod Baldwin and Marvin Smith who gave the final presentations, were able to introduce ideas that had not been mentioned before.

For example, Marvin included, amidst many points made quickly near the end of his talk to stay within his allotted time, the fascinating remark that most air traffic control training has the policy "not to show favoritism to anyone, whether they need it or not." That raises many issues. Is that the policy? Are we assuming that we must not show favoritism in training? Why? What would constitute favoritism? How does this fit various theories of learning and reinforcement? Would forms of computer assistance during training that allow individual students to practice and overcome specific demonstrated weaknesses constitute favoritism? If not, would the same processes become favoritism if an instructor, rather than a machine, was providing the extra individual tuition? His remark raised many important issues, and typified dozens of comparable remarks during this meeting which have also provoked thought, been obviously relevant, and broadened perspectives. I have tried to note them and learn from them. We have all appreciated this never-ending flow of new, fascinating, constructive, and provocative ideas.

A NATO ASI dealing with automation and system issues in air traffic control must look to the future. The anticipated increase of automation in air traffic control systems was a major prompting for the ASI. Although an ASI is intended to gather and review existing knowledge and to identify gaps in it, yet it was surprising how many speakers independently chose to look back as well as forward. Sometimes their purpose was to show that a problem is old and much is already known about it. Sometimes they traced how current practices originated, in order to question their validity. Sometimes they sought to demonstrate that an out-of-date and inapposite approach should be replaced by something better. Looking backwards was an aspect of questioning what has been taken for granted, perhaps for too long.

Recurring Themes

Some issues emerged persistently during this ASI, including the following. It is often quite practical to integrate contributions from disciplines which apparently have nothing in common but which actually are compatible with each other and mutually supportive. A multiplicity of approaches all have something useful to contribute to the solution of problems of automation in air traffic control. Contributions, whether they be from air traffic controllers, from human factors, from computer science, or from other disciplines, have to be integrated so that they can affect the design process. Because many key decisions are taken early during the design process, contributions must influence the design process itself, and the role of subsequent contributions is limited by the stage in the design process at which they are applied. The extent to which design processes predetermine the kinds of error that can occur seems to have been underestimated.

All the disciplines represented at the meeting had encountered some difficulties in providing information in applicable forms to influence design processes appropriately. It is not sufficient to plan and conduct research if the only products are journal articles, standards, or handbooks, essential though these are. The research evidence has to be applied and integrated into the design. Nowhere does this seem to be done satisfactorily, although its necessity is acknowledged and many attempts are made. Lack of appropriate mechanisms to apply research findings to design processes appears to be the main difficulty. This problem is linked to some uncertainty about how valid and general some of the existing data and findings are. Is all the existing evidence actually worth applying to the design processes? If not, how do we determine which should be applied and which should not? What criteria could serve such a purpose? What should the balance of evidence be between previous research and current and future research? What are the best measurements to gain evidence in the form of practical advice at the design stages? How can research findings be made more acceptable to designers, so that they are more willing to adapt design processes of future air traffic control systems to incorporate evidence from research?

Throughout this meeting, there has been caution about where automation might lead, which has taken the forms of withdrawal from the extremes of automation and the eschewal of haste in the introduction of automation. Automation has the potential to resolve many problems but also to introduce new problems which may be difficult to recognize or resolve. There is no antagonism towards automation; indeed it is taken for granted that there will in the future be more automation in air traffic control; but the general view is that the introduction of automation would benefit from greater caution, a slower pace, and more painstaking identification of all its implications beforehand.

Because this ASI encompassed a very broad range of different but ultimately compatible approaches, ranging from the highly theoretical to the very practical, it was able to reveal that the traditional division between theory and practice in the context of automation in air traffic control is an over-simplification that is not helpful in meeting the main needs. Most problems must be addressed broadly because they have both theoretical and practical applications and implications. Satisfactory solutions require their theoretical and practical aspects to be integrated and not separated.

There has been a contrast between what has been said and what has been done at this ASI. The claims about the impact of automation on air traffic control have emphasized that the process is system-driven, yet the topics discussed most have almost all been concerned with its human-driven aspects. The reasons why this has happened are not obvious. If the complaint that most aspects of automation are system-driven is valid, why were the proceedings of this ASI not more system-driven? Perhaps the lesson is that, in the past, most issues have been or have appeared to be system-driven, but in the future, the human-driven issues will be most important. There was the comparable impression, based apparently on considerable substantiating evidence, that the design processes for more automated air traffic control systems are biased towards equipment and technical factors, but the ASI itself did not have this bias. This is either a reflection of those who attended it, or evidence that the bias is beginning to be redressed.

The topic of the adaptability and flexibility of software kept recurring. "Adaptability" and "flexibility" are emotive words which seem to describe advantages, and do so if the system has been optimally designed to accommodate them. However adaptability and flexibility can be difficult to distinguish from "instability" and "unpredictability," emotive words with the opposite connotation. Emotive concepts may lead to unjustified assumptions in such contexts. Discussions of adaptability and flexibility perhaps did not fully address where the balance of advantage lies, and how the adaptability of the human-machine interface in an automated system may need to be limited, so that its true advantages continue to outweigh its potential disadvantages.

Training

This ASI reflected the increasing interest in training in relation to air traffic control and its automation. On the whole, training was viewed as part of a continuum rather than as a separate topic. In the deliberations on training, the processes preceding training received extensive consideration. Not only were recruitment and selection discussed, but so was the public image of air traffic control and its impact on recruitment, and the factors which form and may change its public image. In retrospect, perhaps more attention should have been paid to such topics in the past, but at least they are receiving some attention now.

It emerged very clearly during the ASI that there is a large and challenging gulf between the public image of air traffic control and the controllers' own view of their profession. In one session, all the other participants at the ASI had an opportunity to question a panel of practicing air traffic controllers. It was a revelation to many of those present how closely controllers identify with their profession and how satisfying their job of controlling air traffic is to them. This seemed true of all controllers in all the NATO nations represented, and appears to be almost contrary to the public image of air traffic control as a profession. Studies of attitude formation may suggest how the public image of air traffic control could be brought closer to reality, may measure whether this has been achieved, and may indicate the consequences for recruitment. The need for the general public to be better informed about air traffic control was identified.

One issue was how increased automation of air traffic control should change the nature and content of what is taught to controllers. Currently the emphasis in teaching is on professional knowledge and on procedures. What other aspects of air traffic control can or should be taught, either in their own right or additionally, as automation is introduced? In on-the-job training in particular, there may be a gap between what is taught — such as knowledge and procedures — and what is learned — such as professional attitudes, expectations, norms, ethos, and standards. Attitudes and the factors which influence the formation of attitudes were recognized as important though neglected. What is not so clear is how far we can or should attempt to influence controllers' attitudes. If the goal of engendering positive attitudes among controllers towards the automation were set, how would we try to achieve such a goal and how would we measure our degree of success? This topic needs further thought and research.

The concept of teachability was discussed. To be of practical use, any automated aid must function efficiently and safely, but controllers must be able to learn to use it (preferably all controllers so that no additional selection procedure is entailed). Issues arising include the best ways to teach controllers how to use a new form of automation and the best measures to confirm that its usage has been taught successfully. Perhaps among the first criteria in deciding whether a form of automation is appropriate for air traffic control should be whether it can readily be taught to all the controllers who would use it.

Other Issues

This ASI concentrated almost exclusively on the impact of automation in air traffic control on the controllers themselves, rather than its impact on supervisors, assistants, managers, instructors, engineers, and planning and maintenance staffs, some of whom may be affected as much as the controllers by automation. Since everyone was kept very busy in considering the impact of automation directly on air traffic control systems and on air traffic controllers, it was probably right not to extend this ASI to consider all the other jobs within air traffic control, since the deliberations could have become too diffuse. But the impact of automation on those who work within air traffic control systems but are not themselves controllers is a seriously neglected topic, which must not remain neglected. It could be an apposite theme for a future ASI.

Several of the issues raised merit reemphasis, for various reasons. One concerns measures of performance. Task analysis was discussed, but not always its full range of advantages and limitations. In previous studies, the most appropriate performance measures may not always have been employed. In the future, better measurement tools will be needed to show the consequences of automation not only in terms of performance but also in terms of associated cognitive skills. Some cognitive effects are not currently measured at all, for example on understanding or memory, but they may be more significant than the routinely measured effects on performance.

The concept of over-control was mentioned but not pursued. It has many implications. One is that there is an optimum level of human intervention in the air traffic control process, and that we should be trying to use automation to promote this optimum level. However, we have not yet defined the optimum level and perhaps do not know how to define it or recognize it if we achieve it. A further issue is whether the optimum level, if there is one, is a product of equipment and automation or is primarily or partly a product of training.

The role of models in relation to automation in air traffic control has been considered from many points of view, but with a dearth of firm conclusions. There are many different kinds of model, and further consideration seems needed of what each kind of model can best achieve and of how each kind of model compares with other techniques to achieve the same objectives. It has not always been clear whether a model has been intended to be descriptive or explanatory.

Perhaps there is a need to reexamine so-called non-scientific approaches. An ASI presumes that those who attend can acquire a great deal of useful information by talking to each other, and ample opportunity for discussion is accordingly provided. This ASI seemed very successful in this regard. If this method is satisfactory for an ASI, and it is, what other objectives in relation to automation and air traffic control might it be equally appropriate for? It is necessary to consider what the alternative methods to talking might be, what different kinds of evidence they could gather, and how talking compares with other methods in terms of validity and use of time and resources. Nevertheless, the conduct of this ASI and its outcome seem to provide evidence that talking to people as a method of exchanging and acquiring information remains very productive and important. Perhaps we should be trying to establish in more quantitative terms its actual efficacy as a method. To achieve some research objectives, the best technique may be to talk to other people, rather than conduct laboratory, simulation, or field studies; build models; or follow equally orthodox scientific procedures.

My final thought is of a completely different kind, considering air traffic control as a sub-system. We wish you a safe and pleasant journey home. Many of you will travel in large aircraft, carrying three or four hundred passengers. You know now, if you did not know before, that there will probably be several other aircraft in the vicinity of yours near the departure and terminal airports. If the traffic densities are high, the controller will be trying to land a group of such large aircraft consecutively because of the relationship between aircraft weight category and wake turbulence separation minima, so as to obtain maximum utilization of the limited runway capacity and thereby gain a small but definite increase in traffic handling capacity. But air traffic control is a sub-system in relation to passenger handling. Partly as a result of the controller's efforts, there may be many hundreds of people at immigration, in the baggage hall, in customs, and seeking ground transportation. Is anyone considering air traffic control as a sub-system in relation to other aspects of passenger handling, in order to compare marginal advantages for air traffic control with perhaps major consequent disadvantages in other parts of the larger passenger handling system, and to ascertain where the overall balance of advantage and disadvantage lies? We hope that your return journey does not add to your practical experience of this problem.

Anyone at Home?

Hugh David

High Wickham, Hastings Old Town, Sussex, UK

"Funny how things turn out," mused the beekeeper, staring into his pint of still, off-the-chill bitter (flat, warm, musty) beer. "I started out as an Air Traffic Controller — until they automated the entire system. I wouldn't like to have to do that job now."

"Before I went to North America — it was still the United States then — I used to work for Thingummy Airways, in ticket sales, but they put in that self-check-in system and I was axed. Thanks, miss." The visiting accountant accepted a glass of ice-cold (gassy, icy, tasteless) lager from the barmaid.

"I used to be a flight attendant with that lot, until they decided to automate airside passenger services and my husband — he's the landlord here — used to be an airline captain. We bought this pub with our redundancy payments."

"You know," said the accountant's wife, "I used to do a lot of flying, but nowadays its either the Transonic Train, or I stay at home and pattern-match by video. I went to Heathrow — what do they call it, Thatcher Airport — a couple of years ago to pick up a parcel. It was like Mirabel in the Eighties...a ghost town."

Briefly, a rumble of distant engines rose above the birdsong, as the 1430 London-Tokyo ruled a straight line through the stratosphere.

"I rang to complain about that last month. I could only get an answering machine — said they'd ring back when someone was in the office, but they never did."

"I got the same with the FAA Corporation. But what with the military giving up flying — and contingency insurance making private flying so expensive — without civil airlines, there wouldn't be anyone up there at all. You don't suppose...?"

"Changing the subject, how's the bee business these days?"

"Fine, fine... I had a salesman round trying to sell me an automated beehive last week. Fully automatic. Made by McDonnell-Douglas/Airbus, of all people. I don't reckon the bees would like it. They're very human in some ways."

"How about the other half?"

NATO ASI Series, Vol. F 73
Automation and Systems Issues in Air Traffic Control
Edited by John A. Wise et al.

Participants

Dr. Neçdet Alpaslan
Dokuz Eylül University
Faculty of Engineering and Architecture
Dept. of Environmental Engineering
Bornova, 35100 IZMIR
TURKEY
Phone: (+90) 51-182-108

Dr. Rod Baldwin
Eurocontrol
Intitute of Air Navigation Services
31 Blvd. Conrad Adenauer
L-1115 Luxembourg
LUXEMBOURG
Phone: (+352) 43-60-611
Fax: (+352) 43-86-69

Dr. André Benoît
Eurocontrol
Rue de la Loi 72
B-1040 Brussels
BELGIUM
Phone: (+32) 2-233-0474
Fax: (+32) 2-233-0419

Mr. Peter S. Boag
Canadian Aviation Safety Board
P. O. Box 9120
Alta Vista Terminal
Ottawa, Ontario
CANADA
Phone: (+1) 819-994-4255
Fax: (+1) 819-997-2239

Mr. John N. Bradbury
7096 Randall Rd.
LeRoy, NY 14482
USA
Phone: (+1) 716-768-7153

Dr. Mary C. Chobot
Mary C. Chobot and Associates
4950 Andrea Ave.
Annandale, VA 22003-4166
USA
Phone: (+1) 703-323-9402

Dr. Giovanni Costa
Istituto di Medicina di Lavoro
Universitá degli Studi di Verona
Policlinic Borgo Roma
I-37131 Verona
ITALY
Phone: (+39) 45-933-295
Fax: (+39) 45-584-456

Dr. Hugh David
Eurocontrol
B. P. 15
91220 Brétigny-sur-Orge
FRANCE
Phone: (+33) 1-69-88-7672
Fax: (+33) 1-60-85-1504

Mr. Paul O. Day
Plessey Research Ltd.
Roke Manor
Romsey
Hants SO51 0ZN
UK
Phone: (+44) 794-833-216
Fax: (+44) 794-833-433

Prof. Anthony Debons
Robert Morris College
Narrows Rd.
Pittsburgh, PA 15108-1189
USA
Phone: (+1) 412-262-8465
Fax: (+1) 412-262-4049

Ms. Pamela Della Rocco
Federal Aviation Administration
Civil Aeromedical Institute
P. O. Box 25082
Oklahoma City, OK 73125
USA
Phone: (+1) 405-680-6838
Fax: (+1) 405-680-4813

Dr. Jack Edwards
President
AI Management & Development, Ltd.
206 Keewaiten Ave.
Toronto, Ontario M4P 1Z8
CANADA
Phone: (+1) 416-488-6068
Fax: (+1) 416-486-6013

Dipl.-Psych. H. Eißfeldt
DLR
Abteilung für Luft-und
Raumfahrtpsychologie
Sportallee 54
D-2000 Hamburg 63
GERMANY
Phone: (+49) 40-51-3096 0
Fax: (+49) 40-51-309-660

Dr. J.A.C. Empson
University of Hull
Department of Psychology
Hull HU6 7RX
UK
Phone: (+44) 482-465-283

Dr. John Fearnsides
The MITRE Corp.
Civil Systems Division
7525 Colshire Dr.
McLean, VA 22102-3481
USA
Phone: (+1) 703-883-7622
Fax: (+1) 703-883-5468

Mr. Ioannis Fokas
Aviation/Aerospace Consultant
Kabouroglou 7
Patissia 111 44
Athens
GREECE
Phone: (+30) 1 223 5484 - 85
Fax: (+30)1 223 4262

Mr. Tom Fudakowski
Transport Canada
Air Traffic Services
Place de Ville C
Ottawa, Ontario K1A 0N8
CANADA
Phone: (+1) 613-998-6595
Fax: (+1) 613-991-9451

Dr. Daniel J. Garland
Embry-Riddle Aeronautical University
Humanities and Social Sciences Dept.
Daytona Beach, FL 32114-3900
USA
Phone: (+1) 904-239-6641
Fax: (+1) 904-239-6459

Mrs. Barbara Gibson
Vice-President
Folino Printing Co.
129 Industrial Dr.
Franklin, OH 45005
USA
Phone: (+1) 513-746-3198
Fax: (+1) 513-746-3846

Dr. Barry P. Goettl
Clemson University
Psychology Department
108 Brackett Hall
Clemson, SC 29634-1511
USA
Phone: (+1) 803-656-2831
Fax: (+1) 803-656-0358

Prof. Jan Goodrich
Embry-Riddle Aeronatical University
Humanities and Social Sciences Dept.
Daytona Beach, FL 32114-3900
USA
Phone: (+1) 904-239-6628
Fax: (+1) 904-239-6459

Dr. Peter Hancock
University of Minnesota
164 Norris Hall
172 Pillsbury Dr., S.E.
Minneapolis, MN 55455
USA
Phone: (+1) 612-625-8527
Fax: (+1) 612-626-7496

Dr. Nilgun Harmancioglu
Dokuz Eylül University
Faculty of Engineering and Architecture
Dept. of Civil Engineering
Bornova, 35100 IZMIR
TURKEY
Phone: (+90) 51-181-047

Ms Claudia Harss
Neureutherstr. 25
D-8000 München 40
Germany
Phone: (+49) 89-271-7303
Fax: (+49) 89-470-4702

Mr. Daniel A. Herschler
University of Central Florida
1101 Shaffer Trail
Oviedo, FL 32765
USA
Phone: (+1) 407-275-2216

Mr. V. David Hopkin
RAF Institute of Aviation Medicine
Senior Principal Psychologist
Psychology Division
Farnborough, Hants GU14 6SZ
UK
Phone: (+44) 252-24461 ext 4364
Fax: (+44) 252-377-839

Dr. Esther E. Horne
University of North Texas
School of Library and Info. Science
P. O. Box 13796
Denton, TX 76203
USA
Phone: (+1) 817-565-2347
Fax: (+1) 817-565-3101

Dr. Peter Jorna
National Aerospace Laboratory
P. O. Box 90502
1006BM, Amsterdam
NETHERLANDS
Phone: (+31) 20-511-3476
Fax: (+31) 20-178-024

Dr. Michael Kaplan
ARI
Office of Basic Research
5001 Eisenhower Ave.
Alexandria, VA 22333-5600
USA
Phone: (+1) 703-274-8641
Fax: (+1) 703-274-5616

Wg. Cmdr. John Lane
RAF West Drayton
O. C. Data processing Wing
Middlesex UB7 9AU
UK
Phone: (+44) 895 42 6016
Fax: (+44) 895 42 6987

Mr. John Levesley
Guild of Air Traffic Officers
National Air Traffic Services, ATCEU
Rm 37
Bournemouth International Airport
Christchurch Dorset BH23 8NA
UK
Phone: (+44) 202 472 378
Fax: (+44) 202 472 236

Prof. Morten Lind
Servolaboratriet
Technical University of Denmark
DTH, Bldg. 326
DK-2800 Lyngby
DENMARK
Phone: (+45) 42-88-0199
Fax: (+45) 42-88-2239

Mr. Randahl Lindgren
Air Transportation Systems Division
Systems Control Technology, Inc.
475 School St. SW, W-1030
Washington, D.C. 2024
USA
Phone: (+1) 202-646-2105

Mr. Peter McAlindon
1423 Oxyoke Dr.
Flint, MI 48532
USA
Phone: (+1) 313-732-1570

Mr. Richard Mogford
Hickling Corporation
350 Sparks Street
Ottawa, Ontario K1R 7S8
CANADA
Phone: (+ 1) 613-237-2220
Fax: (+ 1) 613-237-7347

Mr. Kevin Norris
Embry-Riddle Aeronautical University
Center for Instructional Design
Daytona Beach, FL 32114-3900
USA
Phone: (+1) 904-239-6371
Fax: (+1) 904-239-645

Ms Gill Nyfield
Saville and Holdsworth Ltd
3 AC Court
High Street, Thames Ditton
Surrey KT7 0SR
UK
Phone: (+44) 81-398-4170
Fax: (+44) 81-398-9544

Mr. William Sandiford
Air Traffic Control Evaluation Unit
Bournemouth International Airport
Christchurch,
Dorset BH23 6DF
UK
Phone: (+44) 202-472-385
Fax: (+44) 202-472-236

Ms Joy Sargis
126 Solace Ct.
Encinitas, CA 92024
USA
Phone: (+1) 619-942-8348

Col. Peter Segerer
Neuer Graben 9
D-7900 Ulm
GERMANY

Mr. Egbert Seidel
ATC Consultant
Wilhelm Busch Weg 24
D-6072 Dreieich 1
GERMANY
Phone: (+49) 6103-31574
Fax: (+49) 612-180-6271

Mr. Arthur Simolunas
Federal Aviation Administration
WDC-V64
475 School Street
Washington, D.C. 20024
USA
Phone: (+1) 202-646-5649
Fax: (+1) 202-267-9794

Mr. Marvin Smith
Embry-Riddle Aeronautical University
Airway Science Simulation Laboratory
300 S. Clyde Morris Ave.
Daytona Beach, FL 32114-3900
USA
Phone: (+1) 904-239-6448
Fax: (+1) 904-239-6459

Prof. Paul Stager
Department of Psychology
York University
4700 Keele Street
Toronto, Ontario M3J 1P3
CANADA
Phone: (+1) 416-736-5122
Fax: (+1) 416-489-1532

Ms Nicole M. Svátek
Canadian Airlines International
Flight Operations Resource Mgt.
3111 Convair Drive
Mississauga, Ontario L5P 1C2
CANADA
Phone: (+1) 416-612-2348
Fax: (+1) 416-612-2075

Dr. Andrew J. Tattersall
University of Wales-College of Cardiff
School of Psychology
P. O. Box 901
Cardiff CF1 3YG
UK
Phone: (+44) 222-87-4000, X5376
Fax: (+44) 222-87-4858

Mr. Mike Tonner
Canadian Air Traffic Control Assoc.
Suite 1100
400 Cumberland
Ottawa, Ontario K1N 8X3
CANADA
Phone: (+1) 613-232-9413
Fax: (+1) 613-232-9780

Dr. Uwe Völckers
Institute fur Flugfuhrung
DFVLR
Postfach 3267
D-3300 Braunschweig
GERMANY
Phone: (+49) 531-395-2520
Fax: (+49) 531-395-2550

Dr. Ron Westrum
Eastern Michigan University
929 5th Street
Ann Arbor, MI 48103
USA
Phone: (+1) 313-769-8636

Dr. Hilda Wing
APN-200
Federal Aviation Administration
800 Independence Ave. SW
Washington, D.C. 20591
USA
Phone: (+1) 202-267-7841
Fax: (+1) 202-267-5121

Dr. John A. Wise
Embry-Riddle Aeronautical University
Center for Aviation/Aerospace Research
Daytona Beach, FL 32114-3900
USA
Phone: (+1) 904-239-6384
Fax: (+1) 904-239-6459

Photographs of Air Traffic Control Facilities

Control Sector Suite at Maastricht UAC (Eurocontrol Photo)

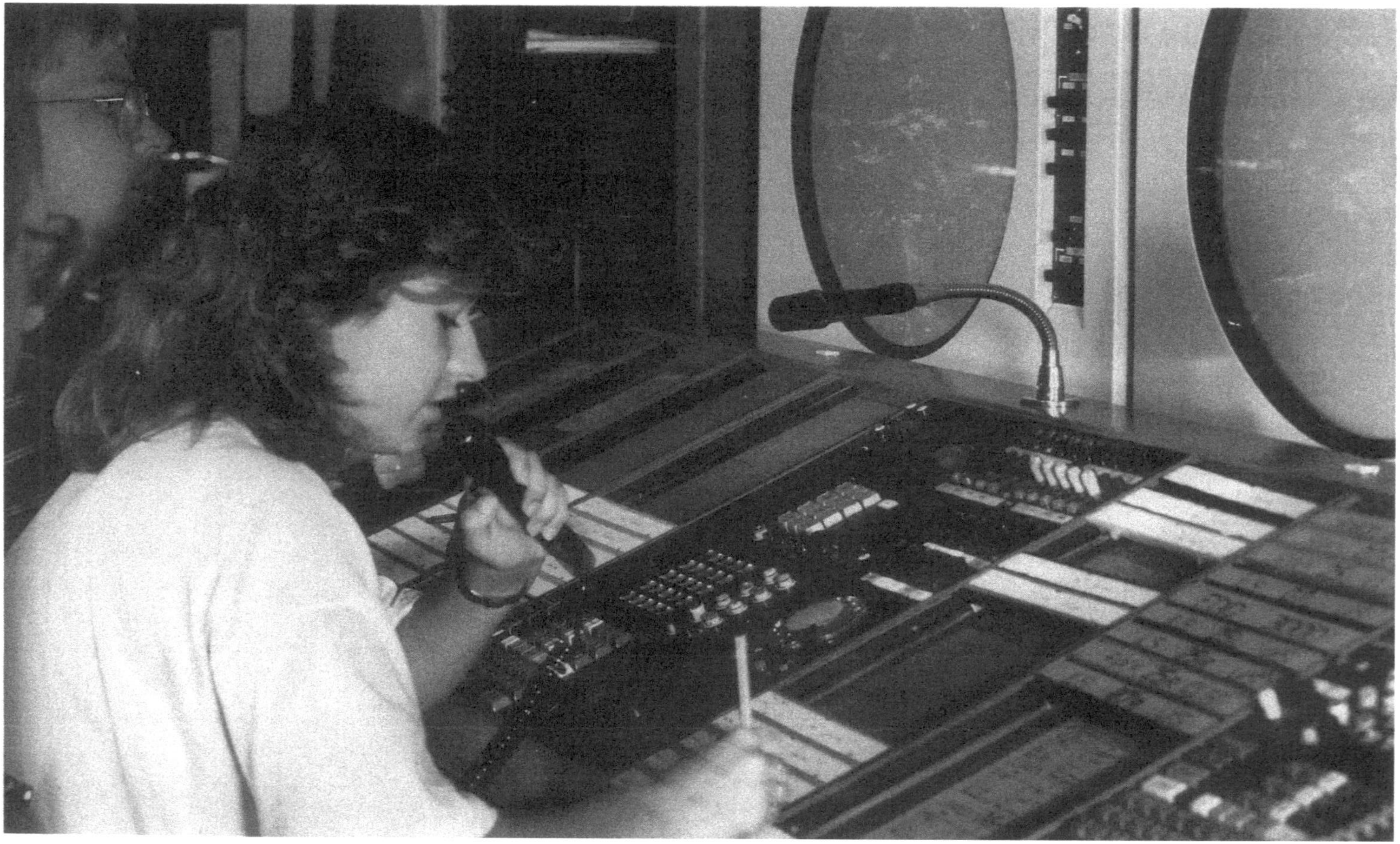

Radar Controller Workstation at Frankfurt Approach Control (© BFS)

Operations Room of London Air Traffic Control Centre (Courtesy of CAA)

Advanced Controller Workstation at BFS, Germany (© BFS)

Two views of the ATC workstations at U. K. Air Traffic Control Evaluation Unit for use in simulations of U.K.'s Central Control Function (Courtesy of CAA)

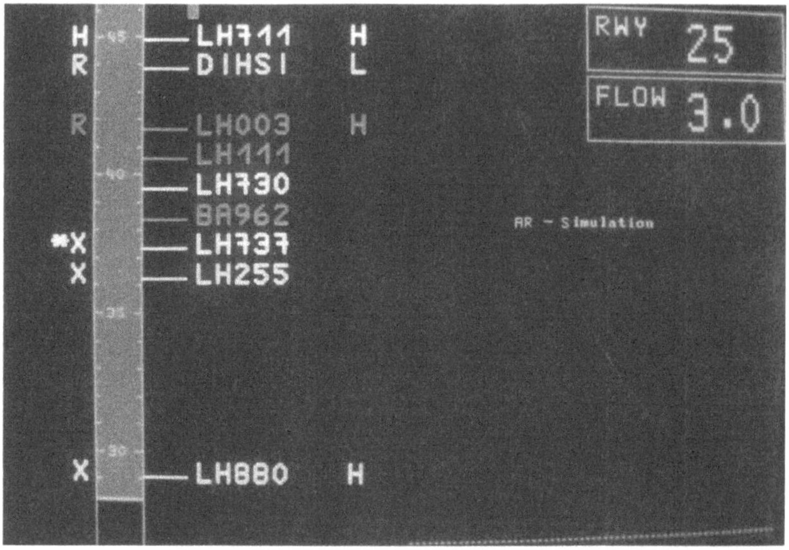

COMPAS Display: Proposed Sequence and Planned Arrival Times
(Left Column=Status, Center Bar= Time Line, Right Column=Aircraft ID) (© DLR)

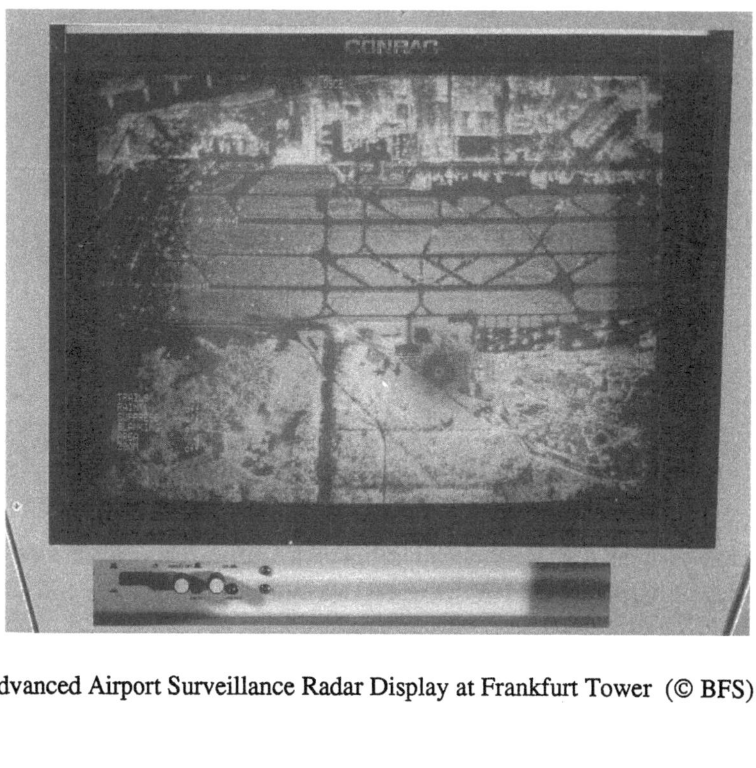

Advanced Airport Surveillance Radar Display at Frankfurt Tower (© BFS)

Radar Control Positions in the Eurocontrol Experimental Centre's Simulations Room (Eurocontrol Photo)

Controller and Pseudo-Pilot Positions During Eurocontrol Experimental Centre Simulation Exercise (Eurocontrol Photo)

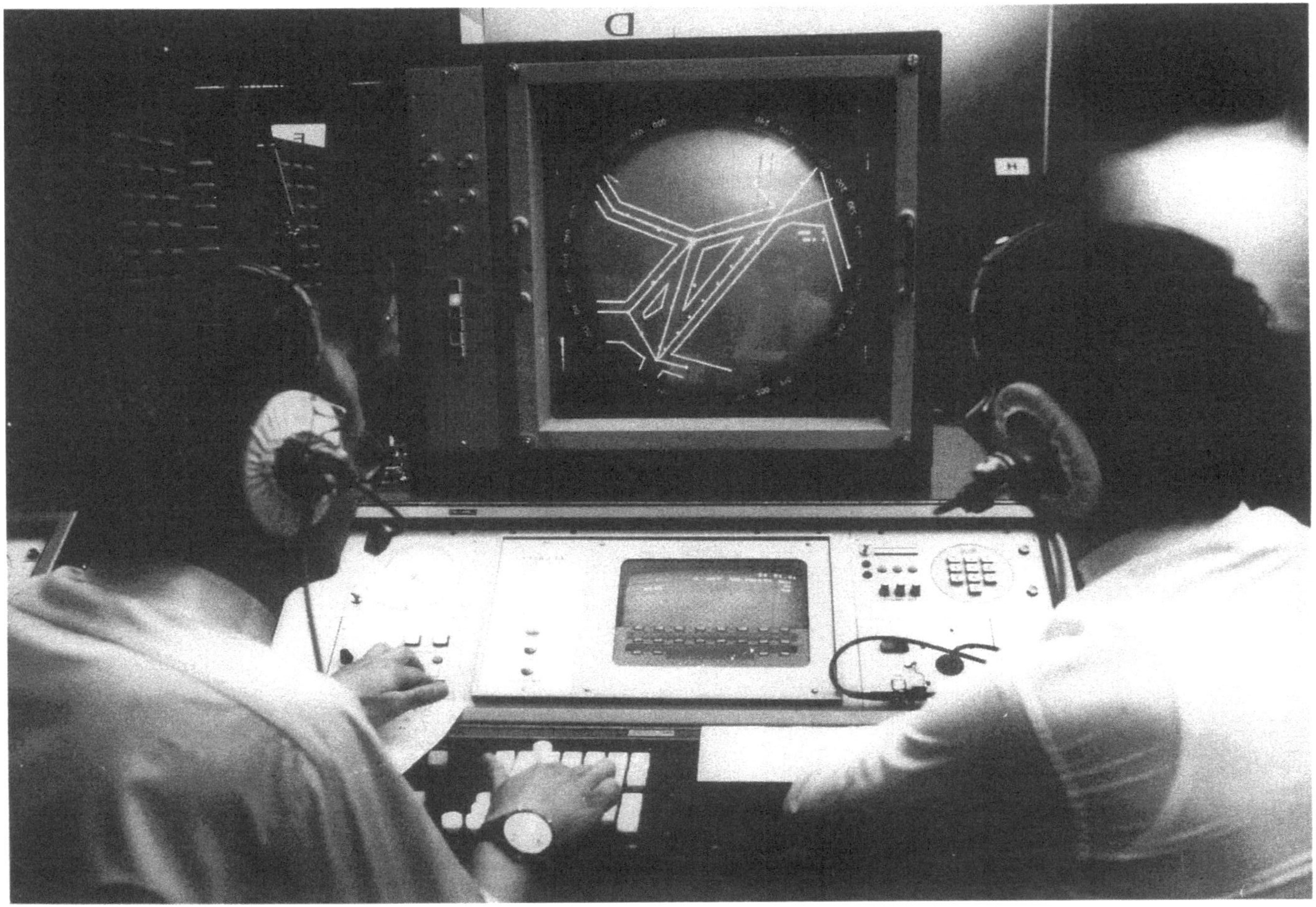

Training at Eurocontrol Institute of Air Navigation Services (Eurocontrol photo)

U.K. Flight Progress Strips (Courtesy of CAA)

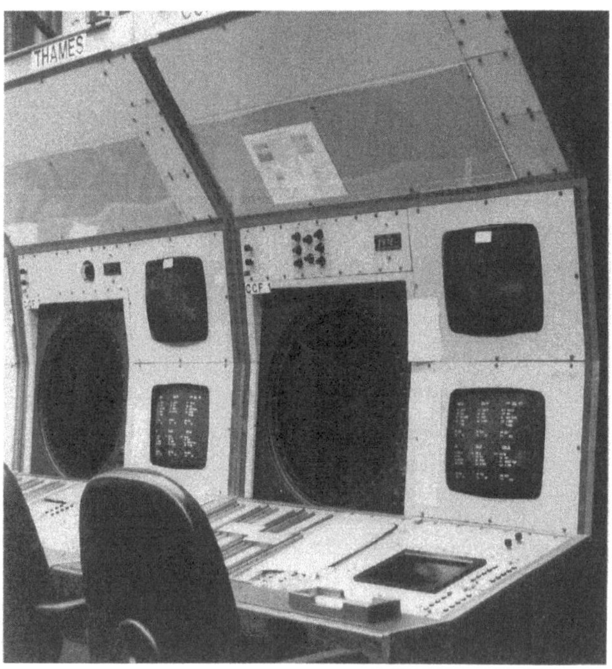

Modern Area Radar Suite at U. K. Air Traffic Control Evaluation Unit (Courtesy of CAA)

Tower Air Traffic Control Training (Courtesy of the FAA)

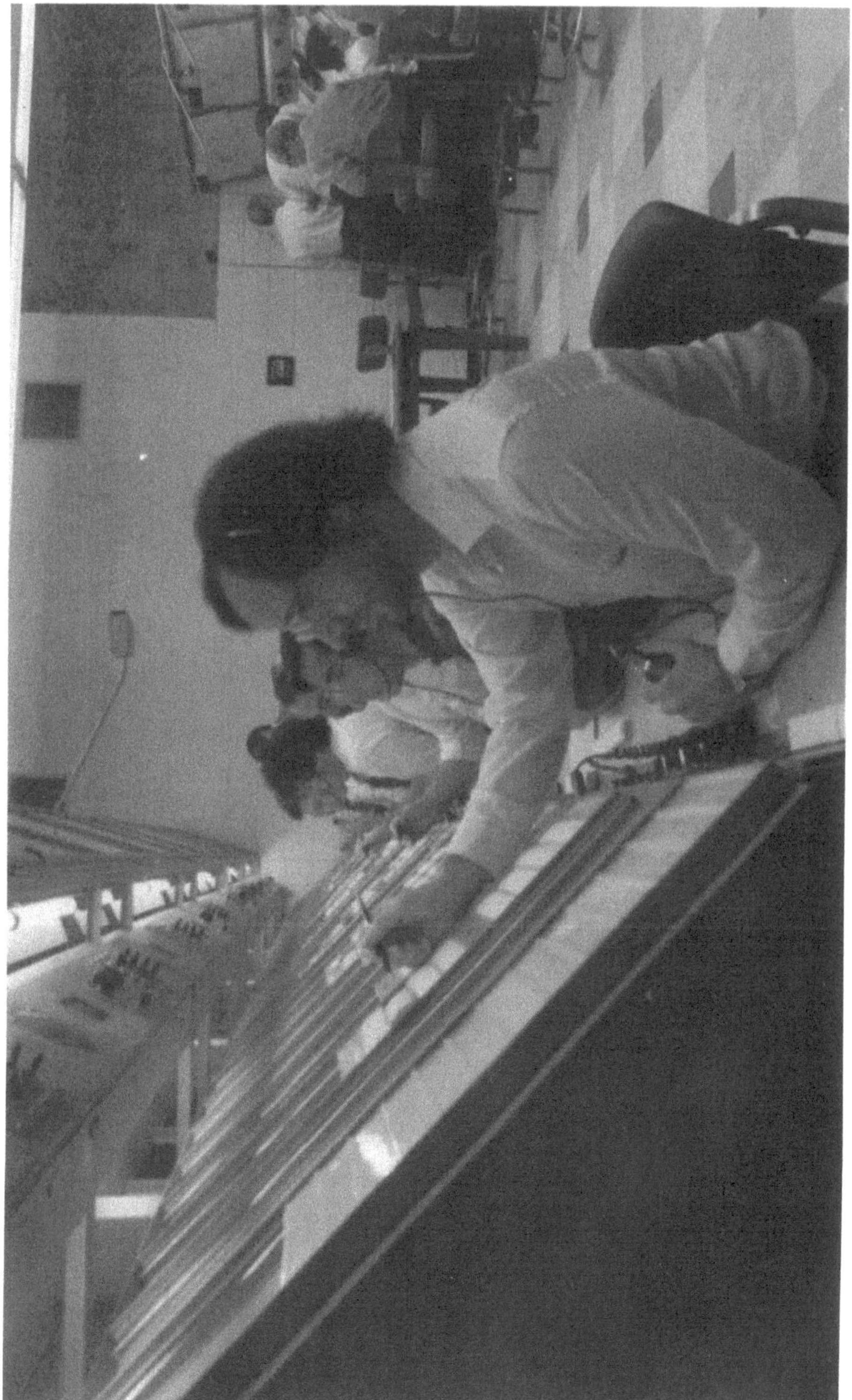

Non-Radar Air Traffic Control Training (Courtesy of FAA)

Index and Glossary

NATO ASI Series F

Including Special Programmes on Sensory Systems for Robotic Control (ROB) and on Advanced Educational Technology (AET)

Vol. 1: Issues in Acoustic Signal – Image Processing and Recognition. Edited by C. H. Chen. VIII, 333 pages. 1983.

Vol. 2: Image Sequence Processing and Dynamic Scene Analysis. Edited by T. S. Huang. IX, 749 pages. 1983.

Vol. 3: Electronic Systems Effectiveness and Life Cycle Costing. Edited by J. K. Skwirzynski. XVII, 732 pages. 1983.

Vol. 4: Pictorial Data Analysis. Edited by R. M. Haralick. VIII, 468 pages. 1983.

Vol. 5: International Calibration Study of Traffic Conflict Techniques. Edited by E. Asmussen. VII, 229 pages. 1984.

Vol. 6: Information Technology and the Computer Network. Edited by K. G. Beauchamp. VIII, 271 pages. 1984.

Vol. 7: High-Speed Computation. Edited by J. S. Kowalik. IX, 441 pages. 1984.

Vol. 8: Program Transformation and Programming Environments. Report on a Workshop directed by F. L. Bauer and H. Remus. Edited by P. Pepper. XIV, 378 pages. 1984.

Vol. 9: Computer Aided Analysis and Optimization of Mechanical System Dynamics. Edited by E. J. Haug. XXII, 700 pages. 1984.

Vol. 10: Simulation and Model-Based Methodologies: An Integrative View. Edited by T. I. Ören, B. P. Zeigler, M. S. Elzas. XIII, 651 pages. 1984.

Vol. 11: Robotics and Artificial Intelligence. Edited by M. Brady, L. A. Gerhardt, H. F. Davidson. XVII, 693 pages. 1984.

Vol. 12: Combinatorial Algorithms on Words. Edited by A. Apostolico, Z. Galil. VIII, 361 pages. 1985.

Vol. 13: Logics and Models of Concurrent Systems. Edited by K. R. Apt. VIII, 498 pages. 1985.

Vol. 14: Control Flow and Data Flow: Concepts of Distributed Programming. Edited by M. Broy. VIII, 525 pages. 1985.

Vol. 15: Computational Mathematical Programming. Edited by K. Schittkowski. VIII, 451 pages. 1985.

Vol. 16: New Systems and Architectures for Automatic Speech Recognition and Synthesis. Edited by R. De Mori, C.Y. Suen. XIII, 630 pages. 1985.

Vol. 17: Fundamental Algorithms for Computer Graphics. Edited by R.A. Earnshaw. XVI, 1042 pages. 1985.

Vol. 18: Computer Architectures for Spatially Distributed Data. Edited by H. Freeman and G.G. Pieroni. VIII, 391 pages. 1985.

Vol. 19: Pictorial Information Systems in Medicine. Edited by K. H. Höhne. XII, 525 pages. 1986.

Vol. 20: Disordered Systems and Biological Organization. Edited by E. Bienenstock, F. Fogelman Soulié, G. Weisbuch. XXI, 405 pages.1986.

Vol. 21: Intelligent Decision Support in Process Environments. Edited by E. Hollnagel, G. Mancini, D.D. Woods. XV, 524 pages. 1986.

NATO ASI Series F

Vol. 22: Software System Design Methods. The Challenge of Advanced Computing Technology. Edited by J.K. Skwirzynski. XIII, 747 pages. 1986.

Vol. 23: Designing Computer-Based Learning Materials. Edited by H. Weinstock and A. Bork. IX, 285 pages. 1986.

Vol. 24: Database Machines. Modern Trends and Applications. Edited by A.K. Sood and A.H. Qureshi. VIII, 570 pages. 1986.

Vol. 25: Pyramidal Systems for Computer Vision. Edited by V. Cantoni and S. Levialdi. VIII, 392 pages. 1986. *(ROB)*

Vol. 26: Modelling and Analysis in Arms Control. Edited by R. Avenhaus, R.K. Huber and J.D. Kettelle. VIII, 488 pages. 1986.

Vol. 27: Computer Aided Optimal Design: Structural and Mechanical Systems. Edited by C.A. Mota Soares. XIII, 1029 pages. 1987.

Vol. 28: Distributed Operating Systems. Theory und Practice. Edited by Y. Paker, J.-P. Banatre and M. Bozyiğit. X, 379 pages. 1987.

Vol. 29: Languages for Sensor-Based Control in Robotics. Edited by U. Rembold and K. Hörmann. IX, 625 pages. 1987. *(ROB)*

Vol. 30: Pattern Recognition Theory and Applications. Edited by P.A. Devijver and J. Kittler. XI, 543 pages. 1987.

Vol. 31: Decision Support Systems: Theory and Application. Edited by C.W. Holsapple and A.B. Whinston. X, 500 pages. 1987.

Vol. 32: Information Systems: Failure Analysis. Edited by J.A. Wise and A. Debons. XV, 338 pages. 1987.

Vol. 33: Machine Intelligence and Knowledge Engineering for Robotic Applications. Edited by A.K.C. Wong and A. Pugh. XIV, 486 pages. 1987. *(ROB)*

Vol. 34: Modelling, Robustness and Sensitivity Reduction in Control Systems. Edited by R.F. Curtain. IX, 492 pages. 1987.

Vol. 35: Expert Judgment and Expert Systems. Edited by J.L. Mumpower, L.D. Phillips, O. Renn and V.R.R. Uppuluri. VIII, 361 pages. 1987.

Vol. 36: Logic of Programming and Calculi of Discrete Design. Edited by M. Broy. VII, 415 pages. 1987.

Vol. 37: Dynamics of Infinite Dimensional Systems. Edited by S.-N. Chow and J.K. Hale. IX, 514 pages. 1987.

Vol. 38: Flow Control of Congested Networks. Edited by A.R. Odoni, L. Bianco and G. Szegö. XII, 355 pages. 1987.

Vol. 39: Mathematics and Computer Science in Medical Imaging. Edited by M.A. Viergever and A. Todd-Pokropek. VIII, 546 pages. 1988.

Vol. 40: Theoretical Foundations of Computer Graphics and CAD. Edited by R.A. Earnshaw. XX, 1246 pages. 1988.

Vol. 41: Neural Computers. Edited by R. Eckmiller and Ch. v. d. Malsburg. XIII, 566 pages. 1988.

NATO ASI Series F

Vol. 42: Real-Time Object Measurement and Classification. Edited by A. K. Jain. VIII, 407 pages. 1988. *(ROB)*

Vol. 43: Sensors and Sensory Systems for Advanced Robots. Edited by P. Dario. XI, 597 pages. 1988. *(ROB)*

Vol. 44: Signal Processing and Pattern Recognition in Nondestructive Evaluation of Materials. Edited by C. H. Chen. VIII, 344 pages. 1988. *(ROB)*

Vol. 45: Syntactic and Structural Pattern Recognition. Edited by G. Ferraté, T. Pavlidis, A. Sanfeliu and H. Bunke. XVI, 467 pages. 1988. *(ROB)*

Vol. 46: Recent Advances in Speech Understanding and Dialog Systems. Edited by H. Niemann, M. Lang and G. Sagerer. X, 521 pages. 1988.

Vol. 47: Advanced Computing Concepts and Techniques in Control Engineering. Edited by M. J. Denham and A. J. Laub. XI, 518 pages. 1988.

Vol. 48: Mathematical Models for Decision Support. Edited by G. Mitra. IX, 762 pages. 1988.

Vol. 49: Computer Integrated Manufacturing. Edited by I. B. Turksen. VIII, 568 pages. 1988.

Vol. 50: CAD Based Programming for Sensory Robots. Edited by B. Ravani. IX, 565 pages. 1988. *(ROB)*

Vol. 51: Algorithms and Model Formulations in Mathematical Programming. Edited by S.W. Wallace. IX, 190 pages. 1989.

Vol. 52: Sensor Devices and Systems for Robotics. Edited by A. Casals. IX, 362 pages. 1989. *(ROB)*

Vol. 53: Advanced Information Technologies for Industrial Material Flow Systems. Edited by S. Y. Nof and C. L. Moodie. IX, 710 pages. 1989.

Vol. 54: A Reappraisal of the Efficiency of Financial Markets. Edited by R. M. C. Guimarães, B. G. Kingsman and S. J. Taylor. X, 804 pages. 1989.

Vol. 55: Constructive Methods in Computing Science. Edited by M. Broy. VII, 478 pages. 1989.

Vol. 56: Multiple Criteria Decision Making and Risk Analysis Using Microcomputers. Edited by B. Karpak and S. Zionts. VII, 399 pages. 1989.

Vol. 57: Kinematics and Dynamic Issues in Sensor Based Control. Edited by G. E. Taylor. XI, 456 pages. 1990. *(ROB)*

Vol. 58: Highly Redundant Sensing in Robotic Systems. Edited by J. T. Tou and J. G. Balchen. X, 322 pages. 1990. *(ROB)*

Vol. 59: Superconducting Electronics. Edited by H. Weinstock and M. Nisenoff. X, 441 pages. 1989.

Vol. 60: 3D Imaging in Medicine. Algorithms, Systems, Applications. Edited by K. H. Höhne, H. Fuchs and S. M. Pizer. IX, 460 pages. 1990.

Vol. 61: Knowledge, Data and Computer-Assisted Decisions. Edited by M. Schader and W. Gaul. VIII, 421 pages. 1990.

Vol. 62: Supercomputing. Edited by J. S. Kowalik. X, 425 pages. 1990.

Vol. 63: Traditional and Non-Traditional Robotic Sensors. Edited by T. C. Henderson. VIII, 468 pages. 1990. *(ROB)*

Vol. 64: Sensory Robotics for the Handling of Limp Materials. Edited by P. M. Taylor. IX, 343 pages. 1990. *(ROB)*

Vol. 65: Mapping and Spatial Modelling for Navigation. Edited by L. F. Pau. VIII, 357 pages. 1990. *(ROB)*

NATO ASI Series F

Vol. 66: Sensor-Based Robots: Algorithms and Architectures. Edited by C.S.G. Lee. X, 285 pages. 1991. *(ROB)*

Vol. 67: Designing Hypermedia for Learning. Edited by D.H. Jonassen and H. Mandl. XXV, 457 pages. 1990. *(AET)*

Vol. 68: Neurocomputing. Algorithms, Architectures and Applications. Edited by F. Fogelman Soulié and J. Hérault. XI, 455 pages. 1990.

Vol. 69: Real-Time Integration Methods for Mechanical System Simulation. Edited by E.J. Haug and R.C. Deyo. VIII, 352 pages. 1991.

Vol. 70: Numerical Linear Algebra, Digital Signal Processing and Parallel Algorithms. Edited by G.H. Golub and P. Van Dooren. XIII, 729 pages. 1991.

Vol. 71: Expert Systems and Robotics. Edited by T. Jordanides and B. Torby. XII, 744 pages. 1991.

Vol. 72: High-Capacity Local and Metropolitan Area Networks. Architecture and Performance Issues. Edited by G. Pujolle. X, 536 pages. 1991.

Vol. 73: Automation and Systems Issues in Air Traffic Control. Edited by J.A. Wise, V.D. Hopkin and M.L. Smith. XIX, 592 pages. 1991.

MIX
Papier aus verantwortungsvollen Quellen
Paper from responsible sources
FSC® C105338

If you have any concerns about our products,
you can contact us on
ProductSafety@springernature.com

In case Publisher is established outside the EU,
the EU authorized representative is:
Springer Nature Customer Service Center GmbH
Europaplatz 3, 69115 Heidelberg, Germany

Printed by Libri Plureos GmbH
in Hamburg, Germany